Integrated Pest Management

David Pimentel • Rajinder Peshin
Editors

Integrated Pest Management

Pesticide Problems, Vol. 3

 Springer

Editors
David Pimentel
Department of Entomology
Department of Ecology and Evolutionary
Biology
Cornell University
Tower Road, East Blue Insectary-Old
Ithaca, New York
USA

Rajinder Peshin
Division of Agricultural Extension
Education, Faculty of Agriculture
Sher-e-Kashmir University of Agricultural
Sciences and Technology of Jammu
Main Campus Chatha, Jammu
India

ISBN 978-94-007-7795-8   ISBN 978-94-007-7796-5 (eBook)
DOI 10.1007/978-94-007-7796-5
Springer NewYork Heidelberg Dordrecht London

Library of Congress Control Number: 2013956045

© Springer Science+Business Media Dordrecht 2014
This work is subject to copyright. All rights are reserved by the Publisher, whether the whole or part of the material is concerned, specifically the rights of translation, reprinting, reuse of illustrations, recitation, broadcasting, reproduction on microfilms or in any other physical way, and transmission or information storage and retrieval, electronic adaptation, computer software, or by similar or dissimilar methodology now known or hereafter developed. Exempted from this legal reservation are brief excerpts in connection with reviews or scholarly analysis or material supplied specifically for the purpose of being entered and executed on a computer system, for exclusive use by the purchaser of the work. Duplication of this publication or parts thereof is permitted only under the provisions of the Copyright Law of the Publisher's location, in its current version, and permission for use must always be obtained from Springer. Permissions for use may be obtained through RightsLink at the Copyright Clearance Center. Violations are liable to prosecution under the respective Copyright Law.
The use of general descriptive names, registered names, trademarks, service marks, etc. in this publication does not imply, even in the absence of a specific statement, that such names are exempt from the relevant protective laws and regulations and therefore free for general use.
While the advice and information in this book are believed to be true and accurate at the date of publication, neither the authors nor the editors nor the publisher can accept any legal responsibility for any errors or omissions that may be made. The publisher makes no warranty, express or implied, with respect to the material contained herein.

Printed on acid-free paper

Springer is part of Springer Science+Business Media (www.springer.com)

Preface

Pests contribute to shortages of food in several ways. They destroy our food and attack us personally. Combined arthropod, disease and weed pests contribute to malnourishment and death to nearly two thirds or more than 66% of the total world population of 7.2 billion people.

Approximately 40% of all the world's food production is lost or destroyed by insects, diseases, and weeds. This loss occurs despite the application of the nearly 3 million tons of pesticides applied to our crops annually. Once the food is harvested an additional 20% of our food is destroyed; in addition to pests, pesticides cause human deaths and damage our environment. Consider there are about 3 million human pesticide poisonings worldwide, with an estimated 220,000 deaths each year.

The widespread use of pesticides is responsible for bird and fish deaths, destruction of many beneficial natural enemies, pesticide residues on and in foodstuffs, loss of vital plant pollinators, ground and surface water contamination, selection for resistance in pests to pesticides, and other environmental problems.

Pesticides can be reduced to *zero* even in the heavily treated crops in the United States—corn and soybeans. A 22-year long experiment carried out in Pennsylvania (see Chap. 6 – this volume) demonstrates this. More research is needed to reduce pesticide use while reducing the negative environmental side-effects of pest control.

The contributors to this book recognize the value of pesticides for pest control and recognize the negative impacts pesticides have on environmental quality and human health. In many instances, they suggest techniques that can be employed to reduce pesticide use while maintaining crop yields. Reducing pesticide use 50% or more while improving pest control economics, public health, and the environment is possible. In fact, successful programs using various techniques in countries like Sweden and Indonesia have reduced pesticide use by close to two-thirds. Clearly we can do better to improve pest control and protect the environment and human health.

Ithaca, New York, USA	David Pimentel
Jammu, India	Rajinder Peshin

Acknowledgements

I wish to express my sincere gratitude to Dr. Rajinder Peshin for inviting me to become his co-editor of this volume and to Springer for agreeing to publish this volume. I thank our authors for their very interesting and informative manuscripts. I would also like to thank the Cornell Association of Professor Emeriti for the partial support of our research through the Albert Podell Grant Program. Finally I wish to thank Michael Burgess for his valuable assistance in proofing and revising these manuscripts for publication.

Ithaca, New York, USA   David Pimentel

Contents

1. Integrated Pest Management and Pesticide Use 1
   Rajinder Peshin and WenJun Zhang

2. Environmental and Economic Costs of the Application of
   Pesticides Primarily in the United States 47
   David Pimentel and Michael Burgess

3. Integrated Pest Management for European Agriculture 73
   Bill Clark and Rory Hillocks

4. Energy Inputs In Pest Control Using Pesticides In New Zealand 99
   Majeed Safa and Meriel Watts

5. Environmental and Economic Benefits of Reducing Pesticide Use 127
   David Pimentel and Michael Burgess

6. An Environmental, Energetic and Economic Comparison of
   Organic and Conventional Farming Systems 141
   David Pimentel and Michael Burgess

7. Pesticides, Food Safety and Integrated Pest Management 167
   Dharam P. Abrol and Uma Shankar

8. Crop Losses to Arthropods 201
   Thomas W. Culliney

9. Crop Loss Assessment in India- Past Experiences and Future
   Strategies 227
   T. V. K. Singh, J. Satyanarayana and Rajinder Peshin

10	Review of Potato Biotic Constraints and Experiences with Integrated Pest Management Interventions	245
	Peter Kromann, Thomas Miethbauer, Oscar Ortiz and Gregory A. Forbes	
11	Biological Control: Perspectives for Maintaining Provisioning Services in the Anthropocene	269
	Timothy R. Seastedt	
12	Herbicide Resistant Weeds	281
	Ian Heap	
13	Strategies for Reduced Herbicide Use in Integrated Pest Management	303
	Rakesh S. Chandran	
14	Herbicide Resistant Crops and Weeds: Implications for Herbicide Use and Weed Management	331
	George B. Frisvold and Jeanne M. Reeves	
15	Integrating Research and Extension for Successful Integrated Pest Management	355
	Cesar R. Rodriguez-Saona, Dean Polk and Lukasz L. Stelinski	
16	Promotion of Integrated Pest Management by the Plant Science Industry: Activities and Outcomes	393
	Keith A. Jones	
17	From the Farmers' Perspective: Pesticide Use and Pest Control	409
	Seyyed Mahmoud Hashemi, Rajinder Peshin and Giuseppe Feola	
18	Evaluation of Integrated Pest Management Interventions: Challenges and Alternatives	433
	K. S. U. Jayaratne	
Index		471

Contributors

D. P. Abrol Professor of Entomology, Faculty of Agriculture, Sher-e-Kashmir University of Agricultural Sciences & Technology of Jammu, Chatha, Jammu - 180 009, Jammu & Kashmir, India.

Michael Burgess Department of Entomology, Cornell University, Horticulture, Research Aide, Greenhouse worker, Tower Road East, Blue Insectary-Old, Ithaca, New York 14853, USA

Rakesh S. Chandran Extension Weed Specialist & Professor, IPM Coordinator, West Virginia University, PO Box 6108, 1076 Agricultural Sciences Building, Morgantown, West Virginia 26506-6108, USA.

Bill Clark Commercial Technical Director, National Institute of Agricultural Botany, Huntingdon Road, Cambridge CB3 0LE, United Kingdom.

Thomas W. Culliney USDA-APHIS, PPQ, Center for Plant Health Science and Technology, Plant Epidemiology and Risk Analysis Laboratory, 1730 Varsity Drive, Suite 300, Raleigh, North Carolina, 27606, USA.

Giuseppe Feola Department of Geography and Environmental Science, University of Reading, Reading, UK.

Greg Forbes CIP-China Center for Asia Pacific, International Potato Center, Room 709, Pan Pacific Plaza, A12 Zhongguancun Nandajie, Beijing 100081, China

George B. Frisvold Professor, University of Arizona, Department of Agricultural & Resource Economics, 319 Cesar Chavez Building, Tucson, Arizona 85721 USA.

Seyyed Mahmoud Hashemi Department of Agricultural Extension and Education, College of Agriculture, University of Tehran, Karaj, Iran.

Ian Heap Director of the International Survey of Herbicide-Resistant Weeds, PO Box 1365, Corvallis, Oregon 97339, USA.

Rory Hillocks European Centre for IPM, Natural Resources Institute, University of Greenwich, Chatham Maritime, Kent, ME4 4TB, United Kingdom.

K. S. U. Jayaratne Associate Professor and the State Leader for Extension Program Evaluation, Department of Agricultural and Extension Education at North Carolina State University, North Carolina State University, Raleigh, NC 27695, USA.

Keith Jones Director of Stewardship & Sustainable Agriculture, CropLife International, 326 Avenue Louise, Box 35, Brussels 1050, Belgium.

Peter Kromann International Potato Center, Post box 17 21 1977, Quito, Ecuador.

Thomas Miethbauer International Potato Center, Apartado 1558, Lima 12, Peru.

Oscar Ortiz International Potato Center, Apartado 1558, Lima 12, Peru

Rajinder Peshin Associate Professor of Agricultural Extension Education at the Sher-e-Kashmir University of Agricultural Sciences and Technology of Jammu, Main Campus : Chatha, Jammu - 180009, India.

David Pimentel Department of Entomology, Department of Ecology and Evolutionary Biology, Cornell University, Tower Road East, Blue Insectary-Old, Ithaca, New York 14853, USA

Dean Polk IPM agent, Rutgers Fruit Research & Extension Center, 283 Route 539, Cream Ridge, New Jersey 08514, USA.

Jeanne M. Reeves Director, Agricultural & Environmental Research Division, Cotton Incorporated, 6399 Weston Parkway, Cary, North Carolina 27513, USA.

Cesar Rodriguez-Saona Associate Extension Specialist, Department of Entomology, Rutgers University, PE Marucci Center for Blueberry & Cranberry Research & Extension, 125A Lake Oswego Rd., Chatsworth, New Jersey 08019, USA

Majeed Safa Lecturer, Department of Agricultural Management and Property Studies, Lincoln University, PO Box 84, Lincoln University, Lincoln 7647, Christchurch, New Zealand.

Jella Satyanarayana Department of Entomology, Acharya N. G. Ranga Agricultural University, Rajendranagar, Hyderabad 500 030, India.

Timothy Seastedt Professor and INSTAAR Fellow, UCB 450, University of Colorado, Boulder, Colorado 80309-0450, USA.

Uma Shankar Division of Entomology, Faculty of Agriculture, Sher-e-Kashmir University of Agricultural Sciences & Technology of Jammu, Chatha, Jammu-180 009, Jammu & Kashmir, India.

T. V. K. Singh Senior Professor, Department of Entomology, Acharya N. G. Ranga Agricultural University, Rajendranagar, Hyderabad 500 030, India.

Lukasz Stelinski Associate Professor, Citrus Research and Education Center, University of Florida, 700 Experiment Station Rd., Lake Alfred, Florida 33850, USA.

Meriel Watts Co-ordinator, Pesticide Action Network, (Aotearoa) New Zealand, PO Box 296 Ostend, Waiheke Island, Auckland 1843, New Zealand.

WenJun Zhang Professor, Sun Yat-sen University, Guangzhou, China; International Academy of Ecology and Environmental Sciences, Hong Kong, China.

About the Authors

D. P. Abrol is working as Professor & Head of the Division of Entomology, Sher-e-Kashmir University of Agricultural Sciences and Technology of Jammu, Faculty of Agriculture, Chatha, India. He has been visiting scholar at: ETH Zurich, Switzerland; Jagiellonian University, Krakow, Poland; Busan, South Korea and Terranagnu, Malaysia. His research addresses pollination biology, honeybee ecology and integrated pest management. He has been honored by several national and international awards. Dr. D. P. Abrol has published more than 200 research papers, 10 chapters of books, 10 review articles and is the author of 10 books published by Springer, CABI, Academic Press and others.

Michael Burgess works as a copy editor for Dr. David Pimentel and works in the Cornell University Greenhouses. He has worked with entomological researchers as an experimentalist, library researcher, copy editor and generally aiding researchers in need at Cornell as a technician for over 25 years.

Rakesh S. Chandran is an Extension Specialist and Professor at West Virginia University, Morgantown, West Virginia, USA. He received a Master of Science degree in Environmental Horticulture from the University of Florida (1993), and a Doctoral degree in Weed Science from Virginia Tech (1997). His primary responsibilities are to carry out an outreach and research program in applied weed science related to agricultural and horticultural commodities in West Virginia and to coordinate the university's Integrated Pest Management (IPM) program. He teaches two courses at West Virginia University and currently serves as the Vice President of the Northeastern Weed Science Society (NEWSS).

Bill Clark is a plant pathologist specializing in cereal disease control strategies. He has worked as an extension pathologist and researcher in plant disease for many years, working in protected crops, ornamentals and arable crops. He has expertise in IPM approaches in a range of cropping systems in the United Kingdom. Bill Clark is currently the Commercial Technical Director at The National Institute of Agricultural Botany (NIAB) in Cambridge, UK. He was formerly the Director of Brooms Barn Research Centre, part of Rothamsted Research and before that worked as a research pathologist for the UK Government Agricultural Advisory Service.

Thomas W. Culliney is an entomologist with the U.S. Department of Agriculture, Animal and Plant Health Inspection Service, Center for Plant Health Science and Technology in Raleigh, North Carolina. He conducts analyses based on standards of the International Plant Protection Convention and the World Organisation for Animal Health, of the risks involved in the importation of agricultural commodities and introduction of alien species. His main interests are in population ecology and biological control of weeds and arthropod pests. He has published more than 40 articles and book chapters on subjects, such as paleoentomology, biological control, ecotoxicology, and sustainable agriculture.

Giuseppe Feola is Lecturer in Environment and Development in the Department of Geography and Environmental Science at the University of Reading, United Kingdom. Giuseppe holds a B.Sc. in Sociology (2002) from the University of Milan-Bicocca, a M.Sc. in Environmental Economics and Management (2003) from Bocconi University in Milan and a Ph.D. in Geography (2010) from the University of Zurich. Giuseppe's research interests include decision-making modeling in social-ecological systems, theories of social-ecological change, and integrated sustainability assessment.

Greg Forbes received his Ph.D. degree in plant pathology from Texas A&M University and spent two years in a postdoctoral position in Montpellier, France at the Institute National de la Recherche Agronomique (INRA). He has worked with the International Potato Center since 1989 with responsibility for research on potato late blight, and more recently for management of other potato diseases. Forbes is interested in disease management strategies appropriate for developing countries and recently has focused on diseases causing degeneration of potato within the context of the roots and tubers and has worked with the bananas CGIAR Research Program.

George B. Frisvold is a Professor and Extension Specialist in the Department of Agricultural and Resource Economics at the University of Arizona. He holds two degrees from the University of California, Berkeley—a B.S. in Political Economy of Natural Resources and a Ph.D. in Agricultural and Resource Economics. He has been Chief of the Resource Policy Branch of USDA's Economic Research Service, a Lecturer at the Johns Hopkins University, and a Senior Economist for the President's Council of Economic Advisers. His research interests include the economics of technological innovation in agriculture, agricultural biotechnologies, and pesticide use.

Seyyed Mahmoud Hashemi is a Ph.D. student in the Department of Agricultural Extension and Education of the University of Tehran, Iran. He received a B.Sc. in Agricultural Extension and Education from Shiraz University and a M.Sc. in Agricultural Extension from the University of Tehran. The areas of his research include management and evaluation of agricultural extension programs.

Ian Heap is the director of the "International Survey of Herbicide-Resistant Weeds" in Corvallis, Oregon. He completed his Ph.D. at the University of Adelaide on "Multiple-resistance in annual ryegrass (*Lolium rigidum*)", the first case of a herbicide-resistant weed in Australia and multiple resistance worldwide. Ian then continued research on herbicide-resistant weeds at the University of Manitoba in Canada, and Oregon State University. Ian has published numerous papers and book chapters on herbicide-resistant weeds and runs the International Survey of Herbicide-Resistant Weeds website at http://www.weedscience.org.

Rory Hillocks is a crop scientist, specializing in integrated crop management and IPM. He received a Master's degree in Applied Plant Sciences (Wye College, University of London) and a Ph.D. in Plant Disease x Nematode interactions (University of Reading, UK, 1984). Before being based permanently at the Natural Resources Institute in the University of Greenwich in the UK, Dr. Hillocks spent 13 years as an agricultural scientist in Sub-Saharan Africa. He continues his research and development interests in smallholder agriculture in Africa and also heads the recently inaugurated European Centre for IPM which aims to promote the wider adoption of IPM for sustainable agriculture in Europe and the Developing World. Website: www.eucipm.org.

K. S. U. Jayaratne is an extension evaluation specialist. He received his B.S. in Agriculture degree from University of Peradeniya, Sri Lanka and M.S. in Extension Education from the University of Illinois, Urbana-Champaign, Illinois, USA. He earned his Ph.D. in Agricultural Education and Studies from Iowa State University, Ames, Iowa, USA in 2001. He is currently an Associate Professor and the State Leader for Extension Program Evaluation at North Carolina State University, Raleigh, North Carolina. His research areas include extension program development, delivery, and evaluation. He teaches Extension Program Planning and Program Evaluation graduate courses at North Carolina State University.

Keith Jones gained his Ph.D. from the University of Reading for research on the persistence of insect baculoviruses. He is currently at CropLife International, where he is responsible for pesticide stewardship programs across the globe. Before, he was at the Natural Resources Institute, UK, where he was head of the Sustainable Agriculture Group and head of the Insect Pathology Section. His research focused on developing microbial insecticides for use in the developing world. He has also run IPM Farmer Field Schools for CARE Sri Lanka and led a team implementing a World Bank-funded cotton IPM program in Uzbekistan.

Peter Kromann is currently working at the International Potato Center as a regional potato scientist. He conducts research and development activities on IPM, seed systems, crop growth and soil-water-plant relations under different climatic and management conditions in Latin America. He has a Bachelors and a Master's in Agricultural Science with specialization in Plant Pathology from the Royal Veterinary and Agricultural University, Copenhagen, Denmark. He received his Ph.D. in Plant Pathology from the University of Copenhagen, Faculty of Life Sciences, working on integrated management of potato late blight.

Thomas Miethbauer is an agricultural engineer, with specialties in agricultural and development economics and did his studies at Kiel University, Institute of Agricultural Economics, Germany. He worked as a World Bank and GIZ consultant in the field of land and production economics as well as in farm-household survey work. For years he was lecturer for cooperate finance and investment at Kiel University of Applied Sciences and Research. Currently he is a senior scientist at the International Potato Center working for the global programs on integrated crop systems research and on social and health sciences, especially in the field of integrated pest management.

Oscar Ortiz graduated from the National University of Cajamarca, Peru; received a Master's degree in Crop Production and Agricultural Extension from the Agrarian University La Molina, Lima, and a Ph.D. from the University of Reading, UK, working on information and knowledge systems for IPM. Ortiz is currently Deputy Director of Research for Regional Programs at the International Potato Center. Before that he lead an interdisciplinary team dealing with global research on potato and sweet potato pest detection methods, risk assessment, synthesizing seed-related lessons, and modeling crop-pathogen-insect-climate interactions. His research includes participatory research for IPM, impact assessment and innovation systems related to crop production.

Rajinder Peshin is an associate professor at the Sher-e-Kashmir University of Agricultural Sciences and Technology of Jammu, India. His Ph.D. is from Punjab Agricultural University, Ludhiana, India. His research expertise is diffusion and evaluation issues associated with sustainable agriculture research and development programs. Dr. Peshin had developed an emperical model for predicting the adoptability of agricultural technologies when put to trial at farmers' fields, and an evaluation methodology for integrated pest management programs. He has published more than 50 scientific papers and chapters of books and has authored three books besides being the editor of two books on integrated pest management published by Springer in 2009.

David Pimentel is a professor of ecology and agricultural sciences at Cornell University, Ithaca, NY 14853. His Ph.D. is from Cornell University. His research spans the fields of energy, ecological and economic aspects of pest control, biological control, biotechnology, sustainable agriculture, land and water conservation, and environmental policy. Pimentel has published over 700 scientific papers and 40 books and has served on many national and government committees including the National Academy of Sciences; President's Science Advisory Council; U.S. Department of Agriculture; U.S. Department of Energy; U.S. Department of Health, Education and Welfare; Office of Technology Assessment of the U.S. Congress; and the U.S. State Department.

Dean Polk is the statewide fruit IPM agent with Rutgers Cooperative Extension, coordinating fruit IPM programming for New Jersey (USA). He received an M.S. in entomology from the University of Idaho in 1979, and worked as a crop consultant in Washington State. He started the Rutgers fruit IPM program in 1981, and worked with the New Jersey Department of Agriculture from 1985–1987, supervising biological control programs. Program interests have included insect mating disruption in tree fruit and blueberries, reduced-risk methods in fruit crops, methods for tracking grower practices and pesticide use, IPM practices for invasive insects, and geo-referenced IPM for fruit pests.

Jeanne M. Reeves is an agricultural economist and Director, Production Economics in the Agricultural and Environmental Research Division of Cotton Incorporated located in Cary, North Carolina (USA). She received her Bachelor's and Master's Degrees from Mississippi State University and Ph.D. from University of Kentucky, all in Agricultural Economics. The areas of her research include factors affecting costs of cotton production, cotton input markets and technologies, and cotton lint marketing.

Cesar Rodriguez-Saona is an Associate Professor and Extension Specialist in Blueberry and Cranberry IPM, Department of Entomology, Rutgers University, P.E. Marucci Center, Chatsworth, New Jersey, USA. He received his M.S. degree from Oregon State University and his Ph.D. from the University of California, Riverside, working on secondary plant compounds for pest control. Dr. Rodriguez-Saona currently conducts basic and applied research on the development and implementation of cost-effective reduced-risk insect pest management practices and delivers educational information to growers. The areas of his research include integrated pest management, insect-plant interactions, tri-trophic interactions, applied chemical ecology, host-plant resistance, and biological control.

Majeed Safa is an agricultural engineer who received his Ph.D. in Modeling Energy Consumption in Agriculture and Environment from Lincoln University, New Zealand. Dr. Safa is currently a lecturer at Lincoln University where he has been a faculty member since 2011. His research interests lie in the area of sustainability, modeling, and energy management in agriculture. Also, he has been involved in several energy auditing and building facility management projects. Dr Safa recently has started to develop artificial neural network (ANN) models to predict energy consumption in agriculture, environment, and residential sectors based on indirect factors.

Jella Satyanarayana is an Entomologist who graduated from Andhra Pradesh Agricultural University (APAU), presently called Acharya N.G. Ranga Agricultural University (ANGRAU), Hyderabad, Andhra Pradesh, India. He received a Master's degree in Agricultural Entomology from the same University, APAU. He completed his Ph.D. at the Indian Agricultural Research Institute (IARI), Pusa, New Delhi with specialization in Integrated Pest Management (IPM) and Insect Toxicology. Dr. Jella is currently a professor at the College of Agriculture, Rajendranagar, Hyderabad, India engaged in teaching Undergraduate & Postgraduate courses and also guiding Postgraduate students. His areas of research include integrated crop management (ICM), environmental ecology with special reference to the impact of climate change on insect population build up and environmental impact assessment.

Tim Seastedt is a Professor of Ecology and Evolutionary Biology and Fellow of INSTAAR at the University of Colorado, Boulder. Much of his research has been conducted as part of the Long-Term Ecological Research (LTER) programs at Konza Prairie and Niwot Ridge. His interests range from plant-consumer-soil interactions to how regional and global environmental changes are affecting and being affected by biotic change. His recent activities have emphasized the ongoing community changes found along the grassland to (melting) glacier gradient that exists in the Colorado Front Range. He has authored over 150 journal articles and book chapters and is the co-editor of the 2013 volume, *Vulnerability of Ecosystems to Climate*.

Uma Shankar is working as an Assistant Professor, Division of Entomology, Sher-e-Kashmir University of Agricultural Sciences and Technology of Jammu, Faculty of Agriculture, Chatha, India. He has expertise in integrated pest management, economic entomology, and pollination of native pollinators. He has published 40 research papers, authored 3 books, 3 manuals and 5 book chapters in national and international publications. He has research projects on IPM of fruits and vegetables and a Network Project on Insect Biosystematics on Native Hymenopteran bees.

T. V. K. Singh an Entomologist who graduated from Andhra Pradesh Agricultural University (APAU), presently known as Acharya N.G. Ranga Agricultural University (ANGRAU), Hyderabad, Andhra Pradesh, India. He received Master's degree in agricultural entomology from the same university, APAU. He completed his Ph.D. from the Indian Agricultural Research Institute (IARI), Pusa, New Delhi with specialization in insect ecology and insect toxicology. Professor Singh is currently working as Senior Professor at the College of Agriculture, Rajendranagar, Hyderabad, India and is engaged in teaching undergraduate & postgraduate courses and has guided more than 25 post graduate students. The areas of his research include insect population dynamics, insecticide resistance and insecticide resistance management to insecticides and Cry toxins. He is project leader of developing stochastic models for predicting *Helicoverpa*. He has published more than 100 papers and 5 books and is the author of *Insect Outbreaks and their Management* (2009) published by Springer.

Lukasz L. Stelinski earned his Master's and Ph.D. degrees in Entomology from Michigan State University. His graduate study research focused on application of semiochemicals for pest management in small fruit and tree fruit. He is currently an Associate Professor of Entomology and Nematology at the University of Florida Citrus Research and Education Center, Lake Alfred, Florida, USA. His research interests include chemical ecology, insect behavior, insect-plant interactions, vector-pathogen-host interactions, and management of insecticide resistance. The majority of Dr. Stelinski's research has an applied aspect focusing on plant protection from insect pests and plant pathogens.

Meriel Watts is a specialist in the adverse effects of pesticides on human health and the environment; and on non-chemical alternatives, with a bachelor degree in agriculture science and a Ph.D. in pesticide policy. She works mainly for community-based organizations like Pesticide Action Network and International POPs Elimination Network, but also undertakes contracts with UN agencies such as UNEP and FAO. She has been a member of numerous New Zealand government boards and committees on pesticides, is a member of Australia's National Toxic

Network, and runs a small organic farm supplying the local market on Waiheke Island, New Zealand.

WenJun Zhang is Professor of ecology at Sun Yat-sen University, China. He completed his Ph.D. in the Northwest A & F University, China. He was the Postdoctoral fellow and project scientist at the International Rice Research Institute (IRRI) during 1997–2000. He is the editor-in-chief of several international journals. He is now working on computational ecology, network biology, modeling, etc.

Chapter 1
Integrated Pest Management and Pesticide Use

Rajinder Peshin and WenJun Zhang

The king is dead: Long live the king

Contents

1.1	Introduction	2
1.2	Pesticides, Pest Management, and Crop Losses	3
1.3	Integrated Pest Management	7
1.4	United States of America	8
	1.4.1 The Huffaker Project and Consortium for IPM (1972–1985)	9
	1.4.2 IPM Initiative of the Clinton Administration (1993–2000)	9
	1.4.3 National IPM Program and Establishment of IPM Centers	10
	1.4.4 Pesticide Use in US Agriculture	10
1.5	Europe	13
	1.5.1 The Netherlands	17
	1.5.2 Denmark	19
	1.5.3 Sweden	21
1.6	India	23
	1.6.1 1975–1990: Operational Research Project	24
	1.6.2 IPM Programs Since 1993	24
	1.6.3 Pesticide Use in Indian Agriculture	28
1.7	China	30
	1.7.1 Development of IPM in China	31
	1.7.2 Pesticide Consumption and Environmental Impact in China	33
1.8	Conclusion	35
References		38

R. Peshin (✉)
Division of Agricultural Extension Education, Faculty of Agriculture,
Sher-e-Kashmir University of Agricultural Sciences and Technology of
Jammu, Main Campus Chatha,
Jammu-180009, India
e-mail: rpeshin@rediffmail.com; rpeshin@gmail.com

W. Zhang
School of Life Sciences, Sun Yat-sen University,
Guangzhou, China
e-mail: wjzhang@iaees.org; zhjwj@mail.sysu.edu.cn

Abstract Worldwide, integrated pest management (IPM) is the policy decision for pest management. It has been five decades since the development of threshold theory and harmonious control strategies were the domain of pest management research in the USA, Canada, and some parts of Europe. In the 1970s the work on development and validation of IPM technologies started in developing countries. The implementation of IPM and pesticide reduction programs has been in place in the developed and developing countries for the last three to four decades. There are plausible questions raised about the objectives of IPM, adoption of IPM practices, and pesticide use. Questions are also being raised on the use of robust indicators to measure the impact of IPM research and extension. Pesticide use by volume, pesticide use by treatment frequency index, reduction in use of more toxic pesticides, and environmental impact quotient have been used as IPM impact evaluation indicators. Low volume pesticides and transgenic crops both decreased and stabilized pesticide use in the 1990s and early 2000s. Since then, the pesticide sales regained an upward trajectory, and pesticide use in agriculture has increased. Transgenetic crops were thus not proven to be a perfect technique in IPM. We propose that the reduction in pesticide use frequency and the environmental impact quotient be the primary indicators to evaluate the success of IPM programs in the future. We have moved full circle from IPM to integrated pest and pesticide management. This chapter analyzes the development and implementation of IPM programs in the developed and developing countries and their impact on pesticide use.

Keywords Integrated pest management · Integrated pesticide management · Pesticides · Crop losses · USA · Europe · Denmark · Netherlands · Sweden · China · India

1.1 Introduction

Though integrated pest management (IPM) is the accepted policy decision worldwide for pest management and large-scale government IPM programs are operational in more than 60 developing and developed countries (FAO 2011), in reality this is often converted into "integrated pesticide management". The strategy of IPM and its implementation has always struggled with interpretation and true progress with ecologically sound IPM being skewed and sketchy. In many countries pesticide use has increased, despite introduction of higher potency, newer pesticides, and transgenic crops.

There are four schools of thought promoting different options in IPM: one promoting the "dominant paradigm," integrated pesticide management, thus training farmers in the right use of pesticides and to target specific pesticides to minimize selection for resistance, conserve beneficials and reduce health and pollution risks (Cooper and Dobson 2007; HGCA 2009; Popp et al. 2013). The second paradigm is IPM incorporating ecologically sound pest management tactics so that pesticides are essentially a last resort (FAO 2011). The third paradigm promotes a pesticide-free pest management (Ramanjaneyulu et al. 2004, 2007, 2009). The fourth paradigm is using transgenic crops to reduce pesticide (insecticide) use (Perlak et al. 2001; Huang et al. 2002; Bannett et al. 2004).

Despite some notable success, the extension of IPM to ensure wider uptake in the future remains a significant challenge in many systems, not the least because each situation and drivers are subtly different. A review of IPM programs and their effectiveness at delivering greater adoption is in most cases not done or not well documented. In many instances IPM technologies developed at the research level have not been effectively scaled up to industry-wide practice because of the lack of a well-conceived and evaluated extension process and buy-in from industry (farmers and their advisors) (Kogan and Bajwa 1999; Pimentel 2005; Peshin et al. 2012; Peshin 2013). The focus of this chapter is to provide a brief account of IPM programs and initiatives and the resultant pesticide use in the USA, Europe (Denmark, the Netherlands and Sweden), and Asia (China and India).

1.2 Pesticides, Pest Management, and Crop Losses

Synthetic pesticides began their development with the discovery of the insecticidal properties of DDT (dichlorodiphenyltrichloroethane) in 1939 by Paul Müller. In 1948, Paul Müller was awarded the Nobel Prize for discovering the pesticidal properties of DDT. The American entomologists proclaimed in 1944, "… never in the history of entomology has a chemical (DDT) been discovered that offers such promise …." (Perkins 1982, p 10) It has been seven decades since the beginning of the synthetic pesticide era. Pesticides have contributed to the saving of crops from ravages caused by pests, thus indirectly contributing to the world's food production (PSAC 1965; Headley 1968; Pimentel et al. 1978), but their use has also been associated with an increasing percentage of losses by insect pests (Pradhan 1964; USDA 1965; Pimentel 1976; Dhaliwal and Arora 1996; Kogan and Bajwa 1999), and potential human health and environmental problems (Pimentel et al. 1978; Pimentel et al. 1993; Pingali and Roger 1995; Waibel and Fleischer 1998; Pretty et al. 2000; Shetty 2004; Pimentel 2005; Shetty and Sabitha 2009). The problems associated with pesticides in agriculture were recognized by the end of 1950s (Pimentel et al. 1951; Brown 1958). Though, "entomologists continued to maintain that insects could be controlled by many different means, but when drawing up their own research plans, they tended to select a chemical as the foundation of the experimental design (Perkins 1982, p 12)." This bonhomie of the plant protection scientists made them ignore the dysfunctional consequences of the pesticide-intensive pest management. This bonhomie led the scientists and farmers onto a "pesticide treadmill" by not anticipating the problems associated with synthetic organic pesticides (van den Bosch 1978). Pesticide use has dysfunctional consequences on human health from residues on food to exposure while applying pesticides by farm workers (Metcalf 1986; WHO 1990; Dinham 1996; Perkins and Patterson 1997).

In the 1950s some voices were being raised about the overreliance on synthetic pesticides. In the 1950s, in response to the development of insecticide resistance and the destruction of natural enemies of insect pests, four entomologists, V.M. Stern, R.F. Smith, R. van den Bosch, and S. Hegan at the University of California, USA, worked on the concept of IPM. In Canada, efforts were taken for "harmonious control," for harmonizing biological and chemical control of orchard pests (Pickett and Patterson 1953; Pickett et al. 1958). The International Organization for Biological Control of Noxious Animals and Plants (IOBC) in Europe was inspired by the work of Stern and his coworkers (Stern et al. 1959) and Pickett et al. (1958) and established a commission for "integrated control" for fruit orchards in 1959 (Frier and Boller 2009). Though at that point in time, environmental pollution from pesticides was not a concern to entomologists, medical and environmental scientists fathomed the possible human health and environmental consequences (Perkins 1982). However, to save the destruction of non-target insect natural enemies the concept of "integrated control," a combination of biological and chemical control based on economic threshold theory, was put forward by Stern et al. in 1959. The environmental problems associated with the synthetic pesticides were brought to center stage for discussion among the public and scientists by Rachel Carson (1962) after publication of the book *Silent Spring*. The book met fierce opposition from pesticide companies, though it led to the rejection of the proposition of the America entomologists, "... never in the history of entomology has a chemical (DDT) been discovered that offers such promise ..." The book firmly argued that uncontrolled and unexamined pesticide use was harming not only animals and birds, but also humans. It evoked strong criticism by biochemists like Robert White Stevons[1] who proclaimed that the world would return to the "Dark Ages," and "the insects and diseases and vermin would once again inherit the earth" if attention was paid to the book of Rachel Carson. van den Bosch (1978, Preface, p. xv) dismissed the claims of the pesticide lobby, ".... Pesticides were big business in 1962 and still big business and pesticides are ideal products like heroin, they promise paradise and deliver addiction .... Pesticide peddlers .... One cure for addiction: use more and more of the product ...."

Pesticide use increased globally in the 1960s. The pesticide market in the 1960 was worth about half a billion dollars (0.58) and experienced steep growth in the 1960s, 1970s, and 1980s (Table 1.1). In the 1960s, the annual sales growth rate was about 30.5% and in the 1970s growth rate increased to 33% annually. Between 1980 and 1993 the pesticide market grew by 9% annually. However, the percent market share of insecticides and fungicides has decreased, whereas herbicide market share has increased (Fig. 1.1). From 1996 onwards, since the commercial cultivation of transgenic insect resistant crops, the pesticide market has been almost static (0.27% annual growth) up to 2001. In fact, pesticide market has been static since the mid-1980s, only increasing in line with inflation (Dinham 2005). The pesticide market declined by 12% between 1998 and 2003, in real terms, according to Allan Woodburn Associates (Dinham 2005). According to pesticide use data of Agrow (2005) Reports/Wood Mechenzie and Cropnosis (Dewar 2005) the world pesticide market declined from $ 31 billion in 1998 to 29.6 in 1999, to 29.2 (2000), to 27.1 (2001),

---

[1] Chemist from American Cyanamid: Source: http://www.pophistorydig.com/?p=11132

Table 1.1 Worldwide pesticide market (billion US $). (Sources: Madhusoodanan (1996) and my own estimates from 1960 to 1993. Anonymous (1998) and own estimates for 1996. Kiely et al. (2004), 2000 and 2001. Allan Woodburn Associates. (2005), 2004. Agranova (2013a), from 2007 to 2012)

Year	Insecticides	Fungicides	Herbicides	Others	Total
1960	0.21	0.23	0.12	0.02	0.58
1970	1.00	0.60	0.94	0.16	2.70
1980	4.03	2.18	4.76	0.64	11.61
1993	7.59	4.73	11.61	1.37	25.30
1996	9.06	6.56	13.75	1.88	31.25
1998	9.10	6.38	14.68	1.88	31.25
2000	9.10	6.38	14.32	2.96	32.77
2001	8.76	6.03	14.12	2.88	31.76
2004	8.98	7.09	14.83	1.77	32.67
2007	9.37	8.29	16.80	1.72	36.18
2008	10.66	10.55	20.79	1.99	43.99
2009	10.20	10.24	17.87	1.85	40.16
2010	11.04	10.57	17.60	1.96	41.16
2011	11.83	11.73	20.46	2.12	46.14
2012	12.78	13.02	21.87	2.27	49.94

Totals may not add due to rounding

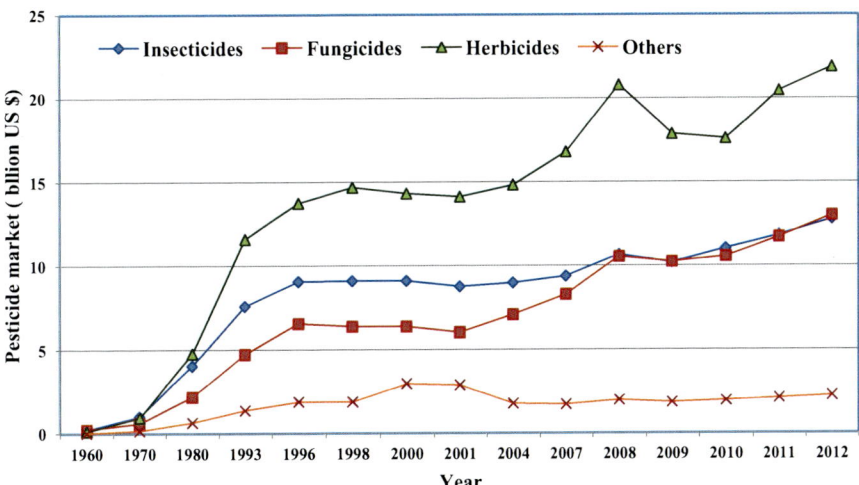

Fig. 1.1 Global insecticide, herbicide, fungicide, and other pesticide markets over time. Between 1960 and 2012, the percent market share of insecticides and fungicides has decreased from 36.2 to 25.6% and 39.7 to 26.1%, respectively, whereas herbicide market has increased from 20.7 to 43.8%

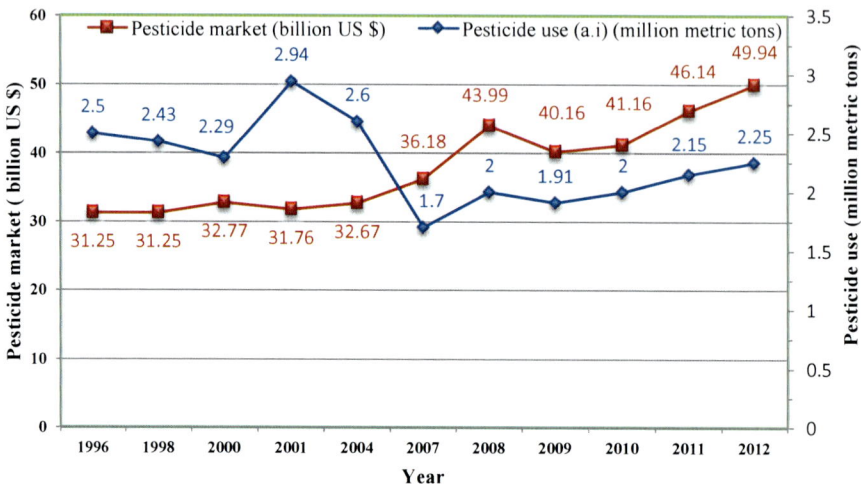

Fig. 1.2 Global pesticide consumption (a.i.) and total pesticide market. The pesticide consumption by mass dropped by 32% between 1996 and 2007 but since then it has increased by 32% by 2012

and to 26.5 (2002). In 2003, it rose to 29.39 (Dewar 2005). The main reason for the decrease in pesticide sales is due to the introduction of transgenic crops. According to the pesticide sales data, the pesticide market did not register growth between 1996 and 2004 (estimates may vary according to source). But since 2004 pesticide market sales started showing an upward movement. In 2004, it increased by 4.6% after inflation (Allan Woodburn Associates 2005). Pesticide use (active ingredients) decreased by 32%, from 2.50 to 1.70 million metric tons, between 1996 (Pimentel 1997) and 2007 (Agranova 2008), (Fig. 1.2). The decrease was driven by many factors, namely the commercial launch of low-volume pesticides (spinosad in 1997; indoxacarb in 2000) replacing some of the organophosphates, growth in cultivation of genetically modified crops which reduced the need for the application of insecticides, and phasing out of insecticide subsidies and development of IPM programs. But since 2007, pesticide use (active ingredients) has increased to 2.25 million metric tons (Agranova 2013a), an increase of 32.35%, of which 24% is consumed in the USA alone, 45% in Europe, and 25% in the rest of the world. The increase in pesticide use has continued since 2007 with the exception of 2009 (Fig. 1.2). The decline in pesticide use by volume in 2009 is attributed to reduced consumption of glyphosate, which constitutes an incredible 20–25% of the total global active ingredient pesticide volume. The estimated pesticide consumption (a.i.) in 2012 was 2.25 million metric tons (Agranova 2013b), an increase of 32.4% over a five-year period with annual average growth rate of 6.5%. In 2011, total volume of pesticide formulations was estimated at 6,985,000 metric tons.[2] This is despite the above-stated facts and mainly driven by increase in herbicide usage. Herbicides account for about 43.80% of the total pesticides sold and the market sales of insecticides and fungicides are almost equal (Fig. 1.3). Pesticides were a big business in the 1960s

---

[2] Personal communication from Dr. R J Bryant, Brychem, UK

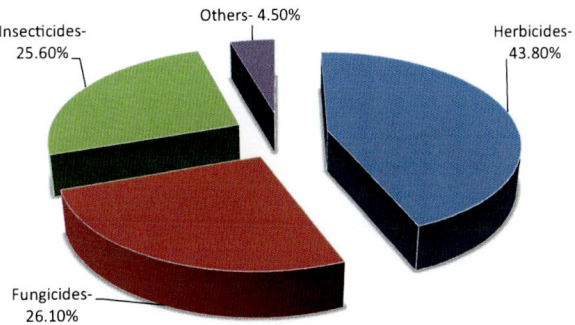

Fig. 1.3 Market share of different groups of pesticides. (Source: Agronova 2013a)

and 1970s (van den Bosch 1978), and continue to be a big business in the twenty-first century, and pesticides are the major pest control paradigm promoted.

However, the crop losses due to pests continue to increase worldwide despite a manifold increase in pesticide use in agriculture since 1960s. For example, crop losses in wheat were estimated at 23.9% in 1964–1965 (Cramer 1967), these losses increased to 34% in 1989–1990 (Oerke et al. 1994). Despite the use of pesticides and implementation of many IPM programs in the last decade of the twentieth century, the crop losses in wheat were estimated at 28.2% (Oerke 2006) which is an increase of 4.3% since 1960s. Similarly, crop losses to pests in cotton crop increased from 24.6% in 1964–1965 (Cramer 1967) to an all-time high of 37.7% in 1988–1990 (Oerke et al. 1994). Since 1996, with the introduction of Bt cotton, the crop losses in cotton declined to 29% for the period 1996–98 (Oerke and Dehne 2004) and 28.8% in 2001–2003 (Oerke 2006). In the rice crop, predominantly cultivated in Asia, the actual losses caused by pests were to the tune of 37% for 2001–2003 period (Oerke 2006).

1.3 Integrated Pest Management

"Integrated Pest Management (IPM)" evolved as a result of the initiatives taken to reduce the complete dependence on synthetic pesticides for managing pests. IPM is, *"A pest management system that, in the context of the associated environment and the population dynamics of the pest species, utilizes all suitable techniques and methods in as compatible a manner as possible, and maintains the pest populations at levels below those causing economically unacceptable damage or loss"*(FAO 1967, p. 19). The term, integrated pest management, was used by Smith and van den Bosch in 1967 (Smith and van den Bosch 1967), and in 1969, the US National Academy of Sciences (1969) formally accepted this term. In 1967, a panel of experts accepted the term "Integrated Pest Control", a synonym for IPM. IPM had been adopted as the main policy, research and extension strategy in the 1970s and 1980s by governments all over the world. The policy decision for research and extension work of IPM was taken by the USA (1972), India (1974), China (1975), Malaysia (1985), the Philippines (1986), Indonesia (1986), Germany (1986),

Denmark (1987), Sweden (1987), and the Netherlands (1991). Billions of tax payers' money has been spent on IPM research and extension since then.

In 2012, the FAO broadened the definition of IPM with stress on the economic, social, and environmental aspects of pest control. It defined IPM as, *"the careful consideration of all available pest control techniques and subsequent integration of appropriate measures that discourage the development of pest populations and keep pesticides and other interventions to levels that are economically justified and reduce or minimize risks to human health and the environment. IPM emphasizes the growth of a healthy crop with the least possible disruption to agro-ecosystems and encourages natural pest control mechanisms"* (FAO 2012).[3]

The definition of IPM though incorporating ecological concerns envisages the use of pesticides as economically justifiable. Therefore, we have moved a full circle to the concept and reality of integrating pesticides with IPM with the caveat to minimize risks to human health and the environment with least possible disruption to agro-ecosystems. According to FAO (2011, p. 76), *"Sustaining IPM strategies requires effective advisory services, links to research that respond to farmers' needs, support to the provision of IPM inputs, and effective regulatory control of chemical pesticide distribution and sale."*

What is the primary quantifiable objective of IPM? Is it to reduce pesticide use in agriculture? If so, would it not be better to state this explicitly as the key objective of IPM and IPM programs so the other elements of IPM would then fall into place automatically (Moss 2010). Therefore, whether pesticide use, either by volume or by treatment frequency index or both, is the most relevant indicator to measure the impact of IPM policies and programs? In the following sections of this chapter we have tried to answer this question and evaluate the impacts of IPM programs on pesticide use.

1.4 United States of America

The history of IPM programs in the USA has been documented by Kogan (1998). With the conclusion of the Integrated Pest Management (IPM) Consortium in 1985 (Frisbie and Adkisson 1985), the United States Department of Agriculture and the Cooperative State Research Stations (USDA-CSRS), a National IPM Coordinating Committee (NIPMCC), was formed (Kogan 1998). This committee provided some funds to support IPM. The NIPMCC and CSRS were the drivers for ushering in the Clinton Administration's National IPM initiative in 1993. Under this initiative IPM practices were to be carried out on 75% of the USA's cropped area by the year 2000 (Sorensen 1994). In the USA, the initiatives for implementing IPM can be divided into three stages: (i) the Huffaker Project and Consortium for IPM, (ii) IPM Initiative of the Clinton Administration, and (iii) National IPM program and the establishment of IPM centers.

---

[3] http://www.fao.org/agriculture/crops/core-themes/theme/pests/ipm/en/

Table 1.2 Extent of adoption of IPM practices in the US agriculture. (Source: USGAO 2001)

Crop	% area estimated by USDA
Cotton	86
Fruits and Nuts	62
Vegetables	86
Soybeans	78
Corn	76
Barley	71
Wheat	65
Alfalfa-hay	40
All other crops and pastures	63

1.4.1 The Huffaker Project and Consortium for IPM (1972–1985)

The Huffaker Project was jointly financed by the United States Department of Agriculture (USDA), National Science Foundation (NSF), and US Environmental Protection Agency (EPA) for a period of five years (1972–1978) (Huffaker and Smith 1972). The Consortium for IPM (CIPM) was the second project funded by the EPA (1979–1981) and USDA and the Cooperative States Research Stations (CSRS) (1981–1985) (Frisbie and Adkisson 1985). Under The Huffaker project, IPM projects were to be carried out on 1.6 million hectares (Kogan 1998) for six crops, namely alfalfa, citrus, cotton, pines, pome and stone fruits, and soybean (Huffaker and Smith 1972). The National IPM Coordination Committee funded through competitive grants short-duration IPM projects after the conclusion of CIPM project in 1985 (Kogan 1998).

The implementation of the Huffaker Project and Consortium for IPM projects led to the reduction by 70–80 % of the use of more environmentally polluting insecticides in 10 years (Huffaker and Smith 1972). The total coverage under these IPM projects was 5.76 million hectares (Frisbie 1985). In 1994, an economic evaluation of 61 IPM programs revealed that IPM methods resulted in lower pesticide use (Norton and Mullen 1994). Earlier, Rajotte et al. (1987) reported about US $ 500 million per year was saved by adoption of IPM practices in the US agriculture by way of reductions in pesticide use.

1.4.2 IPM Initiative of the Clinton Administration (1993–2000)

The National IPM Initiative of the Clinton Administration in 1993 projected the implementation of IPM on 75 % of the US crop area by the year 2000 (Sorensen 1994). The cropland area of IPM adoption in different crops was as high in cotton and vegetables as 86 % (USGAO 2001). The target of achieving implementation of IPM on 75 % of the cropped area was almost achieved (Table 1.2). The use of highly toxic pesticides was reduced by 70–80 % (USGAO 2001). After the review of performance of the National IPM Initiative by the US General Accounting Office (USGAO), the road map for the National IPM program was drawn.

1.4.3 National IPM Program and Establishment of IPM Centers

The federal IPM coordinating committee established in 2003 set overall goals and priorities for the National IPM Program. The goals of this program are (i) to improve the economic benefits of adopting IPM practices, (ii) to reduce the potential risks to human health and the environment caused by pests and the use of IPM practices, and (iii) to minimize adverse environmental effects from pests and the use of IPM practices.

The road map for the National IPM program provided states a grant of US $ 10.75 million annually for IPM extension. Four Regional Pest Management Centers were created in 2000 by the Cooperative Research Education and Extension Service (CSREES) for implementing the IPM in the USA. Four United States Department of Agriculture (USDA) Regional IPM Centers (North Central, North Eastern, Southern, and Western IPM Centers) were established in the USA in 2000 (USDA 2013). In 2004, an interagency national evaluation group was formed to harmonize IPM impact assessment and program evaluation. The logic model of evaluation provides the frame work for assessing the impact. (For details on IPM logic model refer to Peshin et al. 2009a, Chap. 2, Volume 2, and Chap. 18 of this volume. The experiences with three IPM centers are discussed in Chap. 2, 3 and 4, Volume 4 of this series.)

1.4.4 Pesticide Use in US Agriculture

Synthetic pesticides use in the US agriculture started in the 1940s. The near-complete reliance on synthetic insecticides in the USA had arrived by the early 1950s to 1960s (Perkins 1982). Since the implementation of the second large-scale IPM project from 1979 to 1985, known as the Consortium for Integrated Pest Management (Frisbie and Adkisson 1985), pesticide use showed a skewed trend. Though pesticide product formulations changed resulting in lessening human health effects and other risks, pesticide use over time in the USA has increased since the start of first IPM program (Huffaker Project 1972) to the implementation of the fourth Phase of IPM. Pesticide use in US agriculture was about 239 million kg in 1970 (prior to the Huffaker Project) which peaked to 369 million kg (excluding non-conventional pesticides) in 1978 (EPA 1997), mainly on account of widespread use of herbicides. The overall use of all types of pesticides in 1979 was 494 million kg, and at the end of CIPM in 1985 it was reduced to 442 million kg, a decrease of 11 %. The use of conventional pesticides also decreased from 379 to 354 million kg (Table 1.3). This was caused by the fact that newer pesticides were used in dosages which were far less than insecticides and herbicides used in the 1970s. Herbicide and fungicide use

Table 1.3 Pesticide usage (active ingredients) in the US agriculture estimates (million kg). (Sources: EPA (1997, 2004, 2011)-EPA Estimates based on USDA/NASS (www.nass.usd.gov) and EPA Proprietary data)

Year	Herbicides	Insecticides	Fungicides	Total Herbicides, fungicides and insecticides	Other conventional pesticides	Other chemicals	Total
1979	223	82	26	331	48	112	494
1980	229	74	27	330	45	103	479
1981	233	69	28	330	47	98	474
1982	228	64	27	319	46	94	459
1983	206	61	27	294	45	89	429
1984	234	59	25	318	45	88	451
1985	227	57	27	311	43	88	442
1988	204	41	24	269	43	80	393
1990	206	37	23	266	60	74	401
1991	200	35	21	256	65	64	385
1992	204	35	20	259	68	73	401
1993	193	33	21	247	70	75	392
1994	220	36	22	278	74	74	426
1995	209	39	22	270	77	76	423
1996	218	37	23	278	86	69	433
1997	213	36	24	273	75	85	433
1998	211	31	24	266	62	96	425
1999	194	42	20	256	64	113	434
2000	196	41	20	257	71	103	430
2001	196	33	19	248	58	105	411
2002	189	44	18	251	58	108	417
2003	193	36	20	249	54	114	418
2004	193	37	20	250	66	110	425
2005	191	33	21	245	54	102	401
2006	185	31	21	237	55	101	393
2007	200	29	20	249	60	88	398

Totals may not add due to rounding

almost leveled off during this period. However, insecticide use in agriculture was drastically reduced by more than 30% (from 82 to 57 million kg).

Lin et al. (1995) in their working paper on "Pesticide and Fertilizer Use and Trends in U.S. Agriculture" based on USDA pesticide surveys showed an increase in pesticide use on corn, cotton, soybean, wheat, fall potatoes, and other vegetables, citrus, apples, and other fruits from 181 million kg in 1971 (prior to Huffaker Project, 1972) to 242 million kg in 1990 (after the conclusion of CIPM project in 1985), an increase of 33.8%. Insecticide use in these crops decreased to 26 million kg (1990) from 63 million kg (1971), a decrease of 58.69%, whereas herbicide use showed a quantum jump of 88.90% for the same period from 90 to 171 million kg.

Per hectare pesticide use increased from 2.142 to 2.442 kg (USDA Surveys 1964–1992), an increase of 14%. The implementation of the Huffaker Project, Consortium for IPM, and other IPM projects propelled the reduction in the use of insecticides per hectare. The insecticide use per unit area was reduced by a whopping 167% from 0.751 to 0.281 kg/ha between 1971 and 1990 in corn, soybeans, wheat, cotton, potatoes, other vegetables, citrus fruit, apples, and other fruit.[4] The reduction from 1970s is primarily due to replacement of organochlorine insecticides with low dosage highly hazardous insecticides like methyl parathion, terbufos, and chlorpyrifos. The drop in insecticide use is also attributed to banning of the DDT and toxaphene and the use of pyrethroids in cotton (Lin et al. 1994).

The Clinton administration's National IPM Initiative in 1993 is reported to have resulted in adoption of IPM practices on a large scale by the year 2000 (Table 1.2). Conversely, the overall pesticide use estimated for all agricultural crops increased from 401 million kg in 1992 to 430 million kg in 2000 (Table 1.3), an increase of 7%. In this period the use of herbicides decreased from 204 to 196 million kg, and fungicides use also decreased from about 21 to 20 million kg. Despite the introduction (1996) and widespread cultivation of transgenic crops during that period, insecticide use continued to grow. Insecticide use grew by 15%, from 35 to 41 million kg between 1992 and 2000 (Table 1.3). Introduction of herbicide-tolerant crops propelled the consumption of low dosage glyphosate herbicide and decline in the use of other herbicides (Fernandez-Cornejo et al. 2009). Use of herbicides, fungicides, and insecticides decreased by less than 1%, from 259 to 257 million kg for the period from 1992 to 2000 (based on EPA estimates, www.nass.usda.gov).

The United States General Accounting Office (USGAO) in 2001 in its audit of the US IPM programs concluded that the quantities of pesticide used may not be the most appropriate indicator to evaluate the success of the IPM program. Then the question arises: What is a robust indicator to evaluate the IPM success? Have we been searching for evaluation indicators for the last four decades to measure the success of IPM? During these four decades billions of dollars have been spent on development and implementation of IPM programs to reduce the use of toxic pesticides (as originally envisioned). Introduction of low dosage herbicides and insecticides in 1990s lowered the use of pesticides (active ingredients by weight). Therefore, pesticide use frequency is a robust indicator for evaluating the impact of IPM programs. Besides, environmental impact quotient (EIQ) field use rating (Kovach et al. 1992) should be used as an indicator to evaluate the impact of IPM programs in reducing use of more toxic pesticides. The success of IPM programs in terms of extent of adoption of IPM practices by the growers has also been questioned as the rate of adoption of IPM has been slow in the USA (Hammond et al. 2006).

Since the creation of Regional IPM Centers in the USA in 2000, pesticide use has decreased from 430 million kg in 2000 to 398 million kg in 2007. According to EPA (2011) estimates for 2006–2007, the amount of organophosphate insecticides used declined by approximately 63% since 2000, from an estimated 40 million kg in 2000 to 15 million kg in 2007. Glyphosate active ingredient has been the widely used pesticide in agriculture since 2001, and around 84 million kg was used in the

---

[4] USDA, ERS Estimates

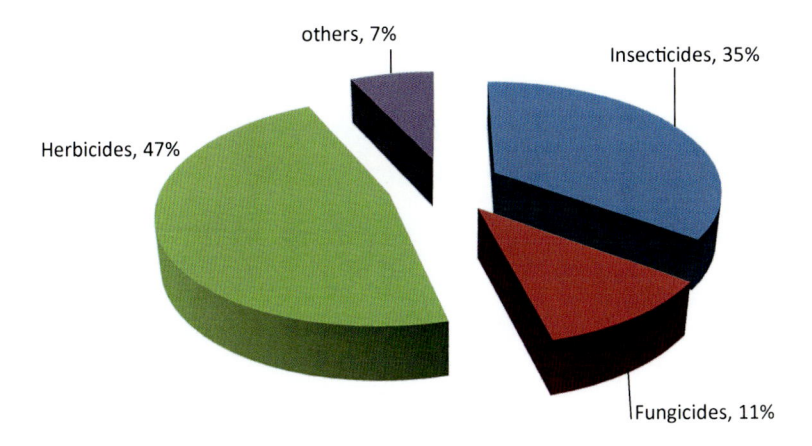

Fig. 1.4 Market share of different types of pesticides in the US-2007

US agriculture in 2007. The total use of herbicides, insecticides, and fungicides in 2007 was 200, 29, and 20 million kg, totaling to 250 million kg, and a skimpy decrease of 2.7% since 2000. Thus the share of glyphosate was more than 33% of the total herbicides, insecticides, and fungicides, and was more than 21% of all types of pesticide use in the US agriculture. However, according to Benbrook (2012), since 1996, transgenic crops increased overall pesticide use by 183 million kg as of 2011. The increase was of 239 million kg in herbicides use, while insecticide use decreased by 56 million kg. (Please refer to Chap. 14, of this volume, and Chap. 2, Volume 4 of this series for the experiences with herbicide resistant and Bt crops in the USA.) Herbicide market share and use by weight is the highest of any pesticide in US agriculture (Figs. 1.4 and 1.5).

To evaluate the progress of IPM programs in achieving their goals, an IPM Performance Measures Working Group by USDA and EPA has been set up. In its first meeting in 2004, the group started work on developing a standard reporting format with common elements for data collection on the outcomes of IPM adoption in the USA. In 2004, the USDA also issued a National Road Map to measure desired outcomes and economic benefits and for reducing the pesticides risks by reducing the use of pesticides (www.goa.gov/products/GAO-01–815).

1.5 Europe

In Europe, IPM programs were developed for orchards (perennial crops). The International Organization for Biological Control of Noxious Animals and Plants (IOBC) established the "Commission on Integrated Control" in 1958 and in 1959 a working group on "Integrated Control in Fruit Orchards" (Freier and Boller 2009). The development and implementation of ecosystem-based technologies in plant protection have been important objectives of the IOBC since its foundation in 1956 (IOBC 2004). The IOBC moved from a biological control concept of pest management to

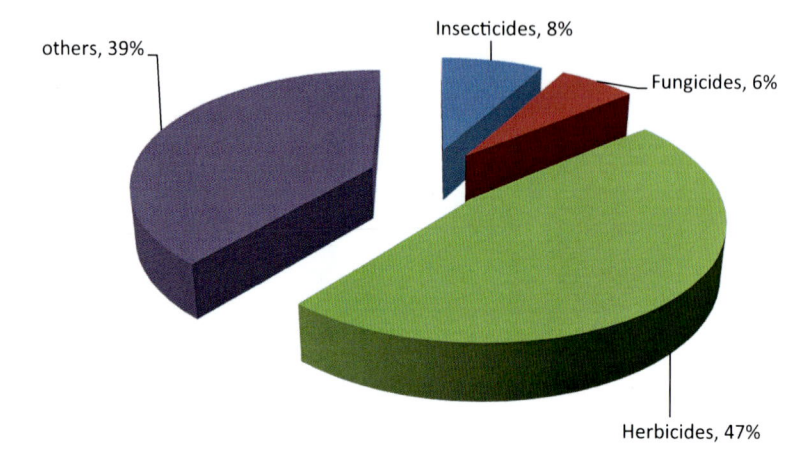

Fig. 1.5 Share of different pesticide groups by volume in the US-2007. The top 10 insecticides used in 2007 were chlorpyrifos, malathion, acephate, naled, dicrotophos, phosmet, phorate, diazinon, dimethoate, and azinphos-methyl. Glyphosate has been the most used pesticide in agriculture since 2001 and around 84 million kg was used in the US agriculture in 2007

IPM to Integrated Production, defined as, "*A farming system that produces high quality food and other products by using natural resources and regulating mechanisms to replace polluting inputs and to secure sustainable farming*" (IOBC 2004, p. 4). In 1974, the IOBC adopted the term "integrated plant protection" and developed IPM systems in all major crops of Europe (Boller et al. 1998). In 1976, integrated production, a concept of sustainable agriculture, was developed by IOBC (IOBS 2004). The European Commission (EC) is promoting low pesticide–input farming in member states, and individual governments will be expected to create the necessary conditions for farmers to adopt IPM. The EC Directive requires member states to establish all necessary conditions for the implementation of IPM by professional pesticide users, and to promote implementation of IPM principles until they become mandatory as of 2014 (EC 2007). In the European Union, IPM is defined through Directive 91/414/EEC: "*The rational application of a combination of biological, biotechnical, chemical, cultural or plant-breeding measures, whereby the use of plant protection products is limited to the strict minimum necessary to maintain the pest population at levels below those causing economically unacceptable damage or loss*" (EC 2007), thereby reducing pesticide use. The EC has been conducting a project called Sustainable Use of Plant Protection Products since 1992. The first phase was concluded in June 1994 with a workshop called "Framework for the Sustainable Use of Plant Protection Products in the European Union." The second phase of the program was initiated in 1994 (EC 2010). The objects of these initiatives were mostly environmental pollution of ground water, surface water, soil, and air by plant protection products; the policy focused on dysfunctional effects of plant protection products themselves and less on use reduction. The Thematic Strategy on the sustainable use of pesticides was adopted in 2006 by the EC, together with a proposal for a Framework Directive on the sustainable use of pesticides (EC 2010).

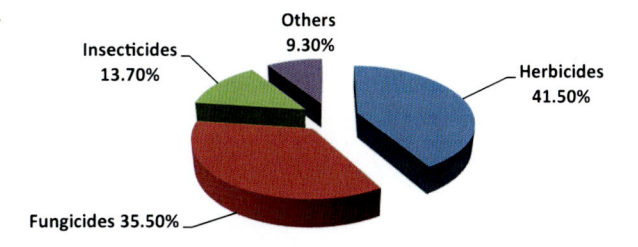

Fig. 1.6 Market share of different types of pesticides in Europe 2010

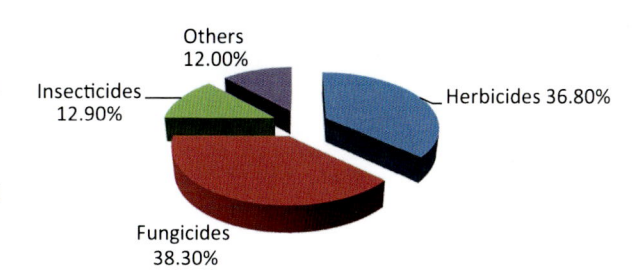

- Herbicides 41.50%
- Fungicides 35.50%
- Insecticides 13.70%
- Others 9.30%

Fig. 1.7 Share of different pesticides by volume in Europe 2010. Annual ECPA crop protection statistical review http://www.ecpa.eu/information-page/industry-statistics-ecpa-total

- Herbicides 36.80%
- Fungicides 38.30%
- Insecticides 12.90%
- Others 12.00%

The member countries in the Organization for Economic Cooperation and Development, commonly known as OECD, conducted a workshop, "OECD Workshop on IPM—Strategies for the adoption and implementation of IPM in agriculture contributing to the sustainable use of pesticides and to pesticide risk reduction," in Berlin, Germany in 2011. The title of the workshop unequivocally confirms that "P" in IPM stands for "Pesticide." Way back in 1998, OECD had also organized a workshop on IPM (OECD 1999). The workshop agreed that IPM can contribute importantly to pesticide risk reduction by: reducing reliance on chemical pesticides and encouraging the use of alternatives, encouraging the use of reduced-risk pesticides when pesticide treatment is necessary, preventing pest problems, to begin with, through better crop management and maintenance of natural resources, and increasing farmer knowledge about agricultural pests and ecosystems.

The European agriculture to date heavily depends on large-scale pesticide use for pest management. Market share of herbicides is the highest (Fig. 1.6) at 41.5 % but fungicide use (a.i.) by volume is the highest (Fig. 1.7), while insecticide use over the years has decreased. Pesticide use pattern since 1992 in EU confirms this trend (Table 1.4). Five countries France (28 %), Spain (14 %), Italy (14 %), Germany (12 %), and the United Kingdom (7 %) accounted for 75 % of the total of 220,000 tons of plant protection chemical consumption in the European Union (Eurostat 2007). Pesticide use by weight is the highest in France and it ranked third in the world as per 2004 data (Aubertot et al. 2005). There has been significant reduction in pesticide use in France, Italy, and United Kingdom in the last 15 years (between 1997 and 2010/2011), with the exception of Germany (Table 1.4). Italy has reduced its use of pesticides in agriculture and horticulture by 56 % followed by the United Kingdom and France (44 %). The increase in pesticide use in Germany during this period is 8 %. There had been a decreasing trend in pesticide use by volume in agriculture in the European Union between 1991 and 1995. The introduction of lower

Table 1.4 Pesticide sale in France, Germany, Italy, and UK (tons active ingredients). (Sources: (1) Eurostat (2013) data source 1997 to 2007 (for Germany 2008 also). (2) FERA (2013) data source UK: 2007 to 2011. (3) BVL (2013) data source Germany: 2009 to 2011. (4) ECPA (2013a) data source France 2008 to 2010. (5) ECPA (2013b) data source Italy 2007 to 2010.)

Country	Pesticide type	1997	1998	1999	2000	2001	2002	2003	2004	2005	2006	2007	2008	2009	2010	2011	Change (%)
Germany	Fungicides	9397	10530	9702	9642	8246	10129	10032	8045	10184	10251	10942	11505	10922	10431	10474	+11.46
	Herbicides	16485	17269	15825	16610	14942	14328	15351	15922	14699	17015	17147	18626	14619	16675	17955	+8.92
	Insecticides	769	1036	953	846	740	742	779	1081	977	813	1092	909	1030	941	883	+14.82
	Other pesticides	4070	4809	3751	3233	3957	4332	4002	3705	3652	3740	3502	3624	3591	3378	3123	−23.27
	Total	30721	33644	30231	30331	27885	29531	30164	28753	29512	31819	32683	34664	30162	31425	33067	+7.64
France	Fungicides	64050	58807	63021	52834	54130	43351	39317	37175	35921	35957	36919	39163	32520	29829	—	−54.43
	Herbicides	33576	36439	42462	30845	32122	28780	24502	26104	29209	23068	26808	27248	22566	22632	—	−32.59
	Insecticides	6074	4672	3612	3103	2487	2308	2224	2460	2505	2140	2100	1254	1212	1033	—	−82.99
	Other pesticides	6092	7835	11406	7912	10896	8009	8481	10360	10630	10447	11428	10912	7470	8410	—	+38.05
	Total	109792	107753	120501	94694	99635	82448	74524	76099	78265	71612	77255	78577	63677	61903	—	−43.62
Italy	Fungicides	52638	53605	52865	52377	48523	63196	54427	52894	53804	50749	20424	23225	19695	18736	—	−64.41
	Herbicides	10536	10665	9741	9507	10063	11829	11587	8947	9206	8924	5829	5027	5578	6259	—	−40.59
	Insecticides	6931	6985	7066	7135	6941	4450	4849	4265	4399	4386	7416	5554	5329	5489	—	−20.81
	Other pesticides	6092	7835	11406	7912	10896	8009	8481	10360	10630	10447	6575	9279	7560	7146	—	+17.30
	Total	84796	84526	82048	79831	76346	94711	86705	84292	85073	81450	40244	43085	38162	37630	—	−55.62
UK	Fungicides	6509	6353	6336	4907	4908	4730	4740	5932	5944	5308	5339	6135	6110	5447	5422	−16.70
	Herbicides	10752	11168	11138	10783	10770	10770	10253	10537	10679	9131	8998	9575	9438	7462	7353	−31.61
	Insecticides	876	965	901	652	650	650	516	557	551	675	537	516	525	436	437	−50.11
	Other pesticides	6351	6896	6924	7259	7198	7198	7054	6437	6427	6037	3829	1879	1757	569	476	−92.51
	Total	24489	25382	25299	23601	23526	23526	22564	23463	23601	21151	18703	18105	17830	13914	13688	−44.11

dose pesticides (especially herbicides, for example, sulfonylurea, glyphosate), was the main driver for this reduction (Ministry of Environment 2003; Lucas and Vall 2003; OECD 2008; Gianessi et al. 2009).

In European countries like Sweden, Denmark, and the Netherlands, programs for reducing pesticide use were initiated in the mid-1980s. The introduction of low-dose pesticides propelled the reduction in pesticide use by volume (26%) between 1991 and 1995, which took place both in the countries implementing pesticide reduction programs (Sweden, Denmark, the Netherlands) and countries with no formal pesticide use reduction programs (Austria, Belgium, France, Germany, Italy, Spain, the United Kingdom) (Urech 1996). This trend reconfirms that pesticide use reduction by weight is not a robust indicator to evaluate the impact of IPM programs. The experiences with IPM and pesticide reduction policies and programs in Europe are covered in Chap. 17, 18, 19, 20, 21 and 22, Volume 4 of this series).

The EU directive 2009/128/EC, to promote IPM for reducing use of chemical pesticides and its adoption by member states, held the international congress on "Pesticide Use and Risk Reduction for future IPM in Europe" in March 2013 in Italy to discuss **regulatory, scientific, and technological information to promote IPM**. Member states of EU have to create the necessary conditions for implementing IPM, which would become mandatory as of 2014 (http://ec.europa.eu/environment/ppps/strategy.htm).

1.5.1 The Netherlands

Pesticide use per unit area is very high in the Netherlands. Between 1984 and 1988, pesticide use per hectare was 20 kg (Proost and Matteson 1997) and the pollution caused by pesticides drew the attention of policy makers. In 1991, a "Multi-Year Crop Protection Plan" (MOANMF 1991) was adopted in the Netherlands, the primary aim of which was a 50% reduction in pesticide use by 2000 (De Jong et al. 2001) and the adoption of non-chemical IPM methods. Annual pesticide sales sank from 21,300 tons in 1985 (David et al. 2000) to about 12,611 tons in 1995 (MOANMF 1996), a decrease of about 41%. Since the implementation of a pesticide reduction program since 1991, pesticide use decreased by 43% in 1996 (PAN Europe 2003) but pesticide use per hectare was still high (Berkhout and van Bruchem 2005). Since 1996 pesticide use has almost stabilized (Table 1.5) and the target of reducing pesticide use by 50% by 2000 was almost achieved.

The targets set for 2004–2010 were 75% reduction in risks by 2005 and 95% by 2010, as expressed by an environmental load indicator (baseline: 1998) (PAN Europe 2007; Statistics Netherlands 2006). These targets were to be achieved by greater adoption of IPM, stricter regulations on pesticide sales and use, improved farmer education, and farm certification. Higher farm gate prices were to be paid to farmers certified as applying "Best Practices" in 2005 on apples, pears, strawberry, parsley, cabbage, and iceberg lettuce (PAN Europe 2007). In 2007, this was expanded to glasshouse production, including tomatoes and sweet peppers. Overall, pesticide use in Dutch agriculture has been reduced by 50% since 1984–1988 but per hectare

Table 1.5 Pesticide use in the Netherlands (tons active ingredients). (Sources: (1) CBS/LEI (n/d), (2) CBS (2004, 2006), (3) CBS (2010), (4) ECPA (2013c))

Pesticide type	1990[1]	1995[2]	1997[1]	1999[2]	2000[2]	2001[2]	2002[2]	2003[2]	2004[2]	2005[3]	2006[3]	2007[3]	2008[3]	2009[4]	2010[4]	Change (%) since 1990
Fungicides	4726	4490	4943	5199	4925	3951	3779	3483	4387	4394	4141	5023	4454	3712	3506	−25.81
Herbicides	4091	3982	3852	3869	3500	3093	4032	3262	3592	3496	3280	3569	3172	2318	2429	−40.62
Insecticides	840	553	486	411	290	276	239	266	248	212	203	214	193	200	198	−64.20
Total (fungicides, herbicides, insecticides)	9657	9025	9821	9479	8715	7320	8050	7011	8227	8102	7624	8806	7819	6230	6113	−36.70
Growth regulators	–	196	–	204	214	181	239	217	218	236	–	225	243	–	–	+23.98
Other pesticides (Including soil disinfectants)	10611	3390	2730	2309	2453	1924	2484	2388	2210	2368	2838	3055	2712	2933	3049	−71.27
Grand total	20268	12611	12001	11992	11382	9425	10773	9616	10655	10704	10462	12086	10744	9163	9182	−54.70 (Pesticide use reduced by 56.89 % since 1985)

Use of soil disinfectants was 44 % of the total pesticide consumption in 1990 (8,938 tons) has reduced to 15 % (1,430 tons) of total pesticide consumption in the Netherlands in 2009. The use of soil disinfectants has reduced by whopping 80 % between 1990 and 2009

use is very high (Berkhout and van Bruchem 2005). Pesticide use was reduced to 7.3 kg/ha in 1998 and it decreased to 6.6 kg/ha in 2004 (CBS 2006), which is still on the high side. Pesticide use in chrysanthemum cultivation is between 40 and 50 kg per hectare, in rose cultivation about 70 kg per hectare, and in potato and onions 10 to 20 kg per hectare (Agricultural Economic Report 2011). In 2000, agriculture and horticulture cut the use of chemical pesticides by over 12 % compared to 1998 (Statistics Netherlands 2006). Compared to 1990, the use of insecticides decreased by more than 64 %. Pesticide use sank from 20,268 tons in 1990 (prior to Multi-Year Crop Protection Plan) to 9,182 tons in 2010, a decrease of 54.70 % (Table 1.5).

1.5.2 Denmark

Denmark was the first country in Europe to formulate an action plan for reducing pesticide use in agriculture. Pesticide use in Denmark increased between 1981 through 1986; there was an increase of about 27 % in 1984 (7,500 tons a.i.) compared to 1981 (6,115 tons a.i.) (PAN Europe 2005). The treatment frequency index of pesticides increased from 1.64 in 1981 to 3.1 in 1985. Between 1981and 1985 it averaged 2.67 (Gianessi et al. 2009) which prompted Denmark to initiate a pesticide use reduction program. In 1986, the first Pesticide Action Plan was put in place (PAN Europe 2003) to target a 25 % reduction in total pesticide consumption by 1992 and 50 % by 1997 (Gianessi et al. 2009). The plan also included measures to encourage the use of less hazardous pesticides. Educating farmers to improve their knowledge and skills in reducing pesticide load was also initiated (Ministry of Environment 2000). Yet, pesticide use increased by 2 % between 1986 and 1992 (Table 1.6). However, between 1993 and 1997 it was reduced by 25.89 %. Pesticide treatment frequency index was reduced from 3.10 in 1985 to 2.45 in 1997 (Jørgensen and Kudsk 2006).

The second Pesticide Action Plan (1997–2003) introduced the indicator treatment frequency index. The Bichel Committee suggested that the treatment frequency index can be reduced from 2.45 in 1997 to between 1.4 and 1.7 by 2007 without adverse economic implications for farmers (Bichel Committee 1999) by adopting IPM practices like damage thresholds, weed harrowing, and other mechanical weed control practices. In 2000, Denmark adopted the Pesticide Action Plan 2. The target was to reach a treatment frequency of less than 2.0 by 2002 (Gianessi et al. 2009) and establish 20,000 ha of pesticide-free zones along key watercourses and lakes. Pesticide use decreased by 2.45 % between 1997 and 2003. The pesticide treatment frequency index was reduced to 2.04 in 2002 (Ministry of Environment 2008). Between 1986 and 2004, farmers had reduced pesticide use by 58 % and the treatment frequency index decreased by 20 % (Jørgensen and Kudsk 2006).

The third Pesticide Action Plan was implemented between 2004 and 2009. The objective of the third Pesticide Action Plan was to lower the treatment frequency below 1.7 by 2009 (Ministry of Environment 2007), to promote pesticide-free cultivation and establish 25,000 ha pesticide-free zones along watercourses and lakes (PAN Europe 2005). This plan included the fruits and vegetables sector for the first time. The plan

Table 1.6 Pesticide use in Denmark (tons a.i.). (Source: The yearly pesticide statistics published by the Danish Environmental Protection Agency http://www.mst.dk/English/)

Pesticide type	1992	1996	1997	1998	1999	2000	2001	2002	2003	2004	2005	2006	2007	2008	2009	2010	2011	Change (%) since 1992
Fungicides	1028	762	869	817	756	650	593	605	580	642	744	576	592	911	484	491	539	−47.57
Herbicides	1739	1869	2726	2619	1892	1982	2164	2105	2205	2087	2308	2479	2583	2813	2012	3172	3512	+101.96
Insecticides	81	46	58	62	60	53	61	56	50	26	38	59	24	40	40	33	30	−62.96
Other pesticides	NA	NA	104	175	135	110	309	146	156	186	209	140	148	287	262	195	158	
Total	2848	2677	3757	3673	2929	2889	3127	2912	2991	2941	3299	3254	3354	4051	2798	3891	4239	+48.84 (−43.48% (since 1984)

provides annual payments of $ 40 million to farmers not using pesticides, $ 24 million for technical assistance, decision support systems, training, and approval procedures, while a pesticide tax was also applied. The Government with this initiative aimed to protect the Danes and nature against undue influence from pesticides and expects the strategy will lead to a reduction in the pesticide load of 40 % over the next three years. Pesticide taxes were increased on insecticides by 54 % and rest by 34 %.

There was a reduction in pesticide use (active ingredients) and the pesticide treatment index, without any reduction in the economic viability of farming (Jørgensen and Kudsk 2006). The reduction in kg active ingredients used was mainly driven by the development of more potent products that are used at lower dosages per ha. As a result of this change, kg active ingredients is not considered to be a valid way of measuring reduction in use of pesticides (Jørgensen and Kudsk 2006).

Nevertheless in 2008, pesticide use jumped to 4,051 metric tons and reached 4,239 tons in 2010 (Table 1.6). Since the end of the first action plan in 1992, pesticide use by weight has increased by 49 % between 1992 and 2010, despite the introduction of low dose pesticides in this period. Pesticide use from 2004 through 2008 stepped up by 37.74 %. Pesticide treatment frequency index increased from 2.51 in 2007 to 2.80 in 2010. It was 2.57 in 2009. Pesticide load per hectare was 2.99 in 2007, 4.48 in 2008, 3.41 in 2009, and 3.92 kg in 2010 (Anonymous 2012). This is a cause of concern. However, if we compare the pesticide use of 1984 (pre-first actions plan) with 2011, Denmark has reduced pesticide use by about 43 % (Table 1.6) but treatment frequency index is almost at the 1984 level.

1.5.3 Sweden

The Swedish Government initiated three pesticide risk-reduction programs since the mid-1980s. The pesticide use reduction program in Sweden can be categorized into three phases: (i) 1986–1990, 50 % reduction (baseline 1981–1985); (ii) 1991–1996, 75 % reduction; and (iii) 1997–2001, no reduction targets but to reduce risks to human health and the environment. In the first phase, 49 % reduction in pesticide use was achieved against a target of 50 %. Against the target of 75 % reduction over a 10-year period (50 %: 1986–1990 and 50 %: 1991–1996) compared to the five-year average during 1981–1985, a reduction of 63 % was achieved against the set target of 75 % (Swedish Board of Agriculture 2009). The average pesticide use by weight between 1981 and 1985 was 4,560 tons and it reduced to averaging 1,690 tons between 1991 and 1995. The reducing trend has been attributed to the reduction in herbicide use due to adoption of low volume herbicides (Pettersson 1994) which has resulted in the reduction of environmental and human health risks.

The pesticides sold for use in agriculture, horticulture, and forestry decreased from a total of 22,800 tons during 1981–1985 to 8,450 tons in 1991–1995, a 63 % reduction (Ekstrom et al. 1996). Pesticide use decreased from 3,120 tons in 1988 to 1,664 tons in 1995. Since 1995, pesticide use in Sweden has stabilized (Table 1.7). The overall pesticide sales have decreased from 5,687 tons in 1988 to 1,652 tons in 2011 (SCB 2012). Per hectare use of pesticide is low in Sweden at 0.390 kg/ha in

Table 1.7 Pesticide use in Swedish agriculture (tons a.i.). (Source: SCB (2002; 2011; 2012) Plant Protection Products in Swedish Agriculture: Statistics, NA: Data not availableNA: Not available)

Pesticide type	1981–1985 (per year average)	1993	1995	1991–1995 (per year average)	1997	1999	2000	2002	2005	2007	2008	2009	2010	2011	Change (%) since 1993
Fungicides	NA	318	200	NA	253	315	233	199	247	219	317	246	221	212	−33.33
Herbicides	NA	1093	975	NA	1303	1285	1364	1447	1280	1345	1472	1090	1205	1404	+28.45
Insecticides	NA	15	17	NA	22	61	20	31	22	38	25	22	19	16	+6.67
Growth regulators	NA	NA	NA	NA	NA	NA	NA	32	17	23	29	27	18	20	
Total	4560	1464	1224	1690	1602	1692	1648	1709	1566	1625	1843	1385	1463	1652	+2.24 (Since 1981–1985: decrease 63.77%)

2011, and ranged between 0.320 and 0.450 kg/ha from 2002 and 2011 (SCB 2012). Herbicide use has always dominated Swedish agriculture. Herbicide use was about 85 % of the total pesticide use in 2011. The quantity of herbicide (a.i.) per hectare decreased from about 1.2 kg in 1981 (Gianessi et al. 2009) to 0.770 kg in 1992 (SCB 2002) and the reduction was attributed to the introduction of low dosage (4 g/ha) sulfonylurea herbicide (Bellinder et al. 1994). In 2002 the quantity further declined to 0.630 kg/ha and was 0.560 kg/ha in 2011 (SCB 2002; SCB 2012).

1.6 India

Many IPM programs have been implemented in India to reduce overreliance on pesticides (mainly insecticides) in cotton, rice and vegetable crops. The first large-scale IPM project was under the Operational Research Project (ORP) in rice and cotton. Besides the ORP, IPM programs implemented were the FAO–Inter-Country Program for IPM in rice crop in 1993, Regional Program on Cotton IPM by Commonwealth Agricultural Bureau International (CABI) in 1993, the Food and Agriculture (FAO)–European Union Program in 2000, the National Agricultural Technology Project for IPM in 2000, and the Insecticide Resistance Management (IRM)-based IPM program in cotton since 2002 (Peshin et al. 2007).

There are multiple public extension systems in India implementing IPM programs. One is the Ministry of Agriculture extension system through its Directorate of Plant Protection Quarantine and Storage, implementing different IPM programs in rice, cotton, vegetables, and oilseeds through 31 Central Integrated Pest Management Centers (CIPMCs) and state departments of agriculture. The second is the Indian Council of Agricultural Research (ICAR) extension system implementing IPM programs through the National Centre for IPM, the Central Institute for Cotton Research (CICR) and its other research institutes and state agricultural universities through ad-hoc IPM projects. Andhra Pradesh Cotton IPM Initiative is another organization implementing IPM. The *Ashta* IPM intervention is implemented in central India. Agriculture Man Ecology (EME) funded by a bilateral agreement between the India and Dutch governments is implementing IPM farmer field schools (FFS) in Karnataka, Andhra Pradesh, and Tamil Nadu. A private sector funded project, Sir Rattan Tata Trust project, in agreement with the Punjab Agricultural University, Ludhiana, has been involved with development, validation, and dissemination of cotton IPM since 2002. Broadly, the IPM implementation in India can be divided into two time periods. These are:

1. IPM programs under Operational Research Project (ORP): 1975–1990
2. IPM programs funded by different agencies since 1993

1.6.1 1975–1990: Operational Research Project

In 1975, the Indian Council of Agricultural Research funded a village level IPM project to test and demonstrate the efficacy, practicability, and economics of IPM. ORP (Operational Research Project) was implemented for the development of location-specific IPM practices in cotton and rice crops. ORP on IPM was implemented in eight states (West Bengal, Orissa, Madhya Pradesh, Andhra Pradesh, Kerala, and Maharashtra in rice and Punjab and Tamil Nadu in cotton) (Krishnaiah 1986). The six ORP projects in rice were implemented under the supervision of the Directorate of Rice Research (DRR), Hyderabad, Kerala Agricultural University, and Department of Agriculture, West Bengal. The highlight of the published literature on the outcomes of ORP in rice and cotton was the reduction in the use of pesticides in cotton and rice (Razak 1986; Sankaran 1987; Dhaliwal et al. 1992; Krishnaiah and Reddy 1989; Simwat 1994; Pasalu et al. 2004). In the rice crop the frequency of pesticide applications decreased from 4–6 in non-ORP areas to an average of 2 in ORP areas (Shankaran 1987). In the cotton crop the insecticide use decreased in the ORP areas compared to non-ORP areas. In Tamil Nadu, farmers covered under the ORP program reduced the insecticide use (a.i.) by weight on cotton by 58.6%, and in Punjab the insecticide applications for sucking pests and bollworms were reduced by 73.3 and 12.4%, respectively (Table 1.8; Dhaliwal et al. 1992; Simwat 1994).

However, the overall trend in the insecticide use in cotton is towards an increasing use of insecticides. In Punjab, the percentage of insecticide costs to the total cost of cultivation increased from 2.1% (1974–1975) to 4.6% (1979–1980). In Indian agriculture total pesticide use between 1973–1974 and 1990–1991 increased from 50,432 to 71,894 tons (Fig. 1.8). Based on the gross cropped area and percentage of total pesticide use in cotton (40–54%) and rice (17–23%), we estimated the per hectare pesticide load in 1990–1991 (at the end of ORP project). Pesticide use (active ingredients) by weight was 4.488 kg/ha in cotton and 0.134 kg/ha in rice (Table 1.9). Though pesticide use per hectare in agriculture in 1970s was only 0.266 kg/ha, this is due to the fact that pesticide use in India varies from crop to crop and region to region. Thus the project reports and published literature on ORP did not reflect on the ground realties on pesticide use in the high pesticide consuming states of Punjab, Andhra Pradesh, and Haryana. However, the ORP project helped in standardizing location-specific technologies in cotton and rice.

1.6.2 IPM Programs Since 1993

The government of India accepted IPM as the main plant protection strategy in mid-1980s. In 1992, Central Integrated Pest Management Centers (CIPMCs) were established by merging Central Plant Protection Stations, Central Surveillance Stations, and Central Biological Stations. The efforts to implement IPM gained momentum in early 1990s. Funded by many donor agencies namely including the FAO, Asian Development Bank (ADB)-CABI, the European Union, and the United Nations De-

1 Integrated Pest Management and Pesticide Use

Table 1.8 Outcomes of the ORP IPM project

Crop	State	IPM tactics	Decrease in pesticide use	Pest problem	Yield	Reference
Cotton	Punjab (1976–1990)	Cultural and mechanical practices Cultivation of short-duration jassid tolerant varieties	(i) 73.3% and 12.4% decrease in average number of insecticide applications for sucking pests and bollworms, respectively (ii) Cost of plant protection increased in case of IPM farmers by 15.8%.	38.5% decrease in bollworm incidence in ORP areas	23.3% increase	Dhaliwal et al. 1992
Cotton	Tamil Nadu (Coimbatore) 1980–1985	IPM practices	(i) Insecticide use (a.i.) by weight decreased from 9.2 to 3.8 kg/ha (ii) Number of insecticide applications was 10.7 in non-IPM area compared to 6.3 in IPM area iii) Expenditure on insecticides reduced by 50.3%, and environmental pollution reduced by 53.4%.	Threefold increase in natural enemy population	IPM area: 2670 kg/ha No-IPM 2,230 kg/ha	Simwat 1994
Rice	Kerala	Cultural practices Pest surveillance Resistant varieties Economic thresholds Conservation of natural enemies	Number of pesticide applications decreased from 4–6 to an average of 2	–	–	Shankaran 1987
Rice	Andhra Pradesh (1981–1986)	Same as in Kerala	–	–	Increase in productivity from 3488 to 4983 kg/ha	Krishnaiah and Reddy 1989

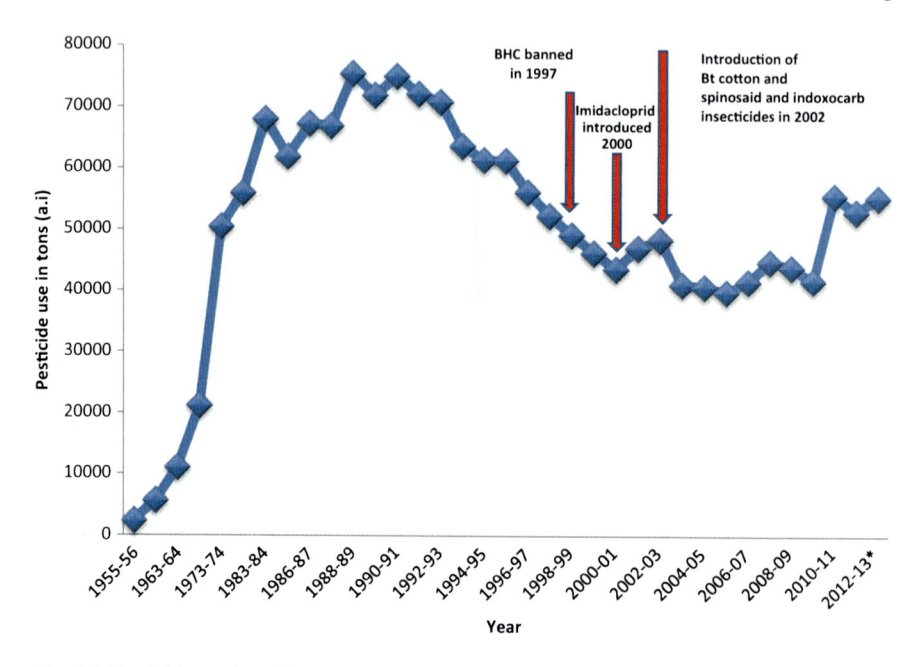

Fig. 1.8 Pesticide use in Indian agriculture

Table 1.9 Pesticides and pesticide formulations banned for manufacture, import, and use

Aldicarb	Chlordane	Endrin	Lindane[a] (Gamma-HCH)	Pentachloro Nitrobenzene	Tetradifon
Aldrin	Chlorofenvinphos	Ethyl Mercury Chloride	Maleic Hydrazide	Pentachlorophenol	Toxaphene (Camphechlor)
Benzene hexachloride (BHC)[b]	Copper Acetoarsenite	Ethyl Parathion	Metoxuron	Phenyl Mercury Acetate	Carbofuron 50% SP
Calcium Cyanide	Dibromochloropropane	Ethylene Dibromide	Nitrofen	Sodium Methane Arsonate	Methomyl 12.5% L and Methomyl 24% formulation
Chlorbenzilate	Dieldrin	Heptachlor	Paraquat Dimethyl Sulphate	TCA (Trichloro acetic acid)	Phosphamidon 85% SL

[a] Banned vide Gazette Notification No S.O. 637 (E) Dated 25/03/2011)-Banned for Manufacture, Import or Formulate from 25th March, 2011 and banned for use from 25th March, 2013.
[b] April 1997

velopment Program (UNDP), IPM programs were implemented in cotton and rice. The Food and Agriculture Organization Inter-Country program for development and application of IPM in rice was operationalized in 1980s. Between 1986 and 1994, a total of 227 demonstrations were organized and 4,951 subject matter specialists were trained (Pawar and Mishra 2004). UNDP IPM program with an outlay of US $ 2.375 million was initiated in 1994 for the development and strengthening of IPM (PAC 1996). This program catered to the development of master trainers, consultancy, and training of trainers. Farmer Centered Agricultural Resource Management (FARM) IPM was started in 1993 for developing the IPM training capacity and building and establishing community based laboratories with the funding of US $ 0.957 million for a period of five years (PAC 1996). Asian Development Bank and Commonwealth Agriculture Bureau International program on cotton IPM for a period of three years was implemented with a budget of US $ 0.153 million. This project helped in the development of human resources and the national IPM program. The National Cotton Research Institute, CICR, was part of this program. Between 1992 and 1997, the Government of India under VIII Plan Period provided budgetary support of US $ 11.25 million (US $1 = Rs. 40 at 2000 rates). Besides the Department of Agriculture and Cooperation, Government of India provided grants to various state governments for IPM training and demonstrations, purchase of bio-pesticides, pheromone traps, and other equipment. FAO-EU cotton IPM project was launched in 1999 in the four states of Karnataka, Maharashtra, Andhra Pradesh, and Tamil Nadu. In the same year, a national plan of action was finalized in consultation with the state departments of agriculture. Under this plan of action, each state has to spend at least 50 % of their agriculture budget on plant protection for promotion of IPM. India adopted the "farmer field school" (FFS) model IPM extension for training the farmers in 1994. The reports on IPM-FFS programs implemented by the Directorate of Plant Protection, Quarantine and Storage, and other agencies reported a decrease in the use of pesticides and an increase in yields (Peshin and Kalra 1998; Peshin 2002). During this period, many pesticides were banned (Table 1.10) and insecticide subsidies were withdrawn and an excise tax of 10 % on pesticides was imposed. This resulted in $60 million annual revenue to Indian Government (Kenmore 1997).

The decade 1990–2000 was most difficult for cotton pest management in India. The IPM programs implemented in cotton in 1990s did not achieve much success in saving the cotton crop from the ravages caused by *Helicoverpa armigera*. The failure of the cotton crop led to farmer suicides. The biological control and bio-pesticide intensive IPM tactics did not work in cotton. With the non-availability of good quality bio-pesticides and biological control organisms, coupled with sub-optimal efficacy under field conditions, cotton cultivators had to depend on insecticides (Kranthi and Russell 2009). IPM packages in cotton were refined to include IRM (Insecticide Resistance Management) as a major component. The Central Institute for Cotton Research (CICR), Nagpur, India, implemented an Insecticide Resistance Management–based IPM (IRM-IPM) program in 10 cotton-growing states of India. These 10 states account for 80 % of insecticide use in cotton (Russell 2004). Between 2002 and 2006, the IRM-IPM project was implemented over 196,000 ha across 1,820 villages in 28 districts of 10 cotton-growing states of India with a funding of US $ 2.58 million from the Ministry of Agriculture, Government of India,

Table 1.10 Pesticide use (a.i.) on cotton and rice in India

Year	Total pesticide use in agriculture[a] (tons)	Pesticide use[a] (kg/ha)	Area under[b] (m ha)		Pesticide use[c] (%)		Estimated pesticide use (tons)[d]		Estimated pesticide use (kg/ha)[d]	
			Cotton	Rice	Cotton	Rice	Cotton	Rice	Cotton	Rice
1984–1985	61881	NA	7.38	126.67	44.5	22.8	27537	14109	3.731	0.111
1990–1991	75033	0.404	7.44	127.84	44.5	22.8	33390	17108	4.488	0.134
1992–1993	70794	0.381	7.54	123.15	40.0	18.0	28318	12743	3.756	0.103
1993–1994	63651	0.341	7.32	122.75	54.0	17.0	34372	10821	4.696	0.088
2006–2007	41510	NA	9.14	123.71	30.0	21.0	12453	8717	1.362	0.070
2010–2011	52979	NA	11.14	125.73	30.0	23.0	15894	12185	1.427	0.097

[a] Pesticide data: Directorate of Plant Protection Quarantine and Storage, Department of Agriculture and Cooperation, Government of India
[b] Directorate of Economics and Statistics, Department of Agriculture and Cooperation, Government of India
[c] Dudani and Sengupta (1991); Unni (1996); Agranova (2008, 2012)
[d] Own estimates

under the Technology Mission on Cotton (TMC) Mini-Mission (MM-II) program (Peshin et al. 2009b). Based on the annual project reports for five years, the overall impact of the project between 2002 and 2006 in terms of the net financial gains to farmers was estimated to be US $ 39.5 million due to US $ 23.0 million from yield increases and US $ 16.5 million from savings on pesticides (ICAR 2007; K.R. Kranthi, unpublished data, CICR). The introduction of Bt cotton in 2002 greatly helped in the revival of cotton productivity in India and reduced per hectare pesticide use (Peshin et al. 2007). (See Chap. 11, Volume 4 of this series for experiences with IPM programs in India based on peer-reviewed evaluation studies.)

1.6.3 Pesticide Use in Indian Agriculture

The total pesticide use in Indian agriculture has declined since 1988–1989 from 75,418 tons to 39,773 tons in 2005–2006 (Fig. 1.8), a reduction of 47.26%. The impact of IPM programs given on the website of the Directorate of Plant Protection Quarantine and Storage, Ministry of Agriculture and Cooperation, Government of India, shows that implantation of IPM programs since 1990s has resulted in reduction of pesticide applications from 50–60% in rice and 19.26–50.50% in cotton, and overall reduction in the use of pesticides from 75,033 to 41,822 tons between 1990–1991 and 2009–2010 (http://ppqs.gov.in/Ipmpest_main.htm). However, the decline in pesticide use was propelled by banning of BHC in 1997 (Peshin et al. 2009c). BHC accounted for about 30% of the total pesticide use. The other drivers for pesticide use reduction are the introduction of low dosage high-potency pesticides, namely imidacloprid in 2000, which replaced the use of the organophosphate group of insecticides for sucking pests in cotton and other crops, spinosad and indoxacarb in 2002, and cultivation of Bt cotton since 2002. The dosage of these insecticides belonging to chloronicotinoids, oxadiazine, and naturalyte is 10–35 times lower than organophosphates. Within a few years since their introduction, the extent of adoption of imidacloprid, spinosad, and indoxacarb (slightly to moderately hazardous) in cotton in 2004–2005 has reached 30, 50, and 35%, respectively (Peshin et al. 2009b). The introduction of Bt cotton in 2002 also contributed to the reduction in pesticide use (Peshin et al. 2007). India is the world's fourth largest grower of GM crops, with a total area of 11.2 million hectares sown in 2012.

Per hectare pesticide use (a.i.) by weight is low in Indian agriculture compared to the many European countries (except Sweden), the USA, Japan, and other developed countries. However, the pesticide consumption varies from region to region and from crop to crop. According to available estimates, out of the total pesticide consumption, 50% and more was used for the cotton crop amounting to $ 340 million, of which $ 254 million (At 2001 prices, 1US$=INR 47) were spent only to control bollworms in non-Bt cotton (Alagh 1988; Mayee et al. 2002). Our estimates show that pesticide use by weight in cotton was and is highest in the pre- and post-IPM era, and 10 years after the introduction of Bt cotton in 2002 (Table 1.10). The introduction of Bt cotton reduced the share of pesticide use in cotton to 30% (Agranova 2008, 2012) from 55%. In a study conducted by Peshin and his coworkers in Indian state of Punjab, the average treatment frequency of insecticide use in non-Bt and Bt cotton was 10.46 and 4.76 respectively and the insecticide use (active ingredients) by weight was 6.440 kg/ha in non-Bt and 2.580 kg/ha in Bt cotton, a difference of 149.86% (Peshin et al. 2007). In vegetable and rice crops, per hectare pesticide consumption by weight was estimated by Birthal and Jha (1997) at 1.108 and 0.306 kg in 1992–1993. Our estimates for different time periods show that pesticide use in cotton was about 4.696 kg/ha in 1993–1994 (pre-Bt cotton era) and it had declined to 1.427 kg/ha in 2010–2011 (Table 1.10).

Despite introduction of low dosage pesticides, banning of high dosage pesticides (Table 1.10), cultivation of Bt cotton, and implementation of many IPM programs, pesticide use after an initial decrease between 2000–2001 and 2005–2006 has increased (Fig. 1.8). Pesticide use for the period from 2005–2006 to 2011–2012 has increased by 39% with annual growth rate of 5.6%. The pesticide sales (at ex-factory level) decreased from 619 in 1998 to 565 million US dollars at May 2013 rates (1US $=Rs. 54) in 2002, a decrease of 8.72%. The pesticide sales stabilized thereafter upto 2007. The sale of pesticides has climbed since then (Fig. 1.9) and reached 1.26 billion US dollars in 2012 (Agranova 2008, 2013a), a steep increase of 128%. There has been a surge in pesticide sales by 21.2% (at constant prices) in 2012 compared to 2011. Over the years there has been a change in the market share of pesticides by type since 1980 (Fig. 1.10). In 1980 the insecticide share was 80% (Phadke 1980) and has reduced to 56% in 2011 (Agranova 2012), and the herbicide market has increased from 7 to 23% for the same period. The promotion of bio-pesticides

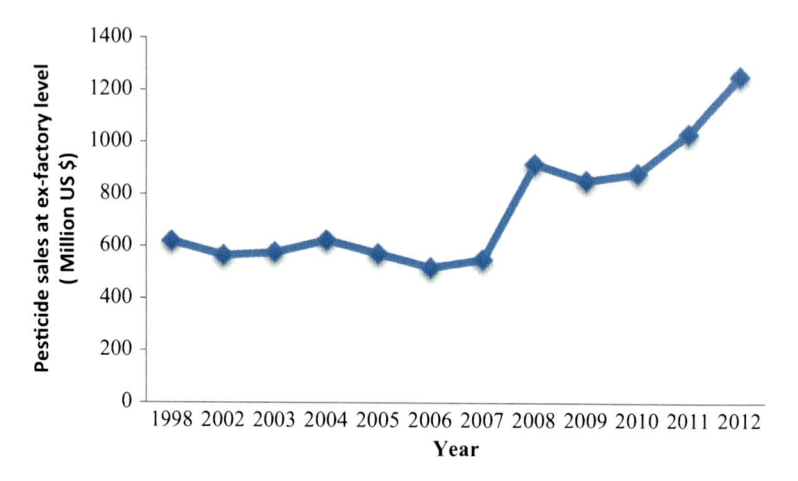

Fig. 1.9 Pesticide sales in India (*at ex-factory level*). The pesticide sales have increased with an average annual growth rate of 21 % since 2007 (*at constant prices*)

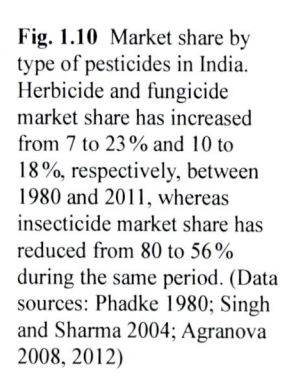

Fig. 1.10 Market share by type of pesticides in India. Herbicide and fungicide market share has increased from 7 to 23 % and 10 to 18 %, respectively, between 1980 and 2011, whereas insecticide market share has reduced from 80 to 56 % during the same period. (Data sources: Phadke 1980; Singh and Sharma 2004; Agranova 2008, 2012)

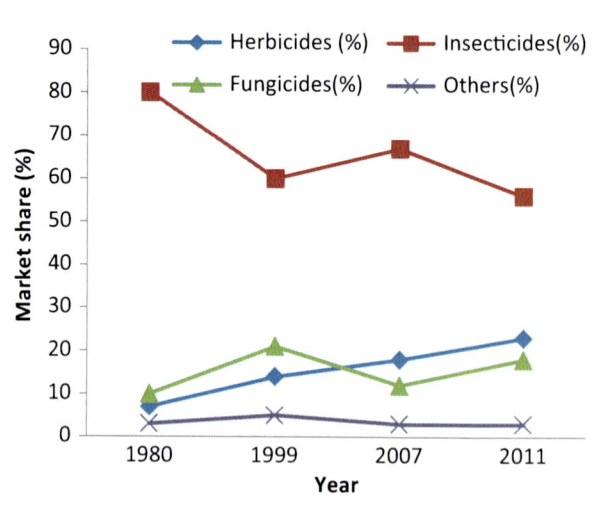

(microbial and botanical pesticides) has led to their increased use in agriculture. In 1996, 219 tons of bio-pesticides were used in Indian agriculture. By 2000, it has increased to 683 tons and has reached 1,262 tons in 2010–2011 (DPPQS 2013).

1.7 China

China is one of the earliest countries to promote integrated control of plant diseases and insect pests. As early as the early 1950s, China put forth the concept of "integrated control" in the research literature (Jing 1997). In 1975, the Chinese plant protection scientists formulated the principle of plant protection "Focus on Prevention and Implement Integrated Control," namely the IPM framework. Lately, this framework was included or realized in some agriculture-related policies, regulations, and provisions in China. Meanwhile the country coordinated and arranged a number of research and promotion programs on IPM and has made great achievements (Zhang et al. 2001).

1.7.1 Development of IPM in China

According to China's level of IPM implementation, the development process of the IPM framework in China can be generally divided into three stages (Wang and Lu 1999):

(i) Pest-centered IPM, that is, the first-generation IPM. For example, during the period of "The Sixth Five Year Plan" (1981–1985), each of the main pests on a certain crop was controlled to a level below the economic threshold using physical, chemical, and biological control methods. In the earlier years of China's reform, pesticide production and imports declined (Table 1.11).

(ii) Crop-centered IPM, that is, the second-generation IPM. For example, during "The Seventh Five Year Plan" (1986–1990), with crop as the focus, a variety of major pests on the crop were controlled. At this stage, IPM emphasized the natural control of pests and IPM systems began to be established. During "The Eighth Five Year Plan" (1991–1995), Many IPM systems have been developed, assembled, improved, and applied. In this period, IPM was demonstrated on more than 200,000 ha of farmland and promoted on more than 6,670,000 ha, and achieved certain positive results.

(iii) Ecosystem-centered IPM is the third-generation IPM. The entire field or regional ecosystem was the focus of IPM; a large quantity of advanced scientific information and data were collected and used, and advanced technologies were developed for IPM practices. Overall, global benefit was expected to be increased with the natural control of ecosystems as the main force. At present, China is in the transition phase between the second- and third-generation IPM.

Migratory locust, *Locusta migratoria manilensis* (Meyen), has historically been a serious insect pest in China. With the focus on environmental conditions and farming systems in the IPM framework, the specific methods to eradicate locust problems were presented in early 1957. The eradication program was organized and invested in by the government. The growth and reproduction of locust was finally inhibited and locust populations was sustainably controlled by transforming habitats, constructing irrigation systems, stabilizing the water table, reclaiming wastelands, implementing crop rotation, planting beans, cotton, sesame, and greening lands (Chen 1979; Ma 1958, 1979).

The rice stem borer, *Scirpophaga incertulas* (Walker), is a serious rice insect pest damaging rice across south China. As early as the 1950s, through investigation and research it was found that adjusting farming systems and selecting appropriate planting dates were the main methods to suppress this pest (Zhao 1958), which has now been applied in IPM practices for the control of this pest.

Table 1.11 Pesticide production, consumption, and import/export of China (10,000 tons). (Source: Ministry of Agriculture of China (http://www.stats.gov.cn/))

Year	Production	Consumption	Import	Export
1983	33.1		6.1	
1984	29.9		5.9	
1985	21.1		1.6	
1986	20.3		0.7	
1987	16.1		1	
1988	17.9		3.4	
1989	20.8		3.7	
1990	22.8		2.8	
1991	25.5	76.1	3.2	
1992	28.1	79.5	3.9	
1993	25.7	84.9	2.3	4.2
1994	29	87.1	3.1	6.1
1995	41.7	108.7	3.4	7.1
1996	42.7	114.1	3.2	7.4
1997	55.2	119.5	4.8	8.8
1998	60.5	123.2	4.4	10.7
1999	62.5	131.2	4.7	14.7
2000	60.7	128	4.1	16.2
2001	78.7	127.5	3.4	19.7
2002	92.9	131.2	2.7	22.2
2003	76.7	132.5	2.8	27.2
2004	87	138.6	2.8	39.1
2005	104	146	3.7	42.8
2006	129.6		4.3	58.3
2007	173.1			
2009	220.0			

In terms of radiation-sterilizing technologies, during the late 1980s about 150,000 radiation-sterilized male *Bactroceraminax* (Enderlein) were released into a citrus orchard with an area of more than 30 ha in Huishui, Guizhou Province, which reduced the citrus injury from 7.5 to 0.005% (Wang and Zhang 1993).

Insect-resistant plant breeding has also been used since the 1950s. Insect-resistant wheat varieties "Xinong6028" and "Nanda 2419" have been bred and planted to successfully control wheat midges (*Sitodiplosis mosellana, Comtarinia tritci* (Kiby)) in north China (Wang et al. 2006). During the 1990s, with government support, transgenic Bt cotton varieties were bred and used to control cotton bollworm and have achieved remarkable success (Zhang et al. 2001; Zhang and Pang 2009). In recent years, the applications of insect-resistant varieties of cotton, rice, wheat, rapeseed, and other crops have also achieved great success in China. According to the statistical data, the total area of transgenic insect-resistant cotton in China has reached 4.667 million hectares, with an average production income of US $ 304.3–342.9/ha (US $1 = 7 RMB Yuan). Annual reduction of chemical pesticide applications has reached 20,000–31,000 tons, equivalent to 7.5% of China's annual total production of chemical insecticides (Zhang and Pang 2009). In general, over the years, IPM programs supported by the Chinese government have demonstrated the positive and significant impact of IPM.

1.7.2 Pesticide Consumption and Environmental Impact in China

China is one of the earliest countries to use pesticides. As early as the Ming Dynasty, the monograph Ben Cao Gang Mu, edited by Li Shizhen, recorded a number of plants and minerals used as pesticides such as veratridine, flavescens, arsenolite, realgar, orpiment and lime, etc. (Chen 2007). China started to manufacture HCH in 1950. In 1957 the first factory in China to produce organophosphorus pesticides was built. During 1960s to 1970s, China mainly manufactured organochlorine, organophosphorus, and carbamate pesticides. Since 1983, China has increased the production of organophosphorus and carbamate pesticides. Meanwhile, pyrethroid and other pesticides were developed (Lin et al. 2000). Since 1994 pesticide exports of China have exceeded its imports. So far, there are more than 2,000 pesticide companies, of which more than 400 companies are manufacturers for original pesticides; more than 300 varieties of original pesticides and 3,000 preparations are being manufactured. China's pesticide production has reached 1.73 million tons (Zhu 2008). China is now the world's largest pesticides producer and exporter, and the second largest consumer of pesticides in the world (Peshin et al. 2009c; See Fig. 1.11 for comparison between China and the developing countries). China has banned the application of high-residual HCH, DDT, and other organochlorine pesticides since 1983. Since 2007, the highly poisonous organophosphorus pesticides, parathionmethyl, parathion, methamidophos, and phosphamidon have been banned for use and sales in China.

In China, rice is the top consumer of pesticides. Rice pesticide sales, accounting for 15% of total sales, reached US $ 538 million in 2006. Vegetable pesticide sales made up 24.2% of total sales.

According to the data from China Customs (Table 1.11; Lan and Bo 2009), in 2008 China imported 44,000 tons of pesticides (US $ 300 million); exported 55,000 tons of fungicides (US $ 240 million), an increase of 5.2% against the export in 2007; exported 136,000 tons of insecticides (US $ 510 million), declining by 1.9%; and exported 277,000 tons of herbicides (US $ 1.23 billion), increasing by 5.1%. In general, herbicides accounted for the greatest portion of total exports.

The excessive pesticide use is serious in China (Table 1.12). Pesticide residues on crop products are a serious problem in China (Table 1.13). Pesticide residues have been detected in grains, rape, vegetables, fruits, tea, and medicinal herbs. A survey on vegetable and fruit markets of China indicated that 41 in 81 vegetable samples were found to have pesticide residues, of which pesticide residues in leek and cabbage exceeded 80 and 60%, respectively, of the national standard.

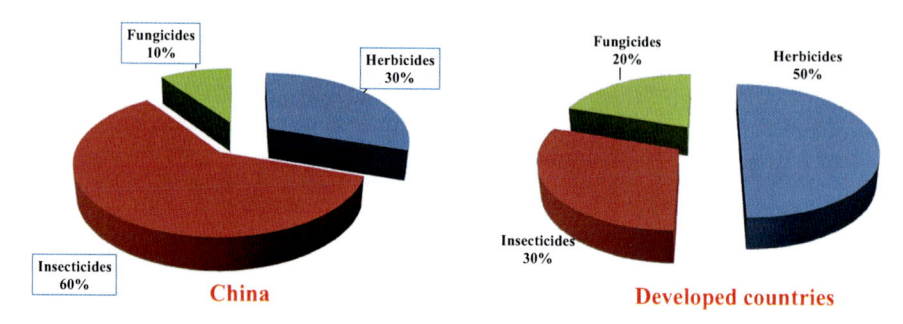

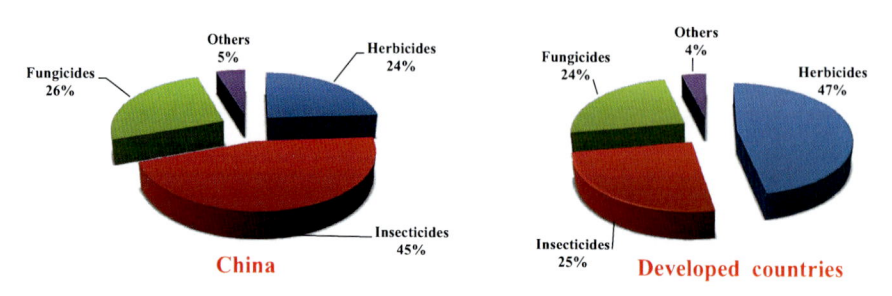

Fig. 1.11 Production and consumption percentages of various pesticides for China (*left*) and developed countries (*right*) in past years. Proportion consumption of insecticides in China was much higher than the developed countries. However, a large number of highly poisonous insecticides have been banned for use in China since 2007. (Peshin et al. 2009c)

Crop pollution has also been caused by pesticide spraying, seed dressing and soaking, and soil treatment with pesticides. Crop pollution could affect crop growth, reducing the yield and quality of crop products. Crop pollution occurs frequently in China, as indicated in Table 1.14.

Growth in pesticide (total pesticides, insecticides, fungicides) use is declining in China due to the implementation of IPM and concerned policies, and the use of low volumes of more toxic pesticides. Moreover, a large number of highly poisonous insecticides have been banned by Chinese government for use in China since 2007, which will greatly promote the development of IPM in China. However, on the whole, the application of IPM technologies in China are still highly localized. Proportionally, the use of insecticides in China was much higher than in the developed countries, although herbicides use in China is greatly increased in recent years. Pesticide misuse is still common and pesticide residue problems are serious. The chemical pesticide use per unit land is 2.6 times that of some developed countries (Liu 2000; Zhang 2001). According to a report in 1999, Anhui Province alone consumed 9,650.89 tons of (active ingredient) pesticides, application dosages reached 0.22 g/m^2, increasing by 43.7 % by weight active ingredients and 24.16% by application dosages over the "The Eighth Five Year Plan" (1991–1995) (Zhang 2001). Exces-

Table 1.12 Application dosage of chemical pesticides in China (kg/ha)

Dosage level	Province or city	Dosage	Dosage level	Province or city	Dosage
I	Shanghai	12.72	(1.5~3.0 kg/ha)	Chongqing	2.47
(>6.0 kg/ha)	Shandong	10.55		Beijing	2.22
	Jiangsu	9.43		Jilin	2.01
	Hubei	7.29		Heilonjiang	1.80
	Hainan	7.12	IV	Shanxi	1.49
	Anhui	7.10	(0.75~1.5 kg/ha)	Sichuan	1.24
	Henan	7.07		Yunnan	0.89
	Zhejiang	6.38		Gansu	0.78
II	Guangdong	5.52	V	Guizhou	0.61
(3.0~6.0 kg/ha)	Jiangxi	5.32	(<0.75 kg/ha)	Shannxi	0.52
	Hunan	5.15		Ninxia	0.34
	Fujian	4.69		Xinjiang	0.26
	Hebei	4.40		Inner Mongolia	0.15
	Liaoning	3.45		Qinghai	0.03
	Tianjing	3.12		Tibet	0.01
III	Guangxi	2.54			

sive use of pesticides in rice and cotton production reached 40 and 50%, respectively (Chen and Han 2005). In recent years the annual number of pesticide poisonings of farmers in Guangdong Province alone has reached 1,500 and is increasing annually.

In general, the application of IPM in China is not widespread. The limited application of IPM in China is attributed to the following reasons: (i) Under the household contract system, agricultural intensification and on-scale operation could not be realized easily, so the farmers have less demand for IPM technologies. (ii) IPM technical extension services systems are insufficient. (iii) Pesticides markets are not ordered and the social environment for IPM application has not yet been established. (iv) We have insufficient theoretical research and application technologies of IPM. At present, IPM technologies are not perfect, and monitoring effectiveness and forecasting accuracy are at a lower level than in the developed countries (Chen and Han 2005).

1.8 Conclusion

From the analysis of IPM and pesticide use in the USA, Europe, and Asia, it is clear that pesticides were and are the primary pest management tools, and the indicators to measure the impact of IPM are not valid, reliable, and robust and raises pertinent questions. First, whether the primary objective of IPM was and is to reduce pesticide use in agriculture? If so, the extent of adoption with respect to the area under an IPM practice is not a reliable and robust indicator to estimate the success of IPM. There are many questions being raised about the measurement of success of IPM pro-

Table 1.13 Pesticide residual situation in China's markets. (Source: Greenpeace (http://www.greenpeace.org/china/zh/))

Sampling markets	Beijing	Shanghai		Guangzhou		Total
	Wal-Mart	Lotus Agriculture	AIB	Vanguard	Trading Market of Crop Products	
No. samples	15	5	5	10	5	45
No. samples with pesticide residuals (percent)	15 (100%)	5 (100%)	3 (60%)	10 (100%)	3 (60%)	40 (89%)
No. samples with more than three kinds of pesticide residuals	13	5	3	9	2	34
No. samples with more than five kinds of pesticide residuals	7	3	3	8	2	25
No. samples with more than 10 kinds of pesticide residuals	2	1	0	2	0	5
No. samples containing illegal or highly toxic pesticides	1	2	1	2	2	9

Sampling markets	Beijing	Shanghai			Guangzhou		Total
	Wal-Mart	Lotus Agriculture	AIB	Trading Market of Crop Products	Vanguard	Trading Market of Crop Products	
No. samples	15	5	5	5	10	5	45
No. samples with pesticide residuals (percent)	15 (100%)	5 (100%)	3 (60%)	4 (80%)	10 (100%)	3 (60%)	40 (89%)
No. samples with more than three kinds of pesticide residuals	13	5	3	2	9	2	34
No. samples with more than five kinds of pesticide residuals	7	3	3	2	8	2	25
No. samples with more than 10 kinds of pesticide residuals	2	1	0	0	2	0	5
No. samples containing illegal or highly toxic pesticides	1	2	1	1	2	2	9

1 Integrated Pest Management and Pesticide Use

Table 1.14 Pesticide pollution to crops and crop products in China

Region	Year	Crop	Area	Crop or Economic Loss	Source
Shannxi	2004	Vegetables	0.0045 million hectares	Yield loss: 7.1%	Zhang 2007
Mudanjiang	2004~2006	Soybean, maize, rice, etc.	0.1223 million hectares	Economic loss: 11.031 million RMB Yuan	Sun et al. 2007
Zhonning, Ningxia	2005~2006	Wolfberry	195 ha	Economic loss: 5.84 million RMB Yuan	Meng et al. 2007
Jiangsu	2000~2005	Rice, wheat, cotton, etc.	0.1285 million hectares	Yield loss: 0.04687 million tons; economic loss: 0.1063 billion RMB Yuan	JSPPS 2006
Anhui	2000~2005	20 kinds of crops, including rice, cotton, soybean, etc.	0.0586 million hectares	Economic loss: 0.223 billion RMB Yuan	Wang et al. 2005
Boxing, Shandong	2004	Cotton	1,000 ha		Li et al. 2005

grams. IPM is aimed to reduce pesticide use in agriculture. However, pesticide use by volume, pesticide use by treatment frequency index, and reduction in use of more toxic pesticides are three commonly used impact indicators. Low volume pesticides and insect-resistant transgenic crops both decreased and stabilized pesticide use in the 1990s and early 2000s. Since then, pesticide sales regained an upward trajectory, and their use in agriculture has increased. Thus, transgenic crops did not prove to be a perfect technique in IPM. On the contrary, cultivation of herbicide-resistant transgenic crops have increased pesticide use. Therefore, reduction in pesticide use frequency and the environmental impact quotient, and not necessarily pesticide use by volume and reduction in use of more toxic pesticides are the primary indicators to evaluate the success of IPM programs in the future. Second, whether IPM stands for "integrated pesticide management" or "integrated pest management"? Pesticides are a much bigger business today than during 1960s and 1970s. Third, what is the way ahead for sustainable pest management? Because of the growing demand for increased crop production worldwide, the intensification of climate change, and other unpredictable factors, IPM will be more important and new guidelines and techniques are needed in the future. It has been four decades since the term IPM was accepted as the primary pest management strategy but we are still struggling with the primary objectives of IPM. Ecologically sound IPM is dead: long live IPM.

Acknowledgments We are grateful to Prof. John Perkins, Member of the Faculty Emeritus,The Evergreen State College, USA, and Dr. Cesar R. Rodriguez-Saona, Department of Entomology, Rutgers University, USA, for reviewing the draft chapter, and for their valuable comments on improving the contents of the chapter.

References

Agranova. (2008). *Agrochemicals-Executive Review* (19th ed.). UK: Agranova.
Agranova. (2012). *Agrochemicals-Executive Review* (23th ed.). UK: Agranova.
Agranova. (2013a). *Agrochemicals-Executive Review* (24th ed.). UK: Agranova.
Agranova. (2013b). *Crop Protection Activities* (online database of global sales and volumes) published by Agranova and updated annually (www.agranova.co.uk)
Agricultural Economic Report. (2011). *Agricultural Economic Report 2011 of the Netherlands Summary*. The Hague, Netherlands: Agricultural Economics Research Institute. www.lei.wur.nl.
Agrow. (2005). First growth in global agrochemical market for a decade. *Agrow, 466,* 18.
Alagh, Y. K. (1988). Pesticides in India agriculture. *Economic & Political Weekly, 20,* 1959–1964.
Allan Woodburn Associates. (1999). *Agrochemicals: Executive Review 1999*. Published by Allan Woodburn Associates, Ltd.
Allan Woodburn Associates. (2005). *Agrochemicals-Executive Review 2005*. Reported In: First growth in global agrochemical market for a decade. *Agrow, 466,* 18.
Anonymous. (1998). *Pesticide Information, 24*(1), 46.
Anonymous. (2012). The agricultural pesticide load in Denmark 2007–2010. *Environmental Review,* 5. http://www2.mst.dk/Udgiv/publikationer/2012/03/978-87-92779-96-0.pdf. Accessed 31 Jan 2013.
Aubertot, J. N., Barbier, J. M., Carpentier, A., Gril, J. J., Guichard, L., Lucas, P., Savary, S., Savini, I., & Voltz, M. (Eds.). (2005). Pesticides, agriculture and the environment: Reducing the use of pesticides and limiting their environmental impact. *Executive Summary of the Expert Report*, INRA et Cemagref (France). http://institut.inra.fr/en/Missions/Inform-public-decision-making/Scientific-Expert-Reports/All-the-news/Pesticides-agriculture-and-the-environment. Accessed 22 Feb 2014.
Bannett, R. M., Ismeal, Y., Kambhampati, U., & Morse, S. (2004). Economic impact of genetically modified cotton in India. *The Journal of Agrobiotechnology Management and Economics, 7,* 96–100.
Bellinder, R., Gummesson, G., & Karlsson, C. (1994). Percentage-driven government mandates for pesticide reduction: The Swedish model. *Weed Technology, 8,* 350–359.
Benbrook, C. M. (2012). Impacts of genetically engineered crops on pesticide use in the U.S.—The first sixteen years. *Environmental Sciences Europe, 24,* 24. doi:10.1186/2190-4715-24-24.
Berkhout, P., & van Bruchem, C. (Eds.). (2005). *Agricultural economic report 2005 of the Netherlands*: English summary. Agricultural Economics Research Institute (LEI), The Hague, The Netherlands. http://ageconsearch.umn.edu/handle/29086. Accessed 22 Feb 2014.
Bichel Committee. (1999). *Report from the Main Committee*. Danish environmental protection agency. http://www2.mst.dk/Udgiv/publications/1998/87-7909-445-7/html/default_eng.htm. Accessed 31 Jan 2012.
Birthal, P. S., & Jha, D. (1997). Socio-economic impact analysis of integrated pest management programmes. In *National Symposium on IPM in India—Constraints and Opportunities* (October 23–24). New Delhi: Indian Agricultural Research Institute.
Boller, E. F., Avilla, J., Gendrier, J. P., Joerg, E., & Malavolta, C. (Eds.). (1998). Integrated production in Europe: 20 years after the declaration of Ovronnaz. *IOBC wprs Bulletin, 21*(1), 41. www.iobc.ch.
Brown, A. W. A. (1958). Insecticide resistance in arthropods. *World Health Organization Monograph Series* 38, Geneva World Health Organization.

BVL (2013). *Absatz an Pflanzenschutzmitteln in der Bundesrepublik Deutschland*. Bundesamt für Verbraucherschutz und Lebensmittelsicherheit (The Federal Office of Consumer Protection and Food Safety), Germany. http://www.bvl.bund.de/SharedDocs/Downloads/04_Pflanzenschutzmittel/meld_par_19_2011_EN.pdf?__blob=publicationFile&v=2. Accessed 16 March 2013.

Carson, R. (1962). *Silent Spring*. New York: Fawcett Crest.

CBS (2004, 2006) Statistical yearbook of the Netherlands 2004 and 2006.Statistics Netherlands, Central Bureau of Statistics (CBS), Wageningen UR and Plant Protection Service; RIVM/CBS (Statistics Netherlands) PrinsesBeatrixlaan 428 2273 XZ Voorburg, The Netherlands.

CBS. (2006). Use of agricultural pesticides stable.*Web Magazine*, 07 February 2006, Statistics Netherlands. http://www.cbs.nl/en-GB/menu/themas/landbouw/publicaties/artikelen/archief/2006/2006-1877-wm.htm. Accessed 22 Feb 2014.

CBS (2010). *Statistical Yearbook 2010*, Statistics Netherlands. www.cbs.nl/statistical yearbook4.

CBS/LEI(n/d). Pesticide use in the Netherlands in the 1990s. http://www.pame.wur.nl/2-plant/envprob/envbx12.htm. Accessed 31 Jan 2013.

Chen, Y. L. (1979). Control and eradicate *Locusta migratoria manilensis* (Meyen). In S. J. Ma (Ed.)., *Integrated Control of Major Insect Pests in China*. Beijing: Science Press.

Chen, Y. X. (2007). *Agricultural Environment Protection*. Bejing: Chemical Industry Press.

Chen, J. L., & Han, Q. X. (2005). Influencing factors and countermeasures of implementing integrated pest management in China. *Journal of Zhongkai University of Agriculture and Technology, 18*(2), 51–58.

Cooper, J., & Dobson, H. (2007). The benefits of pesticides to mankind and environment. *Crop Protection, 26*, 1337–1348.

Cramer, H. H. (1967). Plant protection and world crop production. *Bayer Pflanzenschutz-Nachrichten, 20*, 1–524.

CropLife. (2009). *Facts and Figures—The Status of Global Agriculture*. CropLife International aisbl, 2009. http://www.croplife.org/. Accessed 1 Dec 2012.

David, C., Bernard, C., & Just, F. (Eds.). (2000). *Sustainable Agriculture in Europe State of the Art and Policies in Five European Countries: Denmark, France, Latvia, Netherlands, and Spain*. European research project DG XII environment and climate programme contract no. ENV4-97-0443 and IC20 CT-97–0035. http://adm-websrv3a.sdu.dk/mas/Reports/MASReport1.pdf. Accessed 1 Dec 2012.

De Jong, F. M., De Snoo, G. R., & Loorij, T. P. (2001). Trends of Pesticide Use in the Netherlands. *Mededelingen (Rijksuniversiteit te Gent. Fakulteit van de Landbouwkundige en Toegepaste Biologische Wetenschappen), 66*(2b), 823–834.

Dewar, A. (2005). *Agrow's Top 20: 2005 edition*. T & F Informa UK Ltd.

Dhaliwal, G. S., & Arora, R. (1996). An estimate of yield losses due to insect pests in Indian agriculture. *Indian Journal of Ecology, 23*, 70–73.

Dhaliwal, G. S., Arora, R., & Sandhu, M. S. (1992). Operational research project at the door of cotton growers. *Farmer Parliament, 27*, 15–16.

Dinham, B. (1996). The success of a voluntary code in reducing pesticide hazards in developing countries. *Green Globe Yearbook*. New York: Oxford University Press.

Dinham, B. (2005). Agrochemical markets soar—Pest pressures or corporate design? *Pesticide News*. http://www.pan-uk.org/pestnews/Issue/pn68/pn68p9.htm. Accessed 1 Dec 2012.

DPPQS. (2013). Directorate of plant protection quarantine and storage. Government of India. http://ppqs.gov.in/Ipmpest_main.htm. Accessed 3 March 2014.

Dudani, A. T., & Sengupta, S. (1991). *Status of Banned and Bannable Pesticides*. New Delhi: Voluntary Health Association of India.

EC. (2007). *European Union Policy for a Sustainable use of Pesticides: The Story Behind the Strategy*. Luxembourg: Office for Official Publications of the European Communities. http://ec.europa.eu/environment/ecolabel. Accessed 14 Feb 2013.

EC. (2010). *Thematic Strategy on Sustainable Use of Pesticides*. Europa website, European Commission. http://ec.europa.eu/environment/ppps/synth/contents.htm. Accessed Nov 2012.

ECPA. (2013a). *Annual European Crop Protection Agency Crop Protection Statistical Review: France*. http://www.ecpa.eu/information-page/industry-statistics-france. Accessed 14 Jan 2013.

ECPA. (2013b). ECPA. (2013a). *Annual European Crop Protection Agency Crop Protection Statistical Review: Italy.* http://www.ecpa.eu/information-page/industry-statistics-italy. Accessed 14 Jan 2013.

ECPA. (2013c). *Annual European Crop Protection Agency Crop Protection Statistical Review: Netherlands.* http://www.ecpa.eu/information-page/industry-statistics-netherlands. Accessed 14 Jan 2013.

Ekstrom, G., Hemming, H., & Palmborg, M. (1996). Swedish pesticide risk reduction 1981–1995: Food residues, health hazards, and reported poisonings. *Reviews of Environmental Contaminationand Toxicology, 147,* 119–147.

EPA. (1997). *Pesticides Industry Sales and Usage 1994 and 1995 Market Estimates.* United States Environmental Protection Agency Office of Prevention, Pesticides, and Toxic Substances Washington, D.C. http://www.epa.gov/opp00001/pestsales/95pestsales/market_estimates1995.pdf. Accessed 2 Nov 2012.

EPA. (2004). *Pesticides Industry Sales and Usage 2000 and 2001 Market Estimates.* United States Environmental Protection Agency Office of Prevention, Pesticides, and Toxic Substances. Washington, D.C. http://www.epa.gov/opp00001/pestsales/01pestsales/market_estimates2001.pdf. Accessed 2 Nov 2012.

EPA. (2011). *Pesticide Industry Sales and Usage 2006 and 2007 Market Estimates.* United States Environmental Protection Agency Office of Prevention, Pesticides, and Toxic Substances. Washington, D.C. http://www.epa.gov/opp00001/pestsales/07pestsales/market_estimates2007.pdf. Accessed 2 Nov 2012.

Eurostat. (2007). The use of plant protection products in the European union data 1992–2003. *Eurosat Statistical Book*, European Commission. http://epp.eurostat.ec.europa.eu/cache/ITY_OFFPUB/KS-76-06-669/EN/KS-76-06-669-EN.PDF. Accessed 1 Dec 2012.

Eurostat. (2013). The Statistical Office of the European Union, European Commission. http://epp.eurostat.ec.europa.eu/portal/page/portal/statistics/search_database. Accessed 1 Dec 2012.

FAO. (1967). *Report of the First Session of the FAO Panel of Experts on Integrated Pest Control.* Food and agriculture organization of the United Nations: Rome.

FAO. (2011). *Save and Grow.* Food and agriculture organization. the FAO online catalogue: http://www.fao.org/docrep/014/i2215e/i2215e.pdf. Accessed 21 Dec 2012.

FAO. (2012). *Integrated Pest Management.* http://www.fao.org/agriculture/crops/core-themes/theme/pests/ipm/en/. Accessed 20 June 2012.

FERA. (2013). *Pesticide Usage Survey*, Food & Environment Research Agency, Department for Environment Food and Rural Affairs, Government UK. http://pusstats.csl.gov.uk/myresults.cfm. Accessed 16 March 2013.

Fernandez-Cornejo, J., Nehring, R., Sinha, E. N., Grube, A., & Vialou, A. (2009). Assessing recent trends in pesticide use in US agriculture. *Presented at the Annual Meeting of the Agricultural and Applied Economics Association (AAEA), 26–28 July 2009.* AAEA, Milwaukee, Wisconsin. http://ageconsearch.umn.edu/bitstream/49271/2/Fernandez-Cornejo%20et%20al%20%20-%20AAEA%202009%20-%20Selected%20Paper%20-%20%20April%2019.pdf. Accessed 1 Dec 2012.

Freier, B., & Boller, E. F. (2009). Integrated pest management in Europe—History, policy, achievements and implementation. In R. Peshin & A. K. Dhawan (Eds.)., *Integrated Pest Management: Dissemination and Impact* (Vol. 2) (pp. 435–454). Dordrecht: Springer.

Frisbie, R. E. (1985). Consortium for integrated pest management (CIPM) organization and administration. In R. E. Frisbie & P. L. Adikisson (Eds.)., *Integrated Pest Management on Major Agricultural Systems* (pp. 1–9). College Station, Texas, USA: Texas A & M University, Texas Agriculture Experimental Station.

Frisbie, R. E., & Adkisson, P. L. (Eds.). (1985). *Integrated Pest Management on Major Agricultural Systems.* College Station: Texas A & M University, Texas Agriculture Experimental Station.

Gianessi, L., Rury, K., & Rinkus, A. (2009). Pesticide reduction policies: An evaluation of pesticide use reduction policies in Scandinavia. *Outlooks on Pest Management,* October 2009. doi:10.1564/20oct01.

Hammond, C. M., Luschei, E. C., Boerboom, C. M., & Nowak, P. J. (2006). Adoption of integrated pest management tactics by Wisconsin farmers. *Weed Technology, 20,* 756–767.

Headley, J. C. (1968). Estimating the production of agricultural pesticides. *American Journal of Agricultural Economics, 50,* 13–23.

HGCA. (2009). Pesticide availability for cereals and oilseeds following revision of Directive 91/414/EEC; effects of losses and new research priorities. (2008). *Research Review* No. 70, Home Grown Cereals Authority, UK. http://www.hgca.com/publications/documents/cropresearch/RR70_Research_ReviewX.pdf. Accessed 24 July 2012.

Huang, J., Hu, R., Fan, C., Pray, C. E., & Rozelle, S. (2002). Bt cotton benefits, costs and impact in China. *The Journal of Agrobiotechnology Management and Economics, 5,* 153–166.

Huffaker, C. B., & Smith, R. F. (1972). The IBP program on the strategies and tactics of pestmanagement. *Proceedings of Tall Timbers Ecology and Management Conference, Ecological Animal Control by Habitat Management.* Tallahassee, Florida, USA.

ICAR. (2007). Insecticide resistance management in cotton pests: Success story. *DARE/ICAR Annual Report.* Directorate of Information and Publications of Agriculture (DIPA). New Delhi: Indian Council of Agricultural Research.

IOBC. (2004). Integrated production: Principles and technical guidelines. *IOBC/wprs Bulletin, 27*(2), 1–12.

Jing, D. C. (1997). Sustainable control of agricultural pests: Situation and perspective of IPM. *Guizhou Agricultural Sciences, 25*(Suppl.), 66–69.

Jørgensen, L. N., & Kudsk, P. (2006). Twenty years' experience with reduced agrochemical inputs. *HGCA R & D Conference, Arable Crop Protection in the Balance Profit and the Environment,* 16.1–16.10

JSPPS. (2006). A survey on crop phytotoxicity in Jiangsu province. http://kcxh.jaas.ac.cn/2006/6-13/103026.html. Accessed 1 Dec 2008.

Kenmore, P. E. (1997). A perspective on IPM. *LEISA Newsletter, 13,* 8–9.

Kiely, T., Donaldson, D., & Grube, A. (2004). *Pesticides Industry Sales and Usage 2000 and 2001 Market Estimates,* U.S. Environmental Protection Agency. EPA estimates based on Croplife America annual surveys, Cropnosis limited data, and EPA proprietary data.p6. http://www.epa.gov/opp00001/pestsales/01pestsales/market_estimates2001.pdf. Accessed 6 Dec 2011.

Kogan, M. (1998). Integrated pest management: Historical perspective and contemporary developments. *Annual Review of Entomology, 43,* 243–270.

Kogan, M., & Bajwa, W. I. (1999). Integrated pest management: A global reality? *Anais da Sociedate Entomolgica do Brasil, 28,* 1–25.

Kovach, J., Petzoldt, C., Degnil, J., & Tette, J. (1992). A method to measure the environmental impact of pesticides. *New York's Food and Life Sciences Bulletin, 139,* 1–8. http://dspace.library.cornell.edu/bitstream/1813/5203/1/FLS-139.pdf. Accessed 16 March 2007.

Kranthi, K. R., & Russell, D. (2009). Changing trends in cotton pest management. In R. Peshin & A. K. Dhawan (Eds.)., *Integrated Pest Management: Innovation- Development Process* (Vol. 1) (pp. 499–541). Dordrecht, Netherlands: Springer.

Krishnaiah, K., & Reddy, P. C. (1989). Operational research project on integrated control of rice pests in Medchal area, Ranga Reddy district, A.P.: Achievements and constraints. *Annual Rice Workshop.* Hissar, India: HAU.

Krishnaiah, K. (1986). Status of integrated pest management in rice. *Plant Protection Bulletin, 38*(1), 5–10.

Lan, J. P., & Bo, Y. L. (2009). Pesticide markets of major countries. *World Pesticides, 5,* 8–16.

Li, J. D., Wei, D. A., & Bai, C. J. (2005). Countermeasures of cotton phytotoxicity. In R. Y. Chen, Y. M. Zhao, J. B. Han, et al. (Eds.)., *Proceedings of Production and Development Forum on Cotton in Yellow River Delta* (pp. 236–239).

Lin, B. H., Vandeman, A., Fernandez-Cornejo, J., & Jans, S. (1994). Integrated pest management, how far havewe come? *Agricultural Outlook,* United States Department of Agriculture, Economic Research Service.

Lin, B., Padgitt, M., Bull, M. L., Delvo, H., Shank, D., & Taylor, H. (1995). *Pesticide and Fertilizer Use and Trends in U.S. Agriculture* (AER-717), Agricultural economic report number 717, United States Department of Agriculture, Economic Research Service. http://www.naldc.nal.usda.gov/download/CAT10831382/PDF. Accessed 6 Dec 2011.

Lin, Y. S., Gong, R. Z., & Zhu, Z. L. (2000). *Pesticides and Environmental Protection*. Beijing: Chemical Industry Press.
Liu, S. S. (2000). Opportunity, challenge and strategy of IPM. *Plant Protection, 26*(4), 35–38.
Lucas, S., & Vall, M. P. (2003). *Pesticides in the European Union, Agriculture and Environment and Rural Development: Facts and Figures*. http://ec.europa.eu/agriculture/envir/report/en/pest_en/report_en.htm. Accessed 14 Feb 2013.
Ma, S. J. (1958). Population dynamics of the locust. *Locusta migratoria manilensis* (Meyen), in China. *Acta Entomologica Sinica, 8*(1), 38–40.
Ma, S. J. (1979). *Integrated Control of Major Insect Pests in China*. Beijing: Science Press.
Madhusoodana, S. (1996). Post GATT scenario: Prospects of developing new molecules. *The Pesticide World, March 1996*, 71–73.
Mayee, C. D., Singh, P., Dongre, A. B., Roa, M. R. K., & Raj, S. (2002). *Transgenic Bt cotton* (pp. 1–30). India: Central Institute for Cotton Research.
Meng, Y. J., Tian, W. R., Xie, S. W., Wang, S. D., & Zen, L. (2007). Phytotoxicity of wolfberry and countermeasures in Zhongning, Ningxia. *Journal of Agricultural Sciences, 28*(3), 95–96.
Metcalf, R. L. (1986). Coevolutionary adaptations of rootworm beetles (*Coleoptera*: Chrysomelidae) to cucurbitacins. *Journal of Chemical Ecology, 12*(5), 1109–1124.
Ministry of Environment. (2000). *Background Report for Pesticide Action Plan II*. Copenhagen.
Ministry of Environment. (2003). *Status of the Minister for the Environment's Action plan for Reducing the Consumption of Pesticides*, 1997 updated 24/11/03, Copenhagen.
Ministry of Environment. (2007). *Denmark: Pesticides plan 2004–2009 for Reducing Pesticide Consumption and its Impact on the Environment*. http://www.jki.bund.de/nn_1150582/EN/Home/ReductionofPlantProtection/Denmark.html. Accessed 31 Jan 2012.
Ministry of Environment. (2008). *Bekaempelsesmiddelstatistik 2007, OrienteringfraMiljostyrelsen* Nr. 4 (Denmark). http://www2.mst.dk/udgiv/publikationer/2008/978-87-7052-802-3/pdf/978-87-7052-803-0.pdf. Accessed 31 Jan 2012.
MOANMF. (1991). *The Multi-year Crop Protection Plan*. Summary Ministry of Agriculture, Nature Management and Fisheries. The Hague, Netherlands.
MOANMF. (1996). Information and Knowledge Centre Landbouw/Ede.1996. *Administration Contract Crop Protection Implementation Plan, Progress Report 1995*. Ministry vanLandbouw, Nature Management and Fisheries.
Moss, S. R. (2010). Non-chemical methods of weed control: Benefits and limitations: Keynote address: *Seventeenth Australasian Weeds Conference*, Published by New Zealand Plant Protection Society.
Norton, G., & Mullen, J. (1994). *Economic Evaluation of Integrated Pest Management Programs: A Literature Review* (448–120). Blacksburg, Virginia, USA: Virginia Cooperative Extension Publication.
OECD. (1999). OECD series on pesticides number 8. *Report of the OECD/FAO Workshop on Integrated Pest Management and Pesticide Risk Reduction*.
OECD. (2008). *Environmental Performance of Agriculture in OECD Countries Since 1990*. http://www.oecd.org/document/10/0,3343,en_2649_33793_40671178_1_1_1_1,00.html. Accessed 31 Jan 2012.
Oerke, E. C. (2006). Crop losses to pests. *The Journal of Agricultural Sciences, 144*(1), 31–43. doi:10.1017/S0021859605005708.
Oerke, E. C., & Dehne, H. W. (2004). Safeguarding production—Losses in major crops and the role of crop protection. *Crop Protection, 23*, 275–285.
Oerke, E. C., Dehne, H. W., Schonbeck, F., & Weber, A. (1994). *Crop production and Crop Protection—Estimated Losses in Major Food and Cash Crops*. Amsterdam: Elsevier Science.
PAC. (1996). *India: National IPM Program: A Country Brief*. Programme advisory committee meeting, FAO Intercountry Programme for IPM in Asia, 06–09 February, 1996. Hyderabad, India.
PAN Europe. (2003). *Pesticide Use Reduction is Working: An Assessment of National Reduction Strategies in Denmark, Sweden, The Netherlands and Norway*. http://www.paneurope.info/Resources/Reports/Pesticide_Use_Reduction_is_Working.pdf. Accessed 1 June 2012.

PAN Europe. (2005). *Danish Pesticide Use Reduction Programme- to Benefit the Environment and the Health.* http://www.pan-europe.info/Resources/Reports/Danish_Pesticide_Use_ Reduction_rogramme.pdf. Accessed 6 Dec 2011.

PAN Europe. (2007). *Pesticide Use Reduction Strategies in Europe: Six Case Studies.* Pesticide Action Network Europe. http://www.pan-europe.info/Resources/Reports/Pesticide_Use_Reduction_Strategies_in_Europe.pdf. Accessed 6 Dec 2011.

Pasalu, I. C., Mishra, B., Krrishnaiah, N. V., & Katti, G. (2004). Integrated pest management in rice in India: Status and prospects. In P. S. Birthal & O. P. Sharma (Eds.)., *Integrated Pest Management in Indian Agriculture*, (pp. 237–245). Proceedings 11, National Centre for Agricultural Economics and Policy Reseach and National Centre for Integrated Pest Management, New Delhi, India. http://www.ncap.res.in/upload_files/workshop/wsp11.pdf. Accessed 16 March 2007.

Pawar, A. D., & Mishra, M. P. (2004). Infrastucture incentives and progress in integrated pest management in India. In P. S. Birthal & O. P. Sharma (Eds.)., *Integrated Pest Management in Indian Agriculture* (pp. 237–245). Proceedings 11, National Centre for Agricultural Economics and Policy Reseach and National Cente for Integrated Pest Management, New Delhi. http://www.ncap.res.in/upload_files/workshop/wsp11.pdf. Accessed 16 March 2007.

Perkins, J. H. (1982). *Insects, Experts, and the Insecticide Crisis: The Quest for New Pest Management Strategies.* New York: Plenum Press.

Perkins, J. H., & Patterson, B. R. (1997). Pests, pesticides and the environment: A historical perspective on the prospects for pesticide reduction. In D. Pimentel (Ed.)., *Techniques for Reducing Pesticide Use* (pp. 13–33). Chichester, United Kingdom: Wiley.

Perlak, F. J., Oppenhuizen, M., Gustafson, K., Voth, R., Sivasupramanian, S., Heering, D., Carey, B., Ihrig, R. A., & Roberts, J. K. (2001). Development and commercial use of bollgard ® cotton in the usa-early promises versus today's reality. *Plant Journal, 27,* 489–502.

Peshin, R., & Kalra, R. (1998) Integrated pest management adoption and economic impact at farmers' level. In G. S Daliwal, N. S. Randhawa, R. Arora, & A. K. Dhawan (Eds.)., *Ecological Agriculture and Sustainable Development* (Vol. 2, pp. 624–638). Chandigarh, India: Centre for Research in Rural and Industrial Development.

Peshin, R. (2002) Economic benefits of pest management. In D. Pimentel (Ed.)., *Encyclopedia of Pest Management.* Boca Raton Florida, USA: Marcel and Dekker Inc./CRC Press. doi:10.1201/NOE0824706326.ch92.

Peshin, R. (2013). Farmers' adoptability of integrated pest management of cotton revealed by a new methodology. *Agronomy for Sustainable Development.* doi:10.1007/s13593-012-0127-4.

Peshin, R., Dhawan, A. K., Vatta, K., & Singh, K. (2007). Attributes and socio-economic dynamics of adopting Bt cotton. *Economic and Political Weekly, 45*(51), 73–80.

Peshin, R., Jayaratne, J., & Singh, G. (2009a). Evaluation research: Methodologies for evaluation of management IPM programs (pp. 105–117). In R. Peshin & A. K. Dhawan (Eds.)., *Integrated Pest Management: Dissemination and Impact* (Vol. 2, pp. 31–78). Dordrecht, Netherlands: Springer. doi:10.1007/978-1-4020-8990-9_2.

Peshin, R., Bandral, R. S, Zhang, W., J., Wilson, L., & Dhawan, A. K. (2009b). Integrated pest management: A global overview of history, programs and adoption.In R. Peshin & A. K. Dhawan (Eds.)., *Integrated Pest Management: Innovation- Development Process* (Vol. 1) (pp. 1–49). Dordrecht, Netherlands: Springer. doi: 10.1007/978-1-4020-8992-3_1.

Peshin, R., Dhawan, A. K., Kranthi, K., & Singh, K. (2009c). Evaluation of the benefits of an insecticide resistance management programme in Punjab in India. *International Journal of Pest Management, 55*(3), 207–220.

Peshin, R., Dhwan, A. K., Singh, K., & Sharma, R. (2012). Farmers' perceived constraints in the uptake of integrated pest management practices in cotton crop. *Indian Journal of Ecology, 39*(1), 123–130.

Pettersson, O. (1994). Reduced pesticide use in scandinavian agriculture. *Critical Reviews in Plant Sciences, 13*(1), 43–55.

Phadke, A. D.(1980). Pesticides: Market and marketing in India. In B. V. David (Ed.)., *Indian Pesticide Industry: Facts and Figures* (pp. 147–150). Mumbai: Vishwas Publications.

Pickett, A. D., & Patterson, N. A. (1953). The influence of spray programme on the fauna of apple orchards in Nova Scotia IV: A Review. *Canadian Entomologist, 85*, 472–487.

Pickett, A. D., Putman, W. L., & Le Roux, E. J. (1958). Progress in harmonizing biological and chemical control of orchard pests in Eastern Canada. *Proceedings of Tenth International Congress of Entomology, 3*, 169–174.

Pimentel, D. (1976). World food crisis: Energy and pests. *Bulletin of Entomological Society of America, 22*, 20–26.

Pimentel, D. (1997). Pest management in agriculture. In D. Pimentel (Ed.)., *Techniques for Reducing Pesticides: Environmental and Economic Benefits* (pp. 1–12). Chichester, United Kingdom: Wiley.

Pimentel, D. (2005). Environmental and economic costs of the application of pesticides primarily in the United States. *Environment, Development and Sustainability, 7*, 229–252.

Pimentel, D., Acguay, H., Biltonen, M., Rice, P., Silva, M., Nelson, J., Lipner, V., Giordano, S., Harowitz, A., & D'Amore, M. (1992). Environmental and economic cost of pesticide use. *Bioscience, 42*(10), 750–760.

Pimentel, D., Schwardt, H. H., & Norton, L. B. (1951). New methods of house fly control in dairy barns. *Soap and Sanitary Chemicals, 27*, 102–105.

Pimentel, D., Krummel, J., Gallahan, D., Hough, J., Merrill, A., Schreiner, I., Vittum, P., Koziol, F., Back, E., Yen, D., & Fiance, S. (1978). Benefits and costs of pesticide use in United States food production. *Bioscience, 28, 772*, 778–784.

Pingali, P. L., & Roger, P. A. (1995). *Impact of Pesticides on Farmers' Health and the Rice Environment*. Dordrecht, Netherlands: Kluwer Academic Press.

Popp, J., Peto, K., & Nagy, J. (2013). Pesticide productivity and food security: A review. *Agronomy for Sustainable Development, 33*, 243–255. doi:10.1007/s13593–012-0105-x.

Pradhan, S. (1964). Assessment of losses caused by insect pests of crops and estimation of insect population. In *Entomology in India, 1938–1963*. New Delhi: Entomological Society of India.

Pretty, J. N., Brett, C., Gee, D., Hine, R., Mason, C. F., Morison, J. I. L., Raven, H., Rayment, M., & van der Bijl G. (2000). An assessment of the total external costs of UK agriculture. *Agricultural Systems, 65*(2), 113–136.

Proost, J., & Matteson, P. (1997). Reducing pesticide use in the Netherlands: With stick and carrot. *Journal of Pesticide Reform, 7*(3), 2–8.

PSAC. (1965). *Restoring the Quality of Our Environment. Report of the Environmental Pollution Panel, President's Science Advisory Committee*. Washington, D.C.: The White House.

Ramanjaneyulu, G. V., Kavitha, K., & Hussain, Z. (2004). *No Pesticides No Pests*. Centre for Sustainable Agriculture. http://www.csa-india.org. Accessed 16 March 2012.

Ramanajaneyulu, G. V., & ZakirHussain. (2007). Redefining pest management: A case study of punukula. In *Sustainable Agriculture-A Pathway to Eliminate Poverty*, SUSTAINET, GTZ.

Ramanjaneyulu, G. V., Chari, M. S., Raghunath, T. A. V. S. Hussain, Z., & Kuruganti, K. (2009). Nonpesticidal management: Learning from experiences. In R. Peshin & A. K. Dhawn (Eds.), *Integrated Pest Management: Innovation Development Process* (Vol.1) (pp. 543–573). Dordrecht, Netherlands: Springer.

Rajotte, E. G., Norton, G. W., Kazmierczak, R. F., Lambur, R. F., & Allen, W. A. (1987). *The National Evaluation of Extension's Integrated Pest Management (IPM) Program*. Blacksburg, Virginia, USA: Virginia Polytechnic and State University.

Razak, R. L. (1986). Integrated pest management in rice crop. *Plant Protection Bulletin, 38*, 1–4.

Russell, D. (2004). Integrated pest management in less developed countries. In Horowitz, A. R., Ishaaya, I. (Eds.)., *Insect Pest Management: Field and Protection Crops* (pp. 141–179). New Delhi: Springer.

Sankaran, T. (1987). Biological control in integrated pest control – Progress and perspectives in India. In N. Mohandas, Association for Advancement of Entomology (India) (Eds.)., *Proceedings of the National Symposium on Integrated Pest Control Progress and Perspectives*, (pp. 151–158). October 15–17. Trivandrum, India.

SCB. (2002). *Pesticides in Agriculture in 2001*. Estimated number of doses final statistics, plant protection products in Swedish agriculture. Number of hectare-doses in 2001: Final Statistics. http://www.kemi.se/Start/Statistik. Accessed 1 Dec 2012.

SCB. (2011). *Pesticides in Agriculture in 2010*. Estimated number of doses final statistics, plant protection products in Swedish agriculture. Number of hectare-doses in 2010: Final Statistics. http://www.kemi.se/Start/Statistik. Accessed 1 Dec 2012.

SCB. (2012). *Pesticides in Agriculture in 2011*. Estimated number of doses final statistics, plant protection products in Swedish agriculture. Number of hectare-doses in 2011. http://www.scb.se/Pages/PublishingCalendarViewInfo____259924.aspx?PublObjId=18586. Accessed 1 Dec 2012.

Shetty, P. K. (2004). Socio-ecological implications of pesticide use in India. *Economic and Political Weekly, 39*(49), 5261–5267.

Shetty, P. K., & Sabitha, M. (2009). Economic and ecological externalities of pesticide use in India. In R. Peshin & A. K. Dhawan (Eds.), *Integrated pest Management: Innovation Development Process* (Vol. 1, pp. 113–129). Dordrecht, Netherlands Springer.

Simwat, G. S. (1994). Modern concepts in insect pest management in cotton. In G. S. Dhaliwal & R. Arora (Eds.), *Trends in Agricultural Insect Pest Management* (pp. 186–237). New Delhi: Commonwealth Publisher.

Singh, A., & Sharma, O. P. (2004). Integrated pest management for sustainable agriculture. In P. S. Birthal & O. P. Sharma (Eds.), *Integrated Pest Management in Indian Agriculture* (pp. 237–245). Proceedings 11, National Centre for Agricultural Economics and Policy Research and National Cente for Integrated Pest Management, New Delhi. http://www.ncap.res.in/upload_files/workshop/wsp11.pdf. Accessed 16 March 2007.

Smith, R. F., & van den Bosch, R. (1967). Integrated control. In W. W. Kilgore & R. L. Doutt (Eds.), *Pest Control: Biological, Physical, and Selected Chemical Methods* (pp. 295–340). New York: Academic Press.

Sorensen, A. A. (1994). *Proceedings of National Integrated Pest Management Forum Arlington*. DeKalb, Illinois, USA: American Farmland Trust.

Statistics Netherlands. (2006). *Use of Agricultural Pesticides Stable*. Web magazine, 7 Feb 2006. statistics Netherlands, PrinsesBeatrixlaan 428, 2273 XZ Voorburg, Netherlands. http://www.cbs.nl/en-GB/menu/themas/landbouw/publicaties/artikelen/archief/2006/2006-1877-wm.htm. Accessed 12 Oct 2012.

Stern, V. M., Smith, R. F., van den Bosch, R., & Hagen, K. S. (1959). The integrated control concept. *Hilgardia, 29*, 81–101.

Sun, G. H., Yu, G. S., & Zhang, W. (2007). Situation and causes of crop phytotoxy by herbicides in Mudanjiang during 2004 and 2006. *China Plant Protection, 20*(3), 32–33.

Swedish Board of Agriculture. (2009). *Jordbruks Statistisk Arsbok 2001–2008*. http://www.sjv.se/amnesomraden/47tatistic/ja.4.7502f61001ea08a0c7fff104195.html. Accessed 14 Feb 2013.

Unni, K. K. (1996). Role of agrochemical industry in Ninth Five Year Plan. Paper presented in: *National Seminar on Agricultural Development Perspective for the XIth Five Year Plan* (pp. 13–15). Ahmadabad: Indian Institute of Management.

Urech, P. A. (1996). Is more legislation and regulation needed to control crop protection products in Europe? *Brighton Crop Protection Conference—Pests and Diseases* (pp. 549–557).

US National Academy of Sciences. (1969). *Insect Pest Management and Control*. Washington, D.C: National Academy of Sciences Publication 1695.

USDA. (2013). *National Information System for the Regional IPM Centers*. United States Department of Agriculture, National Institute of Food and Agriculture. http://www.ipmcenters.org/Projects/. Accessed 7 Dec 2012.

USDA Surveys. (1964–1992). National Agricultural Statistics Service (USDA/NASS). Agricultural chemical usage. http://www.ers.usda.gov/media/871516/arei32.pdf. Accessed 2 Nov 2012.

USDA. (1965). Losses in agriculture. *Agriculture Handbook No. 291*. Washington, D.C.: Agricultural Reseach Service, US Government Printing Office.

USGAO (United States General Accounting Office). (2001). *Agricultural Pesticides Management: Improvements Needed to Further Promote Integrated Pest Management*. http://www.gao.gov/new.items/d01815.pdf.

van den Bosch, R. (1978). *The Pesticide Conspiracy*. Berkeley: University of California Press.

Waibel, H., & Fleischer, G. (1998). *Kosten und Nutzen des chemischen Pflanzenschutzes in der deutschen Landwirtschaft aus gesamtwirtschaftlicher Sicht* (Social Costs and Benefits of Chemical Plant Protection in German Agriculture). Kiel, Germany: Vauk Verlag.

Wang, H. L., Wang, B. L., & Li, Z. W. (2006). Advance in research on integrated Pest management. *Journal of Henan Institute of Science and Technology, 34*(3), 40–42.

Wang, H. S., & Zhang, H. Q. (1993). Release effect of radiation sterilized female Bactroceraminax (Enderlein). *Acta Agriculturae Nucleatae Sinica, 14*(1), 26–28.

Wang, M. Y., Han, D. Y., Gen, J. G., Jing, G. L., Zhou, Q. F., & Wang, S. Q. (2005). Situation and countermeasures of crop phytotoxicity by pesticides in Anhui Province. *Anhui Agricultural Science Bulletin, 11*(5), 44–45.

Wang, Q. D., & Lu, G. Q. (1999). A brief analysis on IPM in China. *Hubei Plant Protection, 6*, 30–32.

WHO (World Health Organization). (1990). *Public Health Impact of Pesticides Used in Agriculture*. Geneva: World Health Organization, WHO-UNEP.

Zhang, J. X. (2001). A study on strategy of plant protection development. *Plant Protection, 27*(5), 36–37.

Zhang, Z. L. (2007). Situation and countermeasures of phytotoxicity of facility vegetables in Shannxi province. *Shaanxi Journal of Agricultural Sciences, 1,* 108–111.

Zhang, L. G., Shi, S. B., & Zhang, Q. D. (2001). IPM in Hubei. *Hubei Plant Protection, 3,* 3.

Zhang, W. J., Jiang, F. B., & Ou, J. F. (2011). Global pesticide consumption and pollution: With China as a focus. *Proceedings of the International Academy of Ecology and Environmental Sciences, 1*(2), 125–144.

Zhang, W. J., & Pang, Y. (2009). Impact of IPM and transgenics in the Chinese agriculture. In R. Peshin & A. K. Dhawan (Eds.), *Integrated Pest Management: Dissemination and Impact* (Vol. 2, pp. 525–553). Dordrecht, Netherlands: Springer.

Zhao, S. H. (1958). Preliminary results for control experiment on *Tryporyzaincertulas* (walker). *ScientiaAgriculturaSinica, 3*(1), 25–38.

Zhu, W. J. (2008). Production situation and development trend of Chinese pesticides. *Journal of China Agrochemicals, 5,* 40–44.

Chapter 2
Environmental and Economic Costs of the Application of Pesticides Primarily in the United States

David Pimentel and Michael Burgess

Contents

2.1	Introduction	48
2.2	Public Health Effects	49
	2.2.1 Acute Poisonings	49
	2.2.2 Cancer and Other Chronic Effects	49
	2.2.3 Pesticide Residues in Food	51
2.3	Domestic Animal Poisonings and Contaminated Products	51
2.4	Destruction of Beneficial Natural Predators and Parasites	52
2.5	Pesticide Resistance in Pests	54
2.6	Honeybee and Wild Bee Poisonings and Reduced Pollination	56
2.7	Crop and Crop Product Losses	57
2.8	Ground and Surface Water Contamination	59
2.9	Fishery Losses	59
2.10	Wild Birds and Mammals	60
2.11	Microbes and Invertebrates	62
2.12	Government Funds for Pesticide Pollution Control	63
2.13	Ethical and Moral Issues	64
2.14	Conclusion	65
References		66

Abstract An obvious need for an updated and comprehensive study prompted this investigation of the complex of environmental and economic costs resulting from the nation's dependence on pesticides. Included in this assessment of an estimated $9.6 billion in environmental and societal damages are analyses of: pesticide impacts on public health; livestock and livestock product losses; increased control expenses resulting from pesticide-related destruction of natural enemies and from

---

D. Pimentel (✉)
Department of Entomology, Department of Ecology and Evolutionary Biology, Cornell University, Tower Road East, Blue Insectary-Old, Room 165, Ithaca, New York 14853, USA
e-mail: dp18@cornell.edu

M. Burgess
Department of Entomology, Cornell University, Horticulture, Research Aide, Greenhouse worker, Tower Road East, Blue Insectary-Old, Room 161, Ithaca, New York 14853, USA
e-mail: mnb2@cornell.edu

the development of pesticide resistance in pests; crop pollination problems and honeybee losses; crop and crop product losses; bird, fish, and other wildlife losses; and governmental expenditures to reduce the environmental and social costs of the recommended application of pesticides.

The major economic and environmental losses due to the application of pesticides in the USA were: public health, $1.1 billion year; pesticide resistance in pests, $1.5 billion; crop losses caused by pesticides, $1.4 billion; bird losses due to pesticides, $2.2 billion; and groundwater contamination, $2.0 billion.

Keywords Agriculture · Costs · Crops · Environment · Livestock · Natural resources · Pesticide · Pesticide resistance · Public health

2.1 Introduction

Worldwide, about 3 billion kg of pesticides is applied each year with a purchase price of nearly $40 billion year^{-1} (PAN-Europe 2003). In the USA, approximately 500 million kg of more than 600 different pesticide types are applied annually at a cost of $10 billion (Pimentel and Greiner 1997). Despite the widespread application of pesticides in the United States at recommended dosages, pests (insects, plant pathogens, and weeds) destroy 37% of all potential crops (Pimentel 1997). Insects destroy 13%, plant pathogens 12%, and weeds 12%. In general, each dollar invested in pesticide control returns about $4 in protected crops (Pimentel 1997).

Although pesticides are generally profitable in agriculture, their use does not always decrease crop losses. Despite the more than 10-fold increase in insecticide (organochlorines, organophosphates, and carbamates) use in the United States from 1945 to 2000, total crop losses from insect damage have nearly doubled from 7 to 13% (Pimentel et al. 1991). This rise in crop losses to insects is, in part, caused by changes in agricultural practices. For instance, the replacement of corn-crop rotations with the continuous production of corn on more than half of the corn acreage has resulted in an increase in corn losses to insects from about 3.5 to 12% despite a more than 1000-fold increase in insecticide (organophosphate) use in corn production (Pimentel et al. 1991). Today corn is the largest user of insecticides of any crop in the United States.

Most benefits of pesticides are based on the direct crop returns. Such assessments do not include the indirect environmental and economic costs associated with the recommended application of pesticides. To facilitate the development and implementation of a scientifically sound policy of pesticide use, these environmental and economic costs must be examined. For some time, the US Environmental Protection Agency pointed out the need for such a benefit/cost and risk investigation (EPA 1977). Thus far, only a few scientific papers on this complex and difficult subject have been published.

2.2 Public Health Effects

2.2.1 Acute Poisonings

Human pesticide poisonings and illnesses are clearly the highest price paid for all pesticide use. Although the EPA (1992) estimated that 300,000 pesticide poisoning occurred annually, the National Institute for Occupational Safety and Health states that the total number of pesticide poisonings in the United States is between 10,000–20,000 year^{-1} (NIOSH 2012). Worldwide, the application of 3 million metric tons of pesticides results in more than 26 million cases of non-fatal pesticide poisonings (Richter 2002). Of all the pesticide poisonings, about 3 million cases are hospitalized and there are approximately 220,000 fatalities and about 750,000 chronic illnesses every year (Hart and Pimentel 2002).

2.2.2 Cancer and Other Chronic Effects

Ample evidence exists concerning the carcinogenic threat related to the use of pesticides. The major types of chronic health effects of pesticides include neurological effects, respiratory and reproductive effects, and cancer. There is some evidence that pesticides can cause sensory disturbances as well as cognitive effects such as memory loss, language problems, and learning impairment (Hart and Pimentel 2002). The malady, organophosphate-induced delayed poly-neuropathy (OPIDP), is well documented and includes irreversible neurological damage. In addition to neurological effects, pesticides can have adverse effects on the respiratory and reproductive systems. For example, 15% of a group of professional pesticide applicators suffered asthma, chronic sinusitis, and/or chronic bronchitis (Weiner and Worth 1969). Studies have also linked pesticides with reproductive effects. For example, some pesticides have been found to cause testicular dysfunction or sterility (Colborn et al. 1997). Sperm counts in males in Europe and the United States, for example, declined by about 50% between 1938 and 1990 (Carlsen et al. 1992).

US data indicate that 18% of all insecticides and 90% of all fungicides are carcinogenic (National Research Council et al. 1987). Several studies have shown that the risks of certain types of cancers are higher in some people, such as farm workers and pesticide applicators, who are often exposed to pesticides, see Table 2.1 (Pimentel and Hart 2001). Certain pesticides have been shown to induce tumors in laboratory animals and there is some evidence that suggest similar effects occur in humans (Colborn et al. 1997).

The United Farm Workers of America and others of the cancer registry in California analyzed the incidence of cancer among Latino farm workers and reported that per year, if everyone in the USA had a similar rate of incidence, there would be 83,000 cases of cancer associated with pesticides in the USA (PAN—North America 2002). The incidence of cancer in the US population due to pesticides ranges from about 10,000 to 15,000 cases year^{-1} (Pimentel et al. 1997).

Table 2.1 Estimated economic costs of human pesticide poisonings and other pesticide-related illnesses in the United States each year

Human health effects from pesticides	Total costs ($)
Cost of hospitalized poisonings 5000[a] × 3 days at $2000 per day	30,000,000
Cost of outpatient-treated poisonings 30,000[b] × $1000[c]	30,000,000
Lost work due to poisonings 5000 workers × 5 days × $80	2,000,000
Pesticide cancers 10,000[b] $100,000/case	1,000,000,000
Cost of fatalities 45 accidental fatalities[a] × $3.7 million	166,500,000
Total	1,228,500,000

[a] Estimated.
[b] See text for details
[c] Includes hospitalization, foregone earnings, and transportation

Many pesticides are also estrogenic—they mimic or interact with the hormone estrogen—linking them to an increase in breast cancer among some women. The breast cancer rate rose from 1 in 20 in 1960 to 1 in 8 in 1995 (Colborn et al. 1997). As expected, there was a significant increase in pesticide use during that time period. Pesticides that interfere with the body's endocrine–hormonal system can also have reproductive, immunological, or developmental effects (McCarthy 1993). While endocrine-disrupting pesticides may appear less dangerous because hormonal effects rarely result in acute poisonings, their effects on reproduction and development may prove to have far-reaching consequences (Colborn et al. 1997).

The negative health effects of pesticides can be far more significant in children than adults, for several reasons. First, children have higher metabolic rates than adults, and their ability to activate, detoxify, and excrete toxic pesticides differs from adults. Also, children consume more food than adults and thus can consume more pesticides per unit weight than adults. This problem is particularly significant for children because their brains are more than five times larger in proportion to their body weight than adult brains, making cholinesterase even more vital. In a California study, 40% of the children working in agricultural fields had blood cholinesterase levels below normal, a strong indication of organophosphate and carbamate pesticide poisoning (Repetto and Baliga 1996). According to the EPA, fetuses where the mother is exposed and toddlers under two years of age are 10 times more at risk for cancer than adults and children from 3 to 15 may have at least three times the cancer risk than adults (USA Today 2003).

Although no one can place a precise monetary value on a human life, the economic "costs" of human pesticide poisonings have been estimated (Table 2.1). For our assessment, we use the EPA standard of $3.7 million per human life (Kaiser 2003). Available estimates suggest that human pesticide poisonings and related illnesses in the United States cost about $1 billion year^{-1} (Pimentel and Greiner 1997).

2.2.3 Pesticide Residues in Food

The majority of foods purchased in supermarkets have detectable levels of pesticide residues. For instance, of several thousand samples of food, the overall assessment in 8 fruits and 12 vegetables is that 73 % have pesticide residues (Baker et al. 2002). In five crops (apples, peaches, pears, strawberries, and celery) pesticide residues were found in 90 % of the crops. A study by Groth et al. (1999) detected 37 different pesticides in apples.

Up to 5 % of the foods tested in 1997 contained pesticide residues that were above the FDA tolerance levels. These foods were consumed even though they violated the US tolerance of pesticide residues in foods because the food samples were analyzed after the foods were sold in the supermarkets (Pesticides Residues Committee—UK 2004).

2.3 Domestic Animal Poisonings and Contaminated Products

In addition to pesticide problems that affect humans, several thousand domestic animals are accidentally poisoned by pesticides each year, with dogs and cats representing the largest number (Table 2.2). For example, of 250,000 poison cases involving animals, a large percentage of the cases were pesticide poisonings (Pimentel and Pimentel 2008). Poisonings of dogs and cats are common which is not surprising because dogs and cats usually wander freely about the home and farm and therefore have greater opportunity to come into contact with pesticides than other domesticated animals.

The best estimates indicate that about 20 % of the total monetary value of animal production, or about $4.2 billion, is lost to all animal illnesses, including pesticide poisonings. It is reported that 0.5 % of animal illnesses and 0.04 % of all animal deaths reported to a veterinary diagnostic laboratory were due to pesticide toxicosis. Thus, $21.3 and $8.8 million, respectively, are lost to pesticide poisonings (Table 2.2).

This estimate is considered low because it is based only on poisonings reported to veterinarians. Many animal deaths that occur in the home and on farms go undiagnosed and unreported. In addition, many are attributed to other factors than pesticides. When a farm animal poisoning occurs and little can be done for the animal, the farmer seldom calls a veterinarian but, rather either waits for the animal to recover or destroys it. Such cases are usually unreported.

Additional economic losses occur when meat, milk, and eggs are contaminated with pesticides. In the United States, all animals slaughtered for human consumption, if shipped interstate, and all imported meat and poultry, must be inspected by the USDA. This is to ensure that the meat and poultry products are wholesome, properly labeled, and do not present a health hazard.

Table 2.2 Estimated domestic animal pesticide poisonings in the United States

Livestock	Number × 1000	$ per head × 1000	Number ill[a] × 1000	$ cost per poisoning[b]	$ cost of poisonings × 1000	Number deaths[c] × 1000	$ cost of deaths[d] × 1000	Total $ × 1000
Cattle	99,000[e]	607[e]	100	121.40	12,140	8	4,856	16,996
Dairy Cattle	10,000[e]	900[e]	10	180.00	1,800	1	900	2,700
Dogs	55,000[f]	125[g]	55	25.00	1,375	4	500	1,875
Horses	11,000[h]	1,000[f]	11	200.00	2,200	1	1,000	3,200
Cats	63,000[f]	207	60	4.00	240	4	80	320
Swine	53,000[e]	66.3[e]	53	13.26	703	4	265	968
Chickens	8 × 10[6e,f]	2.5[e]	6000	0.40	2,400	500	1,250	3,650
Turkeys	2.8 × 10[5e]	106	280	2.00	560	25	250	810
Sheep	11,000[e]	82.40[e]	11	16.48	181	1	82.2	63
Total	8.582 × 10[6f]				21,599			30,582

[a] Based on a 0.1 % illness rate (see text)
[b] Based on each animal illness costing 20 % of total production value of that animal
[c] Based on a 0.008 % mortality rate (see text)
[d] The death of the animal equals the total value for that animal
[e] USDA (1989)
[f] USBC (1990)
[g] Estimated
[h] FAO (1986)

Pesticide residues are searched for in animals and their products. However, of more than 600 pesticides in use now, the National Residue Program (USDA, Office of Inspector General 2010) only searches for about 40 different pesticides, which have been determined by FDA, EPA, and FSIS to be of public health concern. While the monitoring program records the number and type of violations, there might be little cost to the animal industry because the meat and other products are sometimes sold and consumed by the public before the test results are available. For example, about 3 % of chickens with illegal pesticide residues are sold in the market (National Research Council et al. 1987).

In addition to animal carcasses, pesticide-contaminated milk cannot be sold and must be disposed of. In some instances, these losses are substantial. In Oahu, Hawaii in 1982, 80 % of the milk supply, worth more than $8.5 million, was condemned by the public health officials because it had been contaminated with the insecticide heptachlor (Baker et al. 2002). This incident had immediate and far-reaching effects on the entire milk industry on the island.

2.4 Destruction of Beneficial Natural Predators and Parasites

In both natural and agricultural ecosystems, many species, especially predators and parasites, control or help control plant-feeding arthropod populations. Indeed, these natural beneficial species make it possible for ecosystems to remain "green."

With the parasites and predators keeping plant-feeding arthropod populations at low levels, only a relatively small amount of plant biomass is removed each growing season by arthropods (Hairston et al. 1960; Pimentel 1988). Like pest populations, beneficial natural enemies and biodiversity (predators and parasites) are adversely affected by pesticides (Pimentel et al. 1993a). The following pests have reached outbreak levels in cotton and apple crops after the natural enemies were destroyed by pesticides:

- cotton—cotton bollworm, tobacco budworm, cotton aphid, spider mites, and cotton loopers;
- apples—European red mite, red-banded leaf roller, San Jose scale, oyster shell scale, rosy apple aphid, wooly apple aphid, white apple aphid, two-spotted spider mite, and apple rust mite (Pimentel et al. 1993a)

Major pest outbreaks have also occurred in other crops due to the destruction of natural enemies. Also, because parasitic and predaceous insects often have complex searching and attack behaviors, sub-lethal insecticide dosages may alter this searching and attack behavior and in this way disrupt effective biological controls (Pimentel et al. 1993a).

Fungicides also can contribute to pest outbreaks when they reduce fungal pathogens that are naturally parasitic on many insects. For example, the use of benomyl reduces populations of entomopathogenic fungi, resulting in increased survival of velvet bean caterpillars and cabbage loopers in soybeans. This eventually leads to reduced soybean yields (Pimentel et al. 1993a).

When outbreaks of secondary pests occur because their natural enemies are destroyed by pesticides, additional and sometimes more expensive pesticide treatments have to be made in efforts to sustain crop yields. This raises the overall costs and contributes to pesticide-related problems. An estimated $520 million can be attributed to costs of additional pesticide application and increased crop losses, both of which follow the destruction of natural enemies by various pesticides applied to crops (Table 2.3).

Natural enemies are being adversely affected by pesticides worldwide. Although no reliable estimate is available concerning the impact of this in terms of increased pesticide use and/or reduced crop yields, entomologists often observe a severe impact due to the loss of natural enemies where pesticides are heavily used in many parts of the world. From 1980 to 1985 insecticide use in rice production in Indonesia drastically increased (Oka 1991) which caused the destruction of beneficial natural enemies of the brown plant hopper and causing the brown plant hopper population to explode. Rice yield decreased to the extent that rice had to be imported to Indonesia. The estimated cost of rice loss in just a 2-year period was $1.5 billion (Soejitno 1999).

After this incident, Dr. I.N. Oka, who had previously developed a successful low-insecticide program for rice pests in Indonesia, was consulted by the Indonesian President Suharto's staff to determine what should be done to rectify the situation. Oka's advice was to substantially reduce insecticide use and return to a sound "treat-when-necessary" program that protected the natural enemies. Following Oka's advice, President Suharto mandated in 1986 on television that 57 of

Table 2.3 Losses due to the destruction of beneficial natural enemies in US crops ($ millions)

Crops	Total expenditures for insect control with pesticides[a]	Amount of added control costs
Cotton	320	160
Tobacco	5	1
Potatoes	31	8
Peanuts	18	2
Tomatoes	11	2
Onions	1	0.2
Apples	43	11
Cherries	2	1
Peaches	12	2
Grapes	3	1
Oranges	8	2
Grapefruit	5	1
Lemons	1	0.2
Nuts	160	16
Other	500	50
Total ($)	1,120	257.4 (520)[b]

[a] Pimentel et al. (1991)
[b] Because the added pesticide treatments do not provide as effective control as the natural enemies, we estimate that at least an additional $260 million in crops are lost to pests. Thus the total loss due to the destruction of natural enemies is estimated to be at least $520 million year^{-1}

64 pesticides would be withdrawn from use on rice, and sound pest management practices implemented. Pesticide subsidies were also reduced to zero. By 1991, pesticide applications had been reduced by 65% and rice yields increased by 12%.

Dr. David Rosen (Hebrew University of Jerusalem, PC, 1991) estimates that natural enemies account for up to 90% of the control of pest species in agroecosystems. I estimate that at least 50% of the control of pest species is due to natural enemies. Pesticides provide an additional control, while the remaining 40% is due to host–plant resistance in agroecosystems (Pimentel 1988).

Parasites, predators, and host–plant resistance are estimated to account for about 80% of the nonchemical control of pest arthropods and plant pathogens in crops (Pimentel et al. 1991). Many cultural controls including crop rotations, soil and water management, fertilizer management, planting time, crop-plant density, trap crops, and polyculture provide additional pest control. Together these non-pesticide controls can be used to effectively reduce US pesticide use by more than 50% without any reduction in crop yields or cosmetic standards (Pimentel et al. 1993a).

2.5 Pesticide Resistance in Pests

In addition to destroying natural enemy populations, the extensive use of pesticides has often resulted in the development and evolution of pesticide resistance in insect pests, plant pathogens, and weeds. An early report by the United Nations Environmental Program (UNEP 1979) suggested that pesticide resistance ranked as one of the top 4 environmental problems of the world. About 520 insect and mite species, nearly 150 plant pathogen species, and about 273 weeds species are now resistant to pesticides (Stuart 1999).

Increased pesticide resistance in pest populations frequently results in the need for several additional applications of the commonly used pesticides to maintain crop yields. These additional pesticide applications compound the pesticide resistance problem by increasing environmental selection of pest populations for resistance. The pesticide resistance problem continues to increase despite all efforts and is spreading to other pest species. Over time extremely high pesticide resistance had developed in the tobacco budworm population on cotton in northeastern Mexico and the Lower Rio Grande of Texas (NAS 1975). Finally approximately 285 000 ha of cotton had to be abandoned, because the insecticides used were totally ineffective due to extreme resistance in the budworm. The economic and social impact on these Texan and Mexican farmers dependent on cotton was devastating. The study by Carrasco-Tauber (1989) reported a yearly loss of \$45– \$120 ha^{-1} to pesticide resistance in California cotton. A total of 4.2 million hectares of cotton were harvested in 1984; thus, assuming a loss of \$82.50 ha^{-1}, approximately \$348 million of the California cotton crop was lost due to pesticide resistance. Since \$3.6 billion of US cotton was harvested in 1984 (USBC 1990), the loss due to resistance for that year was approximately 10 %. Assuming a 10 % loss in other major crops that receive heavy pesticide treatments in the United States, crop losses due to pesticide resistance are estimated to be about \$1.5 billion year^{-1}.

Efforts to control resistant *Heliothus* spp. (corn ear worm) exact a cost on other crops when large, uncontrolled populations of *Heliothus* and other pests disperse onto other crops. In addition, the cotton aphid and the whitefly populations exploded as secondary cotton pests because of their pesticide resistance and their natural enemies' exposure to high concentrations of insecticides (Pimentel et al. 1993a).

The total external cost attributed to the development of pesticide resistance is estimated to range between 10 and 25 % of current pesticide treatment costs (Harper and Zilberman 1990), or more than \$1.5 billion each year in the United States. In other words, at least 10 % of pesticide used in the USA is applied just to combat increased resistance that has developed in several pest species.

Although the costs of pesticide resistance are high in the United States, the costs in tropical developing countries are significantly greater, because pesticides are not only used to control agricultural pests, but also vital for the control of arthropod disease vectors. One of the major costs of resistance in tropical countries is associated with malaria control. By 1985, the incidence of malaria in India after early pesticide use declined to about 1.86 million cases from a peak of 70 million cases. However, because mosquitoes developed resistance to pesticides, as did malarial parasites to drugs, the incidence of malaria in India has now ranges between 1.5–2.0 million cases year^{-1} (Reid 2000; Kakkilaya 2012). Problems are occurring not only in India but also in the rest of Asia, Africa, and South America. The total number of people at risk of malaria in 2010 in the world is now 3.3 billion (WHO 2011).

2.6 Honeybee and Wild Bee Poisonings and Reduced Pollination

Honeybees and wild bees are vital for pollination of fruits, vegetables, and other crops. Bees are essential to the production of about one-third of US and world crops. Their benefits to US agriculture are estimated to be about $40 billion year^{-1} (Pimentel et al. 1997). Because most insecticides used in agriculture are toxic to bees, pesticides have a major impact on both honeybee and wild bee populations. D. F. Mayer (Washington State University, PC, 1990) estimates that approximately 20% of all honeybee colonies are adversely affected by pesticides. He includes the approximately 5% of US honeybee colonies that are killed outright or die during winter because of pesticide exposure. Mayer calculates that the direct annual loss reaches $13.3 million year^{-1} (Table 2.4). Another 15% of the honeybee colonies are either seriously weakened by pesticides or suffer losses when apiculturists have to move colonies to avoid pesticide damage. According to Mayer, the yearly estimated loss from partial honeybee kills, reduced honey production, plus the cost of moving colonies totals about $25.3 million year^{-1}. Also, as a result of heavy pesticide use on certain crops, beekeepers are excluded from 4 to 6 million ha of otherwise suitable apiary locations, according to Mayer. He estimates the yearly loss in potential honey production in these regions is about $27 million (Table 2.4).

In addition to these direct losses caused by the damage to honeybees and honey production, many crops are lost because of the lack of pollination. In California, for example, approximately 1 million colonies of honeybees are rented annually at $55 per colony to augment the natural pollination of almonds, alfalfa, melons, and other fruits and vegetables (Burgett 2001). Since California produces nearly half of our bee-pollinated crops, the total cost for honeybee rental for the entire country is estimated at $40 million year^{-1}. Of this cost, I estimate that at least one-tenth or $4 million is attributed to the effects of pesticides (Table 2.4). Estimates of annual agricultural losses due to the reduction in pollination caused by pesticides may be as high as $4 billion year^{-1} (J. Lockwood, University of Wyoming, PC, 1990). For most crops, both yield and quality are enhanced by effective pollination. Several investigators have demonstrated that for various cotton varieties, effective pollination by honeybees resulted in yield increases of from 20 to 30%.

Mussen (1990) emphasizes that poor pollination will not only reduce crop yields, but also equally important, it will reduce the quality of some crops, such as melons and fruits. In experiments with melons, E.L. Atkins (University of California at Davis, PC, 1990) reported that with adequate pollination melon yields increased 10% and melon quality was raised 25% as measured by the dollar value of the melon crop.

Based on the analysis of honeybee and related pollination losses from wild bees caused by pesticides, pollination losses attributed to pesticides are estimated to represent about 10% of pollinated crops and have a cost of about $210 million year^{-1} (Table 2.4). Clearly, the available evidence confirms that the yearly cost of direct

Table 2.4 Estimated honeybee losses and pollination losses from honeybees and wild bees

Colony losses from pesticides	$13.3 million year^{-1}
Honey and wax losses	$25.3 million year^{-1}
Loss of potential honey production	$27.0 million year^{-1}
Bee rental for pollination	$8.0 million year^{-1}
Pollination losses	$210.0 million year^{-1}
Total	$283.6 million year^{-1}

honeybee losses, together with reduced yields resulting from poor pollination, is significant.

2.7 Crop and Crop Product Losses

Basically, pesticides are applied to protect crops from pests in order to increase yields, but sometimes crops are damaged by the pesticide treatments. This damage occurs when (1) the recommended dosages suppress crop growth, development, and yield; (2) pesticides drift from the targeted crop to damage adjacent crops; (3) residual herbicides either prevent chemical-sensitive crops from being planted; and/or (4) excessive pesticide residue accumulates on crops, necessitating the destruction of the harvested crop. Crop losses translate into financial losses for growers, distributors, wholesalers, transporters, retailers, food processors, and others. Investments as well as potential profits are lost. The costs of crop losses increase when the related costs of investigations, regulation, insurance, and litigation are added to the equation. Ultimately the consumer pays for these losses in higher marketplace prices. Data on crop losses due to pesticides are difficult to obtain. Many losses are never reported to the state and federal agencies because the parties settle privately (Pimentel et al. 1993a).

Damage to crops may occur even when recommended dosages of herbicides and insecticides are applied to crops under normal environmental conditions. Recommended dosages of insecticides used on crops have been reported to suppress growth and yield in both cotton and strawberry crops (ICAITI 1977; Reddy et al. 1987; Trumbel et al. 1988). The increase in susceptibility of some crops to insects and diseases following normal use of 2,4-D and other herbicides has been demonstrated (Oka and Pimentel 1976; Pimentel 1994). Furthermore, when weather and/or soil conditions are inappropriate for pesticide application, herbicide treatments may cause yield reductions ranging from 2 to 50 % (Pimentel et al. 1993a).

Crops are lost when pesticides drift from the target crops to non-target crops located as much as several miles downwind (Barnes et al. 1987). Drift occurs with most methods of pesticide application including both ground and aerial equipment; the potential problem is greatest when pesticides are applied by aircraft. With aircraft, from 50 to 75 % of the pesticide applied never reaches the target area (Akesson and Yates 1984; Mazariegos 1985; Pimentel et al. 1993a). In contrast, 10 to 35 % of the pesticide applied with ground application equipment misses the target area

Table 2.5 Estimated loss of crops and trees due to the use of pesticides

Impacts	Total Costs in millions of US dollars
Crop losses	136
Crop applicator insurance	245
Crops destroyed because of excess pesticide contamination	1000
Government investigations and testing	10
Total	1391

(Hall 1991). The most serious drift problems are caused by "speed sprayers" and ultra-low-volume (ULV) equipment, because relatively concentrated pesticide is applied. The concentrated pesticide has to be broken into small droplets to achieve adequate coverage.

Crop injury and subsequent loss due to drift are particularly common in areas planted with diverse crops. Because of the drift problem, most commercial applicators carry insurance that costs about \$245 million year^{-1} (Pimentel et al. 1993a; Table 2.5).

When residues of some herbicides persist in the soil, crops planted in rotation are sometimes injured. This has happened with a corn and soybean rotation. When atrazine or Sceptor herbicides were used in corn, the soybean crop planted after was seriously damaged by the herbicides that persist in the soil.

If the herbicide treatment persists in the soil and prevents another crop from being grown, soil erosion may be intensified (Pimentel et al. 1993a) assuming the soil is left exposed to the elements.

Losses due to pesticides average 0.1% in annual US production of corn, soybeans, cotton, and wheat, together these crops account for about 90% of the herbicides and insecticides used in US agriculture; this 0.1% loss was valued at \$35.3 million in 1987 (National Research Council et al. 1989). Assuming that only one-third of the incidents involving crop losses due to pesticides are reported to authorities, the total value of all crop lost because of pesticides could be as high as three times this amount or \$106 million annually.

However, this \$106 million does not take into account other crop losses, nor does it include major events such as the large-scale losses that have occurred in one season in Iowa (\$25–30 million), in Texas (\$20 million), and in California's aldicarb/watermelon crisis (\$8 million) (Pimentel et al. 1993a). These recurrent losses alone represent an average of \$30 million year^{-1}, raising the estimated average crop loss value from the use of pesticides to approximately \$136 million each year.

Additional losses are incurred when food crops exceed the FDA and EPA regulatory tolerances for pesticide residue levels and have to be disposed of. Assuming that all the crops and crop products that exceed the FDA and EPA regulatory tolerances (reported to be 1–5%) were disposed of as required by law, then about \$1 billion in crops would be destroyed because of excessive pesticide contamination. Special investigations and testing for pesticide contamination are estimated to cost the nation more than \$10 million each year (Pimentel et al. 1993a).

2.8 Ground and Surface Water Contamination

Certain pesticides applied at recommended dosages to crops eventually end up in ground and surface waters. The three most common pesticides found in groundwater are aldicarb, alachlor, and atrazine (Trautmann et al. 2012). Estimates are that nearly one-half of the groundwater and well water in the United States is or has the potential to be contaminated (Holmes et al. 1988; USGS 1996). EPA (1990) reported that 10% of community wells and 4% of rural domestic wells have detectable levels of at least one pesticide of the 127 pesticides tested for in a national survey. Estimated costs to sample and monitor well and groundwater for pesticide residues costs $1,100 well^{-1} year^{-1} (USGS 1995). With 16 million wells in the US, the cost of monitoring all the wells for pesticides would cost $17.7 billion year^{-1} (Stone and American Ground Water Trust 1998; Pimentel and Pimentel 2008). Two major concerns about groundwater contamination with pesticides are that about one-half of the human population obtains its water from wells and once groundwater is contaminated, the pesticide residues remain for long periods of time. Few microbes are present in groundwater that can degrade the pesticides and the groundwater recharge rate is less than 1% year^{-1} so even dilution of the contaminant pesticide will be a slow process (CEQ and Barney 1980).

Monitoring pesticides in groundwater is only a portion of the total cost of groundwater contamination. There is also the high cost of cleanup. For instance, at the Rocky Mountain Arsenal near Denver, Colorado, the removal of pesticides from the groundwater and soil was estimated to cost approximately $2 billion (Greene 1994). If all pesticide-contaminated groundwater was to be cleared of pesticides before human consumption, the cost would be about $500 million year^{-1}. Note the cleanup process requires a water survey to target the contaminated water for cleanup. Thus, in addition to the monitoring and cleaning costs, the total cost regarding pesticide-polluted groundwater is estimated to be about $2 billion annually. The $17.7 billion figure shows how impossible it would be to expect the public to pay for pesticide-free well water or even to test for pesticide contamination (Pimentel and Pimentel 2008).

2.9 Fishery Losses

Pesticides are washed into aquatic ecosystems by water runoff and soil erosion. About 13 ha^{-1} year^{-1} of soil is washed and/or blown from pesticide-treated cropland into adjacent locations including rivers and lakes (Unnevehr et al. 2003). Pesticides also can drift during application and contaminate aquatic systems. Some soluble pesticides are easily leached into streams and lakes. Gilliom et al. (2007) analyzed stream water from 1992 to 2001 for pesticides and their degradates in US streams found that one or more pesticides or their degradates occurred over 90% of the time in streams in agricultural areas, urban areas and mixed land uses areas and 65% of the time in stream water from undeveloped areas. At least one pesticide was detected in water samples in 33% of major aquifers located in mixed land use areas

(Gilliom et al. 2007). Organochlorine pesticides (most of which are banned for use in the United States) were detected in the tissue of over 90% of fish sampled from streams in agricultural areas, urban areas, and mixed land use areas and in 57% of fish sampled from streams in undeveloped areas (Gilliom et al. 2007).

Once in aquatic ecosystems, pesticides cause fishery losses in several ways. These include high pesticide concentrations in water that directly kill fish; low doses that may kill highly susceptible fish fry; or the elimination of essential fish foods like insects and other invertebrates. In addition, because government safety restrictions ban the catching or sale of fish contaminated with pesticide residues, such fish are unmarketable and are an economic loss.

Only 6–14 million fish are reported killed by pesticides each year (Pimentel et al. 1993a). However, this is an underestimate because fish kills cannot be investigated quickly enough to determine accurately the cause of the kill. Also, if the fish are in fast-moving waters in rivers, the pesticides are diluted and/or the pesticides cannot be identified. Many fish sink to the bottom and cannot be counted.

The best estimate for the value of a fish is $10. This is based on EPA fining Coors Beer $10 per fish when they polluted Clear Creek in Colorado (US Water News 2002). Thus, the estimate of the value of fish killed each year is only $10–24 million year^{-1}. This is an under estimate and I estimate $100 million year^{-1} minimum.

2.10 Wild Birds and Mammals

Wild birds and mammals are damaged and destroyed by pesticides and these animals make excellent "indicator species." Deleterious effects on wildlife include death from the direct exposure to pesticides or secondary poisonings from consuming contaminated food; reduced survival, growth, and reproductive rates from exposure to sub-lethal dosages; and habitat reduction through the elimination of food resources and refuges. In the United States, approximately 3 kg of pesticide is applied per hectare on about 160 million hectares of cropland each year (Pimentel et al. 1993a). With such heavy dosages of pesticides applied, it is expected that wildlife would be significantly impacted.

The full extent of bird and mammal kills is difficult to determine because birds and mammals are often secretive, camouflaged, highly mobile, and live in dense grass, shrubs, and trees. Typical field studies of the effects of pesticides often obtain extremely low estimates of bird and mammal mortality (Mineau et al. 1999) since bird and small mammal carcasses disappear quickly, well before they can be found and counted. Even when known numbers of bird carcasses were placed in identified locations in the field, from 62 to 92% of the animals disappeared overnight due to vertebrate and invertebrate scavengers (Balcomb 1986). In addition, field studies seldom account for birds that die a distance from the pesticide treated areas. Finally, birds often hide and die in inconspicuous locations.

Nevertheless, many bird kills caused by pesticides have been reported. For instance, 1200 Canada geese were killed in one wheat field that was sprayed with a 2:1 mixture of parathion and methyl parathion at a rate of 0.8 kg ha^{-1} (White et al. 1982). Carbofuran applied to alfalfa killed more than 5000 ducks and geese in five incidents, while the same chemical applied to vegetable crops killed 1400 ducks in a single application (Flickinger et al. 1980, 1991). Carbofuran is estimated to kill 1–2 million birds each year (American Bird Conservancy 2010a). Another pesticide, diazinon, applied to three golf courses killed 700 Atlantic brant geese of the wintering population of just 2500 birds (Stone and Gradoni 1985). In 1988, the US EPA cancelled diazinon use on sod farms and golf courses due to numerous bird kills; acute lethal and reproductive effects for birds occur at levels below those detected in the field (EPA 2004).

American Bird Conservancy reports that an estimated 67 million birds are killed each year by twelve pesticides that are particularly harmful to birds (i.e., fenthion, chlorfenapyr, ethyl parathion and several rodent poisons) in the United States as of 1992 (American Bird Conservancy 2010). Birds are not only killed in the US but also killed as they migrate from North America to South America. For example, more than 4000 carcasses of Swainson's hawks were reported poisoned by pesticides in late 1995 and early 1996 in farm fields of Argentina (CWS 2012). Although it was not possible to know the total kill, conservatively it was estimated to be more than 20,000 hawks.

Several studies report that the use of some herbicides has a negative impact on some young birds. Since the weeds would have harbored some insects in the crops, the weeds nearly total elimination by herbicides is devastating to particular bird populations (Potts 1986; R. Beiswenger, University of Wyoming, PC, 1990). This has led to significant reductions in the gray partridge in the United Kingdom and in the common pheasant in the United States. In the case of the partridge, population levels have decreased more than 77% because the partridge chicks (also pheasant chicks) depend on insects to supply them with needed protein for their development and survival.

Frequently the form of a pesticide influences its toxicity to wildlife (Hardy 1990). Pesticide-treated seed and insecticide granules, including carbofuran, fensulfothion, fonofos, and phorate, are particularly toxic to birds. Estimates are that from 0.23 to 1.5 birds ha^{-1} were killed in Canada, while in the United States the estimates ranged from 0.25 to 8.9 birds killed ha^{-1} year^{-1} by these pesticides (Mineau 1988).

Pesticides also adversely affect the reproductive potential of many birds and mammals. Exposure of birds, especially predatory birds, to chlorinated insecticides caused reproductive failure, sometimes attributed to eggshell thinning (Elliot et al. 1988). Most of the affected predatory birds, like the bald eagle and peregrine falcon, have recovered since the banning of DDT and most other chlorinated insecticides in the US (Unnevehr et al. 2003). Although the US and most other developed countries have banned DDT and other chlorinated insecticides, countries such as India and China are still producing, exporting, and using DDT (Asia Times 2001). Pesticide-caused habitat alteration and destruction can be expected to reduce mammal and bird populations. When glyphosate (Roundup) was applied to forest clear-cuts to eliminate low-growing vegetation like shrubs and small trees, the southern red-backed vole population was greatly reduced because its food source and cover were practically eliminated (D'Anieri et al. 1987). Similar effects from herbicides have been reported on other mammals (Isenring 2010). Overall, the impacts of pesticides on mammal populations have been inadequately investigated.

Although gross values for wildlife are not available, expenditures involving wildlife made by humans are one measure of the monetary value. Non consumptive users (i.e., tourists, sightseers) of wildlife spent an estimated $14.3 billion on their activity (USFWS 1988). Yearly, US bird watchers spend an estimated $600 million on their hobby and an additional $500 million on birdseed, for a total of $1.1 billion (USFWS 1988). For bird watching, the estimated cost is about 40¢ per bird. The money spent by hunters to harvest 5 million game birds was $1.1 billion, or approximately $216 per bird (USFWS 1988). The estimated cost of replacing a bird of an affected species to the wild, as in the case of the Exxon Valdez oil spill, ranged from $170 to $6,000 for sea birds and eagles (Cleveland et al. 2012).

If damages that pesticides inflict on birds occur primarily on the 160 million ha of cropland that receives the most pesticides, and the bird population is estimated to be 4.4 birds ha^{-1} of cropland (Boutin et al. 1999), then 720 million birds are directly exposed to pesticides. Also, if it is conservatively estimated that only 10 % of the bird population is killed by the pesticide treatments, it follows that the total number of birds killed is 72 million birds. Note this estimate is at the lower range of 0.25–8.9 birds killed ha^{-1} year^{-1} mentioned earlier for the US.

The American bald eagle and other predatory birds suffered high mortalities because of DDT and other chlorinated insecticides. The bald eagle population declined primarily because of pesticides and was placed on the endangered species list. After DDT and the other chlorinated insecticides were banned in 1972, it took nearly 30 years for these bird populations to recover. The American bald eagle was recently removed from the endangered species list (Millar 1995).

I assumed a value of a bird to be about $30 based on the information presented. Thus, the total economic impact of pesticides on birds is estimated to be $2.1 billion year^{-1}. This estimate does not include birds killed due to the death of one or both of the nesting parents and in turn causes the deaths of the nestlings. It also does not include nestlings killed because they were fed contaminated arthropods and other foods.

2.11 Microbes and Invertebrates

Pesticides easily find their way into soils, where they may be toxic to arthropods, earthworms, fungi, bacteria, and protozoa. Small organisms are vital to ecosystems because they dominate both the structure and function of ecosystems (Pimentel et al. 1992). An estimated 4.5 t ha^{-1} of fungi and bacteria exist in the upper 15 cm of soil. They and the arthropods make up 95 % of all species and 98 % of the biomass in the upper 15 cm of soil (excluding vascular plants). Microbes are essential to the proper functioning of the terrestrial ecosystem because they break down organic matter, enabling the vital chemical elements to be recycled (Atlas and Bartha 1987; Pimentel et al. 1997). Equally important is the ability of some microorganisms to "fix" nitrogen, making it available to plants and ecosystems (Pimentel et al. 1997).

Earthworms and insects aid in bringing new soil to the surface at a rate of up to 200 t ha^{-1} year^{-1} (Pimentel et al. 1993a). This soil movement improves soil formation and structure for plant growth and makes various nutrients more available for absorption by plants. The holes (up to 10,000 holes m^{-2}) in the soil made by earthworms and insects also facilitate the percolation of water into the soil (Edwards and Lofty 1982).

Insecticides, fungicides, and herbicides reduce species diversity in the soil as well as the total biomass of these biota. Stringer and Lyons (1974) reported that where earthworms had been killed by pesticides, the leaves of apple trees accumulated on the surface of the soil and increased the incidence of scab in the orchards. Apple scab, a disease carried over from season to season on fallen leaves, is commonly treated with fungicides. Some fungicides, insecticides, and herbicides are toxic to earthworms which would otherwise remove and recycle the fallen leaves.

On golf courses and other lawns, the destruction of earthworms by pesticides results in the accumulation of dead grass or thatch in the turf (Potter and Braman 1991). To remove this thatch special equipment must be used and it is expensive.

Although these microbes and invertebrates are essential to the vital structure and function of both natural and agricultural ecosystems, it is impossible to place a monetary value on the damage caused by pesticides to this large group of organisms. To date, no relevant quantitative data on the value of microbe and invertebrate destruction by pesticides are available.

2.12 Government Funds for Pesticide Pollution Control

A major environmental cost associated with all pesticide use is carrying out state and federal regulatory actions, as well as pesticide-monitoring programs needed to control pesticide pollution. Specifically, these funds are spent to reduce the hazards of pesticides and to protect the integrity of the environment and public health.

About $10 million is spent each year by state and federal governments to train and register pesticide applicators (Pimentel and Pimentel 2008). Also, more than $60 million is spent each year by the EPA to register and re-register pesticides. In addition, about $400 million is spent to monitor pesticide contamination of fruits, vegetables, grains, meat, milk, water, and other items for pesticide contamination. Thus, at least $470 million is invested by state and federal governmental organizations. Although enormous amounts of government funds are being spent to reduce pesticide pollution, many costs of pesticides are not taken into account. Also, many serious environmental and social problems remain to be corrected by improved government policies.

2.13 Ethical and Moral Issues

Although pesticides provide about $40 billion year^{-1} in saved US crops, the data of this analysis suggest that the environmental and social costs of pesticides to the nation totaled approximately $10 billion. From a strict cost/benefit approach, pesticide use is beneficial. However, the nature of the environmental and public health costs of pesticides has other trade-offs involving environmental quality and public health.

One of these issues concerns the importance of public health versus pest control. For example, assuming that pesticide-induced cancers numbered more than 10,000 cases year^{-1} and that pesticides returned a net agricultural benefit of $32 billion year^{-1}, each case of cancer is "worth" $3.2 million in pest control. In other words, for every $3.2 million in pesticide benefits, one person falls victim to cancer. Social mechanisms and market economics provide these ratios, but they ignore basic ethics and values.

In addition, pesticide pollution of the global environment raises numerous other ethical questions. The environmental insult of pesticides has the potential to demonstrably disrupt entire ecosystems. All through history, humans have felt justified in removing forests, draining wetlands, and constructing highways and housing in various habitats. White (1967) has blamed the environmental crisis on religious teachings of mastery over nature. Whatever the origin, pesticides exemplify this attempt at mastery, and even a noneconomic analysis would question justification of pesticide use. A careful and comprehensive assessment of the environmental impacts of pesticides on agriculture and natural ecosystems is very much needed.

In addition to the ethical status of ecological concerns are questions of economic distribution of costs. Although farmers spend about $10 billion year^{-1} for pesticides, little of the pollution costs that result are borne by them or the pesticide-producing chemical companies. Rather, most of the costs are borne off-site by public illnesses and environmental degradation. Standards of social justice suggest a need for a more equitable allocation of responsibility.

These ethical issues do not have easy answers. Strong arguments can be made to support pesticide use based on social and economic benefits. However, evidence of these benefits should not cover up the public health and environmental problems. One goal should be to maximize the benefits while at the same time minimizing the health, environmental and social costs. A recent investigation pointed out that US pesticide use could be reduced by one-half without any reduction in crop yields and that systems of organic agriculture can produce corn and soybeans yields equivalent to conventional agriculture over a 22 year period without any pesticides (Pimentel et al. 1993b; Pimentel et al. 2005). The judicious use of pesticides could reduce the environmental and social costs, while it benefits farmers economically in the short term and supports sustainability of agriculture in the long term.

Public concern over pesticide pollution confirms a national trend toward environmental values. Media emphasis on the issues and problems caused by pesticides has contributed to a heightened public awareness of ecological concerns. This awareness is encouraging research in sustainable agriculture and in nonchemical pest management.

Granted, substituting nonchemical pest controls in US agriculture would be a major undertaking and would not be without costs. The direct and indirect benefits and costs of implementation of a policy to reduce pesticide use should be researched in detail. Ideally, such a program should both enhance social equitability and promote public understanding of how to better protect public health and the environment, while supplying abundant, safe food. Clearly, it is essential that the environmental and social costs and benefits of pesticide use be considered when future pest control programs are being considered and developed. Such costs and benefits should be given ethical and moral scrutiny before policies are implemented, so that sound, sustainable pest management practices are available to benefit farmers, society, and the environment.

2.14 Conclusion

An investment of about $10 billion in pesticide control each year saves approximately $40 billion in US crops, based on direct costs and benefits. However, the indirect costs of pesticide use to the environment and public health need to be balanced against these benefits. Based on the available data, the environmental and public health costs of recommended pesticide use totaled an estimated $9.6 billion each year (Table 2.6). Users of pesticides pay directly only about $3 billion, which includes problems arising from pesticide resistance and destruction of natural enemies. Society eventually pays this $3 billion plus the remaining $9 billion in environmental and public health costs (Table 2.6).

Our assessment of the environmental and health problems associated with pesticides was made more difficult by the complexity of the issues and the scarcity of data. For example, what is an acceptable monetary value for a human life lost or a human illness due to pesticides? Equally difficult is placing a monetary value on killed wild birds and other wildlife; on the dearth of information on the value of invertebrates lost, or microbes lost; or on the price of contaminated food and groundwater.

In addition to the costs that cannot be accurately measured, many costs are not included in the $9.6 billion figure. If the full environmental, public health and social costs could be measured as a whole, the total cost might be nearly double the $9.6 billion figure. Such a complete and long-term cost/benefit analysis of pesticide use would reduce the perceived profitability of pesticides. The efforts of many scientists to devise ways to reduce pesticide use in crop production while still maintaining crop yields have helped but a great deal more needs to be done. Sweden, for example, as of from 1991–1996 reduced pesticide use by 64% without reducing crop yields and/or cosmetic standards, Denmark by 1997 reduced pesticide use by 47%, the Netherlands from 1990–2000 reduced pesticide use by 43% and Norway has from 1985–1996 reduced pesticide use by 54% (PAN-Europe 2003). At the same time, public pesticide poisonings have been reduced by 77%. It would be helpful, if the United States adopted a similar goal to that of Sweden or Denmark.

Table 2.6 Total estimated environmental and social costs from pesticides in the United States

	Costs in Millions of US dollars
Public health impacts	1140
Domestic animal deaths and contaminations	30
Loss of natural enemies	520
Cost of pesticide resistance	1500
Honeybee and pollination losses	334
Crop losses	1391
Fishery losses	100
Bird losses	2160
Groundwater contamination	2000
Government regulations to prevent damage	470
Total	9645

Unfortunately with some groups in the USA, IPM is being used as a means of justifying pesticide use.

Acknowledgement This research was supported in part by the Podell Emeriti Award at Cornell University.

References

Akesson, N. B., & Yates, W. E. (1984). Physical parameters affecting aircraft spray application. In W. Y. Garner & J. Harvey (Eds.)., *Chemical and Biological Controls in Forestry* (pp. 95–115). Washington, D.C.: American Chemical Society (ACS Symposium Series 238).

American Bird Conservancy. (2010). Pesticides and birds. American bird conservancy. http://www.abcbirds.org/abcprograms/policy/toxins/pesticides.html. Accessed 12 July 2012.

American Bird Conservancy. (2010a). Pesticide profile—carbofuran. Washington, D.C.: American Bird Conservancy. http://www.abcbirds.org/abcprograms/policy/toxins/profiles/carbofuran.html. Accessed 6 July 2012.

Asia Times. (2001). India's industrial pest. *Asia Times Online*, India/Pakistan, June 14, 2001. http://www.atimes.com/ind-pak/CF14Df01.html. Accessed 6 July 2012.

Atlas, R. M., & Bartha, R. (1987). *Microbial Biology: Fundamentals and Applications* (2nd ed.). Menlo Park, California, USA: Benjamin Cummings Co.

Baker, B. P., Benbrook, C. M., Groth III, E., & Lutz Benbrook, K. (2002). Pesticide residues in conventional, integrated pest management (IPM)-grown and organic foods: Insights from three US data sets. *Food Additives and Contaminants, 19*, 427–446.

Balcomb, R. (1986). Songbird carcasses disappear rapidly from agricultural fields. *Auk, 103*, 817–821.

Barnes, C. J., Lavy, T. L., & Mattice, J. D. (1987). Exposure of non-applicator personnel and adjacent areas to aerially applied propanil. *Bulletin of Environmental Contamination and Toxicology, 39*, 126–133.

Boutin, C., Freemark, K. E., & Kirk, D. E. (1999). Spatial and temporal patterns of bird use of farmland in southern Ontario. *Canadian Field-Naturalist, 113*, 430–460.

Burgett, M. (2001). Pacific Northwest honey bee pollination survey—2000. *The Bee Line: The Newsletter of the Oregon State Beekeepers Association, 26*(3), 1, 3–6. http://www.orsba.org/htdocs/download/apr01.PDF. Accessed 6 July 2012.

Carlsen, E., Giwercman, A., Keilding, N., & Skakkebaek, N. E. (1992). Evidence for decreasing quality of semen during the past 50 years. *British Medical Journal, 305,* 609–613.

Carrasco-Tauber, C. (1989). *Pesticide productivity revisited.* M.S. Thesis, Amherst, Massachusetts: University of Massachusetts - Amherst.

CEQ, & Barney, G. O. (1980). *The Global 2000 Report to the President of the US: Entering the 21st Century: A Report/Vol. 1, The summary Report. Council on Environmental Quality (CEQ).* New York: Pergamon Press.

Cleveland, C., NOAA, & Saundry, P. (2012). Exxon Valdez oil spill. In C. Cleveland (Ed.)., *Encyclopedia of earth.* Washington, D.C.: Environmental Information Coalition, National Council for Science and the Environment. http://www.eoearth.org/article/Exxon_Valdez_oil_spill. Accessed 6 July 2012.

Colborn, T., Dumanoski, D., & Myers, J. P. (1997). *Our Stolen Future: Are We Threatening Our Fertility, Intelligence, and Survival?–A Scientific Detective Story.* New York: Plume Penguin Group.

CWS. (2012). Pesticides and wild birds. Hinterland weho's who. Canadian Wildlife Service (CWS). http://hww.cwf-fcf.org/hww2.asp?id=230. Accessed 6 July 2012.

D'Anieri, P., Leslie, D. M. Jr., & McCormack, M. L. Jr. (1987). Small mammals in glyphosate-treated clearcuts in northern Maine. *Canadian Field-Naturalist, 101,* 547–550.

Edwards, C. A., & Lofty, J. R. (1982). Nitrogenous fertilizers and earthworm populations in agricultural soils. *Soil Biology Biochemistry, 14,* 515–521.

Elliot, J. E., Norstrom, R. J., & Keith, J. A. (1988). Organochlorines and eggshell thinning in northern gannets (Sula bassanus) from Eastern Canada, 1968–1984. *Environmental Pollution, 52,* 81–102.

EPA. (1977). *Minutes of administrator's pesticide policy advisory committee.* Washington, D.C.: US Environmental Protection Agency.

EPA. (1990). *National pesticide survey–summary results of EPA's national survey of pesticides in drinking water wells.* Office of water, office of pesticides and toxic substances. Washington, D.C.: US Environmental Protection Agency. http://nepis.epa.gov/Exe/ZyNET.exe/10003H1X.TXT?ZyActionD=ZyDocument&Client=EPA&Index=1986+Thru+1990&Docs=&Query=&Time=&EndTime=&SearchMethod=1&TocRestrict=n&Toc=&TocEntry=&QField=&QFieldYear=&QFieldMonth=&QFieldDay=&IntQFieldOp=0&ExtQFieldOp=0&XmlQuery=&File=D%3A%5Czyfiles%5CIndex%20Data%5C86thru90%5CTxt%5C00000005%5C10003H1X.txt&User=ANONYMOUS&Password=anonymous&SortMethod=h%7C-. Accessed 6 July 2012.

EPA. (1992). *Hired farmworkers: Health and well-being at risk.* HRD-92–46, February 14, 1992. U.S. Government Accountability Office. http://www.gao.gov/assets/160/151490.pdf. Accessed 6 July 2012.

EPA. (2004). *Interim reregistration eligibility decision.* US Environmental Protection Agency, Prevention, Pesticides and Toxic Substances (7508C) EPA 738-R-04-006, May 2004. http://www.epa.gov/oppsrrd1/REDs/diazinon_red.pdf. Accessed 13 Sept 2012.

FAO. (1986). *FAO Production Yearbook* (Vol. 40). Rome: Food and Agriculture Organization of the United Nations.

Flickinger, E. L., King, K. A., Stout, W. F., & Mohn, M. M. (1980). Wildlife hazards from furadan 3G applications to rice in Texas. *Journal of Wildlife Management, 44,* 190–197.

Flickinger, E. L., Juenger, G., Roffe, T. J., Smith, M. R., & Irwin, R. J. (1991). Poisoning Canada geese in Texas by parathion sprayed for control of Russian wheat aphid. *Journal of Wildlife Diseases, 27,* 265–268.

Gilliom, R. J., Barbash, J. E., Crawford, C. G., Hamilton, P. A., Martin, J. D., Nakagaki, N., Nowell, L. H., Scott, J. C., Stackelberg, P. E., Thelin, G. P., & Wolock, D. M. (2007). Pesticides in the nation's streams and ground water, 1992–2001. US Geological Survey, USGS Circular 1291 (Revised 2007). http://pubs.usgs.gov/circ/2005/1291/. Accessed 13 Sept 2012.

Greene, M. (1994). Rocky mountain arsenal: States' rights and the cleanup of hazardous wastes. Conflict Resolution Consortium. http://www.colorado.edu/conflict/full_text_search/AllCRC-Docs/94-58.htm. Accessed 13 Sept 2012.

Groth, E. III, Benbrook, C. M., & Lutz, K. (1999). Do you know what you're eating? An analysis of US government data on pesticide residues in foods. Consumers Union of United States, Inc., Public Service Projects Department, Technical Division. http://www.consumersunion.org/pdf/Do_You_Know.pdf. Accessed 6 July 2012.

Hairston, N. G., Smith, F. E., & Slobodkin, L. B. (1960). Community structure, population control and competition. *American Naturalist, 94,* 421–425.

Hall, F. R. (1991). Pesticide application technology and integrated pest management (IPM). In D. Pimentel (Ed.)., *Handbook of Pest Management in Agriculture* (Vol. II, pp. 135–170). Boca Raton, Florida, USA: CRC Press.

Hardy, A. R. (1990). Estimating exposure: The identification of species at risk and routes of exposure. In L. Somerville & C. H. Walker (Eds.)., *Pesticide Effects on Terrestrial Wildlife* (pp. 81–98). London: Taylor and Francis.

Harper, C. R., & Zilberman, D. (1990). Pesticide regulation: Problems in trading off economic benefits against health risks. In D. Zilberman & J. B. Siebert (Eds.)., *Economic Perspectives on Pesticide Use in California: A Collection of Research Papers* (pp. 181–208). Berkeley, California, USA: Department of Agricultural and Resource Economics, University of California.

Hart, K., & Pimentel, D. (2002). Public health and costs of pesticides. In D. Pimentel (Ed.)., *Encyclopedia of Pest Management* (pp. 677–679). New York: Marcel Dekker.

Holmes, T., Nielsen, E., & Lee, L. (1988). Managing groundwater contamination in rural areas. *Rural Development Perspectives, 4,* 35–40. http://naldc.nal.usda.gov/download/IND89047077/PDF. Accessed 12 July 2012.

ICAITI. (1977). An environmental and economic study of the consequence of pesticide use in Central American cotton production: Final report. Guatemala: Central American Research Institute for Industry (ICAITI).

Isenring, R. (2010). Pesticides reduce biodiversity. *Pesticide News, 88,* 4–7. http://www.pan-uk.org/pestnews/Issue/pn88/PN88_p4-7.pdf. Accessed 16 Oct 2013.

Kaiser, J. (2003). Economics: How much are human lives and health worth? *Science, 299,* 1836–1837.

Kakkilaya, B. S. (2012). Malaria site. All about malaria: History, aetiology, pathophysiology, clinical features, diagnosis, treatment, complications and control of malaria. http://www.malaria-site.com/malaria/MalariaInIndia.htm. Accessed 19 Dec 2012.

Mazariegos, F. (1985). The use of pesticides in the cultivation of cotton in Central America. *Industry and Environment (published by United Nations Environment Programme), 8*(3), 5–7.

McCarthy, S. (1993). Congress takes a look at estrogenic pesticides and breast cancer. *Journal of Pesticide Reform, 13,* 25.

Millar, J. G. (1995). Fish and wildlife service's proposal to reclassify the bald eagle in most of the lower 48 states. *Journal of Pesticide Reform, 29,* 71.

Mineau, P. (1988). Avian mortality in agroecosystems. I. The case against granule insecticides in Canada. In M. P. Greaves, P. W. Greig-Smith, B. D. Smith, & British Crop Protection Council (Eds.)., *Field Methods for the Study of Environmental Effects of Pesticides* (pp. 3–12). London: Smith. BPCP Monograph No. 40. British Crop Protection Council BPCP.

Mineau, P., Fletcher, M. R., Glaser, L. C., Tomas, N. J., Brassard, C., Wilson, L. K., Elliott, J. E., Lyon, L. A., Henny, H., Bolinger, T., & Porter, S. L. (1999). Poisoning of raptors with organophosphorus and carbamate pesticides with emphasis on Canada, US, and UK. *Journal of Raptor Research, 33,* 1–37.

Mussen, E. (1990). California crop pollination. *Gleanings in Bee Culture, 118,* 646–647.

NAS. (1975). *Pest Control: An Assessment of Present and Alternative Technologies* (5 volumes). Washington, D.C.: National Academy of Sciences.

National Research Council, Committee on Scientific and Regulatory Issues Underlying Pesticide Use Patterns and Agricultural Innovation. (1987). *Regulating Pesticides in Food: The Delaney Paradox.* Washington, D.C.: National Academy Press.

National Research Council, Committee on the Role of Alternative Farming Methods in Modern Production Agriculture. (1989). *Alternative Agriculture.* Washington, D.C.: National Academy Press.

NIOSH. (2012). A story of impact: NIOSH pesticide poisoning monitoring program protects farmworkers. National Institute for Occupational Safety and Health (NIOSH), Centers for Disease Control and Prevention, US Department of Health and Human Services. http://www.cdc.gov/niosh/docs/2012-108/pdfs/2012-108.pdf. Accessed 4 Oct 2012.

Oka, I. N. (1991). Success and challenges of the Indonesian national integrated pest management program in the rice based cropping system. *Crop Protection, 10,* 163–165.

Oka, I. N., & Pimentel, D. (1976). Herbicide (2,4-D) increases insect and pathogen pests on corn. *Science, 193,* 239–240.

PAN-Europe. (2003). Pesticide use reduction is working: An assessment of national reduction strategies in Denmark, Sweden, the Netherlands and Norway. Pesticides Action Network—Europe. http://www.epha.org/IMG/pdf/Pure_is_Working.pdf. Accessed 19 Dec 2012.

PAN—North America. (2002). PANNA: Latino farmworkers face greater risk of cancer. Pesticide Action Network—North America. http://www.panna.org/legacy/panups/panup_20020719.dv.html. Accessed 12 July 2012.

Pesticides Residues Committee—U. K. (2004). Pesticide residues in food: Facts not fiction. Pesticides residues committee—United Kingdom. PRC Leaflet. http://www.pesticides.gov.uk/guidance/industries/pesticides/advisory-groups/PRiF/PRC-Pesticides-Residues-Commitee/Other_PRC_Information/prc-leaflet-pesticide-residues-in-food-facts-not-fiction. Accessed 14 March 2013.

Pimentel, D. (1988). Herbivore population feeding pressure on plant host: Feedback evolution and host conservation. *Oikos, 53,* 289–302.

Pimentel, D. (1994). Insect population responses to environmental stress and pollutants. *Environmental Reviews, 2,* 1–15.

Pimentel, D. (1997). Pest management in agriculture. In D. Pimentel (Ed.)., *Techniques for Reducing Pesticide Use: Environmental and Economic Benefits* (pp. 1–11). Chichester, United Kingdom: Wiley.

Pimentel, D., & Greiner, A. (1997). Environmental and socio-economic costs of pesticide use. In D. Pimentel (Ed.), *Techniques for Reducing Pesticide Use: Environmental and Economic Benefits* (pp. 51–78). Chichester, United Kingdom: Wiley.

Pimentel, D, & Hart, K. (2001). Pesticide use: Ethical, environmental, and public health implications. In W. Galston & E. Shurr (Eds.), *New Dimensions in Bioethics: Science, Ethics and the Formulation of Public Policy* (pp. 79–108). Boston: Kluwer Academic Publishers.

Pimentel, D., & Pimentel, M. (2008). *Food, Energy, and Society, third edition.* Boca Raton, Florida, USA: CRC Press.

Pimentel, D., McLaughlin, L., Zepp, A., Latikan, B., Kraus, T., Kleinman, P., Vancini, F., Roach, W. J., Graap, E., Keeton, W. S., & Selig, G. (1991). Environmental and economic impacts of reducing US agricultural pesticide use. In D. Pimentel (Ed.), *Handbook on Pest Management in Agriculture* (Vol. I, pp. 679–718). Boca Raton, Florida, USA: CRC Press.

Pimentel, D., Stachow, U., Takacs, D. A., Brubaker, H. W., Dumas, A. R., Meaney, J. J., O'Neil, J. A. S., Onsi, D. E., & Corzilius, D. B. (1992). Conserving biological diversity in agricultural/forestry systems. *Bioscience, 42,* 354–362.

Pimentel, D., Acquay, H., Biltonen, M., Rice, P., Silva, M., Nelson, J., Lipner, V., Giordana, S., Horowitz, A., & D'Amore, M. (1993a). Assessment of environmental and economic impacts of pesticide use. In D. Pimentel & H. Lehman (Eds.), *The Pesticide Question: Environment, Economics and Ethics* (pp. 47–84). New York: Chapman & Hall.

Pimentel, D., McLaughlin, L., Zepp, A., Lakitan, B., Kraus, T., Kleinman, P., Vancini, F., Roach, W. J., Graap, E., Keeton, W. S., & Selig, G. (1993b). Environmental and economic effects of reducing pesticide use in agriculture. *Agriculture, Ecosystems & Environment, 46,* 273–288.

Pimentel, D., Wilson, C., McCullum, C., Huang, R., Dwen, P., Flack, J., Tran, Q., Saltman, T., & Cliff, B. (1997). Economic and environmental benefits of biodiversity. *Bioscience, 47,* 747–757.

Pimentel, D., Hepperly, P., Hanson, J., Douds, D., & Seidel, R. (2005). Environmental, energetic, and economic comparisons of organic and conventional farming systems. *Bioscience, 55,* 573–582.

Potter, D. A., & Braman, S. K. (1991). Ecology and management of turfgrass insects. *Annual Review of Entomology, 36,* 383–406.

Potts, G. R. (1986). *The Partridge: Pesticides, Predation and Conservation.* London: Collins.

Reddy, V. R., Baker, D. N., Whisler, F. D., & Fye, R. E. (1987). Application of GOSSYM to yield decline in cotton. I. Systems analysis of effects of herbicides on growth, development and yield. *Agronomy Journal, 79,* 42–47.

Reid, C. (2000). Malaria in India. In Implications of climate change on malaria in Karnataka, India. Senior Honors Thesis in Environmental Science. Providence: Brown University. http://www.brown.edu/Research/EnvStudies_Theses/full9900/creid/malaria_in_india.htm. Accessed 12 July 2012.

Repetto, R., & Baliga, S. S. (1996). Pesticides and the immune system: The public health risks. Washington, D.C.: World Resources Institute. http://www.wri.org/publication/pesticides-and-the-immune-system. Accessed 12 July 2012.

Richter, E. D. (2002). Acute human pesticide poisonings. In Pimentel, D. (Ed.)., *Encyclopedia of Pest Management* (pp. 3–6). New York: Marcel Dekker.

Soejitno, J. (1999). *Integrated Pest Management in Rice in Indonesia: A Success Story.* Asia-Pacific Association of Agricultural Research Institution. Bangkok: FAO Regional Office for Asia & the Pacific. http://www.apaari.org/wp-content/uploads/2009/05/ss_1999_02.pdf. Accessed 14 March 2013.

Stone, A. W., American Ground Water Trust. (1998). Ground water for household water supply in rural America: Private wells or public systems? In *Gambling with Groundwater—Physical, Chemical, and Biological Aspects of Aquifer-stream Relations:* Proceedings of the Joint Meeting of the Twenty Eighth Congress of the International Association of Hydrogeologists and the Annual Meeting of the American Institute of Hydrologists. Las Vegas, Nevada, USA. http://www.agwt.org/events/Education_Papers/RuralWaterInUSA.pdf. Accessed 12 July 2012.

Stone, W. B., & Gradoni, P. B. (1985). Wildlife mortality related to the use of the pesticide diazinon. *Northeastern Environment Science, 4,* 30–38.

Stringer, A., & Lyons, C. (1974). The effect of benomyl and thiophanate-methyl on earthworm populations in apple orchards. *Pesticide Science, 5,* 189–196.

Stuart, C. (1999). Development of resistance in pest populations. In M. Lieberman (Ed.)., *Report on Genetically Modified Food Crops. Chemistry and Public Policy [Chem 191].* South Bend, Indiana, USA: Notre Dame University. http://www.nd.edu/~chem191/TOC.html. Accessed 12 July 2012.

Trautmann, N. M., Porter, K. S., & Wagenet, R. J. (2012). Pesticides and groundwater: A guide for the pesticide user. Cornell University Cooperative Extension, Pesticide Safety Education Program (PSEP), Fact Sheet. http://psep.cce.cornell.edu/facts-slides-self/facts/pest-gr-gud-grw89.aspx. Accessed 6 July 2012.

Trumbel, J. T., Carson, W., Nakakihara, H., & Voth, V. (1988). Impact of pesticides for tomato fruit worm (Lepidoptera: Noctuidae) suppression on photosynthesis, yield, and non-target arthropods in strawberries. *Journal of Economic Entomology, 81,* 608–614.

UNEP. (1979). The State of the Environment: Selected Topics – 1979: Report of the Executive Director United Nations Environmental Programme. *Environment International, 2*(3), 187–200.

Unnevehr, L. J., Lowe, F. M., Pimentel, D., Brooks, C. B., Baldwin, R. L., Beachy, R. N., Chornesky, E. A., Hiler, E. A., Huffman, W. E., King, L. J., Kuzminski, L. N., Lacy, W. B., Lyon, T. L., McNutt, K., Ogren, W. L., Reginato, R., & Suttie, J. W. (2003). *Frontiers in Agricultural Research: Food, Health, Environment, and Communities.* Washington, D.C.: National Academies of Science. http://www.nap.edu/openbook.php?record_id=10585&page=217. Accessed 12 July 2012.

USA Today. (2003). Child cancer risk cited in chemical rules. *USA Today,* March 3, 2003. http://usatoday30.usatoday.com/news/health/2003-03-03-children-cancer_x.htm. Accessed 13 Dec 2012.

USBC. (1990). *Statistical Abstract of the United States: 1991.* Washington, D.C.: US Bureau of the Census.

USDA. (1989). *Agricultural Statistics 1989*. Washington, D.C.: US Department of Agriculture.
USDA, Office of Inspector General. (2010). FSIS national residue program for cattle. Audit Report 24601-08-KC, March 2010. http://www.usda.gov/oig/webdocs/24601-08-KC.pdf. Accessed 12 July 2012.
USFWS. (1988). *1985 National Survey of Fishing, Hunting and Wildlife-associated Recreation*. Washington, D.C.: US Fish and Wildlife Service, US Department of the Interior.
USGS. (1995). Pesticides in public supply wells of Washington state. US Geological Survey, USGS Fact Sheet 122–96. http://wa.water.usgs.gov/pubs/fs/fs122-96/. Accessed 13 Sept 2012.
USGS. (1996). Pesticides found in ground water below orchards in the quincy and pasco basin. USGS fact sheet 171–96. http://wa.water.usgs.gov/pubs/fs/fs171–96/. Accessed 19 Dec 2012.
U.S. Water News. (2002). Clear creek beer spill in 2000 costs Coors $500,000. *U.S. Water News Online*. http://www.uswaternews.com/archives/arcrights/2clecre1.html. Accessed 6 July 2012.
Weiner, B. P., & Worth, R. M. (1969). Insecticides: Household use and respiratory impairment. *Hawaii Medical Journal, 28*(4), 283–285.
White, L. (1967). The historical roots of our ecological crisis. *Science, 155,* 1203–1207.
White, D. H., Mitchell, C. A., Wynn, D., Flickinger, E. L., & Kolbe, E. J. (1982). Organophosphate insecticide poisoning of Canada geese in the Texas Panhandle. *Journal of Field Ornithology, 53,* 22–27.
WHO. (2011). *World Malaria Report 2011*. FACT SHEET. Global Malaria Programme, World Health Organization. http://www.who.int/malaria/world_malaria_report_2011/WMR2011_factsheet.pdf. Accessed 12 Sept 2012.

Chapter 3
Integrated Pest Management for European Agriculture

Bill Clark and Rory Hillocks

Contents

3.1	Introduction	75
3.2	European Union Pesticide Reduction Strategy	76
3.3	Impact of Pesticide Withdrawals	77
3.4	United Kingdom Government Initiatives	79
	3.4.1 United Kingdom Agri-Environment Schemes (AES)	79
3.5	Implementation of EU Legislation in the UK	81
	3.5.1 Measures for Prevention and/or Suppression of Harmful Organisms	81
	3.5.2 Tools for Monitoring	84
	3.5.3 Threshold Values as Basis for Decision-Making	85
	3.5.4 Non-Chemical Methods to be Preferred	89
	3.5.5 Target-Specificity and Minimization of Side Effects	90
	3.5.6 Reduction of Use to Necessary Levels	91
	3.5.7 Application of Anti-Resistance Strategies	92
	3.5.8 Records, Monitoring, Documentation and Check of Success	93
3.6	Conclusions	95
References		96

Abstract Legislation to decrease pesticide use in European agriculture has given renewed importance to integrated pest management (IPM). The adoption of IPM on all farms in Member States by 2014 is the main pillar of the European Union (EU) strategy to mitigate the negative impact of rapid pesticide removals on European food production. Legislation under the EU's 'Thematic Strategy for the Sustainable Use of Pesticides' is directed primarily at minimizing the impact of pesticide use

---

B. Clark (✉)
National Institute of Agricultural Botany, Huntingdon Road,
Cambridge CB3 0LE, United Kingdom
e-mail: bill.clark@niab.com

R. Hillocks
European Centre for IPM, Natural Resources Institute,
University of Greenwich, Chatham Maritime,
Kent ME4 4TB, United Kingdom
e-mail: r.j.hillocks@gre.ac.uk

on the environment and human health and does not promote IPM directly. In the absence of IPM technologies which can deliver significant decreases in pesticide use while maintaining the productivity and profitability of agricultural and horticultural enterprises, further action will be required to develop and promote IPM and to ensure that IPM adoption results in a decrease in total pesticide use. All Member States are required to develop National Action Plans (NAPs) for pesticide reduction and implementation of IPM. The approach in the NAPs differs between EU countries and the United Kingdom (UK); the UK Government favors voluntary measures and aims to decrease the non-target effects of pesticides, not necessarily a decrease in pesticide use. In the EU Framework Directive on the sustainable use of pesticides, there are eight general principles for integrated pest management and this chapter describes how each of these is being addressed in UK agriculture.

Keywords European Union · Pesticides · IPM · Framework Directive · Sustainable use

List of abbreviations

ADAS	Agricultural Development Advisory Service (UK)
AHDB	Agriculture and Horticulture Development Board
AES	Agri-environment Schemes
AFS	Assured Food Standards
BAP	Biodiversity Action Plan
BBRO	British Beef Research Organization
CPMP	Crop Protection Management Plan
CRD	Chemicals Regulation Directorate
CRP	Crop Protection Association
DEFRA	Department of the Environment, Food and Rural Affairs (UK)
EC	European Commission
ELC	European Landscape Convention
ELS	Entry Level Stewardship
ES	Environmental Stewardship
EU	European Union
FERA	Food and Environment Research Agency
FRAC	Fungicide Resistance Action Committee
INRA	The Institut National de la Recherche Agronomique
IPM	Integrated Pest Management
IRAC	Insecticide Resistance Action Committee
HLS	Higher Level Stewardship
HGCA	Home Grown Cereals Authority
MS	Member State (of the EU)
NAP	National Action Plan
NIAB	National Institute of Agricultural Botany
NRoSO	National Register of Spray Operators

NSTS   National Sprayer Testing Scheme
PGRO   Pulse Growers Research Organization
RDPE   Rural Development Programme for England
SSSI   Sites of Special Scientific Interest
SUD   Sustainable Use Directive
TSSP   Thematic Strategy on the Sustainable Use of Pesticides
VI   Voluntary Initiative
WBM   Wheat Blossom Midge
WFD   Water Framework Directive
WRAC   Weed Resistance Action Committee

Part I: European pesticide policy and integrated pest management

3.1 Introduction

Integrated pest management (IPM) as a concept for decreasing the reliance of farming on pesticides has been around for more than 30 years. With the exception of a few major crops such as cotton, where IPM has been implemented in response to heavy insecticide dependency and the development of resistance in target pest populations, the adoption of IPM globally has been limited. The first major obstacle to wider adoption has been the absence of sufficient IPM technologies and systems which are practical and economic to implement on-farm. The second impediment to adoption is the knowledge-intensive nature of IPM, requiring significant changes to farming practice and some understanding of pest/natural enemy dynamics. Thirdly, most biopesticides and other IPM component technologies used on their own are less effective than the conventional pesticides they replace and have to be used in combination, sometimes with decreased application of a conventional pesticide. The approach to crop protection becomes one of 'pest management' rather than 'pest control' and IPM becomes a component of integrated crop management.

Legislation has been introduced by the European Parliament aimed at decreasing the use of conventional pesticides in European agriculture (Hillocks 2011). As part of a suite of legislation collectively known as the Thematic Strategy on the Sustainable Use of Pesticides (TSSP), the European Commission (EC) has introduced a statutory requirement for crop protection to be conducted under a system of IPM on all farms in European Union (EU) Member States. EU Member States are expected to deliver action plans to encourage IPM and facilitate a decrease in pesticide use.

Because IPM is a principle rather than a specific technology, there are numerous definitions, depending on the context for its application and the desired outcomes. For the purpose of this review we are using the definition provided by the EU in EC Directive 91/414/EEC: "*IPM is the rational application of a combination of biological, biotechnical, chemical, cultural or plant-breeding measures, whereby the use of plant protection products is limited to the strict minimum necessary to maintain the pest population at levels below those causing economically unacceptable damage or loss."*

3.2 European Union Pesticide Reduction Strategy

The TSSP encompasses a number of EC Directives aimed at decreasing risks to the environment and to public health, related to pesticide use. The Directives seek to promote a more sustainable use of pesticides and target a significant reduction in risks associated with pesticide use and a significant decrease in overall pesticide use 'consistent with the necessary protection of crops'.

When the strategy was being developed in 2007, the aim was to:

- Minimize the hazards and risks to health and environment from the use of pesticides
- Improve controls on the use and distribution of pesticides
- Reduce the levels of harmful active substances, including substituting safer alternatives for the most dangerous ones
- Encourage conversion to low-input or pesticide-free cultivation
- Establish a transparent system for reporting and monitoring the program.

The TSSP was developed and is being implemented within the EC by the Directorates General for Health (DG SANCO) and Environment. As a consequence, the drivers and indicators are all concerned with health benefits, environmental protection and biodiversity enhancement, rather than being set in the context of actively promoting IPM and sustaining farm productivity and profitability.

The component of the TSSP where IPM is introduced as a statutory requirement in EU Member States is the 'Sustainable Use Directive' (SUD). Wider adoption of IPM is seen by the EC as the means to mitigate the negative impacts of rapid pesticide withdrawal on farming livelihoods and food production. Member States (MS) are required to publish, by end of 2012, National Action Plans (NAPs) for pesticide reduction in agriculture. IPM-compliance is to be achieved by the beginning of 2014. NAPs are expected to achieve a reduction in the risks and effects of pesticide use on human health and the environment. This does not necessarily require a decrease in total pesticide use. Similarly, a farm could become IPM compliant without an associated decrease in total pesticide use, provided there was a reduction in the non-target effects of pesticide application.

Each MS is adopting a different approach to pesticide reduction within their NAPs. The British Government for instance, believes that existing measures under the 'Voluntary Initiative' (see Part 2) in some respects, already go beyond what is required under the SUD and only minor adjustments will be needed to meet the terms of compliance. The assumption is that actions encouraged under the Initiative to protect the environment, especially water courses and to enhance biodiversity, will be sufficient to significantly decrease pesticide use. Farmers are able to access payments for compliance with environmental protection measures under the Voluntary Initiative.

The French Government has more ambitious targets, introducing in 2008, a plan known as ECOPHYTO 2018 which aims to decrease pesticide use by 50% over the ten year period to 2018. The Institut National de la Recherche Agronomique (INRA 2010) concludes from their farm surveys that improved pesticide application management, based on the widespread use of existing decision support tools and field observations, could lead to reductions in pesticide use of between *e.g.* 3% (pea) and 40% (grain maize), compared with intensive management, in most cases without affecting production levels. The results of the survey demonstrated that the commitment to a 50% reduction of pesticide use will be a difficult target to achieve. During an average year similar to 2006, if all French farms switched to a system of integrated production the INRA model estimated that the reduction in pesticide use would be 50% in arable crops, 37% in viticulture, 21% in fruit orchards and 100% in grasslands.

3.3 Impact of Pesticide Withdrawals

While the EU Parliament and policy makers within DG SANCO might wish to accelerate the withdrawal of pesticides under the precautionary principle, experts among the agricultural stakeholders have published reports warning that further rapid decline in available active ingredients for conventional pesticides will threaten European food production. The Agriculture Development Advisory Service in the UK estimated that there would be a decline in food production of between 25% and 53%, depending on which of a range of pesticide withdrawal criteria were approved by the European Parliament (ADAS 2008). In France, INRA (2010) estimates that conversion to low input production systems to meet the 'Ecophyto' targets by 2018, would result in a 15–20% decrease in production for oilseed rape and potato. The INRA model predicted that full adoption of 'integrated production' in French agriculture would be associated with production decreases (in value terms) estimated at 12% for arable crops, 24% for viticulture and 19% for fruits (based on 2006 prices).

Rapid withdrawal of conventional pesticides and resulting decreases in food production and/or increased costs of production would be at odds with European policy on food security which requires food production per capita to be maintained or increased in the coming years. Furthermore, the EC does not wish to see Europe's farmers suffering declining incomes as a result of no longer having access to key pesticides.

The farming community has been skeptical about IPM and in the main, has preferred to retain their pesticide regimes. However, pesticide withdrawals that have already occurred, have forced some farmers to reconsider IPM although they are finding that the necessary technologies are just not available. In some EU countries, there is insufficient technical support to help farmers to identify the best IPM component technologies and how to integrate them into an IPM system adapted to their own farms.

The crisis in pest management brought about by pesticide withdrawals under EC registration rules, has so far had the greatest impact in so called 'minor use' horticultural crops, where the scale of production does not justify the expense of pesticide registration. Field vegetables, soft fruits and orchard fruits, are also the crops grown over smaller areas with higher returns per unit area than for arable crops. It is in these crops where alternatives to conventional pesticides are most widely used. There is still a long way to go before sufficient alternative crop protection technologies and IPM solutions are available for arable crops, such that pesticide reduction will not result in major decreases in yield and/or quality. Because arable crops, particularly cereals and oilseeds, occupy the majority of farmland in the EU and in total, use the most pesticide, major decreases in total pesticide use will have to take place in these farming systems. If such decreases are to be achieved without substantial production losses, economically viable IPM solutions for the arable sector will be required.

Excluding the horticulture sector and with the possible exception of control of *Septoria* in wheat, arable farmers may not yet be experiencing much of an economic impact with respect to management of diseases and insect pests but, they are already facing acute problems in weed management. A number of important weeds have become more difficult to control as a result of herbicide withdrawals and the worst case is that of black grass in cereal/oilseed rape rotations. As farmers have to rely on fewer active ingredients, the rate at which resistance to those few active ingredients builds up in the target weed population will increase (Roteveel et al. 2011).

Most farms in UK and many across Europe use some form of non-chemical pest control and would be able to say they were IPM-compliant simply, for instance, because they use pest-resistant crop varieties or sometimes use mechanical weeding (Bailey et al. 2009). The likely outcome of the TSSP is that initially, the principle of minimal compliance will result in little change to present practice, with many farms able to say they are using IPM. However, as time goes on and pesticides become still fewer, the negative impact on food production will become increasingly apparent. More widespread adoption of IPM as a system, rather than as ad-hoc pest management components, will become necessary if farmers are to protect their livelihoods. The shortage of alternative crop protection technologies and IPM systems that can sustain farm productivity at its present levels will then become apparent.

Now is the time to introduce a rural livelihood perspective to the TSSP and to bring the farmer back to the center of research and development of IPM systems 'fit for purpose'. Conflict between environmental and economic goals is not necessary; the policy and practice must be developed to make these goals compatible. The health and environmental outcomes through decreased use of conventional pesticides will be achieved much more quickly and sustainably if the farming community is positively engaged in the process and there is harmony between EU policy on European food security and on pesticide reduction/IPM implementation.

Part II: Implementing integrated pest management in the United Kingdom

3.4 United Kingdom Government Initiatives

Several schemes to promote the safe use of pesticides have been implemented in the UK, including training initiatives and registration schemes, mostly voluntary but these have been very successful. In 2006 the UK Government published the "UK Pesticides Strategy: A Strategy for the Sustainable Use of Plant Protection Products". This strategy had a number of aims including:

- Protecting consumers by minimising risks from pesticides residues in food;
- Protecting users and workers by minimizing exposure to pesticides;
- Protecting residents and bystanders by minimizing exposure from spray operations;
- Reducing water pollution caused by pesticides;
- Reducing the impact of pesticides on biodiversity;
- Maintaining the availability of sufficient methods of crop protection particularly for minor crops;
- Encouraging the introduction of cost-effective alternative approaches and greater use of integrated crop and pest management.

Under these broad aims are specific initiatives including the registration and training of spray operators (NRoSO), the implementation of agri-environment schemes (AES) and the Voluntary Initiative, details of which are given below. The UK approach has been one of reducing the impact of pesticides on the environment, rather than the simple reduction of pesticide use which can have unintended consequences.

3.4.1 United Kingdom Agri-Environment Schemes (AES)

Over two decades, UK AES have helped make arable land not just a source of food, but a haven for the country's wildlife and a source of beneficial insects. Farmers have joined with conservationists to maintain production while safeguarding the countryside. (See "Agri-environment schemes in action", at http://www.naturalengland.org.uk/publications/)

AES are voluntary agreements that pay farmers and other land managers to manage their land in an environmentally friendly way. The first AES in the UK, Environmentally Sensitive Areas, was launched in 1987. The schemes are run by Natural England, on behalf of the Department of the Environment, Food and Rural Affairs (DEFRA). AES are supported through the Rural Development Programme for England 2007–2013 (RDPE), with EU funding from the European Agricultural Fund for Rural Development. The main schemes are Environmental Stewardship (ES): Entry-Level Stewardship (ELS) and Higher-Level Stewardship (HLS).

3.4.1.1 Environmental Stewardship

Environmental Stewardship (ES) is an agri-environment scheme open to all farmers and funded by the UK Government and the European Union (EU). Farmers and land managers across England enter into voluntary management agreements with Natural England in order to deliver the scheme. In return for ES payments, the farmer agrees to protect and enhance certain features of the landscape including wildlife, landscapes, historic features and natural resources. There are two main elements to Environmental Stewardship:

- Entry Level Stewardship (ELS) provides a basic approach to supporting the good stewardship of the countryside. This is done through simple and effective land management that goes beyond the Single Payment Scheme requirement.
- Higher Level Stewardship (HLS) involves more complex types of management and agreements are tailored to local circumstances. HLS rewards much higher standards of environmental management and is targeted at land and features of greatest environmental value. It is the main delivery mechanism to achieve targets for the condition of Sites of Special Scientific Interest (SSSIs), Biodiversity Action Plan (BAP) targets a range of other national and international targets. These include, for example, the protection and management of landscape character and features under the European Landscape Convention (ELC) and the Water Framework Directive (WFD) (Natural England 2009).

3.4.1.2 Voluntary Initiative

In 2001 the Government accepted proposals put forward by the farming and crop protection industry to minimize the environmental impacts from pesticides. This set of measures and initiatives became known as the Voluntary Initiative (VI) which began in April 2001. It is a UK-wide package of measures, designed to reduce the environmental impact of the use of pesticides in agriculture, horticulture and outdoor amenities. Initially a list of 27 proposals, the program finally included over 40 different projects covering research, training, communication and stewardship. The package was developed as a better and more effective means of fulfilling the Government's environmental objectives of improving water quality and biodiversity on arable farmland, than would be achieved by a proposal to introduce a tax on crop protection products. The VI was developed by the Crop Protection Association together with leading national agricultural and farming organizations, and in consultation with the main environmental groups (Goldsworthy 2006). The VI has continued since as a voluntary program of work promoting responsible pesticide use. The VI has four major schemes operating nationwide: the National Sprayer Testing Scheme (NSTS) organized and run by AEA; the National Register of Sprayer Operators (NRoSO) set-up and run by City and Guilds; the Crop Protection Management Plan (CPMP) operated by NFU and the BASIS BETA programme for practicing agronomists to gain further understanding of the interactions between production and the environment. NSTS tested machines now cover 86.8% of the sprayed area, the NRoSO has over 20,000 members on its CPD training programs and there are over 850 agronomists who have completed BETA training.

There are over 58,000 voluntary AES agreements, covering over 6 million ha—about 66% of agricultural land in England. AES have largely been successful in halting the loss and deterioration of the highest priority habitats on farmland, and are now restoring or enhancing many of these. Habitat creation has had variable results but improved techniques have resulted in notable successes. Populations have been increased of certain nationally scarce farmland birds. Bumble bee abundance increased 15–35 times on AES sown wildflower mixes in arable areas, compared to control areas.

3.5 Implementation of EU Legislation in the UK

In the EU Framework Directive on the sustainable use of pesticides and the accompanying document, the "Development of guidance for establishing Integrated Pest Management (IPM) principles", eight general principles for integrated pest management were identified as:

1. Measures for prevention and/or suppression of harmful organisms
2. Tools for monitoring
3. Threshold values as basis for decision-making
4. Non-chemical methods to be preferred
5. Target-specificity and minimization of side effects
6. Reduction of use to necessary levels
7. Application of anti-resistance strategies
8. Records, monitoring, documentation and check of success

This section discusses how these eight principles have been approached within UK cropping systems. The emphasis here is on arable cropping systems and wheat as the major arable crop within UK rotations.

3.5.1 *Measures for Prevention and/or Suppression of Harmful Organisms*

3.5.1.1 Crop Rotation

The basic technique of crop rotation forms the basis of pest and disease suppression for the majority of arable crops in the UK. The growing of different crops in a sequence ensures that pests and diseases are avoided or reduced in subsequent crops. The major cause of yield loss in wheat grown consecutively is 'take-all', caused by the soil-borne fungus *Gaeumannomyces graminis*. In some soils this fungus would prevent a second wheat crop being grown but in most second wheat crops it would cause some yield loss (c. 10–15%). This disease can almost be prevented completely by employing a crop rotation that avoids wheat crops being grown in succession or in close rotation. Virtually every wheat grower in the UK employs this simple rotational technique. Where wheat is grown in succession or in very close rotations, seed treatments containing the fungicide silthiofam, are used to reduce the severity of the disease but control is much more effective using rotational techniques.

Some fungal pathogens can survive and overwinter on crop debris and pose a threat to subsequent crops. Examples of this include *Fusarium graminearum* which affects both wheat and maize and survives on straw between crops. Wheat following maize in the rotation would be at much higher risk from the disease than if a non-host crop was grown before. Consequently in high risk areas wheat following wheat or maize would be avoided wherever possible to control this disease. Where wheat and maize are grown in close rotations, disease pressure from *F. graminearum* can be very high, leading to high levels of fungicide use on the ears of wheat to prevent ear blight.

Soil-borne pests such as beet cyst nematode (BCN) (*Heterodera schachtii*) are also routinely controlled by rotation. BCN is a persistent soil-borne pest which can reduce root yields by up to 60%. If susceptible crops are grown in close rotation, the pest will build up over time and render fields unusable for sugar beet and some other crops. The recent increased incidence of the problem in the UK is thought to be caused by the concentration of the beet growing area in the UK and a trend to closer rotations.

3.5.1.2 Cultivation Techniques

Cultivation practices are an integral part of crop protection measures in most arable cropping systems. Examples include the use of deep cultivation or plowing to remove crop debris from the soil surface. This has the effect of removing the source of many fungal diseases such as eyespot (*Tapesia* spp.), *Fusarium graminearum*, and tan spot (*Drechslera tritici-repentis*) which survive and overwinter on straw from the previous crop. Plowing is also an integral component of weed control strategies, particularly against grass weeds such as black-grass (*Alopecurus myosuroides*) and brome (*Bromus* spp.). In parts of the UK control of such grass weeds is not possible using chemical control methods alone. The use of shallow tine cultivation in conjunction with broad spectrum herbicides (referred to as a 'stale seedbed' technique) is also widely practiced. Cultivation techniques such as the use of stale seedbeds, reduced tillage and direct sowing are integrally linked to agronomic practices such as adjusting sowing dates and seed rates. These techniques can reduce pest and disease pressure, allowing reduced doses or numbers of applications of chemical pesticides.

3.5.1.3 Use of Insect Pest and Disease Resistant Varieties/Use of Certified Seed

The use of resistant varieties and certified seed and planting material are a routine part of UK cropping systems. Cereal and oilseed rape variety testing is undertaken by a levy-funded organization in the UK (The Home Grown Cereals Authority, HGCA, part of the Agricultural and Horticultural Development Board, AHDB). Varieties are tested by the HGCA in national trials and varieties are compared against a list of criteria which aim to improve the overall performance of varieties, including pest and disease resistance. Varieties not achieving minimum standards will not be recommended or will be removed from the recommended list if, after recommendation, they subsequently fall below the minimum standards. Results of variety testing are published by the AHDB (www.hgca.com). Similar systems operate for many crops in the UK such as the Processors and Growers Research Organization (PGRO) who publish the PGRO Pulse Levy funded Recommended Lists for peas and beans and the British Beet Research Organization (BBRO), who publish the Recommended List of sugar beet varieties.

In the UK, Seed Certification Schemes exist to protect farmers and their customers by ensuring that the seed they buy meets certain quality standards. All certified seed must meet prescribed standards of varietal identity and purity, germination and freedom from weed seeds. The directives define standards of purity and germination that the seed must meet in order to be certified. The National Institute of Agricultural Botany (NIAB) provides the government with technical services to implement the seed certification scheme as well as national listing and UK Plant Breeders Rights. NIAB monitors the quality of all seed stocks passing through the certification system and maintains pedigree records for all seed lots and seed crops in England and Wales. The Official Seed Testing Station for England & Wales is operated by NIAB in Cambridge. In Scotland the Official Seed Testing Station is operated by SASA in Edinburgh. Although not part of the cereal seed certification system, many seed stocks are tested for seed-borne diseases such as *Microdochium nivale, Fusarium graminearum* and *Tilletia tritici*.

3.5.1.4 Balanced Fertilizer Use

Major nutrients, particularly nitrogen, are a major cost to crop production but optimizing fertilizer inputs is difficult to achieve. The UK government publishes guidelines for fertilizer use and the majority of farmers follow these national guidelines (DEFRA 2010). It is well-established that fertilizer use in cereal crops, particularly nitrogen fertilizer, has a direct effect on disease pressure and the risk of lodging, so it is in the farmers' interest to try and optimize use. Higher than optimal use can lead to higher disease pressure from certain biotrophic fungal pathogens such as yellow rust (*Puccinia striiformis*) and brown rust (*Puccinia triticina*) as well as the main UK disease septoria (*Mycosphaerella graminicola*), leading to higher fungicide use and increased costs. Higher than optimal use also increases the risk of lodging and additional costs are incurred by having to apply plant growth regulators to try and prevent lodging. Hence, there are real incentives for farmers not to over apply nitrogen fertilizer.

Optimizing nitrogen fertilizer use is not a precise science but is important both economically and environmentally. Measurement of the amount of nitrogen available in the soil pre-planting is carried out on some fields although an approximate figure can be used where previous cropping is known. The amount of soil mineral nitrogen (SMN) is used to estimate the amount of additional fertilizer required by the crop.

3.5.1.5 Prevention of Spread of Harmful Organisms

Where disease organisms are soil-borne, such as *Soil-borne wheat mosaic virus* (SBWMV), or *Beet necrotic yellow vein virus* (Rhizomania) appropriate hygiene measures to reduce the movement of soil spread are undertaken. Movement of soil on farm machinery from field to field or to adjacent farms would be avoided wherever possible. There are relatively few examples of this in arable cropping systems, mostly where soil-borne non-indigenous pathogens have been introduced into the UK.

3.5.1.6 Protection and Enhancement of Important Beneficial Organisms

This is done largely through the Agro-environment Schemes which give financial encouragement to farmers to make their land not just a source of food, but a haven for the country's wildlife and a source of beneficial insects. This is most obviously manifested in the countryside by field margins or 'conservation headlands' which are often sown with wildlife-enhancing seed mixtures to provide plant species beneficial to insects and birds. Other management practices include buffer strips to protect watercourses and overwintered stubble to provide food sources for wildlife.

3.5.2 Tools for Monitoring

In the UK there are a number of monitoring systems in place to give information to farmers and advisers. For example the HGCA funds a number of schemes that aim to give information to aid decision making. The HGCA funds research on disease control, including independent fungicide performance testing, disease monitoring and variety testing. Major fungicide active ingredients are tested at a range of doses on winter wheat, winter barley and oilseed rape. The results of these tests give growers and advisers independent information on dose and product choice on a wide range of diseases. General guidance on disease control practices is provided in booklets such as the HGCA Wheat Disease Management Guide (HGCA 2012a) which includes information on decision making in relation to growth stage, disease levels, etc. Similar guides are published for barley and oilseed rape (HGCA 2003).

Similar schemes exist for pest management options. Pests can reduce yield by 10 % or more and need to be carefully managed. Pest management is essential to:

- Prevent virus transmission
- Apply pesticides in a timely manner and only where economically justified
- Minimize the development of insecticide resistance
- Reduce the impact on the environment, including beneficial insects.

Major arable crop pests include aphids, slugs, orange wheat blossom midge, wheat bulb fly, cabbage stem flea beetle and pollen beetle.

Changes in pesticide availability (e.g., due to regulation), efficacy (e.g., due to pesticide resistance) and crop susceptibility (e.g., changes in varieties grown) mean that approaches to pest management have to be constantly adapted. With increasing concerns about the environment, it is recognized that pest control needs to be balanced against encouraging other insects which can benefit the crop. Integrated strategies seek to use cultural control options, encourage natural enemies and only use chemical crop protection methods when they are fully justified usually by the use of thresholds.

In addition to these 'historic' surveys, i.e., recording what has happened, in-season information is collected and translated into near real-time warnings and alerts for growers and advisers. An example of such a service is 'CropMonitor'. This project has several collaborators and is funded by HGCA and DEFRA, employing independent researchers and advisers who monitor reference crops at sites located throughout England. The sites are inspected weekly during the growing season and up-to-date measurements of crop pest and disease activity in a range of arable crops are reported on an open access website (www.cropmonitor.co.uk). All data gathered are analyzed to identify disease and pest risks, seasonal variation in disease development and the effectiveness of control strategies. Users are alerted to emerging threats during the growing season and advised on appropriate courses of action. The service also runs in-season risk models for diseases such as septoria (*M. graminicola*), yellow rust (*P. striiformis*) and eyespot (*Tapesia* spp.). Some models are season specific e.g., Septoria risk model, others are generic, reflecting average disease risk due to climatic differences—e.g.,phoma stem canker. Both types of model/risk forecasts are useful in helping to determine risk in particular areas. There is a higher risk from Septoria in the west of the UK where the climate is wetter, than in the east, where the climate is much drier (Fig. 3.1). This general disease risk map is useful for making strategic decisions on variety choice and base fungicide programs. In the case of phoma stem canker in canola, the risk is higher in the east of England (Fig. 3.2).

3.5.3 Threshold Values as Basis for Decision-Making

The use of thresholds for decision making is an integral part of integrated pest and disease management and thresholds are widely used in the industry (Ellis et al. 2009). Farmers and advisers make regular crop visits throughout the growing season to monitor pest, disease and weed levels in crops. Threshold figures and

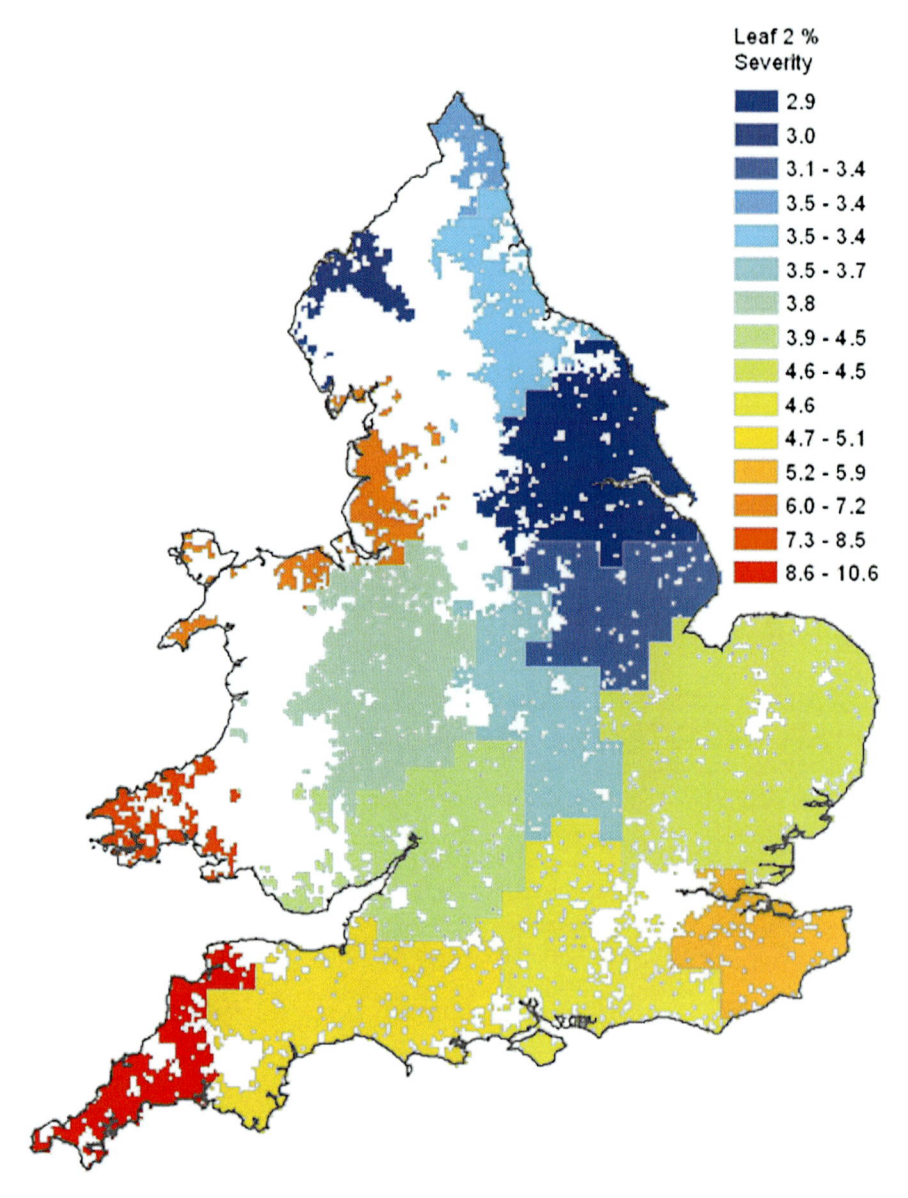

Fig. 3.1 Example of risk model prediction for risk from septoria (*M. graminicola*). (Numbers are predicted disease levels on leaf 2). (CropMonitor 2012)

monitoring methods are published for all major pests in most arable crops. Most decisions on insecticide use are based on some form of monitoring or sampling and then the application of some threshold value which determines the economic need for treatment. Examples include:

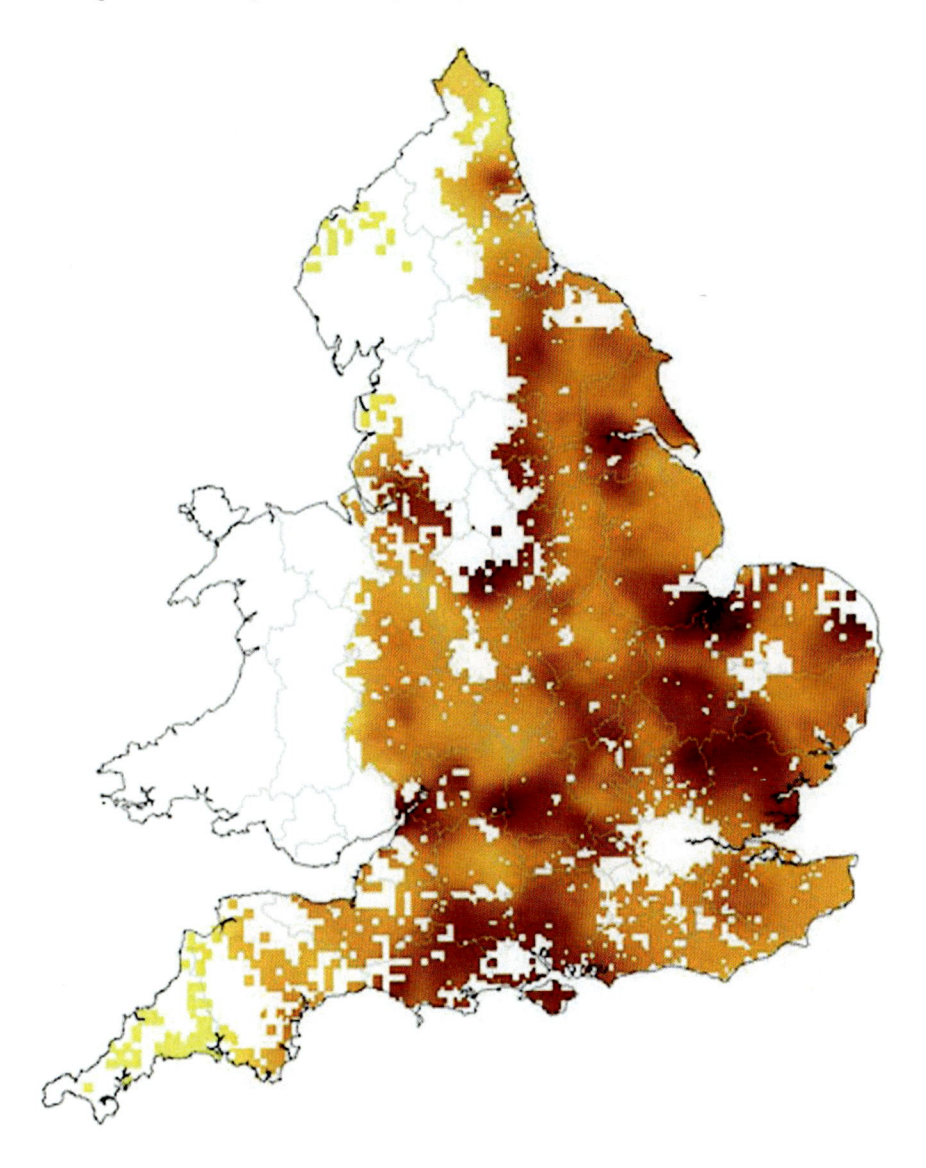

Fig. 3.2 Incidence of phoma stem canker (*Leptosphaeria maculans*) 1997–2006. (CropMonitor)

3.5.3.1 Wheat Bulb Fly (*Delia coarctata*)

Wheat bulb fly is most prevalent insect pest in eastern England. Adult flies lay eggs on bare soil from August until early September and these remain dormant throughout late autumn and early winter. In risk areas, soil is sampled and eggs present are counted. Egg numbers above 250/m^2 present a risk of economic damage to autumn-drilled wheat crops. Egg numbers above 100/m^2 justify the use of seed treatment on the latest-drilled crops of wheat or barley. In very high risk situations, foliar sprays may need to be applied and they too are determined by the incidence of damage to the crop. Thresholds vary with growth stage:

- 10% of tillers attacked at GS20
- 15% of tillers attacked at GS21
- 20% of tillers attacked from GS22 onwards.

(HGCA 2012b)

3.5.3.2 Pollen Beetle (*Meligethes aeneus*)

Pollen beetles migrate into winter oilseed rape crops from mid-March and throughout April to feed on pollen and lay eggs. If flowers are not open, beetles bite into and kill the flower buds. Crops are most at risk when the weather is warm (above 15 °C). Monitoring traps and online forecasts of pollen beetle migration help to focus monitoring activities. Crops with low plant populations have a higher pollen beetle threshold than more dense plantings so the threshold for treatment varies from 10–30 pollen beetles per plant (HGCA 2012c).

3.5.3.3 Orange Wheat Blossom Midge (*Sitodiplosis mosellana*)

The emergence of orange wheat blossom midge (WBM) in the late 1990s in the UK led to the rapid development of a pheromone-based decision support system. WBM has a very patchy spatial distribution and also varies from year to year depending on climatic conditions. In the UK, precipitation causing moist soil conditions at the end of May, followed by warm still weather in late May/early June can lead to serious midge outbreaks. The egg-laying female is small and remains hidden in the crop canopy. The larvae are also hidden within the wheat ear, presenting a difficult spray target. The numbers of males caught in pheromone traps are correlated to the level of egg-laying by females. When the male midges are caught, this typically indicates a 2-day window before eggs are laid in the crop by female midges. The use of pheromone traps for WBM helped in spray timing decisions although the rapid discovery and breeding of varieties with genetic resistance to WBM in the early 2000s, reduced the need to monitor midge numbers and flight activity. In susceptible varieties, midge numbers tend to be monitored at night, along with pheromone traps to target monitoring times and thresholds based on the number of midges per year are used. The majority of wheat varieties in the UK are now resistant to this pest.

3.5.3.4 Cereal Disease Control

Decisions on the need for fungicide use in cereal crops are complex as usually there are several target diseases. Also, the severity of a disease in any season depends on the amount of disease inoculum, weather patterns and the varieties' genetic ability to resist the disease pressure. A higher fungicide dose is needed when disease pressure is high and varietal resistance is low. Conversely, a resistant variety facing low disease pressure may not require any treatment. Because of the complexity of diseases and weather/variety interactions, disease forecasting is not very precise. Risk assessment is often reduced to estimating major categories of risk such as nil, low, moderate or high. For foliar diseases the risk assessment is normally based on a visual assessment in the crop at 10–20 locations throughout a field. The crop would normally be assessed every 7–10 days starting in the spring and finishing at the end of flowering. If the crop has been sprayed with a fungicide, an interval of around 10–14 days can normally be allowed before the crop has to be monitored again. Often the effective threshold for treatment is very low so presence or absence is often the trigger for treatment. Decision support tools or thresholds for soil-borne or stubble-borne diseases such as eyespot (*Tapesia* spp.) are often based on a combination of average historic or regional risk combined with soil type, previous cropping, cultivation method and sowing date alongside a visual assessment of the incidence and severity of the disease. Consequently, there are several interacting factors that determine the likely risk from a single disease. When the decision to spray is made there will usually be more than one disease to consider so a combination of thresholds may be used to make the final decision.

3.5.4 Non-Chemical Methods to be Preferred

Crop production in arable crops in the UK has a high dependency on pesticides although pest, weed and disease control strategies do incorporate non-chemical control methods.

Before the widespread use of chemical pesticides, a 'systems approach' was often applied to pest control. Farmers made more use of cultural control, host-plant resistance and some aspects of biological control. Cultural methods such as crop rotation, the manipulation of sowing and harvesting dates, the use of resistance to pests and diseases and the use of farming systems that encouraged natural biological control. This was not, however, a utopian agricultural paradise as crop failures due to diseases such as bunt (*Tilletia tritici*) occurred frequently and yields were low. However, with the proliferation of artificial insecticides and fungicides in the late 20th century, the emphasis changed from a reliance on a systems approach, to a reliance on chemical pesticides, alongside a breeding program which emphasized yield and quality, rather than pest or disease resistance. The new approach of using chemical control methods allowed farmers to utilize new varieties and achieve higher and more consistent yields, without the threat of crop failure from pests and diseases. Consumer demand for blemish-free produce also fuelled the demand for high levels of pest and disease control, often unachievable using biological control methods. Consequently, there are still relatively few biological control agents that are used successfully in arable crops. However, the development of insecticide, fungicide and herbicide resistance in the late 20th century has re-invigorated interest in more sustainable systems which utilize and integrate techniques such as pest and disease-resistant crops, cultural techniques, the use of pheromones, encouraging natural predators, etc. These are all concepts which were familiar to researchers in the mid 20th century as fundamental components of IPM (Stern et al. 1959).

Biological control of insect pests, particularly in protected crops, outdoor horticultural crops and fruit, has been widely and very successfully practiced for decades (Bale et al. 2008). The first successes of biological control in protected crops were the control of the glasshouse whitefly *Trialeurodes vaporariorum* using the parasitic wasp *Encarsia formosa*, and the control of the glasshouse spider mite *Tetranychus urticae* using the predatory mite *Phytoseiulus persimilis*, both still widely used today. Most pests of protected crops can now be managed with biological control agents (Lenteren 2000).

There have been few studies to assess the adoption of IPM component technologies in the arable sector but one postal survey conducted in the UK (Bailey et al. 2009) revealed that crop rotation and seed treatment were the two most popular measures. Resistant varieties were used by 60–70 % of respondents but is also at the top of the list of measures which had been dropped, reflecting the more profitable practice of using higher yielding disease-susceptible varieties and protecting them with fungicides. If major reductions in total pesticide use in European agriculture are to be achieved, more will have to be done to develop economically viable IPM systems for arable crops which consume the bulk of the pesticides used in food production.

3.5.5 *Target-Specificity and Minimization of Side Effects*

Plant protection products (pesticides) have many benefits, including helping to ensure we have access to sufficient quantities of good quality, reasonably priced foodstuffs. The UK Government believes that the best way to minimize the risk of adverse impacts is through a range of statutory and voluntary controls. The Government has therefore published the UK Pesticides Strategy which does this (DEFRA 2006a).The Strategy provides a framework for plant protection product legislation, policies and initiatives that contribute to promoting sustainable development. The strategy aims to help protect the countryside and natural resources, supporting sustainable food and farming and sustainable consumption and production. It also aims to minimize the adverse impacts of using plant protection products. The UK Strategy foreshadows the requirements of the EU Thematic Strategy for Pesticides.

By law, everyone who uses pesticides professionally in the UK must have received adequate training in using pesticides safely and be skilled in the job they are carrying out. This applies to users, operators and technicians (including contractors), managers, employers, self-employed people and people who give instruction to others on how to use pesticides. A qualification called a 'certificate of competence' is needed before it is legal to supply, store or use agricultural pesticides.

The Plant Protection Products (Sustainable Use) Regulations 2012 came into force on 18 July 2012, replacing the previous UK legislation governing the use of pesticides. Guidance on the safe use and storage of plant protection products already existed in Codes of Practice (DEFRA 2006b).

3.5.6 Reduction of Use to Necessary Levels

There is an economic imperative to optimizing pesticide use. There is no incentive for farmers to over-use pesticides as this would have a negative impact on financial returns for their business. However, any pesticide use has to be balanced against risk of losses due to pests or diseases. Financial losses due to pests and diseases can be so large that a single seasonal loss of crop could lead to financial ruin. Consequently, the cost of pesticide use has to be balanced against the potential financial loss if pest or disease control fails. This situation makes farmers and advisers risk-averse and may lead to pesticide use above the optimal level (which cannot be known in advance). Clearly there is a place for decision support tools to help the farmer make the decision as to whether pesticide use is warranted and what the appropriate intervention should be. This would include information on pesticide choice, dose and timing. Let us illustrate this dilemma in the UK with wheat production. In the UK, the climate, particularly mild winters, favors the development of foliar diseases and disease levels are often higher than in other EU countries. Data from the UK variety testing program (HGCA Winter Wheat UK Recommended List—www.hgca.com) show that these diseases, on average, cause a 20% yield loss in crops with an average fungicide treated yield of just over 10 t(11 US t) per hectare. Varieties that are very responsive to fungicide use can give yield responses of up to 28% whereas, most disease-resistant varieties gives a yield response of only 11%. Also, at any spray timing, the farmer is trying to control up to five or six different diseases, some of which are potentially very damaging. Fungicide use is therefore very profitable. A typical fungicide input in wheat would cost c. € 80 ha^{-1} (US $ 130 ha^{-1}) and this would give an average yield response of almost 2.0 t (2 US t) per hectare, giving a profit of c. € 300/ha)($ 380/ha). Growing wheat varieties that are disease resistant rather than disease-prone may seem advisable when attempting to devise a reduced inputs wheat production system. However, the disease-prone varieties tend to be higher yielding and so give higher financial margins than lower yielding disease-resistant varieties. Even accounting for differential fungicide requirements, it is still more cost-effective to grow the higher yielding disease-prone varieties. Farmers have access to many sources of independent and commercial advice to help them make such decisions. Most farmers would employ a specialist professional agronomic adviser to help make decisions on pesticide use on the farm.

National pesticide usage surveys of arable crops are carried out every two years by the Food and Environment Research Agency (FERA) on behalf of DEFRA, to give general information about national and regional trends in pesticide use. Similar surveys are carried out for top and soft fruit, ornamentals, protected crops and grassland (DEFRA 2012).

3.5.7 Application of Anti-Resistance Strategies

The UK Resistance Action Groups (RAGs) are UK-based groups consisting of experts from the Crop Protection Association (CPA) member companies, other representatives from the agrochemical industry, a range of independent organizations, including public-sector research institutes, and the Chemicals Regulation Directorate (CRD). The groups are completely independent of CRD and work to produce guidance on pesticide resistance issues. There are four autonomous groups dealing with issues relating to herbicides, fungicides, insecticides and rodenticides. These groups work closely with their international equivalents, the Insecticide Resistance Action Committee (IRAC), the Weed Resistance Action Committee (WRAC) and the Fungicide Resistance Action Committee (FRAC). Each resistance action group publishes guidelines on pesticide use to avoid resistance problems (e.g., FRAG 2011).

There are issues of fungicide resistance in the majority of the modern fungicide groups currently available. Consequently, manufacturers take the issue of fungicide resistance development very seriously as a threat to their long term business. The Fungicide Resistance Action Committee (FRAC) and the UK Fungicide Resistance Action Group (FRAG-UK) are very active in trying to devise and promote strategies to avoid resistance development. Debate is on-going as to whether there is such a thing as a successful anti-resistance strategy as there are few examples where a planned or reactive strategy has been successful in slowing or preventing the further development of resistance. Where such a strategy has worked it is often unclear why—and so the industry continues to apply the general principles promoted by FRAC and FRAG-UK. These principles being primarily:

- Limiting the exposure of the pathogen population to the fungicide, mainly by reducing the number of applications per season.
- Avoiding the use of fungicides where the target pathogen is already well established in the crop.
- Mixing or alternating fungicides with different modes of action.
- Manipulating dose (generally described as avoiding multiple low doses and promoting the use of high doses).

Some of these principles are based on general assumptions, some are impracticable, and others contradicted by experimental evidence. The issue of dose is contentious and there is no general agreement as to the effect of dose on selection. Experimental work with strobilurin fungicides and *Septoria tritici* (Fraaije et al. 2003) clearly showed that high doses posed a greater selection pressure. In terms of sustainable disease control, we have a medium-term set of problems in managing pesticide resistance:

3 Integrated Pest Management for European Agriculture

- No anti-resistance strategy presently exists that can prevent resistance development.
- Inevitable resistance development to remaining single-site active ingredients.
- Increasing development costs leading to a falling number of active ingredients, most of which have single-site modes of action.
- Varieties lacking durable resistance to the major pathogens.

3.5.8 Records, Monitoring, Documentation and Check of Success

In 2001 the Government accepted proposals put forward by the farming and crop protection industry to minimize the environmental impacts from pesticides. These became known as the Voluntary Initiative. By 2006 the program had met or exceeded the vast majority of its targets. In the light of this, the VI Steering Group proposed to Ministers that the Voluntary Initiative should continue as a rolling two year program. These proposals were welcomed by the Government and the VI has continued since as a voluntary program of work promoting responsible pesticide use. Some of the successes of the VI include measurable benefits for biodiversity through:

1. The investment of £ 5 million (US $ 8 million) by the crop protection industry in research projects and the implementation of a Biodiversity Strategy Action Plan. For example the SAFFIE Sustainable Arable Link project demonstrated that skylark plots, which are now recognized in the Entry Level Scheme, can deliver an almost 50% increase in skylark fledglings survival.
2. Improved surface water quality (Environment Agency data). This work showed a significant reduction in the number of samples exceeding 0.1 ppb of pesticides, compared with the average for 1998–2002, achieved partly through the H_2OK campaign of the Voluntary Initiative aimed at decreasing water pollution from agriculture.
3. Reductions in pesticide residues in water in pilot catchments where up to 60% reductions were achieved.
4. Improved awareness by farmers of the potential environmental risks from pesticide use through widespread communication of VI messages. This has led to improved competence by advisers and improved field and handling practice by sprayer operators.
5. The establishment of Crop Protection Management Plans (CPMPs)—a self-audited assessment of crop protection activities across the whole farm. CPMPs have now been included in the Entry Level Stewardship scheme.
6. The establishment of a nationwide National Sprayer Testing Scheme (NSTS). Tested machines now account for almost 80% of the sprayed area. NSTS is now part of the audit for most farm crop assurance schemes.
7. The establishment of a National Register of Sprayer Operators (NRoSO) who are committed to adopt best practice in pesticide handling and application. NRoSO

membership is now included in the audit procedures of the major farm crop assurance schemes.
8. The creation of Environmental Information Sheets as an aid to risk management for all products sold by members of the Crop Protection Association.
9. The development of technical solutions and risk assessment tools in water catchment areas for communicating best practice advice for reducing pesticide residues in water. The VI is now working with DEFRA to provide this expertise in the England Catchment Sensitive Farming Delivery Programme.
10. The establishment of a Biodiversity and Environmental Training for Advisers qualification.

The Voluntary Initiative has produced and promoted many best practice guidelines on topics such as container cleaning, pesticide disposal, pesticide container disposal, sprayer washings, etc.

3.5.8.1 Quality Assurance Schemes

The majority of arable crops produced in the UK are grown to Assured Food Standards (AFS): AFS is an organization that promotes and regulates food quality in the UK. It licenses the Red Tractor quality mark, a product certification program that comprises a number of farm assurance schemes, including livestock schemes. The Red Tractor Farm Assurance Combinable Crops & Sugar Beet scheme sets out to maintain, develop and promote Assurance standards within the industry. The aim is to provide consumers and retailers with confidence about product quality attributes including food safety and environmental protection. Certification to Red Tractor Farm Assurance Combinable Crops and Sugar Beet demonstrates that the high standards of production meet nationally agreed levels of best agricultural practice and that crops grown on farms are managed by well-qualified and highly professional farmers. Crops covered by the scheme include wheat, barley, rye, oilseeds, linseed and pulses such as peas and beans.

3.5.8.2 Adviser Standards

Professionally qualified advisers are an integral part of UK advice on farms. There are many independent advisers as well as advisers linked with agricultural distribution and agrochemical manufacturing companies. All advisers must be qualified and accredited by a company called BASIS (Registration) Ltd. BASIS is an independent standards setting and auditing organization for the pesticide, fertilizer and allied industries. The company plays a key role in the training and certification of people who work in the pesticide and fertilizer sectors. It sets the syllabuses and examination standards and maintains a list of approved trainers, colleges and training providers who offer suitable training modules. BASIS administers the statutory certificate of competence for the pesticide industry, the BASIS Certificate in Crop Protection, which is the recognised statutory qualification for pesticide sellers, suppliers and advisers. It is available in a range of specialties including agriculture, amenity horticulture, commercial horticulture, aquatics, grassland and forage crops. BASIS administers the Professional Registers for qualified pesticide and fertilizer advisers and for public health pest control professionals. BASIS was established by the pesticide industry in 1978 to develop standards for the safe storage and transport of agricultural and horticultural pesticides and to provide a recognised means of assessing the competence of staff working in the sector. The growing range of BASIS qualifications has allowed people in the industry to develop their skills and demonstrate their professionalism. BASIS standards have been adopted by key organizations such as the environment agencies, county councils, supermarkets, crop assurance schemes and farmers. Over 4,000 people have become members of the BASIS Professional Register.

3.6 Conclusions

Pesticides are fundamental to the way combinable crops are grown in the UK. They provide a relatively cheap and efficient way of controlling the major pests, weeds and diseases. The UK Government has implemented several schemes to promote the safe use of pesticides including training initiatives and registration schemes including Agri-Environment Schemes and The Voluntary Initiative. These are mostly voluntary schemes but have been very successful. Many EU policy makers are attracted by the concept of IPM and see it as a means to further reduce pesticide use in crops. However, many IPM principles are already normal farm practice in the UK. The majority of farmers currently adopt many practices that are regarded as integral components of IPM—crop rotation, use of disease and pest-resistant varieties, use of thresholds and decision support tools, etc. Successful UK farmers are highly innovative and the majority will rapidly adopt new technologies and incorporate them into their farming systems. As a consequence, wheat yields in the UK are some of the highest in Europe and were increased from 3.5 t/ha in the 1960s to today's average of over 8 t/ha due in large part to science-led advances and innovations in agricultural production. However, strong evidence indicates that yields of wheat in most European countries have reached a plateau. Europe under climate change will be a major global food producer. Currently, EU regulations are putting pressure on the agricultural and horticultural industries to reduce pesticide use, either directly through pesticide reduction plans or indirectly through water quality regulations. This combined pressure will almost certainly reduce food production in the EU. In the context of Global Food Security challenges, it seems incongruous that EU policy presently aims at reducing pesticide use which would almost certainly reduce average yields of the major combinable crops (HGCA 2009).

Some policy makers associate IPM with reduced pesticide use and although this has been the case with protected crops, it has not yet been successful in most cereal and oilseed rape based arable cropping systems. Techniques or legislation which seek to reduce pesticide use *per se* are rarely successful unless farmers are compensated significantly for the loss of yield which often results. Farming must be profitable otherwise it is not sustainable.

References

ADAS. (2008). *Evaluation of the Impact on UK Agriculture of the Proposal for a Regulation of the European Parliament and of the Council Concerning the Placing of Plant Protection Products on the Market.* UK: Agricultural Development & Advisory Services.

Bailey, A. S., Bertaglia, M., Fraser, I. M., Sharma, A., & Douarin, E. (2009). Integrated pest management portfolios in UK arable farming: Results of a farmer survey. *Pest Management Science, 65*(4), 1030–1039.

Bale, J. S., van Lentern, J. C., & Bigler, F. (2008). Biological control and sustainable food production. *Philosophical Transactions of the Royal Society, 363*(1492), 761–776.

CropMonitor (2012). http://www.cropmonitor.co.uk/. Accessed 1 March 2014.

DEFRA. (2006a). *UK Pesticides Strategy: A Strategy for the Sustainable Use of Plant Protection Products.* Crown Copyright. Department for Environment, Food and Rural Affairs. http://www.pesticides.gov.uk/Resources/CRD/Migrated-esources/Documents/U/Updated_National_Strategy.pdf. Accessed July 2012.

DEFRA. (2006b). *The Code of Practice for Using Plant Protection Products.* Crown copyright, Department for Environment, Food and Rural Affairs. http://www.pesticides.gov.uk/farmers_growers_home.asp #Codes_of_Practice. Accessed July 2012.

DEFRA. (2010). *Fertilizer Manual RB209* (8th ed., p. 257). UK: The Stationery Office.

DEFRA. (2012). *Pesticide Usage Survey Reports.* http://www.fera.defra.gov.uk/scienceResearch/science/lus/pesticideUsageFullReports.cfm. Accessed July 2012.

Ellis, S., Berry, P., & Walters, K. (2009). A review of invertebrate pest thresholds. *HGCA Research Review No.73*, Home Grown Cereals Authority, UK. http://www.hgca.com/publications/documents/cropresearch/RR73.pdf. Accessed July 2012.

Fraaije, B. A., Lucas, J. A., Clark, W. S., & Burnett, F. J. (2003). QoI resistance development in populations of cereal pathogens in the UK. *Proceedings British Crop Protection International Congress, 2,* 689–694.

FRAG. (2011). *Fungicide Resistance Management in Cereals.* Fungicide Resistance Action Group, UK. http://www.pesticides.gov.uk/Resources/CRD/Migrated-Resources/Documents/F/FRAG_Cereals_Resistance_guidelines_v4_March_2011.pdf. Accessed July 2012.

Goldsworthy, P. (2006). *The Voluntary Initiative: Successes and Future Challenges.* Crop Protection Association. http://www.hgca.com/publications/documents/cropresearch/Paper_17_Patrick_Goldsworthy.pdf. Accessed July 2012.

HGCA. (2003). *Pest Management in Cereals and Oilseed Rape—A Guide.* Home Grown Cereals Authority. http://adlib.everysite.co.uk/resources/000/250/907/HGCA_pest_mgt_cereals_and_OSR.pdf. Accessed July 2012.

HGCA. (2009). Pesticide availability for cereals and oilseeds following revision of Directive 91/414/EEC; effects of losses and new research priorities. (2008). *Research Review No. 70*, Home Grown Cereals Authority, UK. http://www.hgca.com/publications/documents/cropresearch/RR70_Research_ReviewX.pdf. Accessed July 2012.

HGCA. (2012a). *Wheat Disease Management Guide.* Home Grown Cereals Authority. http://adlib.everysite.co.uk/resources/000/160/665/G54_Wheat_disease_management_guide_2012.pdf. Accessed July 2012.

HGCA. (2012b). *Wheat Bulb Fly: Risk Assessment and Control.* Topic Sheet No. 118, Home Grown Cereals Authority. http://www.hgca.com/cms_publications.output/2/2/Publications/On-farm%20information/Wheat%20bulb%20fly%E2%80%93%20risk%20assessment%20and%20control.mspx?fn=show&pubcon=9062. Accessed July 2012.

HGCA. (2012c). *Controlling Pollen Beetle and Combating Insecticide Resistance in Oilseed Rape*. Information Sheet No.13, Home Grown Cereals Authority. http://www.hgca.com/cms_publications.output/2/2/Publications/Publication/Controlling%20pollen%20beetle%20and%20combating%20insecticide%20resistance%20in%20oilseed%20rape.mspx?fn=show&pubcon=8521. Accessed July 2012.

Hillocks, R. J. (2011). Farming with fewer pesticides: EU pesticide review and resulting challenges for UK agriculture. *Crop Protection, 31*(1), 85–93.

INRA. (2010). Ecophyto R & D: Which options to reduce pesticide use? http://www.international.inra.fr/the_institute/advanced_studies/ecophyto_r_d. Accessed June 2012.

Lenteren, J. C. van (2000). A greenhouse without pesticides: Fact or fantasy? *Crop Protection, 19*(6), 375–384.

Natural England. (2009). Agri-environment schemes in England 2009—a review of results and effectiveness. http://www.naturalengland.org.uk/publications. Accessed July 2012.

Roteveel, T., Jorgensen, L. N., & Heimbach, U. (2011). Resistance management in Europe: A preliminary proposal for the determination of a minimum number of active substances necessary to manage resistance. *EPPO Bulletin, 41*(3), 432–438.

Stern, V. M., Smith, R. F., van den Bosch, R., & Hagen, K. S. (1959). The integration of chemical and biological control of the spotted alfalfa aphid. *Hilgardia, 29*(1), 81–101.

Chapter 4
Energy Inputs In Pest Control Using Pesticides In New Zealand

Majeed Safa and Meriel Watts

Contents

4.1	Introduction	100
	4.1.1 Agriculture in New Zealand	100
	4.1.2 Main Production	101
4.2	Pest Control Methods	102
	4.2.1 Non-Chemical Methods	102
	4.2.2 Chemical Pest Control Methods	104
4.3	Environmental and Health Impacts of Pesticide Application	106
4.4	Pesticide Consumption in New Zealand	108
4.5	Spraying Systems and Technologies	109
	4.5.1 Ground Spraying Systems	110
	4.5.2 Aerial Spraying Systems	111
4.6	Energy Consumption in Pest Control in New Zealand	112
	4.6.1 Main Energy Inputs in Spraying	112
	4.6.2 Fuel	113
	4.6.3 Tractors and Field Machines	114
	4.6.4 Labor	115
	4.6.5 Pesticides	116
4.7	Energy Inputs in Pesticides Application in New Zealand	117
	4.7.1 Operational Energy	117
	4.7.2 Energy Consumption of Pesticides in Different Agriculture Sectors	118
	4.7.3 Total Energy Inputs in Pest Control Using Pesticides in New Zealand	119
	4.7.4 CO_2 Emission in Pesticide Application	120
4.8	Summary	122
References		122

M. Safa (✉)
Department of Agricultural Management and Property Studies, Lincoln University, PO Box 84, Lincoln, Christchurch 7647, New Zealand
e-mail: Majeed.Safa@lincoln.ac.nz

M. Watts
Pesticide Action Network Aotearoa New Zealand, PO Box 296 Ostend, Waiheke Island, Auckland 1843, New Zealand
e-mail: merielwatts@xtra.co.nz

D. Pimentel, R. Peshin (eds.), *Integrated Pest Management*,
DOI 10.1007/978-94-007-7796-5_4,
© Springer Science+Business Media Dordrecht 2014

Abstract This chapter examines the energy consumption of pesticide applications in New Zealand. Energy use in pest control using pesticides is investigated in the horticultural, arable, pastoral and forestry sectors, based on 30 different groups of farm products.

On average, total energy consumption in pesticide applications in New Zealand was estimated at about 2,350,757 GJ. Energy use in pesticide applications was about 160 MJ/ha. The pastoral and horticultural sectors are ranked first and second in terms of total energy usage for pesticide applications, at about 1,109,389 GJ and 704,511 GJ, respectively. The horticultural sector has the most intensive pesticide consumption at around 5,855 MJ/ha.

The total operational energy was about 20 % of total energy use in pest control of which 90 % was for fuel. Herbicides are applied more than other pesticide in most agricultural sectors; therefore, the energy equivalent of herbicides is ranked first with 1,353,503 GJ and 58 % of total energy use. Fungicides and insecticides are mostly used in the horticultural sector.

Total CO_2 emissions from pesticide applications was estimated at around 145857 t, or approximately 10 kg/ha.

Keywords Energy inputs · CO_2 emission · Pest Control · New Zealand

Abbreviations:
ATV   All Terrain Vehicles
FAO   Food and Agriculture Organization of the United Nations
GDP   Gross domestic product
PTO   Power Take off
SVFC   Specific volumetric fuel consumption
SVFE   Specific volumetric fuel efficiency

4.1 Introduction

4.1.1 Agriculture in New Zealand

New Zealand's economy is heavily dependent on exports from agricultural production, accounting for nearly 51 % of New Zealand's exports by value (Statistics New Zealand 2008). Farms cover about 50 % of New Zealand's land area (Ministry of the Environment 2006). In general, New Zealand farmers practice a form of 'industrialized' agriculture that relies on relatively high inputs of fossil fuels, not only to power machinery and irrigation directly but also embedded in artificial fertilizers and agrichemicals from their manufacture (Wells 2001; Safa and Samarasinghe 2011). Consequently, New Zealand is one of the countries with the highest energy input per unit weight of agricultural output in the world (Conforti and Giampietro 1997). In New Zealand, the agricultural sector produces about 4.6 % of total GDP, while its proportion of greenhouse gas (GHG) emissions is, surprisingly, over 54% of total national emissions (Kelly 2007; Energy & Environment 2009).

New Zealand's climate is not extremely cold or hot; therefore, high energy modifications are not widely used in agriculture, such as animal housing or heating. Moreover, 99% of cows and sheep graze directly on pasture and farmers don't need to spend energy on harvesting this land. Pigs and poultry are often intensively housed, but these are relatively small sectors of New Zealand's livestock industries. Other high intensive energy use farming activities, such as greenhouses production, are not very significant. According to Stout (1990), compared to some developed countries, the productivity of New Zealand farms was low and there was potential to increase yields and improve energy efficiency. However, comparing the figures for agricultural production through recent FAO statistics, the yield of many farm products in New Zealand has increased during the last few years and more increase is predicted in the future.

4.1.2 Main Production

The relative proportion of different types of production in New Zealand varies with technical and financial factors. Following increases in the prices of dairy products in 2007 and 2008, many farms were converted from arable to dairy production and some existing dairy farmers increased their stocking rates. However, the high exchange rate for the New Zealand dollar and increasing oil prices are also likely to affect farming patterns. In addition, investigation into the effects of economic changes on farm production in the short term is difficult and farmers' reactions to price changes are always slower than those in other sectors. They cannot easily change capital equipment, establish orchard trees, or change crops after sowing, and they cannot convert their farms from dairy to arable use quickly. According to Statistics New Zealand (2007), 12,279,599 ha (82%) of farm land is in pastoral farms used for livestock and dairy production. Forestry ranks second with 13% of land, and horticulture and arable farms have similar proportions at 2% of total land.

New Zealand is the world's eighth largest milk producer (MAF 2011) and exports about 95% of its milk production, making it the world's largest milk exporter (FAO 2011). Continued demand for beef, lamb and wool is expected to hold prices at a level higher than for the previous five years. New Zealand meat and wool production in 2011 was reduced by undesirable weather in the second half of 2010 (MAF 2011). However, due to high international prices, meat production is expected to return to previous levels (FAO 2011).

Wheat, barley, maize, oats, potatoes and peas are the main arable crops in New Zealand. Favorable weather, sufficient precipitation during the growing season, and high agricultural production prices at sowing time increased farm production during 2009 (Statistics New Zealand 2010; FAO 2012). However, average growth was predicted for recent years (MAF 2011). One of the fastest growing arable sectors is the production of seed from vegetable and grain crops mainly for export to the Northern Hemisphere, estimated to be around 50,000 ha in recent years.

Forests can provide social and environmental services, including the sequestration of carbon, combating desertification and rehabilitating degraded lands. Manmade forests have been developed in recent years and farmers have been encouraged to develop forests to improve soil conservation and carbon fixation (Rhodes and Novis 2002). The continued demand from China, the need for reconstruction after the earthquakes in New Zealand and Japan, and floods in Australia, all raise the international timber demand. Around 24.8 million cubic meters of logs and timber were harvested from New Zealand's exotic forest in the year ending 31 December 2010, up 19% from the previous year (MAF 2011).

New Zealand's diverse geography and climate allows the production of a wide range of fruits and vegetables. Grapes, kiwifruit, apples and pears are the main horticultural products in New Zealand (MAF 2011). Kiwifruit, apples and avocados are the main fresh fruit produced, and wine is the main indirect product; all are export focused. Other vegetables and fruits are grown to supply the domestic market and for export. Flower, bulb and seed production for export and supply to the domestic markets has increased in recent years. Compared with other agricultural sectors such as dairy, the price of the main horticultural crops has not increased significantly. As with other agricultural sectors, several factors can change horticultural production. For example, a new disease caused a sharp reduction in production of kiwifruit, but production is forecast to grow steadily over the coming year (MAF 2011).

4.2 Pest Control Methods

Pests destroy an estimated 37% (insects 13%, plant pathogens 12%, and weeds 12%) of all potential agricultural production every year. When the post-harvest losses are added to the pre-harvest losses, total agricultural production losses due to pests increase to 52% (Pimentel and Pimentel 2008). Three different methods of pest control—chemical, mechanical and biological- are usually applied to control or eliminate fungi, insects and weeds on modern farms. On small farms, organic farms, and in areas with cheap labor sources, farmers use more mechanical methods. Other pest controls methods used for specific crops or conditions include insect collectors, thermal weeding and soil disinfectors. However, most farmers choose chemical methods because they are perceived to be faster, cheaper and more effective than non-chemical methods.

4.2.1 Non-Chemical Methods

4.2.1.1 Simple Pest Control Methods

Pest control is as old as agriculture itself. The first generation of farmers soon found there was huge competition between their plants and other plant species in the use of space, light and soil nutrition. Additionally, animals, birds and insects were interested in eating their produce. Therefore, they started to develop simple methods to protect their products. Some of the initial pest control methods are still in use and effective on small farms, organic farms and in areas with cheap labor sources. Using a scarecrow to scare birds would be a good example of a simple and initial pest control method with minimum energy consumption that is still in use. Fencing, hunting and burning are other examples of initial pest control methods developed based on nature, material availability and the characteristics of the pests.

Weeding by hand (hand weeding) is an initial method for providing more space and resources for plants. On many farms in developing countries, home gardens and small farms in developed countries, weeds are still taken out by hand or with simple tools. Cultivation with tools is often carried out when the weeds are very small. Hand weeding is generally delayed until weeds grow large enough to be grasped easily. The method requires sufficient soil moisture to ensure that weeds can be easily and completely pulled out of the ground while minimizing damage to the crop.

4.2.1.2 Mechanical and Cultivational Pest Control

Since the beginning of agricultural production, pests have also been managed without chemicals: by hand, cultivational practices and mechanical means. For example, hand removal is still used on many small farms and in kitchen gardens. The use of light traps, attractant lures including pheromone lures, mating disruption with pheromones, companion planting, trap crops, rotational planting, intercropping, multi-cropping, and effective natural plant nutrition are all techniques used by modern organic farmers and by those practicing agroecological farming. Some farmers use cultivational practices such as deep plowing to expose soil pests to sunlight and predators. Others encourage birds to remove pests. In New Zealand, mob-stocking pastures at a particular stage in the lifecycle of the grass grub was used effectively to drastically reduce populations of this pest (East and Pottinger 1975).

Developing technology enabled farmers to use more mechanized methods. It is very difficult to establish when and where tools were first developed to control weeds. Mechanical weeding equipment is used on straight-row planted farms including cereals, vegetable, vineyards and orchards (CIGR 1999); in this system weeds within the crop rows are not easily removed. Mechanical weeding methods are always faster, and cheaper than hand weeding, but less effective because weeds within the rows are not removed. However, compared with chemical weed control, mechanical weeding is generally less effective, unless resistance problems occur, and more time consuming, although can offer additional benefits to soil and plant health.

Mechanical weeders root out or cut weeds; therefore sufficient soil moisture is necessary to achieve maximum efficiency. The process can improve the soil condition and also help to aerate the soil. Mechanical weeding achieves maximum efficiency when carried out during warm, dry and sunny days as weeds will dry out before they can re-root, and soil compaction will be minimized. Also, under these conditions that allow for maximum efficiency for mechanical weeders, the main crops are more flexible and will be injured less when the implement passes by. Note that weeds with extensive spreading root systems cannot be killed by cutting them off at the surface (Kubik 2005).

Hoes, blade and rotary cultivators, brushes and harrows are the most common mechanical weeding machines. Several mechanical weed control methods can destroy soil structure and increase soil compaction. To achieve the best results from mechanical weeding, crop plants should be planted in straight rows. With most mechanical weeding implements, driver skill is crucial to achieving the highest efficiency. If plants were not sown in straight rows, or the driver does not have enough experience, implements can damage the crops.

4.2.1.3 Biological Pest Control

Biological pest control is attractive from an environmental stand point (Lucas 2005; Vincent et al. 2009). In biological pest control methods, many pest species may be partially or completely controlled through the use of natural enemies such as predators, parasites and pathogens. However, other pest species may still need to be managed. Successful biological control can decrease the population of the target species over successive years, or act very quickly within a season in the case of mass inundation. In addition, regeneration and re-establishment programs can aid the recovery of naturally occurring natural enemy species.

Biological control is a well-established practice in agriculture, and biological control programs are used successfully all over the world. For example, *Trichogramma* wasps, minute endoparasitoids of insect eggs and the most widely augmented species of natural enemies, have been mass-produced and field released for almost 70 years. Worldwide, over 32 million ha of agricultural crops and forests are treated annually with *Trichogramma* spp. in 19 countries, especially in China and the republics of the former Soviet Union (Li 1994). In 2010, 230 species of natural enemies were being used in pest management in all regions of the world (Van Lenteren 2012).

However, a number of practical challenges to biological control exist such as retaining, distributing and applying stocks of viable biological agents. An introduced species can change the biodiversity of an agroecosystem dramatically. It should be confirmed that the introduced species does not target crops, beneficial insects, or native species both in the short term and long term. Improving biological methods can be expensive and time consuming in the short term, but with long-term dividends. However, much can be done to conserve naturally occurring pest enemies within the crop, at little or no cost, to achieve long-term sustainability, and drastically reduce pesticide usage (Landis and Orr 2002).

4.2.2 Chemical Pest Control Methods

The average worldwide growth in the use of agrichemicals is around 4.4% per year (Vlek et al. 2004). The first recorded chemical use on farms was by the Sumerians who used sulfur compounds as insecticides, around 4500 years ago. However, chemical pest control did not become widespread until the 18th and 19th centuries. In modern farming systems most farmers choose chemical methods because they believe them to be faster, cheaper, and more effective than other pest control methods. Global pesticides use is about 3 billion kg, costing nearly 40 billion US $ per year (Pimentel and Pimentel 2008). Significant challenges to plant protection by chemical pest control methods include pesticide resistance, pest resurgence, new pests and diseases, cost, and environmental and health issues.

In agriculture, a wide range of pesticides are used for a variety of purposes. Pesticides should control weeds, insects and fungi without causing serious harm to the crops (Smil 2008). Their responsibilities are prevention, avoidance, monitoring, and suppression of weeds, insects, diseases and other pests. Pesticide use generally reduces crop losses. However pesticide use creates a number of adverse effects, including human and animal poisonings, cancer and other chronic health effects, reduced biological diversity, and soil and water contamination. These adverse effects should be balanced against the benefits from pesticides. Some studies show that through appropriate management, it is possible to reduce pesticide use without reducing crop yields (Pimentel and Pimentel 2008).

Untrained home gardeners can access all kinds of pesticides but have little awareness about their toxic effects and no training in how to apply them.

Nevertheless, the use of pesticides is increasing rapidly in some countries. Pesticides have become a major environmental hazard, the main source of pollution in agriculture (Lal 2004), and a major hazard to health in some countries.

Because of public and scientific concern about the environmental effects of agrichemical use, new components have been introduced into spray programs to reduce pesticide losses from runoff and leaching and reduce pesticide residues in crops. Research is being carried out to introduce new natural methods and to improve traditional methods. For example, several studies have been undertaken to improve the genetic resistance of crops to pests, encourage pests' biological enemies, employ crop rotation, combinations with conservation tillage and the use of natural forages and trees (CIGR 1999; Lal 2004; Pimentel and Pimentel 2008). Some government programs in Canada, Sweden, Denmark and Indonesia have reduced pesticide use in some crops by 50–65% with minimum impact on yields and quality (Pimentel et al. 2005). There is now also high-level support for the replacement of pesticides with what is increasingly known as agroecological methods that use pesticides only as a last resort when other approaches are not sufficient. For example, the International Code of Conduct on the Distribution and Use of Pesticides (the Code), a global guidance document on pesticide management for all public and private entities associated with the distribution and use of pesticides was adopted by the FAO in 1985, has a guidance document on Pest and Pesticide Management Policy Development (2010) that promotes the adoption of integrated pest management (IPM). Its definition of IPM involves an ecosystem approach to pest management and states that "Pesticides are only used in those cases where there are no effective or economically viable alternatives" (FAO 2010).

In terms of energy, using pesticides is much more energy intensive than mechanical pest control methods. For example, in organic farms, energy used for weed control using cultivators takes half the energy used for herbicide weed control (Pimentel 2009).

4.3 Environmental and Health Impacts of Pesticide Application

Drift, volatilization, and runoff are the three important routes for environmental impacts from pesticide applications. Drift is defined as the movement of pesticide droplets or solid particles outside the area being treated. Drift can reduce the efficiency of the operation and waste farmers' money and time. Drift is also an important concern for human health and untargeted animals and plants. Losses due to spray drift can vary between 1–30% of pesticides applied (Briand et al. 2002). Selecting and calibrating appropriate sprayer equipment nozzles and line pressures, using correct spraying techniques, working in the right environmental conditions (moisture, temperature, and wind), defining an appropriate buffer zone, using the largest droplet size consistent with acceptable pest control and improving driving skills can reduce drift significantly (Davis and Williams 1990; Briand et al. 2002; Ramaprasad et al. 2004; Tsai et al. 2005). Liquid pesticides that are applied to crops can volatilize very quickly and may be blown by wind into nearby areas, potentially posing a threat to wildlife, livestock and humans. Volatilization mostly depends on the type of pesticide, application technique and environmental and soil conditions (Haenel and Siebers 1995; Reichman et al. 2013). Volatilization can result in the transfer of pesticides from tropical and temperature zones to the Arctic and Antarctic regions. Runoff occurs when rain events follow pesticide application and can be a major cause of contamination of surface waters.

Pesticide use frequently results in residues in food, posing additional risks to human health in the long term, something that is impossible for consumers to recognize. Many horticultural producers apply significant quantities of pesticides over a short period of time, and often close to harvest. Some vegetable growers apply pesticides more than 20 times over 30 days. The New Zealand government carries out two types of food residue monitoring: the Food Residue Surveillance Programme and intermittent Total Diet Surveys. A range of residues are always found, usually on a par with other developed countries (NZFSA 2009).

Global pesticide poisoning figures are not exact because of the difficulty in gathering data; many people who are poisoned do not report the event to medical centers and, even of those incidences reported to doctors, many are not reported to central databases. Additionally, many people and medical personnel do not recognize the symptoms of pesticide poisoning so it is often misdiagnosed. Surveillance exercises indicate that the rate of underreporting is about 98% in Central America (Murray et al. 2002). Nevertheless, an estimated 355,000 people are killed through unintentional exposure to pesticides each year (World Bank 2008). Global estimates of nonfatal poisoning of agricultural workers range from 1–100 million per year (Watts 2010). Surveys have found that between 10–94% of agricultural workers applying pesticides in developing countries experience acute symptoms of pesticide poisoning (Watts 2010). The figures will not be nearly so high in New Zealand, but reporting is poor and there is no real indication of the extent of either acute or chronic effects.

Chronic effects of pesticides on humans, not taken into account in the figures above, include birth defects, altered birth outcomes such as changes in head circumference and body weight, cancer, neurodevelopmental problems including reduced cognitive ability, reproductive problems, immune suppression, chronic neurological problems such as Parkinson's disease, and metabolic problems such as obesity and diabetes (Watts 2010). One study in New Zealand showed the risk of cancer in female workers using insecticides and herbicides in horticulture and the fruit growing sector is significantly higher than for the general public (Dryson et al. 2008). 'tMannetje et al. (2008) found an elevated risk of non-Hodgkin's lymphoma (NHL) for field crop and vegetable growers (Odds Ratio 2.74), for horticulture and fruit growing (OR 2.28) and particularly for women (OR 3.44). Sheep and dairy farming was not associated with increased risk of NHL.

Environmental impacts include acute and chronic poisoning of animals (terrestrial and aquatic), reduced biological diversity including loss of insects beneficial to agricultural pest control and to the wider ecosystem, and contamination of soil, surface and ground water, the marine ecosystem, air, rain, snow and fog. Persistent organic pollutants, including some still in use, have evaporated from the fields on which they were used and travelled through the atmosphere to be deposited in the Arctic and Antarctic environments where levels are now significant. The impacts of pesticides on the New Zealand environment are not well understood. However, limited monitoring has revealed contaminants in soil, groundwater, surface water (Buckland et al. 1998), in the air over the Southern Alps (Lavin et al. 2012), in marine sediments (Milne 2010) and marine mammals (Stockin et al. 2010). For example, in New Zealand the Sixth National Survey of pesticides in groundwater found residues in 23 % of wells sampled, and in nine of 14 regions of the country. Twenty-two different pesticide active ingredients were found, most commonly herbicides (Close and Skinner 2012). Residues of historic use pesticides in soil include copper, lead, arsenic, DDT, dieldrin, lindane, and endosuflan (Love et al. 2005). Hot spots of contamination include soils around spray storage sheds and equipment wash down areas, historic orchards and sheep dip sites (Gaw 2003; Love et al. 2005). One investigation of pastoral soil samples found that 58 % of samples taken in the Bay of Plenty region contained residues of total DDT above the permitted value for conversion to dairy of 0.2 mg/kg (Love et al. 2005). Soil samples from around the country contained DDT and its degradation products DDE and DDD (<0.03–289 mg/kg), endosulfan (BDL-0.39 mg/kg), copper (7–523 mg/kg), dieldrin (<0.005–56 mg/kg), metolachlor (0.005–0.22 mg/kg), hexaclorobenzene (BDL-0.31 mg/kg), arsenic (<2–58 mg/kg) and lead (<3–1250 mg/kg) (Love et al. 2005). Copper, arsenic, and arsenate of lead were historically applied in orchards (Gaw 2003); copper is still used in some. Mercury from mercuric fungicides has been found in horticultural soils (median in Auckland region of 0.1 mg/kg) (Gaw 2003). Residues of current use pesticides in soil include organophosphate and organonitrogen insecticides, fungicides and herbicides (Gaw 2003).

4.4 Pesticide Consumption in New Zealand

Comparing pesticide consumption in different years is not easy. Because of the wide range of companies and products, and the broad classifications of the later, together with the lack of government collection of usage figures, and the difficulty in obtaining data from the companies, comprehensive and reliable data on the use of agrichemicals in New Zealand was difficult to obtain. Also, every year new, more concentrated products are introduced that can reduce later sales volume without reducing pesticidal activity (Ministry of the Environment 2006). The value of pesticide products, moreover, depends on several financial factors that make the analyses more difficult. Non-agricultural use is one of the major limitations in estimating usage figures. There are few records available of this kind of use in New Zealand, but it can be substantial. For example, some estimations show the annual use of glyphosate to manage roadside weeds in the Auckland region alone to be about 25,000 L of formulated product (pers.Comm. Burton, Biothermal, 2012)[1]. There is no information available on home use, but this can also be substantial, especially the use of herbicides.

Agcarm (the agrichemical industry's organization in New Zealand) provides a report from members only, which is estimated to cover about 80 % of agrichemical sales (Manktelow et al. 2005). Over the last ten years (2000–2010), according to FAOSTAT, the value of imported agrichemicals in New Zealand has increased by approximately 62 %. Between 2000 and 2006 the value of imported herbicides, fungicides and insecticides increased 61 %, 50 %, and 42 %, respectively (Manktelow et al. 2005). Unfortunately, data for these individual categories were not available after 2006. Most herbicides are used in the pastoral and forestry sectors, and the horticulture sector accounts for the largest volume of fungicides and insecticides.

Total pesticide consumption for each farming sector depends on the total land area cropped and the sum of diseases, pests and weeds for a particular farm production system. There is no recent reliable data on the total pesticide consumption in New Zealand. It is estimated that the value of agrichemical sales in New Zealand is around $ 200 million NZ$ (too difficult to verify) (Ministry of the Environment 2006). Compared with world standards, the pesticides market in New Zealand is very small, which reduces opportunities for establishing new products because of the high cost of development and registration relative to sales.

Over the last few years new and more effective pesticides have been introduced and the pesticide products being used have been changed considerably. For example, the quantity of biological materials sold has increased and there has been a decrease in the quantities of miticides sold (Manktelow et al. 2005). The fluctuating trends of herbicide, fungicide and insecticide use during the last few years would be mostly because of major changes in the pesticide types used. Because of new pesticides and fluctuating prices of most pesticides (mostly downwards), neither the value nor the quantity are good indicators of pesticide consumption.

---

[1] Biothermal is a roadside weed maintenance contractor in Auckland.

Table 4.1 Areas and pesticide used in sector group. (Manktelow et al. 2005; Statistics New Zealand 2007)

Sector group	Total New Zealand area (ha)(2007)	Areas as % of total (%)	Total tonnes (a.i/yr)	Mean pesticide loading (kg a.i./ha/yr)(2005)	Percentage of total use (%)
Horticulture	246,748	1.7	3,254,606	13.19	49
Arable	318,416	2.2	773,751	2.43	12
Pastoral Farming	12,279,599	84	2,087,532	0.17	32
Forestry	1,849,897	13	499,472	0.27	8

The wet maritime climate of New Zealand creates suitable conditions for disease development, which affects fungicide use in New Zealand. As shown in Table 4.1, based on pesticides loading (Manktelow et al. 2005) and area of agriculture sectors (Statistics New Zealand 2007), horticulture is the most intensive pesticide using sector, using 49% of total pesticides on about, 1.7% of land, and this must be taken into consideration.

Manktelow et al. (2005) estimated the proportion of herbicide, insecticide and fungicide use in each agricultural sector. Table 4.2 shows herbicides are applied mostly on pastoral farms and the horticulture sector ranked the highest for insecticide and fungicide consumption at 88.6% and 61.9%, respectively.

4.5 Spraying Systems and Technologies

There are several techniques and technologies for applying pesticides on farms, including aerial spraying, air-blast spraying, boom spraying, in-furrow spraying, soil injection, dust and granular application. It appears the technology of farm spraying has not significantly changed during the last few decades. However, the new large commercial sprayers use GPS guidance to prevent overlap, misses and reduce drift (Kubik 2005; Kondo et al. 2011).

Spraying liquid is more common than dust and granular applications. Sprayed chemicals are mixed with water and broken down into droplets by forcing the liquid under pressure through an orifice, injecting the liquid into a fast moving air stream, or spraying the liquid off the surface of a rapidly rotating disc (Hawker and Keenlyside 1985; Culpin 1986). Tank, pump, filter, spraybar (boom), nozzles and mixing devices are the most important parts of liquid sprayers. Selecting appropriate spraying size, selecting right tractor speed, spraying pressure, nozzle size, and spraying boom height can increase the efficiency of pesticide use (Bell and Cousins 1991; CIGR 1999; Hunt 2001; Bell 2005; Kubik 2005). Furthermore, the shape and size of paddocks, barriers in borders, environmental conditions, availability of clean water and fuel, and driving skill can influence field efficiency by sprayers.

Spraying techniques and technologies can be categorized based on pesticides properties, injection system, amount of liquid applied per hectare, power sources,

Table 4.2 The percentage of pesticide use by agricultural sector. (Manktelow et al. 2005)

Sector group	Herbicides (%)	Insecticides (%)	Fungicides (%)
Horticulture	13.2	88.6	61.9
Arable	12.3	3.5	3.7
Pastoral Farming	55.5	6.3	34.4
Forestry	19	1.8	0

targeted pests and targeted species. In this study, to analyze energy use in pesticide application, the applications are categorized into ground and aerial spraying.

4.5.1 Ground Spraying Systems

Ground sprayers range in size from simple hand-carried (knapsack) sprayers to advanced self-propelled units, which spray a mixture of water and chemical droplets, through a spray nozzle under pressure. The size of droplets depends on the nozzles and pressure. There is a wide range of sprayers based on available technology, crop, and size of farm. Usually, sprayers used on arable and pastoral farms are wider than horticultural sprayers.

During pesticide application losses to the air vary from a few percent to 20–30 %, although it can reach as high as half the total amount (Van den Berg et al. 1999). The amount of atmospheric loss is influenced by several factors like the physicochemical properties of the compounds, the environmental conditions and the application techniques (Bedos et al. 2002). The transport of pesticide droplets to adjacent areas (the influence of weather conditions, microclimate, topography, and product types) and the amount of polluting agents released into the atmosphere have been studied by scientists.

Hand-carried sprayers or knapsacks are used to apply small amount of pesticides. They can also be used to spray livestock, greenhouses, nurseries and areas that are difficult to reach like valleys and areas with high slopes, as well as trees and crops on small farms and home gardens. In other countries they are also used in plantations and field crops. For the compressed air sprayer, the air pressure in the tank is increased by a hand operated pump or small diaphragm or piston pumps and when the valve is opened, pesticides are freely delivered to the nozzle. In some other knapsacks, the pumps directly deliver high pressure pesticides to the nozzles (Hawker and Keenlyside 1985; Bell and Cousins 1991; CIGR 1999). Most small tank knapsacks are carried in a backpack. Barrow sprayers use a similar mechanism to spray. They usually have two or four wheels and can carry greater amounts of pesticides; however, their maneuverability is much less than hand-carried or knapsack sprayers. The pumps of some barrow sprayers have a motor to drive the pumps and some of their pumps are powered by All Terrain Vehicles (ATV) or small tractors (Bell and Cousins 1991).

Trailed and 3-point boom sprayers are the most common sprayers on arable and pastoral farms, ranging in length from 6–36 m (CIGR 1999). The pumps of Trailed and 3-point boom sprayers are driven by Power Take off (PTO) systems and long booms are folded by tractors' hydraulic systems. Self-propelled sprayers are popular on large arable farms or in some specific farm production systems (Culpin 1986; Hunt 2001; Bell 2005). Self-propelled sprayers are the most common sprayers used by contractors as they can spray a large area over a short time.

In orchards, both surfaces of leaves and fruits should be covered completely by pesticides. In orchards the width of rows is around 2–4 m and trees can grow up to six meters; therefore, sprayers need higher pressure to reach the top of trees. In spraying orchards, vertical booms and airblast sprayers are used to spray pesticides along a narrow vertical band giving better coverage of leaves, fruits, and branches (CIGR 1999; Bell 2005). In air blast sprayers, nozzles may be arranged around the fan to increase atomization of the spray, which can improve coverage; however, it increases the risk of volatilization and drift in dry and windy conditions. Tower sprayers have better coverage and less drift than other horticultural sprayers in orchards (Culpin 1986; Bell and Cousins 1991; CIGR 1999) but require better driving skills. Due to conditions in orchards, the coverage width and speed of sprayers are less than on pastoral and arable farms and that reduces the capacity of the farm's spraying operation.

Ground spraying methods are more accurate as there is less drift because spray is applied only in the areas where needed, but drift still occurs. Ground spraying can increase soil compaction and spread disease and weeds; it also limits work during wet months and in hilly areas.

4.5.2 Aerial Spraying Systems

From the early years of the twentieth century, aircraft were used for seed sowing and dusting fertilizers; however, after the First World War (1920s) many of the war aircraft were used for spraying farms. Aerial spraying is the fastest method to apply baits, fertilizers and pesticides on large farms, high country farms and forests. One of the main limitations of using aircraft is drift, which affects useful and untargeted plants, animals and humans (Culpin 1986; CIGR 1999); this reduces farmers' interest in using aerial applications on multi crop farms and orchards. Moreover, aerial spraying on farms with overhead power lines, trees and hedges is very difficult and dangerous. The proximity of waterways, roads and houses also limits the use of aerial spraying.

Helicopters and aircraft sprayers fly at speeds of 15–240 km/h with effective spray widths of 12–20 m (CIGR 1999). Aerial sprayers can apply agrichemicals much faster than ground sprayers so fuel consumption and machinery use per hectare in aerial applications are lower than for ground applications. Also, quick applications are very useful when pest control is required immediately over large areas.

Helicopters have several features that make them more attractive for some types of spraying operations. According to the Civil Aviation Authority (CAA) of New Zealand, 80 aircraft and approximately 200 helicopters on the New Zealand register are involved in agricultural work. Helicopters have become more popular than fixed-wing aircraft in New Zealand over last few years and spray 70–80% of the total agrichemicals used on New Zealand farms. Some of the advantages of helicopters are:

- greater manoeuvrability
- ability to operate at lower speeds thus allowing greater precision for product placement
- better penetration of dense and deep foliage
- ability to hover in place, reverse, and take off and land vertically.

Aerial spraying is more effective on hilly and large farms, but effectiveness depends very much on weather conditions. In addition, highly trained pilots are required for aircraft and more vehicles and trained staffs are required to load chemical and fuel onto aircraft and helicopters.

4.6 Energy Consumption in Pest Control in New Zealand

4.6.1 Main Energy Inputs in Spraying

Studies by McChesney et al. (1982); Nguyen and Hignett (1995); Wells (2001); Barber (2004); Barber and Glenys (2005) and Saunders et al. (2006) have estimated energy consumption in different farming systems and farm productions. Most of these studies are interesting and provide useful information; however, agriculture is a complex system and it is not easy to estimate the average energy use for the whole country from a limited number of farms. Many energy studies in New Zealand use a small number of farms or do not mention the sample size. Also, some of them do not indicate the location of the farms on which they estimated energy use. Moreover, most studies only estimated energy consumption in a particular agricultural system while some compared energy use of different methods and in different countries.

Due to different farming systems, spraying techniques and agricultural products, it is very difficult to have definite energy consumption figures for pesticide applications. Main energy inputs in pesticide applications include fuel, labor, machinery and pesticides. Farmers use knapsacks (which do not consume significant amounts of fuel and machinery energy) in many greenhouses and small nurseries, but the labor costs per hectare for knapsacks are much higher than for boom sprayers. Many factors, such as environmental and soil conditions, driving skill, shape and size of paddocks also affect energy use in pesticide applications.

In terms of energy, using pesticides is much more energy intensive than mechanical pest control methods. For example, in organic farms, energy used for weed control using cultivators takes half the energy used for herbicide weed control (Pimentel 2009). The energy component in agrichemicals comes mainly from its manufacture, packaging and transport (Stout 1990; CIGR 1999). Fuel consumption and machinery use depend on the number of applications, farm production and farming system. The number of applications would be increased in high pest or disease pressure situations. Due to large potential variations in the number of applications, finding actual use patterns is extremely difficult. For example, in the pastoral and forestry sectors, a percentage of farms are usually sprayed but it varies with area and farming system.

4.6.2 Fuel

Fuel consumption in specific operations depends on soil conditions, crop type, groundspeed and rolling resistance (Smil 1991). Also, fuel consumption in spraying depends on the tractor, sprayer, shape and size of farm, and driver skill. The energy component in fuel comes mainly from the heat of combustion; furthermore, the energy required to drill, transport and refine the petroleum should be added to this figure (Stout 1990). Fuel consumption expressed as litres per hectare (L/ha), is a better measurement of fuel consumption than that expressed as litres per hour (L/h), as it uses the same bases to compare different inputs and operations (McLaughlin et al. 2008). Specific volumetric fuel consumption (SVFC) is the most common method used to estimate energy efficiency of a tractor using the units of L/kWh. However, sometimes instead of SVFC, specific volumetric fuel efficiency (SVFE), with units of kWh/l, is used (Grisso et al. 2004).

There are several methods to estimate the fuel consumption of tractors based on the power of the tractors; nevertheless, due to the effect of parameters such as altitude above sea level, soil conditions (soil type, moisture, density and residue cover), barometric pressure, humidity and temperature on tractor power and fuel consumption, most of these methods work only in specific areas (McLaughlin et al. 2002; Serrano et al. 2007; Bertocco et al. 2008; Safa et al. 2010). Furthermore, these methods can only predict fuel usage of diesel engines under full loads, but under partial loads and conditions when engine speeds are decreased from full throttle these methods do not work (Siemens et al. 1999; Safa et al. 2010).

For an accurate estimation, fuel consumption is measured before and after any farm operation by filling the fuel tank of the equipment (tractor, combine, or pump) and recording the difference in volume. After sampling several different farms and conditions, a formula was arrived at using mathematical modelling methods (Safa and Tabatabaeefar 2002). The energy input is determined from fuel consumption per operation for one hectare times the fuel equivalent energy per litre, as shown in Eq 4.1.

$$\text{Energy (input) /hectare} = \text{Operation fuel consumption (L/ha)} \times \text{Fuel energy (MJ/L)} \tag{4.1}$$

The formula for fuel consumption depends significantly on field efficiency. The efficiency of tractors and self-propelled sprayers is analyzed with respect to engine, power transmission and wheel soil system (Pellizzi et al. 1988; Serrano et al. 2007; Safa et al. 2010). Matching of tractor and implement, using hydraulic 3-point linkage equipment, using Power-Take-Off (PTO) equipment, selecting the right travel pattern, having large paddocks, regular servicing, adjusting tire inflation pressure, matching engine speed and gear selection, improving traction efficiency, using turbochargers and improving farmers' awareness, are all methods that could reduce fuel usage and improve field efficiency (Barber 2004; Grisso et al. 2004; Safa et al. 2010) and can reduce fuel consumption around 10 % in crop production (Pimentel 2009).

Diesel is the main fuel for tractors and other agricultural machinery because diesel engines are stronger, and have a higher efficiency and longer life than gasoline-powered engines (Safa et al. 2010). McChesney (1981) estimated diesel consumption for spraying at approximately 3 l/ha in New Zealand. However, because of developing technology and the use of more efficient machines and methods, the current rate is much lower than his estimation. There are large differences between different estimations of diesel consumption in ground spraying: Witney (1988): 1 L/ha, Dalgaard et al. (2001): 1.2 L/ha, CIGR (1999) 1.5 L/ha to 3 L/ha, and Wells (2001): 3 L/ha.

Estimations of fuel consumption for aerial spraying range from 0.035 L/ha in New Zealand (Barber 2004) to 1.85 L/ha in southern Queensland, Australia (Ghareei Khabbaz 2010). Comparing catalogs of helicopters and aircraft and data collection from contractors shows that fuel consumption for most aircraft and helicopters ranges from 58 L/h to 200 L/h, which means that due to high field capacity, fuel consumption per hectare is much lower and per hour is much higher than ground applications.

According to Saunders et al. (2006), an extra 23 % of energy consumption beyond the energy contents of diesel fuel and gasoline consumed accounts for processing, refining, and transport of crude oil and final products to, and within, New Zealand. Thus the total energy consumption for diesel and gasoline were taken to be 43.6 MJ/ha and 39.9 MJ/ha, respectively.

4.6.3 Tractors and Field Machines

Most commercial energy in agriculture is used in agricultural machinery manufacture and operation (Stout 1990). This energy can be categorized into energy required for manufacturing, maintenance and repair (Fluck and Baird 1980). Estimating the energy consumption of field machinery is much more complicated than determining energy consumption of other farm inputs (Smil 2008) because of the wide range of different tractors and sprayers and also different companies use different processes for producing machinery.

To compare energy use for producing and repairing tractors and equipment, energy use per kg has usually been used. Due to different technologies and different components, weight is not a good estimation index to compare energy consumption in producing machinery. There are large differences between different estimations: 75 MJ/kg (Roller et al. 1975), 90 MJ/kg (McChesney et al. 1978), 80.23 MJ/kg (Hornacek 1979), 27 MJ/kg (Fluck and Baird 1980), 85 MJ/kg (Stout 1990),129 MJ/kg for sprayers and 138 MJ/kg for tractors (CIGR 1999),132 MJ/kg for sprayers and 144 MJ/kg for tractors (Lague and Khelifi 2001), and 80 MJ/kg for sprayers and 160 MJ/kg for tractors (Wells 2001). Comparing the above rates, it appears that improving technology does not change the energy consumption for producing agricultural machinery. CIGR (1999) considered several steps in calculating these energy coefficients: first, the energy required for producing the raw materials; second, the energy used in the manufacturing process; third, the energy consumption for transporting the machine to the consumer; and fourth, the energy used in repairs and maintenance.

To calculate the energy input of tractors and other field equipment, it was necessary to know the weight, working life span, and the average surface area on which they were used annually (Safa et al. 2011). The estimated life can be taken from the ASAE Standard D497.6 (2009) and the estimated weight of different machines and equipment can be taken from companies' catalogues.

To calculate the energy used in producing and repairing agricultural machinery, the following formula was used:

$$ME = (G \times E) / (T \times Ca) \qquad (4.2)$$

where ME is machine energy (MJ/ha); G is the weight of the implement (kg); E is the energy sequestered in agricultural machinery (MJ/kg); T is the economic life of the machine (h) and; Ca was effective field capacity (ha/h).

For calculation of Ca, the following equation was used:

$$Ca = (s \times w) \times FE/10 \qquad (4.3)$$

where s is ground speed(km/h); w is the width of the machine (m) and; FE is field efficiency(%), which was taken from the ASAE Standard D497.6 (ASAE 2009).

4.6.4 Labor

Before the invention of the tractor, hand and draught domestic animals were the only choices for power generation needed for agricultural operations. Even now, human power is the main source (73 %) of energy in agricultural operations in many developing countries (Stout 1990). In the future, human labor on fully mechanized (mechatronic) farms could be reduced to almost nil. Nevertheless, some scientists believe that organic agriculture, one of the important choices for future farming, needs more manual work for harvesting and weeding (WCED 1987; Pimentel et al. 2005; Wallgren and Höjer 2009) and, in some crops, this could be up to 35 % (WCED 1987; Pimentel et al. 2005; Wallgren and Höjer 2009).

There are several different thermodynamic and sequestered methods for analyzing human energy (Fluck and Baird 1980). Human energy is analyzed through measuring heart rates and recording oxygen consumption (Stout 1990). The energy output of humans depends on gender, weight, body size, age, activity and climate (Smil 1994); therefore, there are a number of different estimations of energy output used for human labor.

Human energy is used less than other energy inputs in modern agriculture (sometimes less than 1 % of all energy inputs) so it has not been calculated in many recent agricultural energy studies. The energy output for a male worker is 1.96 MJ/hr and 0.98 MJ/hr for a female worker (Singh and Mittal 1992; Mani et al. 2007). One must recognize that human energy, especially in developed countries, is the most expensive form of energy in field operations which encourages farmers to use better machinery and cultivate crops with minimum need for labor.

Most physical activities in pesticide application involved driving, adjusting, and servicing tractors and sprayers, which consumed significantly less energy than physical weed control. However, estimating human energy use in operations such as tractor servicing is difficult as this also contributes to other farm products. Farmers clearly expended different amounts of energy per hour for each operation and several factors, such as gender, weight and age can influence their energy use.

4.6.5 *Pesticides*

Pesticides are the most energy intensive of all farm inputs (Stout 1990). Most ingredients used in pesticide production come from petrochemical products such as ethylene, methane and propylene (Safa et al. 2011); and transportation on the farm uses significant amounts of fuels. Energy used in formulation, packaging and transport, as well as manufacturing active ingredients, inert ingredients and adjuvants should be considered as a part of pesticide energy equivalents.

Studies such as Helsel (1992) and CIGR (1999) estimated energy use of some pesticides. but these studies did not cover all products, especially new ones. Pesticide products vary between brands and, because of patent and commercial issues, it is impossible to access the details of active ingredients, inert ingredients, and manufacturing processes. Exact documented energy consumption in manufacturing is not available and would be very difficult to estimate, especially for newer pesticides which are introduced continuously and labelled for use at very low rates. In this study, the energy coefficients for herbicides, insecticides and fungicides were taken from Saunders et al.'s (2006) report and these were 310, 315, and 210 MJ/kg, respectively.

Using mechanical pest control, biological pest control, resistant varieties, crop rotation, cover crops, and optimal planting spaces and dates can cut the quantities of pesticides required. New technologies such as precision agriculture and integrated pest management (IPM) can reduce pesticide usage and the operator's workload with minimum if any reduction in agricultural production.

4 Energy Inputs In Pest Control Using Pesticides In New Zealand

Table 4.3 Operational energy consumption and energy intensity in agricultural sectors in New Zealand (GJ)

	Machinery	Fuel	Labor	Total Energy use (%)	Energy use per hectare (MJ/ha)[a]
Horticulture	18,615.1	95,170.2	935.6	114,721.0 (24%)	465
Arable	4,289.9	62,473.2	182.9	66,946.0 (14%)	210
Pastoral	17,539.7	262,348.5	871.1	280,759.3 (60%)	23
Forestry	66.8	7167.9	6.4	7,241.2 (2%)	4
Total (%)	40,512 (10%)	427,160 (90%)	1,996 (0.5%)	469,667.5	32

[a] It is notable that only a proportion of farms are sprayed each year (especially on forestry and pastoral farms); therefore, the energy use per hectare on land on which applications are made should be more than this

4.7 Energy Inputs in Pesticides Application in New Zealand

Energy inputs in pesticide applications can be separated into the energy content of the pesticides and the operational energy including fuel, machinery and labor. As mentioned previously, various direct and indirect factors can influence pesticide consumption and operational energy use in pesticide applications.

4.7.1 Operational Energy

Operational energy includes the labor, fuel, and machinery used in pesticide applications. In this study, the main groups of agricultural production and the area of each production system were collected from Statistics New Zealand (2007). Statistics New Zealand (2007) categorizes farm production into 30 main groups and provides the average cultivated area for each group. The most varied sector is horticulture with 14 production groups, including vegetables, floriculture, nurseries and fruit trees.

The average number of applications and other operational information were collected for each group through interviews with farmers, scientists and contractors, with all possible tools and methods being used to collect data with the highest accuracy. As was expected, the methods of pesticide application, spraying frequency, and machinery used in some production types was completely different from other production types in the same group. Therefore, a reasonable average of the required data for each production group was selected.

The total operational energy consumption is estimated as 469,667 GJ in New Zealand (Table 4.3). Fuel is the most important operational energy input (at 90%) and machinery is ranked second with 427,160 and 40,511 GJ, respectively. As expected, labor energy was less than 1%. The proportion of labor energy and fuel in pastoral farming and forestry due to more aerial spraying was lower than for other production types. In horticultural production due to the use of smaller size sprayers, the proportion of labor and fuel is more than other sectors.

Pastoral farming, due to large areas, and horticulture, due to intensive applications, ranked as the highest energy consumers in the agriculture sectors with 280,759 GJ (60%) and 114,721 GJ (24%), respectively. Notably the number of applications in some horticultural production is much higher than for other production groups such as more than 20 applications on average in onion production and eight applications in some olive orchards. Moreover, the average speed and width of coverage of sprayers in fruit orchards is lower than that of typical boom sprayers on crop and pastoral farms, which reduces the field capacity during orchard pesticide applications. In some nurseries, greenhouses and small orchards the operational energy use is very low due to use of knapsack sprayers.

Due to the higher numbers of applications for some horticultural production, the energy use per hectare of operational energy was significantly higher than for other agricultural sectors, at 465 MJ/ha. The arable sector ranked second at approximately 210 MJ/ha. On average, energy consumption during operations for all agricultural sectors was estimated at approximately 32 MJ/ha in New Zealand. As the forestry and pastoral farming sectors have the highest proportion of aerial applications this reduces their operational energy use per hectare.

Operational energy use for aerial applications was estimated at around 18,859 GJ, mostly on forestry and pastoral farms. As mentioned before, fuel consumption and machinery use per hectare during aerial spraying is lower than ground spraying per hectare. The proportion of aerial spraying is around 4% of total operational energy. However this does not include the energy embedded in fixed wing planes or helicopters.

With the fuel consumption in spraying (90%) and other farm operations representing a high proportion of operational energy, fuel consumption should be considered more than other operational energy inputs for reducing energy consumption. As mentioned before, there are several technical ways to reduce fuel consumption in agricultural operations. Due to different farm conditions and product properties, it would be very difficult to provide a general plan to reduce fuel use in all different farming activities. A fuel conservation plan should be developed based on farm production, farmers' knowledge, available technology and the most common pests.

4.7.2 *Energy Consumption of Pesticides in Different Agriculture Sectors*

The total energy component of pesticide use in New Zealand was estimated to be 1,881,408 GJ. Herbicides are the main pesticide energy use with 1,353,503 GJ. As shown in Table 4.4 and Fig. 4.1, the intensive use of pesticides in the horticultural sector affects energy consumption as well. However, pastoral farming, because of the large area and high herbicide use, has the highest proportion of total pesticide use.

Pesticide energy use in each agricultural sector depends on the energy equivalent of pesticides and the volume of agrichemicals used. Herbicide energy ranked high-

Table 4.4 Energy use (GJ) of pesticides and energy use per hectare in agriculture sectors in New Zealand

	Herbicides	Insecticides	Fungicides	Total Energy(%)	Energy use per hectare (MJ/ha)
Horticulture	178,662	128,990	282,138	589,790 (31%)	2390
Arable	166,481	7,710	11,183	185,374 (10%)	582
Pastoral	751,194	71,684	5,751	828,629 (44%)	67
Forestry	257,165	0	20,130	277,295 (15%)	149
Total	1,353,503 (70%)	208,384 (10%)	319,522 (20%)	1,881,409	128

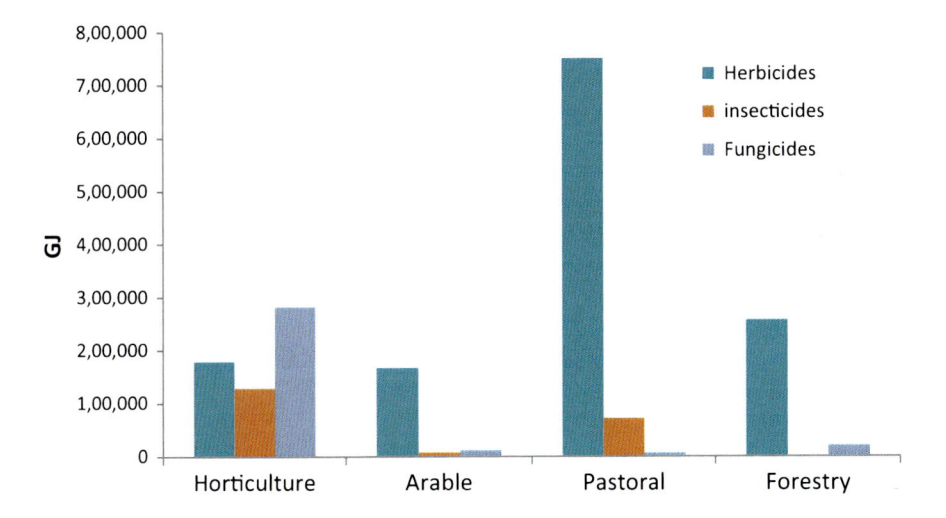

Fig. 4.1 Energy use (GJ) in agriculture sectors

est in all agricultural sectors except the horticultural sector, with 70% (1,353,503 GJ) of total energy, while fungicide energy is around 20% of total pesticide energy.

The high energy intensity in the horticulture sector was expected; it is four times more than the arable sector and around 35 times more than pastoral farming per hectare. The average energy use per hectare of all pesticides was estimated at around 128 MJ/ha, which is around five times more than the operational energy.

4.7.3 Total Energy Inputs in Pest Control Using Pesticides in New Zealand

The total energy use for pesticide applications in New Zealand was estimated at around 2,350,757 GJ. As expected, pastoral farming has the highest proportion of total energy use for pesticide applications, with 1,109,389 GJ, due to the large area;

Table 4.5 Total energy use (GJ) and energy use per hectare of pesticides and operational energy inputs in New Zealand

	Herbicides	Insecticides	Fungicides	Operational Energy	Total Energy (%)	Energy use per hectare (MJ/ha)
Horticulture	178662	128990	282138	114721	704511 (30%)	2855
Arable	166481	7710	11183	66946	252320 (11%)	792
Pastoral	751194	71684	5751	280759	1109389 (47%)	90
Forestry	257166	0	20130	7241	284537 (12%)	154
Total (%)	1353503 (58%)	208384 (9%)	319202 (14%)	469667 (20%)	2350757	160

and the horticultural sector ranked second due to intensive pesticide applications, with 704,511 GJ (Table 4.5).

Horticultural and arable farming have the most intensive energy use per hectare with 2,855 and 792 MJ/ha, respectively. Average total energy use per hectare was estimated to be 160 MJ/ha (Table 4.5 and Fig. 4.2). Comparing Tables 4.3 and 4.5 shows that operational energy in arable and pastoral farming is higher than other sectors, at 27% and 25%, respectively. The percentage of operational energy to total energy in orchard and vegetable production is higher than for other groups but is very low in nurseries and greenhouses, which are estimated to be around 16% of the horticultural sector. The percentage of operational energy, mostly from aerial applications, of total energy use in forestry sector was estimated at only 2.5%.

As shown in Table 4.5, energy use in each farming sector depends on pesticide use, the area of that sector, type of pesticide application and spraying frequency. For example, fungicides and most herbicides are applied by aerial application in only some forests; therefore, the proportion of insecticide and operational applications per hectare are lower than in other sectors. Another example is the high usage of fungicides in vegetable farms, nurseries and orchards in humid areas of the North Island.

4.7.4 CO_2 Emission in Pesticide Application

The direct link between energy use and CO_2 emissions in agricultural production results in a similar pattern for both factors. Compared with emissions from other farm activities, CO_2 emissions from pesticide applications are not very large. The CO_2 emissions from pesticides and operations were calculated based on estimated energy use in pesticide applications (Table 4.6). Total CO_2 emission was estimated

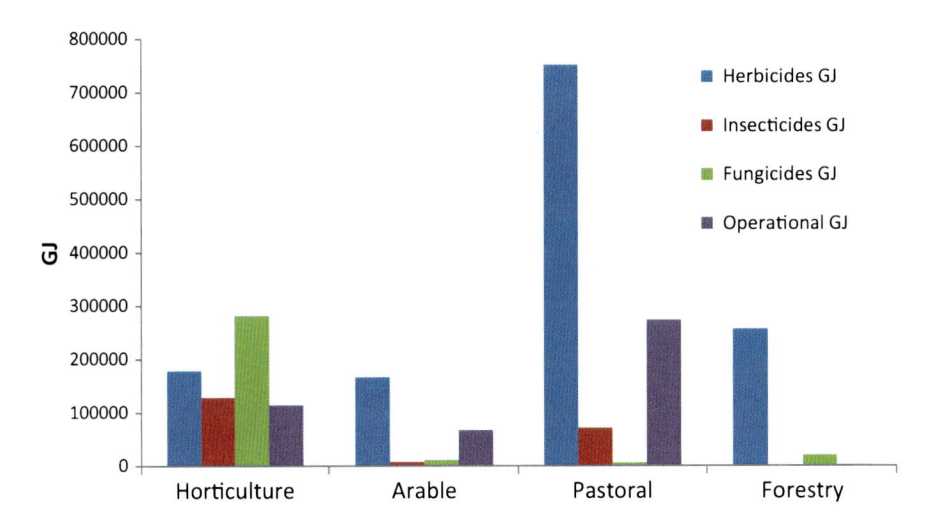

4 Energy Inputs In Pest Control Using Pesticides In New Zealand

Fig. 4.2 Energy use (GJ) of farm inputs in agriculture sectors in New Zealand

Table 4.6 Total CO_2 (tonnes of CO_2) and $kgCO_2$ per hectare of pesticides and operational energy inputs in New Zealand

	Herbicides	Insecticides	Fungicides	Operational CO_2	Total CO2 (%)	CO_2 Emission per hectare (kg CO_2/ha)
Horticulture	10720	7739	16928	8214	43601 (30%)	177
Arable	9989	463	671	4678	15800 (11%)	50
Pastoral	45072	4301	345	19602	69320 (48%)	6
Forestry	15430	0	1208	498	17136 (12%)	9
Total (%)	81210 (56%)	12503 (9%)	19152 (13%)	32992 (23%)	145857	10

at 145,857 t, with pastoral and horticulture farming having the highest proportions at 48% and 30%, respectively.

As expected, the horticulture sector has the most intensive CO_2 emissions with 177 kg CO_2/ha, which is three times more than CO_2 emissions on arable farms and 29 times more than average CO_2 emissions on pastoral farms per hectare. CO_2 emissions from fuels are around 90% of operational emissions in spraying; therefore, better farm management can reduce CO_2 emissions significantly. Comparing CO_2 emissions in different farming sectors shows using aerial applications can significantly reduce CO_2 emission per hectare, which is obvious in the forestry and pastoral farming sectors.

4.8 Summary

There is a wide range of active ingredients and techniques used in pesticide applications and every year new agrichemicals with different prices are introduced to markets, which make it difficult to compare annual pesticide consumption. In this study, energy use in farm production in New Zealand was investigated for the pastoral, arable, forestry, and horticultural sectors.

Based on available data, pesticide consumption is more intensive, and energy use and CO_2 emissions higher, in the horticultural sector compared with other sectors, which would increase environmental and health costs of fruit and vegetables. The high proportion of operational energy shows the importance of developing new techniques and machinery to reduce energy use and CO_2 emissions. For example, the results show that operational energy use in aerial applications is much lower than for ground spraying. However, to compare different pest control techniques, other factors such as environmental and health impacts, cost, effectiveness and maintenance availability should be considered. Energy use and CO_2 emissions, as well as health and environmental costs, are likely to be even lower for agroecological techniques of pest management, such as light traps, attractant lures, mating disruption with pheromones, companion planting, trap crops, rotational planting, intercropping, biological controls, and effective natural plant nutrition.

References

ASAE. (2009). *ASAE D497.6:2009. Agricultural Machinery Management Data*. St. Joseph, Michigan, USA: American Society of Agricultural and Biological Engineers (ASAE).

Barber, A. (2004). *Seven Case Study Farms: Total Energy & Carbon Indicators for New Zealand Arable & Outdoor Vegetable Production*. AgriLINK New Zealand Ltd. http://www.agrilink.co.nz/Portals/Agrilink/Files/Arable_Vegetable_Energy_Use_Main_Report.pdf. Accessed 2 Jan 2013.

Barber, A., & Glenys, P. (2005). *Energy Use and Efficiency Measures for the New Zealand Arable and Outdoor Vegetable Industry*. Climate change office and energy efficiency and conservation authority, Auckland. AgriLINK New Zealand Ltd. http://www.agrilink.co.nz/Portals/agrilink/Files/Arable_Vege_Energy_Efficiency_Stocktake.pdf. Accessed 2 Jan 2013.

Bedos, C., Cellier, P., Calvet, R., & Barriuso, E. (2002). Occurrence of pesticides in the atmosphere in France. *Agronomie, 22*(1), 35–49.

Bell, B. J. (2005). *Farm Machinery* (5th ed.). Ipswich United Kingdom: Old Pond Publishing.

Bell, B. J., & Cousins, S. (1991). *Machinery for Horticulture*. Ipswich, United Kingdom: Farming Press Books.

Bertocco, M., Basso, B., Sartori, L., & Martin, E. C. (2008). Evaluating energy efficiency of site-specific tillage in maize in NE Italy. *Bioresource Technology, 99*(15), 6957–6965.

Briand, O., Bertrand, F., Seux, R., & Millet, M. (2002). Comparison of different sampling techniques for the evaluation of pesticide spray drift in apple orchards. *Science of The Total Environment, 288*(3), 199–213.

Buckland, J., Jones, P., Ellis, H., & Salter, R. (1998). *Ambient Concentrations of Selected Organochlorines in Rivers*. Wellington, New Zealand: Organochlorines Program, Ministry for the Environment.

CIGR. (1999). *CIGR Handbook of Agricultural Engineering*. International Commission of Agricultural Engineering. St. Joseph, Michigan, USA: American Society of Agricultural Engineers.

Close, M., & Skinner, A. (2012). Sixth national survey of pesticides in groundwater in New Zealand. *New Zealand Journal of Marine and Freshwater Research, 46*(3), 443–457.

Conforti, P., & Giampietro, M. (1997). Fossil energy use in agriculture: An international comparison. *Agriculture, Ecosystems & Environment, 65*(3), 231–243.

Culpin, C. (1986). *Farm Machinery* (11th ed.). London: Collins.

Dalgaard, T., Halberg, N., & Porter, J. R. (2001). A model for fossil energy use in Danish agriculture used to compare organic and conventional farming. *Agriculture, Ecosystems & Environment, 87*(1), 51–65.

Davis, B. N. K., & Williams, C. T. (1990). Buffer zone widths for honeybees from ground and aerial spraying of insecticides. *Environmental Pollution, 63*(3), 247–259.

Dryson, E., 't Mannetje, A., Walls, C., McLean, D., McKenzie, F., Maule, M., et al. (2008). Case-control study of high risk occupations for bladder cancer in New Zealand. *International Journal of Cancer, 122*(6), 1340–1346.

East, R., & Pottinger, R. P. (1975). Starling (Sturnus vulgaris L.) predation on grass grub (Costelytra zealandica (White), Melolonthinae) populations in Canterbury. *New Zealand Journal of Agricultural Research, 18*(4), 417–452.

Energy, & Environment. (2009). *New Zealand Energy & Environment Business Week*. In (ed., Vol. 6, pp. 1), Media Information, Christchurch, New Zealand.

FAO. (2010). *International Code of Conduct on the Distribution and Use of Pesticides, Guidance on Pest and Pesticide Management Policy Development*. Rome: Food and Agriculture Organization of the United Nations. http://www.vegetableipmasia.org/docs/FAO-Pest&Pesticide-ManagementPolicy_June2010.pdf. Accessed 2 Jan 2012.

FAO. (2011). *FAO Food Outlook Report 2011*. Food and Agriculture Organization of the United Nations. http://www.fao.org/docrep/014/al978e/al978e00.pdf. Accessed 2 Jan 2013.

FAO. (2012). *FAOSTAT*. Rome: Food and Agriculture Organization of the United Nations. http://faostat.fao.org/site/339/default.aspx. Accessed 28 Feb 2014.

Fluck, R. C., & Baird, C. D. (1980). *Agricultural Energetics*. Westport, CT, USA: AVI Pub. Co.

Gaw, S. (2003). *Pesticides in horticultural Soils in the Auckland Region*. ARC Working Report No. 96. Auckland Regional Council, Auckland, New Zealand.

Ghareei Khabbaz, B. (2010). *Life cycle energy use and greenhouse gas emissions of Australian cotton: Impact of farming systems* [Thesis]. University of Southern Queensland, Toowoomba. http://eprints.usq.edu.au/19556/. Accessed 2 Jan 2013.

Grisso, R. D., Kocher, M. F., & Vaughan, D. H. (2004). Predicting tractor fuel consumption. *Applied Engineering in Agriculture, 20*(5), 553–561.

Haenel, H.-D., & Siebers, J. (1995). Lindane volatilization under field conditions: Estimation from residue disappearance and concentration measurements in air. *Agricultural and Forest Meteorology, 76*(3–4), 237–257.

Hawker, M. F. J., & Keenlyside, J. F. (1985). *Horticultural Machinery* (3rd ed.). London: Longman.

Helsel, Z. R. (1992). Energy and alternatives for fertilizer and pesticide use. In R. Fluck (Ed.)., *Energy in World Agriculture* (Vol. 6, pp. 177–202). Amsterdam: Elsevier Sci. Publ. Co.

Hornacek, M. (1979). Application de l'analyse énergetique à 14 exploitations agricoles. [Energetic analysis of 14 agricultural practices]. *Étude du CNEEMA (Centre National d'Etude et d'Expérimentation de Machinisme Agricole), (France)* 457.

Hunt, D. (2001). *Farm Power and Machinery Management* (10th ed.). Ames, Iowa, USA: Iowa State University Press.

Kelly, G. (2007). Renewable energy strategies in England, Australia and New Zealand. *Geoforum, 38*(2), 326–338.

Kondo, N., Monta, M., & Noguchi, N. (2011). *Agricultural robots: Mechanisms and Practice*. (English Ed.). Sakyo-ku, Japan: Kyoto University Press & Trans Pacific Press.

Kubik, R. (2005). *How to Use Implements on Your Small-scale Farm*. St. Paul, Minnesota, USA: Motorbooks.

Lague, C., & Khelifi, M. (2001). Energy use and time requirements for different weeding strategies in grain corn. *Canadian Biosystems Engineering, 43*, 2.13–2.21.

Lal, R. (2004). Carbon emission from farm operations. *Environment International, 30*(7), 981–990.

Landis, D. A., & Orr, D. B. (2002). Biological control: Approaches and application. In E. B. Radcliffe & W. D. Hutchinson (Eds.), *Radcliffe's IPM World Textbook*. St. Paul, Minnesota, USA: University of Minnesota. http://ipmworld.umn.edu/. Accessed 3 Jan 2013.

Lavin, K. S., Hageman, K. J., Marx, S. K., Dillingham, P. W., & Kamber, B. S. (2012). Using trace elements in particulate matter to identify the sources of semivolatile organic contaminants in air at an alpine site. *Environmental Science & Technology, 46*(1), 268–276.

Li, L.-Y. (1994). Worldwide use of Trichogramma for biological control on different crops: A survey. In E. Wajnberg & S. A. Hassan (Eds.), *Biological Control with Egg Parasitoids* (pp. 37–51). Wallingford United Kingdom: CAB International.

Love, B., Gaw, S. K., & SEM NZ Limited. (2005). *Background Levels of Agrichemical Residues in Bay of Plenty Soils. A Preliminary Technical Investigation*. Whakatane: Solutions in Environmental Management (SEM), Whakatane, New Zealand.

Lucas, R. (2005). *Managing Pests and Diseases: A Handbook for New Zealand Gardeners*. Nelson, New Zealand: Craig Potton Pub.

MAF. (2011). *Situation and Outlook for New Zealand Agriculture and Forestry*. Ministry of Agriculture And Forestry, Wellington, New Zealand. http://www.mpi.govt.nz/news-resources/publications.aspx?title=Situation%20and%20Outlook%20for%20New%20Zealand%20Agriculture%20and%20Forestry. Accessed 3 Jan 2013.

Mani, I., Kumar, P., Panwara, S. G., & Kanta, K. (2007). Variation in energy consumption in production of wheat–maize with varying altitudes in hilly regions of Himachal Pradesh, India. *Energy, 32*(12), 2336–2339.

Manktelow, D., Stevens, P., Zabkiewicz, J., Walker, J., Gurnsey, S., Park, N., Zabkiewicz, J., Teulon, D., & Rahman, A. (2005). *Trends in Pesticide Use in New Zealand: 2004*. Report to the New Zealand Ministry for the Environment, Auckland, New Zealand.

McChesney, I. G. (1981). *Field Fuel Consumption of Tractors*. Internal Report 3, Joint centre for environmental sciences, Lincoln College, University of Canterbury, New Zealand.

McChesney, I. G., Pearson, R. G., Bubb, J. W., & Joint Centre for Environmental Sciences (N.Z.). (1978). *Energy use on Canterbury mixed cropping farms: A pilot survey*. Occasional paper (Joint centre for environmental sciences (NZ)0, no. 5, Lincoln College, Lincoln, New Zealand.

McChesney, I. G., Sharp, B. H. M., & Hayward, J. A. (1982). Energy in New Zealand agriculture: Current use and future trends. *Energy in Agriculture, 1,* 141–153.

McLaughlin, N. B., Gregorich, E. G., Dwyer, L. M., & Ma, B. L. (2002). Effect of organic and inorganic soil nitrogen amendments on mouldboard plow draft. *Soil and Tillage Research, 64*(3–4), 211–219.

McLaughlin, N. B., Drury, C. F., Reynolds, W. D., Yang, X. M., Li, Y. X., Welacky, T. W., & Stewart, G. (2008). Energy inputs for conservation and conventional primary tillage implements in a clay loam soil. *Transactions of the ASABE, 51*(4), 1153–1163.

Milne, J. R. (2010). *Wellington Harbour Marine Sediment Quality Investigation. Supplementary Report*. Environmental monitoring and investigations department, Greater Wellington Regional Council, Wellington, New Zealand. http://www.gw.govt.nz/assets/council-publications/Wellington%20Harbour%20Marine%20Sediment%20Quality%20Investigation%20Supplementary%20Report.pdf. Accessed 3 Jan 2013.

Ministry of the Environment. (2006). Study of the New Zealand *Product Stewardship Scheme for Agrichemical Containers*. Prepared for the Ministry for the Environment by responsible resource recovery Ltd. Ministry for the Environment, Wellington, New Zealand. http://www.mfe.govt.nz/publications/waste/product-stewardship-agrecovery-may06/index.html. Accessed 3 Jan 2013.

Murray, D., Wesseling, C., Keifer, M., Corriols, M., & Henao, S. (2002). Surveillance of pesticide-related illness in the developing world: Putting the data to work. *International Journal of Occupational and Environmental Health, 8*(3), 243–248.

Nguyen, M. L., & Hignett, T. P. (1995). Energy and labour efficiency for three pairs of conventional and alternative mixed cropping (pasture-arable) farms in Canterbury, New Zealand. *Agriculture, Ecosystems and Environment, 52,* 163–172.

NZFSA. (2009). Making sure New Zealand's food is safe. New Zealand Food Safety Authority, Wellington, New Zealand. http://www.foodsmart.govt.nz/elibrary/making-sure-zealand-nz-food-is-safe/food-is-safe.pdf. Accessed 3 Jan 2013.

Pellizzi, G., Guidobono Cavalchini, A., Lazzari, M., & Commission of the European Communities. (1988). *Energy Savings in Agricultural Machinery and Mechanization.* London: Elsevier Applied Science.

Pimentel, D. (2009). Reducing energy inputs in the agricultural production system. *Monthly Review, 61*(3), 92–101.

Pimentel, D., Hepperly, P., Seidel, R., Hanson, J., & Douds, D. (2005). Environmental, energetic, and economic comparisons of organic and conventional farming systems. *Bioscience, 55*(7), 573–582.

Pimentel, D., & Pimentel, M. (2008). *Food, Energy, and Society* (3rd ed.). Boca Raton, Florida, USA: CRC Press.

Ramaprasad, J., Tsai, M.-Y., Elgethun, K., Hebert, V. R., Felsot, A., Yost, M. G., et al. (2004). The Washington aerial spray drift study: Assessment of off-target organophosphorus insecticide atmospheric movement by plant surface volatilization. *Atmospheric Environment, 38*(33), 5703–5713.

Reichman, R., Yates, S. R., Skaggs, T. H., & Rolston, D. E. (2013). Effects of soil moisture on the diurnal pattern of pesticide emission: Comparison of simulations with field measurements. *Atmospheric Environment, 66*(February), 52–62.

Rhodes, D., & Novis, J. (2002). The impact of incentives on the development of plantation forest resources in New Zealand. *MAF Information Paper No. 45.* Ministry of Agriculture and Forestry (MAF), Wellington, New Zealand. http://maxa.maf.govt.nz/forestry/publications/impact-of-incentives-on-plantation-forest-resources/information-paper-45.pdf. Accessed 3 Jan 2013.

Roller, W. L., Keener, H. M., Kline, R. D., Mederski, H. J., & Curry, R. B. (1975). *Grown Organic Matter as a Fuel Raw Material Resource.* Report No NASA CR-2608. Prepared by the Ohio agricultural research and development center, Wooster, Ohio. http://ntrs.nasa.gov/search.jsp?R=19760003481. Accessed 3 Jan 2013.

Safa, M., & Samarasinghe, S. (2011). Determination and modelling of energy consumption in wheat production using neural networks: "A case study in Canterbury province, New Zealand". *Energy, 36*(8), 5140–5147.

Safa, M., & Tabatabaeefar, A. (2002). *Energy consumption in wheat production.* Paper presented at the IACE 2002, Wuxi, China. IACE.

Safa, M., Samarasinghe, S., & Mohssen, M. (2010). Determination of fuel consumption and indirect factors affecting it in wheat production in Canterbury, New Zealand. *Energy, 35*(12), 5400–5405.

Safa, M., Samarasinghe, S., & Mohssen, M. (2011). A field study of energy consumption in wheat production in Canterbury, New Zealand. *Energy Conversion and Management, 52*(7), 2526–2532.

Saunders, C., Barber, A., & Taylor, G. (2006). *Food Miles: Comparative Energy/Emissions Performance of New Zealand's Agriculture Industry.* Lincoln, New Zealand: Agribusiness & Economics Research Unit, Lincoln University.

Serrano, J. M., Peça, J. O., Marques da Silva, J., Pinheiro, A., & Carvalho, M. (2007). Tractor energy requirements in disc harrow systems. *Biosystems Engineering, 98*(3), 286–296.

Siemens, J. C., Bowers, W., Holmes, R. G., & Deere & Company. (1999). *Machinery Management: How to Select Machinery to Fit the Real Needs of Farm Managers.* East Moline, Illinois, USA: John Deere Publishing.

Singh, S., & Mittal, J. (1992). *Energy in Production Agriculture.* New Delhi: Mittal Publications.

Smil, V. (1991). *General energetics: Energy in the biosphere and civilization.* New York: Wiley.

Smil, V. (1994). *Global Ecology: Environmental Change and Social Flexibility.* London: Routledge.

Smil, V. (2008). *Energy in Nature and Society: General Energetics of Complex Systems.* Cambridge, United Kingdom: The MIT Press.

Statistics New Zealand. (2007). Agricultural areas in hectares by farm type (ANZSIC06). *Statistics New Zealand*, Wellington, New Zealand. http://www.stats.govt.nz/browse_for_stats/industry_sectors/agriculture-horticulture-forestry/2007-agricultural-census-tables/land-use-farm-counts.aspx. Accessed 3 Jan 2013.

Statistics New Zealand. (2008). Agricultural production statistics (Final): June 2007. *Statistics New Zealand*, Wellington. http://www.stats.govt.nz/browse_for_stats/industry_sectors/agriculture-horticulture-forestry/AgriculturalProduction_HOTPJun07final.aspx. Accessed 3 Jan 2013.

Statistics New Zealand. (2010). Harvests increase for wheat, barley, and maize grain. *Statistics New Zealand*, Wellington, New Zealand. http://www.stats.govt.nz/browse_for_stats/Corporate/Corporate/CorporateCommunications_MRJun09prov.aspx. Accessed 3 Jan 2013.

Stockin, K. A., Law, R. J., Roe, W. D., Meynier, L., Martinez, E., Duignan, P. J. Bridgen, P., Jones, B. (2010). PCBs and organochlorine pesticides in Hector's (Cephalorhynchus hectori hectori) and Maui's (Cephalorhynchus hectori maui) dolphins. *Marine Pollution Bulletin, 60*(6), 834–842.

Stout, B. A. (1990). *Handbook of Energy for World Agriculture*. London: Elsevier Science Pub. Co.

t Mannetje, A., Dryson, E., Walls, C., McLean, D., McKenzie, F., Maule, M., et al. (2008). High risk occupations for non-Hodgkin's lymphoma in New Zealand: Case-control study. *Occupational and Environmental Medicine, 65*(5), 354–363.

Tsai, M.-Y., Elgethun, K., Ramaprasad, J., Yost, M. G., Felsot, A. S., Hebert, V. R., et al. (2005). The Washington aerial spray drift study: Modeling pesticide spray drift deposition from an aerial application. *Atmospheric Environment, 39*(33), 6194–203.

Van den Berg, F., Kubiak, R., Benjey, W. G., Majewski, M. S., Yates, S. R., Reeves, G. L., et al. (1999). Emission of pesticides into the air. *Water Air and Soil Pollution, 115*(1–4), 195–218.

Van Lenteren, J. C. (2012). The state of commercial augmentative biological control: Plenty of natural enemies, but a frustrating lack of uptake. *Biocontrol, 57*(1), 1–20.

Vincent, C., Weintraub, P., & Hallman, G. (2009). Physical control of insect pests. In V. H. Resh & R. T. Cardé (Eds.), *Encyclopedia of Insects a second edition*. San Diego, California, USA: Academic Press.

Vlek, P. L. G., Rodríguez-Kuhl, G., & Sommer, R. (2004). Energy use and CO_2 production in tropical agriculture and means and strategies for reduction or mitigation. *Environment, Development and Sustainability, 6*(1–2), 213–233.

Wallgren, C., & Höjer, M. (2009). Eating energy—Identifying possibilities for reduced energy use in the future food supply system. *Energy Policy, 37*(12), 5803–5813.

Watts, M. (2010). *Pesticides: Sowing Poison, Growing Hunger, Reaping Sorrow* (2nd ed.). Pesticide Action Network Asia and the Pacific. Penang: Jutaprint. http://www.pananz.net/resources/Div_Loaded_Files/Documents/Sowing_Poison/sowingpoisongrowinghunger_2ndedition.pdf. Accessed 3 Jan 2013.

WCED. (1987). *Our Common Future*. World Commission on Environment and Development. Oxford: Oxford University Press. http://www.un-documents.net/wced-ocf.htm. Accessed 3 Jan 2013.

Wells, C. (2001). *Total Energy Indicators of Agricultural Sustainability: Dairy Farming Case Study*. Ministry of Agriculture and Forestry, Wellington, New Zealand. http://maxa.maf.govt.nz/mafnet/publications/techpapers/techpaper0103-dairy-farming-case-study.pdf. Accessed 3 Jan 2013.

Witney, B. (1988). *Choosing and Using Farm Machines*. Harlow, United Kingdom: Longman Scientific & Technical; Wiley.

World Bank. (2008). *Agriculture for Development*. Washington, D.C.: World Bank. http://web.worldbank.org/WBSITE/EXTERNAL/EXTDEC/EXTRESEARCH/EXTWDRS/0,,contentMDK:23062293~pagePK:478093~piPK:477627~theSitePK:477624,00.html. Accessed 3 Jan 2013.

Chapter 5
Environmental and Economic Benefits of Reducing Pesticide Use

David Pimentel and Michael Burgess

Contents

5.1	Introduction	128
5.2	Extent of Pesticide Use in the U.S.	128
5.3	Crop Losses and Changes in Agricultural Technologies	129
5.4	Estimated Benefits/Costs From Reduced Pesticide Use	131
5.5	Techniques to Reduce Pesticide Use	132
	5.5.1 Pesticide Application Technologies	132
	5.5.2 Insecticides	133
	5.5.3 Herbicides	134
5.6	Overall Pesticide-Reduction Assessment	135
5.7	Environmental and Public Health Costs of Pesticide Use	135
5.8	Conclusion	136
References		136

Abstract Pesticides cause serious public health problems and considerable damage to agricultural and natural ecosystems. We confirm previous reports that it is feasible to reduce pesticide use by 50% or more. The Swedish Government achieved a 61% reduction in pesticide use and the Indonesian Government achieved a 65% reduction in pesticide use without a reduction in crop yields. In fact in Indonesia the result of the reduction in pesticide use was a 12% increase in rice yield.

Keywords Pesticides · Environment · Agriculture · Foods · Economic benefits

---

D. Pimentel (✉)
Department of Entomology, Department of Ecology and Evolutionary Biology,
Cornell University, Tower Road East, Blue Insectary-Old, Room 165,
Ithaca, New York 14853, USA
e-mail: dp18@cornell.edu

M. Burgess
Department of Entomology, Cornell University, Horticulture, Research Aide,
Greenhouse worker, Tower Road East, Blue Insectary-Old, Room 161,
Ithaca, New York 14853, USA
e-mail: mnb2@cornell.edu

D. Pimentel, R. Peshin (eds.), *Integrated Pest Management*,
DOI 10.1007/978-94-007-7796-5_5,
© Springer Science+Business Media Dordrecht 2014

5.1 Introduction

Several studies suggest that it is technologically feasible and desirable to reduce pesticide use in the U.S. by 50% (OTA 1979; NAS 1989; Pimentel et al. 1991).

Denmark had a major increase in the volume of pesticides used in agriculture in the early 1980s that resulted in a serious decline in farm wildlife. In response to this decline and to protect consumers and farm workers, Denmark developed an action plan in 1985 to reduce pesticide use by 50% within 12 years (PAN-Europe 2005). Sweden also approved a program in 1988 to reduce pesticide use by more than 50%. Actually Sweden reduced pesticide use by 61% from 2001 through 2009 (Ekström and Bergkvist 2008). U.S. farmers in 2007 used an estimated 512 million kg per year at a cost of $ 10 billion per year (Pimentel 2005; Grube et al. 2011). Pesticide benefits are estimated to be about $ 4 for every dollar invested in pesticides (Pimentel 2005). However, these costs do not reflect the public health and environmental costs (See Pimentel and Burgess In Press.). Assessments of the direct and indirect costs of using pesticides in the U.S. are difficult to determine because of the complexity of pest problems.

The objective of this chapter is to estimate the potential agricultural, public health, and environmental benefits of reducing pesticide use in the U.S. by approximately 50% by examining the costs and benefits of current pesticide use patterns on about 40 U.S. major crops.

5.2 Extent of Pesticide Use in the U.S.

Of the estimated 500 million kg of pesticides applied annually in the U.S., about 15–19% are insecticides, 69–74% are herbicides, and 11–12% is fungicides (Benbrook, 2009; ISIS 2010; Grube et al. 2011; Pimentel et al. 1993a). The 500 million kg of pesticides used in U.S. agriculture are applied at an average rate of approximately 3 kg/ha to the 114 million ha. Thus a significant cropland area (38%) receives no pesticides (Pimentel et al.1993b).

The application of pesticides for pest control is not evenly distributed among the crops. Overall, 93% of the hectarage of row crops like corn, soybeans, and cotton is treated with pesticides (Pimentel et al. 1993b).

In contrast, less than 10% of forage crops are treated with pesticides.

Pesticides are applied to about 62% of all US crop acreage and about 93% of all row crop acreage (Muir 2012).

Over 90% of all corn acreage and 98% of soybean acreage is treated with herbicides (USDA 2009). The treated hectarage with insecticides is less than herbicides or between 8% and 57% of the acreage of these two crops is treated with insecticides (Table 5.1).

Of the approximately 314 million kg of the insecticides applied most are applied to corn and soybeans (Table 5.1).

Table 5.1 U.S. hectarage treated with pesticides. (Pimentel et al. 1993b)

Land-Use Categories	Total hectares	All Pesticides		Herbicides		Insecticides		Fungicides	
		Treated hectares	Quantity (x 10^6 kg)	Treated hectares	Quantity (x 10^6 kg)	Treated hectares	Quantity (x 10^6 kg)	Treated hectares	Quantity (x 10^6 kg)
Agricultural	472	114	396	90	69	22	314	4	1
Gov. + Industrial	150	28	50	30	40	–	10	–	–
Forest Lands	290	2	4	2	3	1	1	–	–
Household Lands	4	4	50	3	25	3	24	1	–
Total	916	148	500	125	137	26	349	5	1

Totals for hectarage treated with various pesticide types exceeds the total treated hectares because the same
Land area can be treated with several classes of pesticide chemicals

However, apples and cotton may be treated with insecticides as many as 20 times per year compared with corn and wheat which may be treated only once per year.

Insecticide use varies considerably among geographic regions. Warm regions of the U.S. often suffer severe insect pest problems. For example, although only 13% of the total alfalfa area is treated with insecticides, 18% of alfalfa hectarage in Texas is treated with insecticides (Bade et al. 2002), nearly 90% of the alfalfa hectarage in the Southern Plains has to be treated to control insect pests (Pimentel et al. 1993a). In the Mountain region of the U.S. where large quantities of potatoes are grown, 65% of the potatoes are treated with insecticides, but in the Southeast, where only early potatoes are grown, about 100% of the potato hectarage is treated (USDA 1975; Pimentel et al. 1993b). Cotton insect pests such as the boll weevil are also more of a problem in the Southeastern U.S. than in other cotton growing regions (Ridgway et al. 1983).

Fungicides are primarily used on fruit and vegetable crops. Approximately 77–95% of the grapes and 97% of the potato areas are treated with fungicides (Gianessi and Reigner 2005; USDA 2009), whereas corn is not treated with fungicides and wheat only rarely is treated (USDA 2009).

5.3 Crop Losses and Changes in Agricultural Technologies

Since 1945, the use of synthetic pesticides in the U.S. has grown about 35-fold (Pimentel et al. 1993a; USDA 2009). The increase in pesticide use is largely due to changes in agricultural practices and cosmetic standards (Pimentel et al. 1977). At the same time, some of the newer pesticides have at least a 10-fold greater

Table 5.2 Average annual pest losses in the United States (1904–1989). (Pimentel et al. 1993a)

Date	Percentage of crops lost to pests per year			
	Insects	Diseases	Weeds	Total
Current	13.0	12.0	12.0	37.0
1989	13.0	12.0	12.0	37.0
1974	13.0	12.0	8.0	33.0
1951–1960	12.9	12.2	8.5	33.6
1942–1951	7.1	10.5	13.8	31.4
1910–1935	10.5	NA	NA	NA
1904	9.8	NA	NA	NA

effectiveness than the older pesticides. For example, in 1945 DDT was applied at a rate of approximately 2 kg/ha. Today, similarly effective insect control is achieved with pyrethroids and aldicarb applied at only 0.1 kg/ha and 0/05 kg/ha, respectively.

Currently an estimated 37% of all crop production is lost annually to pests (13% to insects, 12% to weeds, and 12% to plant pathogens) in spite of the use of pesticides and non-chemical controls. Although pesticide use has increased over the past 5 decades, crop losses to pests have not shown a concurrent decline. According to survey data collected from 1942 to present losses from weeds have fluctuated with an overall slight decline, due to improved mechanical, chemical, and cultural weed control practices, from 14 to 12% (Table 5.2). During the same period, U.S. losses from plant pathogens including nematodes, increased slightly from 10.5 to 12%. This increase results in part from reduced sanitation, higher cosmetic standards, and abandonment of crop rotation practices.

The share of crop yields lost to insects and mites has nearly doubled during the past 40 years (Table 5.2), despite more than a 10-fold increase in both the amount and toxicity of synthetic insecticides used. The increase in crop losses due to insects per hectare has been offset by increased crop yields obtained with higher-yielding varieties and greater use of fertilizers and irrigation. Crop losses have increased despite intensified insecticide usage due to several major changes in agricultural practices (Pimentel et al. 1991). These changes include:

1. The planting of some crop varieties more susceptible to insect pests.
2. The destruction of natural enemies by insecticides.
3. Increased insecticide resistance of pests.
4. The increase in crop monocultures and reduced crop diversity.
5. The reduction of Food and Drug Administration tolerances for insects and insect parts in foods.
6. Increased use of aircraft application of pesticides.
7. Reduction in crop sanitation, including infected crop and fruit materials.
8. Reduced tillage and more crop residues left on the soil surface.
9. Planting crops in climatic regions where potential insect pests are more abundant.
10. The use of herbicides that alter the physiology of crop plants, making them more vulnerable to insect attack (Pimentel et al. 1991).

5.4 Estimated Benefits/Costs From Reduced Pesticide Use

A reduction in U.S. pesticide use would require substituting non-chemical alternatives for chemical pest control and improving the efficiency of pesticide application technology. Such changes, if done properly, would improve pest control technology.

Losses from pests for the 40 major crops grown with pesticides have been estimated by examining data on current crop losses, by reviewing loss data based on experimental field tests, and by consulting pest control specialists. Pimentel et al. (1993b) took this collection of current pesticide use and crop loss data and estimated the costs if pesticides were reduced (suggesting how much they could be reduced) and several alternatives were employed. Combining these data, however, has often been difficult. For example, data from published experimental field tests usually emphasize the benefits of pesticide use. Thus loss data associated with pesticide treatments usually emphasize benefits over costs (Pimentel et al. 1978).

In addition, field tests often exaggerate total crop losses because assessments of insect, disease, and weed losses are carried out separately and then combined. For untreated apples, insects are reported to cause a 50–100% crop loss, diseases a 50–60% crop loss, and weeds about a 6% loss (Ahrens and Cramer 1986; Pimentel et al. 1991). This approach yields an estimated total loss of approximately 140% from all pests combined! A more accurate estimate of the losses in the absence of pesticides ranges from 80 to 90% based on current cosmetic standards (Ahrens and Cramer 1986). While Ahrens and Cramer (1986) cited crippling losses in the 1980s, a trip to the supermarket when apples are in season will reveal that organic apples and other fruit are being successfully produced in various locations, organic apples are being grown successfully in eastern Washington state and other countries certified by the USDA (Zerbe 2009). Jim Travis, professor emeritus of plant pathology has been quoted concerning organic apple production in the northeastern U.S., "We live in a lush environment with beneficial insects and organisms that could help us grow organic apples here even better. Someday, it may actually shift, and the East Coast may be the best place for organic [apples]" (Zerbe 2009). Increasing interest in organic fruit production by fruit growers even in the northeastern U.S., which has many apple arthropod and disease pests, has resulted in experts offering advice to produce apples profitably using organic systems as long as cosmetic standards concessions are made (Peck et al. 2009).

Exactly how much overlap exists among insect, disease, and weed loss figures for apples and other crops is not known.

Our analysis has other important limitations (See Pimentel et al. 1993b). The figures for current crop losses to pests, despite heavy pesticide use, are based on U.S. Department of Agriculture data and other estimates obtained from pesticide specialists. We emphasize that these are estimates. For certain crops, little or no experimental data are available concerning yields with pesticide use and various alternative pest control systems (Pimentel et al. 1993b). In addition, for some, recent crop data is not available. With these crops, our data were generally extrapolated from available data on closely related crops. Although we recognize the limitations of the data used in this analysis, we believe in the need to assemble available information to provide a first approximation of the potential for reducing pesticide use by one half. We hope that better data will be available in the future so a more complete analysis of pesticide costs and benefits can be made.

Reduction of the risks associated with pesticides is in itself a complicated issue, particularly because of environmental and health trade-offs; they could not be included in the analysis (See Pimentel et al. 1993b). One example, however, includes the conflict between reducing pesticide use and promoting soil conservation through the use of no-till culture. No-till and reduced-till culture can greatly reduce soil erosion, but these practices also significantly increase the need for herbicides, insecticides, and fungicides (Taylor et al. 1984; Pimentel et al. 1991).

However, although reducing pesticide use may require reducing the use of some no-till systems, highly cost-effective soil conservation alternatives to no-till are available. These include ridge-till, crop rotations, strip cropping, contour planting, terracing, windbreaks, mulches, cover crops, and green mulches (Moldenhauer and Hudson 1988). Ridge till can be employed without the use of herbicides, and it controls soil erosion more effectively than no-till (Russnogle and Smith 1988).

5.5 Techniques to Reduce Pesticide Use

The increase in crop losses associated with the recent changes in agricultural practices suggests that some alternative practices exist that might reduce pesticide use. Two important agricultural practices that apply to all agricultural crops include the widespread use of monitoring (scouting) and improved pesticide application equipment. Currently a significant number of pesticide treatments are applied unnecessarily and at improper times due to a lack of treat-when-necessary programs which scouting by either a professional scout or the farmer (with some training) would remedy. Furthermore, the mode of application can result in much pesticide being unnecessarily lost (i.e., 75% is lost during aerial application while only 25% reaches the target area). By increasing monitoring and improving application equipment, more efficient pest control can be achieved.

5.5.1 *Pesticide Application Technologies*

The amount of pesticides reaching target areas could be increased by changing the type of application equipment employed, especially reducing the use of aircraft ultra-low volume application equipment, which wastes about 75% of the pesticides applied (Pimentel et al. 1991). The amount of pesticide waste could be reduced by 25% if ground application instead of air application were used (Mazariegos 1985; Pimentel and Levitan 1986; Pimentel et al. 1991). In addition, covering the spray boom with a plastic shroud can further reduce drift 85% (Ford 1986), thereby allowing for an additional reduction in pesticide use (Pimentel et al. 1991).

5.5.2 Insecticides

Corn and soybeans account for more than 92% of the total insecticide use in agriculture (Pimentel 2012, unpublished). Thus reducing insecticide use in these two crops by substituting non-chemical alternatives would contribute significantly to a reduction in insecticide use.

5.5.2.1 Corn

During the early 1940s, little or no insecticide was applied to corn and the losses to insects were only 3.5% (USDA 1954). Since then, insecticide use on corn has increased more than 1000-fold, whereas losses due to insects have increased more than 3.4-fold (Ridgeway 1980) which is primarily due to the abandonment of crop rotations (Pimentel et al. 1991). Today more than 50% of the corn is grown in continuous monoculture with 2.2 million kg of insecticide applied annually (USDA 2009). By reinstituting crop rotation, nearly 95% of the insecticide use in corn production could be eliminated. Rotating corn with soybeans or similar high value crops will increase yields and net profits (Helmers et al. 1986; Pimentel et al. 1991). From a more comprehensive perspective, rotating corn with other crops also has other advantages, including reducing weed and plant pathogen losses and decreasing soil erosion and rapid water runoff problems (Helmers et al. 1986; Pimentel et al. 1991).

Combining crop rotations with corn varieties resistant to the corn borer and chinch bug could reduce insecticide use on corn by 80% while concurrently reducing corn losses due to insects (Schalk and Ratcliff 1977; Pimentel et al. 1991). Such a move might increase the cost of corn production only an estimated $ 10 per hectare above current costs of corn grown continuously (Pimentel et al. 1991). Using an attractant combined with insecticides has been reported to reduce insecticide use by 99% (Paul 1989; Pimentel et al. 1991).

5.5.2.2 Soybean

According to the USDA (2011) in 2005, 1.04×10^6 kg of insecticide active ingredient was used on soybean hectarage in the United States for the states that reported data. For the states reporting, the soybean crop area to which insecticides were applied in 2005 ranged from a low of 2% in Kentucky to a high of 44% for Louisiana with the average per state being 18.6% (USDA 2011).

5.5.2.3 Cotton

The potential for reducing insecticide us in U.S. cotton is well illustrated in the following in Texas: Since 1966, insecticide use in Texas cotton has been reduced by almost 90% (OTA 1979). The technologies adopted to reduce insecticide use were: monitoring pest and natural enemy populations to determine when to treat, biological control, host-plant resistance, stalk destruction (sanitation), uniform planting date, water management, fertilizer management, rotations, clean seeds, and altered tillage practices (OTA 1979; King et al. 1986; Pimentel et al. 1991).

Currently, a total of about 6.7 million kg of insecticides are applied to cotton, and it is estimated that that this amount could be reduced by approximately 40% through the use of the aforementioned technologies (Pimentel et al. 1991; USDA 2009).

5.5.3 Herbicides

Corn and soybeans account for about 71% of the total herbicides applied in U.S. agriculture (Osteen and Livingston 2006; USDA 2009). I use these crops to illustrate the potential for decreasing herbicide use.

5.5.3.1 Corn

About (31%) of the herbicides used on crops are applied on corn (Osteen and Livingston 2006; USDA, 2009; Grube et al. 2011). More than 3 kg of herbicides are applied per hectare of corn, and more than 96% of the corn hectarage is treated (USDA 2009). Schweizer et al. (1988) found that a reduced or moderate level of herbicides can control weeds in irrigated corn and still control weed seed reserves. Hanna et al. (1996) have found that despite herbicide treatment of 95% of corn hectarage, over 70% of corn hectares are also cultivated to control weeds. As of 1993 91% of the corn land is also cultivated to help control weeds (Pimentel and Lehman 1993).

The average costs and returns per hectare for no-till, reduced-till, and conventional-till culture have actually been found to be quite similar (Duffy and Hanthorn 1984). For example, added labor, fuel, and machinery costs for conventional-till practices for corn were approximately $ 24/ha higher than those for no-till. However, the costs for the added fertilizers, pesticides, and seeds for the no-till system were $ 22/ha higher than conventional-till (Duffy and Hanthorn 1984).

It might be possible to reduce herbicide use on corn by about 60% if the use of mechanical cultivation and rotations were increased (Forcella and Lindstrom 1988). Corn and soybean rotations have been found to provide substantially higher returns than either crop grown alone and continuously (Helmers et al. 1986).

5.5.3.2 Soybeans

The second-largest amount of herbicides is applied to soybeans, with approximately 96% of soybean hectarage receiving 47 million kg herbicide treatments for weed control, more than half the amount used on corn (USDA 2009). About 96% of the hectarage also receives some tillage and mechanical cultivation for weed control (Duffy 1983). Several techniques have been developed that increase the efficiency of chemical applications. The rope-wick applicator has been used in soybeans to reduce herbicide use approximately 90%, and this applicator was found to increase soybean yields 51% over conventional herbicide treatments (Dale 1980). Also, a new model of recirculating sprayer saves 70–90% of the spray emitted that is not trapped by the weeds themselves (Matthews 1985). Spot treatments are a third method of reducing unnecessary pesticide treatments.

In addition, alternative techniques are available to reduce the need for herbicides in soybeans. These include ridge-till tillage, mechanical cultivation, row spacing, planting dates, weed-tolerant varieties, crop rotations, spot treatments, and reduced dosages (Russnogle and Smith 1988). Employing several of these techniques in combination might reduce herbicide use in soybeans by about 60% (D. Pimentel, per. comm. 2012).

5.6 Overall Pesticide-Reduction Assessment

By substituting non-chemical alternatives for some insecticides and herbicides used on 40 major crops, I estimate that total agricultural pesticide use can be reduced by approximately 50% (note, fungicides are a small percentage of the pesticides used). The added costs for implementing these alternatives are estimated to be approximately $ 500 million per year. These alternatives would increase total pest-control costs approximately 15% and would increase total food production costs at the farm only 0.3%.

5.7 Environmental and Public Health Costs of Pesticide Use

The public now pays a high price for its use of pesticides. Pesticide-control measures now cost approximately $ 4 billion annually, not including the indirect environmental and public health costs, which total more than $ 2.2 billion annually (Pimentel and Burgess In Press). Perhaps the most serious social and environmental costs related to pesticide use are the human pesticide poisonings. Annually in the US approximately 20,000 accidental poisonings occur, mostly from agricultural pesticides, with 2,000 cases requiring hospitalization. These pesticide poisonings result in approximately 50 fatalities per year. Pesticides are also implicated in numerous human diseases, including cancer and sterility. An estimated 6,000 cases of pesticide induced cancer occur each year (EPA 1987).

5.8 Conclusion

Pesticides cause serious public health problems and considerable damage to agricultural and natural ecosystems. This article confirms previous reports that it is feasible to reduce pesticide use by 50% or more at a cost of $ 500 million per year (Pimentel and Cilveti 2003). Such a finding supports the estimates of the Office of Technology Assessment (OTA 1979) and the National Academy of Sciences (1989) as well as the policies adopted by Swedish Government and the Indonesian Government. The Ekström and Bergkvist (2008) report that, focusing the reduction of pesticide use on crops receiving the heaviest pesticide treatments, Sweden achieved a 61% reduction in pesticide use. The Indonesian Government has achieved a 65% reduction in pesticide use (Oka 1991; Resosudarmo 2001). In both cases the results have proven beneficial to pest control, the environment, and public health.

Acknowledgement We wish to express our sincere gratitude to the Cornell Association of Professors Emeriti for the partial support of our research through the Albert Podell Grant Program.

References

Ahrens, C. H. H., & Cramer, H. H. (1986). Improvement of agricultural production by pesticides. In F. P. W. Winteringham. (Ed.)., *Environment and Chemicals in Agriculture* (pp. 151–162). New York: Elsevier.

Bade, D., Bean, B., Black, M., Downing, S., Grichar, J., Muegge, M., Patrick, C., & Stichler, C. (2002). *Crop Profile for Alfalfa in Texas*. National Information System for the Regional IPM Centers, National Institute of Food and Agriculture, US Department of Agriculture. http://www.ipmcenters.org/cropprofiles/docs/txalfalfa.pdf. Accessed 20 Sept 2012.

Benbrook, C. (2009). Impacts of *Genetically Engineered Crops on Pesticide Use in the United States: The First Thirteen Years*. The Organic Center. http://www.organic-center.org/reportfiles/GE13YearsReport.pdf. Accessed 27 June 2012.

Dale, J. E. (1980). Roppe wick applicator—tool with a future. *Weeds Today, 11*, 3–4.

Duffy, M. (1983). Pesticide use and practices, 1982. Economic Research Service., *Agricultural Information Bulletin*, No. 462. USDA, Washington, D.C. http://naldc.nal.usda.gov/download/CAT87212367/PDF. Accessed 28 June 2012.

Duffy, M., & Hanthorn, M. (1984). *Returns to Corn and Soybean Tillage Practices*. Economic Research Service, Agricultural Economic Research Service. Agricultural Economic Report No. 508. USDA. Washington, D.C. http://naldc.nal.usda.gov/download/CAT87202233/PDF. Accessed 28 June 2012.

EPA. (1987). *Unfinished Business: A Comparative Assessment of Environmental Problems*. Appendix #1. Report of the Cancer Risk Work Group. Washington, D.C.: United States Environmental Protection Agency. http://nepis.epa.gov/Exe/ZyNET.exe/2000BZS0.txt?ZyActionD=ZyDocument & Client=EPA&Index=1986%20Thru%201990&Docs=&Query=FNAME%3D2000BZS0.TXT%20or%20%28%20unfinished%20or%20business%20or%20a%20or%20comparative%20or%20assessment%20or%20environmental%20or%20problems%20or%20appendix%20or%20i%29&Time=&EndTime=&SearchMethod=1&TocRestrict=n&Toc=&TocEntry=&QField=&QFieldYear=&QFieldMonth=&QFieldDay=&UseQField=&IntQFieldOp=1&ExtQFieldOp=1&XmlQuery=&File=D%3A%5CZYFILES%5CINDEX%20DATA%5C86THRU90%5CTXT%5C00000001%5C2000BZS0.txt&User=ANONYMOUS&Password=anonymous&SortMethod=h%7C-&MaximumDocuments=10&FuzzyDegree=0&ImageQuality=r7

5g8/r75g8/x150y150g16/i425&Display=p%7Cf&DefSeekPage=x&SearchBack=ZyActionL& Back=ZyActionS&BackDesc=Results%20page&MaximumPages=1&ZyEntry=3. Accessed 27 June 2012.

Ekström, G., & Bergkvist, P. (2008). Swedish pesticide reduction: 1986–2006. In D. Pimentel (Ed.)., *Encyclopedia of Pest Management*. New York: Taylor and Francis. http://dx.doi.org/10.1081/E-EPM-120044723. Accessed 23 Feb 2014.

Forcella, F., & Lindstrom, M. J. (1988). Movement and germination of weed seeds in ridgetill crop production systems. *Weed Science, 36*, 56–59.

Ford, R. J. (1986). Field trials of a method for reducing drift from agricultural sprayers. *Canadian Agricultural Engineering, 28*, 81–83.

Gianessi, L. P., & Reigner, N. (2005). *The Value of Fungicides in U.S. Crop Production*, CropLife International, September, 2005. http://www.croplifefoundation.org/upload/137%20CropLife%20Foundation%20Fungicide%20Benefits.pdf. Accessed 20 Sept 2012.

Grube, A., Donaldson, D., Kiely, T., & Wu, L. (2011). *Pesticide Industry Sales and Usage: 2006 and 2007 Market Estimates.* Biological and Economic Analysis Division, Office of Pesticide Programs, Office of Chemical Safety and Pollution Prevention. Washington, D.C.: US Environmental Protection Agency. http://epa.gov/oppfead1/cb/csb_page/updates/2011/sales-usage06-07.html. Accessed 19 Sept 2012.

Hanna, M., Hartzler, R., Erbach, D., Paarlberg, K., & Miller, L. (1996). *Cultivation: An Effective Weed Management Tool.* Iowa State University, University Extension. Pm-1623. http://www.extension.iastate.edu/Publications/PM1623.pdf. Accessed 27 June 2012.

Helmers, G. A., Langemeit, M. R., & Atwood, J. (1986). An economic analysis of alternative cropping systems for east-central Nebraska. *American Journal of Alternative Agriculture, 4*, 153–158.

ISIS. (2010). *GM Crops Increase Herbicide Use in the United States*. Institute of Science in Society (ISIS). ISIS Report. 18/01/10. http://www.i-sis.org.uk/GMcropsIncreasedHerbicide.php. Accessed 28 June 2012.

King, E. G., Phillips, J. R., & Head, R. B. (1986). Thirty-ninth annual conference report on cotton insect research and control. In *Proceedings of the Beltwide Cotton Production Research Conference* (pp. 126–135). Memphis, Tennessee, USA: National Cotton Council.

Matthews, G. A. (1985). Application from the ground. In P. T. Haskell (Ed.)., *Pesticide Application: Principles and Practice* (pp. 95–117). Oxford: Claredon Press.

Mazariegos, F. (1985). The use of pesticides in the cultivation of cotton in Central America. *Industry and Environment: A Quarterly Review, 8*, 5–8.

Moldenhauer, W. C., & Hudson, N. W. (Eds.)., (1988). *Conservation Farming on Steep Lands.* Ankeny, Iowa, USA: Soil and Water Conservation Society, World Association of Soil and Water Conservation. http://pdf.usaid.gov/pdf_docs/PNABB290.pdf. Accessed 28 June 2012.

Muir P. (2012). *Pesticide Use in the US.* Oregon State University. http://people.oregonstate.edu/~muirp/uspestic.htm. Accessed 3 Oct 2013.

NAS. (1989). *Alternative Agriculture.* Committee on the role of alternative farming methods in modern agriculture. Board on Agriculture, National Research Council, National Academy of Sciences (NAS). Washington, D.C.: National Academy of Sciences Press. http://www.nap.edu/openbook.php?isbn=0309039851. Accessed 28 June 2012.

Office of Technology Assessment (OTA). (1979). *Pest Management Strategies* (Vol. II). *Working Papers*. Washington, D.C.: Office of Technology Assessment.

Oka, I. N. (1991). Success and challenges of the Indonesian national integrated pest management program in the rice based cropping system. *Crop Protection, 10*, 163–165.

Osteen, C., & Livingston, M. (2006). Chapter 4.3: Pest management practices. In K. Wiebe & N. Gollehon (Eds.)., *Agricultural Resources and Economic Indicators (AREI)*, 2006 Edn. Economic Research Service. US Department of Agriculture, *Economic Information Bulletin*, No. EIB-16. http://www.ers.usda.gov/publications/arei/eib16/. Accessed 28 June 2012.

PAN-Europe. (2005). *Danish Pesticide Reduction Programme—to Benefit the Environment and the Health*. Pesticide Action Network—Europe. http://www.pan-europe.info/Resources/Reports/Danish_Pesticide_Use_Reduction_Programme.pdf. Accessed 19 Sept 2012.

Paul, J. (1989). Getting tricky with rootworms. *Agricultural Age, 33*, 6–7.

Peck, G. M., Merwin, I. A., Agnello, A., Caldwell, B., Curtis, P., Gardner, R., Helms, M., Rosenberger, D., Thomas, E., & Watkins, C. (2009). *A Grower's Guide to Organic Apples*. NYS IPM Publication No. 223. Cornell University Cooperative Extension, NYS Integrated Pest Management, New York State Department of Agriculture and Markets. http://nysipm.cornell.edu/organic_guide/apples.pdf. Accessed 7 Feb 2013.

Pimentel, D. (2005). Environmental and economic costs of the application of pesticides primarily in the United States. *Environment, Development and Sustainability, 7*, 229–252.

Pimentel, D., & Levitan, L. (1986). Pesticides: Amounts applied and amounts reaching pests. *Bioscience, 36*, 86–91.

Pimentel, D., & Lehman, H. (Eds.)., (1993). *The Pesticide Question: Environment, Economics, and Ethics*. New York: Chapman and Hall.

Pimentel, D., & Cilveti, M. V. (2003). Reducing pesticide use, successes in. In D. Pimentel (Ed.)., *Encyclopedia of Pest Management* (Vol. 2, pp. 551–553). Boca Raton, Florida, USA: CRC Press (Taylor & Francis Group).

Pimentel, D., & Burgess, M. (2014). Environmental and economic costs of the application of pesticides primarily in the United States. In D. Pimentel & R. Peshin (Eds.)., *Integrated Pest Management: Pesticide Problems* (Vol. 3, pp. 47–71). Dordrecht Netherlands: Springer.

Pimentel, D., Terhune, E., Dritschilo, W., Gallahan, D., Kinner, N., Nafus, D., Peterson, R., Zareh, N., Misiti, J., & Haber-Schaim, O. (1977). Pesticides, insects in foods, and cosmetic standards. *Bioscience, 27*, 178–185.

Pimentel, D., Shoemaker, C., Whitman, R. J., Bellotti, A., Beyer, N., Brick, A., Brodel, C., Caunter, I., Cornell, H., Dritschilo, W., Gunnison, D., Habte, M., Hurd, L., Johnson, P., Krummel, J., Liebherr, J., Loye, M., Mackenzie, D., Nafus, D., Oka, I., Rao, R., Saari, D., Smith, J., Stack, R., Udovic, D., Yip, C., & Zareh, N. (1978). Systems management program for corn pest control in New York State. *Search Cornell (University), 8*(1), 1–16.

Pimentel, D., McLaughlin, L., Zepp, A., Lakitan, B., Kraus, T., Kleinman, P., Vancini, F., Roach, W. J., Graap, E., Keeton, W. S., & Selig, G. (1991). Environmental and economic effects of reducing pesticide use. *Bioscience, 41*, 402–409.

Pimentel, D., McLaughlin, L., Zepp, A., Lakitan, B., Kraus, T., Kleinman, P., Vancini, F., Roach, W. J., Graap, E., Keeton, W. S., & Selig, G. (1993a). Environmental and economic effects of reducing pesticide use in agriculture. *Agriculture, Ecosystems and Environment, 46*, 273–288.

Pimentel, D., McLaughlin, L., Zepp, A., Lakitan, B., Kraus, T., Kleinman, P., Vancini, F., Roach, W. J., Graap, E., Keeton, W. S., & Selig, G. (1993b). Environmental and economic impacts of reducing U.S. agricultural pesticide use. In D. Pimentel & H. Lehman (Eds.)., *The Pesticide Question: Environment, Economics, and Ethics* (pp. 223–278). New York: Chapman and Hall.

Resosudarmo, B. P. (2001). The economy-wide impact of integrated food crop pest management in Indonesia. Economy and environment program for Southeast Asia, Singapore. *Research Report 2001–RR11*.

Ridgeway, R. (1980). Assessing agricultural crop losses caused by insects. In *Assessment of Losses which Constrain Production and Crop Improvement in Agriculture and Forestry* (pp. 229–233). Proceedings of the E.C. Stakman Commemorative Symposium. Minneapolis, Minnesota, USA: University of Minnesota.

Ridgway, R. L., Lloyd, E. P., & Cross, W. H. (1983). Cotton insect management with special reference to the boll weevil. Agricultural Handbook No. 589. US Department of Agriculture. Washington, D.C.: Agricultural Research Service.

Russnogle, J., & Smith, D. (1988). More dead weeds for your dollar. *Farm Journal, 112*(2), 9–11.

Schweizer, E. E., Lybecker, D. W., & Zimdahl, R. L. (1988). Systems approach to weed management in irrigated crops. *Weed Science, 36*, 840–845.

Schalk, J. M., & Ratcliffe, R. H. (1977). Evaluation of the United States Department of Agriculture program of alternative methods of insect control: Host plant resistance to insects. *FAO Plant Protection Bulletin, 25*, 9–14.

Taylor, F., Raghaven, G. S. V., Negi, S. C., McKyes, E., Viger, B., & Watson, A. K. (1984). Corn grown in a Ste. Rosalie clay under zero and traditional tillage. *Canadian Agricultural Engineering, 26*, 91–95.

USDA. (1954). *Losses in Agriculture*. US Department of Agriculture, Agricultural Research Service. ARS-20–1.
USDA. (1975). *Farmer's Use of Pesticides in 1971- Extent of Crop Use*. US Department of Agriculture. Washington, D.C.: Economic Research Service (Agricultural Economic Report No. 268).
USDA. (2009). *Agricultural Statistics 2009*. Washington, D.C.: US Government Printing Office.
USDA. (2011). *Agricultural Statistics 2011*. Washington, D.C.: US Government Printing Office.
Zerbe, L. (2009). *Grow Organic Apples at Home*. Rodale: Where Health Meets Life. http://www.rodale.com/growing-organic-apples?page=0,1. Accessed 7 Feb 2013.

Chapter 6
An Environmental, Energetic and Economic Comparison of Organic and Conventional Farming Systems

David Pimentel and Michael Burgess

Contents

6.1	Introduction	142
6.2	Methods and Materials	143
	6.2.1 Conventional (Synthetic Fertilizer and Herbicide-Based)	143
	6.2.2 Organic, Animal Manure and Legume-Based	145
6.3	Measurements Recorded in the Experimental Treatments	146
	6.3.1 Data Collection	146
	6.3.2 Analytical Methods	147
6.4	Results	148
	6.4.1 Crop Yields under Normal Rainfall	148
	6.4.2 Energy Inputs	150
	6.4.3 Economics	151
	6.4.4 Soil Carbon	152
	6.4.5 Soil Nitrogen	153
	6.4.6 Nitrate Leaching	153
	6.4.7 Herbicide Leaching	155
6.5	Discussion	157
	6.5.1 Soil Organic Matter and Biodiversity	157
	6.5.2 Oil and Natural Gas Inputs	158
	6.5.3 Crop Yields and Economics	159
	6.5.4 Challenges for Organic Agriculture	160
	6.5.5 Policy Needs	161
6.6	Conclusion	161
References		162

D. Pimentel (✉)
Department of Entomology, Department of Ecology and Evolutionary Biology,
Cornell University, Tower Road East, Blue Insectary-Old, Room 165,
Ithaca, New York 14853, USA
e-mail: dp18@cornell.edu

M. Burgess
Department of Entomology, Cornell University, Horticulture, Research Aide,
Greenhouse worker, Tower Road East, Blue Insectary-Old, Room 161,
Ithaca, New York 14853, USA
e-mail: mnb2@cornell.edu

D. Pimentel, R. Peshin (eds.), *Integrated Pest Management*,
DOI 10.1007/978-94-007-7796-5_6,
© Springer Science+Business Media Dordrecht 2014

Abstract Various organic technologies have been utilized for about 6,000 years to make agriculture sustainable while at the same time conserving soil, water, energy and biological resources. Benefits of organic technologies include higher soil organic matter and nitrogen, lower fossil energy inputs, yields similar to conventional systems, and conservation of soil moisture and water resources, especially advantageous under drought conditions. Traditional organic farming technologies may be adopted by conventional agriculture to make it more sustainable and ecologically sound.

Keywords Cover crops · Soybeans · Corn · Soil organic matter

6.1 Introduction

Heavy agricultural reliance on synthetic-chemical fertilizers and pesticides is having serious impacts on public health and the environment (Colburn et al. 1997). The estimated environmental and health costs of the recommended use of pesticides costs the nation about $ 10 billion per year (Pimentel 2005). In the United States over 90% of corn farmers rely on herbicides for weed control (Pimentel et al. 1993). Atrazine, one of the most widely used herbicides on corn, is also one of the most commonly found pesticides in streams and groundwater (USGS 2001). The allowable atrazine level in municipal water systems is 3 ppb and this is 30 times the biological threshold level that Hayes et al. (2002) have demonstrated alters developmental processes in frogs.

Fertilizer and animal manure-nutrient losses have been associated with deterioration of some large fisheries in North America (Frankenberger and Turco 2003). Doughty (2003) relates the runoff of soil and nitrogen fertilizer from US Corn Belt corn production to the anaerobic "dead zone" that has developed in the Gulf of Mexico. The National Academy of Sciences (NAS 2003) reports that excessive fertilizer use is responsible for $ 2.5 billion in annual losses in agricultural inputs. Modern agricultural practices are responsible for increased likelihood of soil erosion. The estimate of public and environment health costs related to soil erosion exceed $ 45 billion yearly (Pimentel et al. 1995).

Integrated pest and nutrient management systems and certified organic agriculture can reduce reliance on agrichemical inputs as well as make agriculture environmentally and economically sound. Pimentel and Pimentel (1996) and the National Academy of Sciences (NAS 2003) have demonstrated that sound management practices can reduce pesticide inputs while maintaining high crop yields and improving farm profitability. Some government programs in Sweden, Ontario, and Indonesia have demonstrated that pesticide use can be reduced 50–65% without sacrificing high crop yields and quality (NAS 2003; Surgeoner and Roberts 1993).

Organic agriculture seeks to augment ecological processes that foster plant nutrition while conserving soil and water resources. Organic systems eliminate agrichemicals and reduce other external inputs to improve the environment as well as farm profitability. The National Organic Standards Program (USDA-AMS 2002) codifies organic production methods that are based on certified practices verified by independent third party reviewers. These systems give consumers assurance of how their food is produced and for the first time give them the ability to select foods based on food production methods. The National Organic Standards Program prohibits the use of synthetic chemicals, genetically modified organisms, and sewage sludge in organically certified production.

While starting from a small base, organic agriculture is now the fastest growing agricultural sector in the U.S. Dimitri and Greene (2002) report a doubling of hectar-age in organic production (from cropland and pasture) from 1992 to 1997 to more than 500,000 ha and increasing to 1.95 million ha in 2008 (ERS 2012). Organic food sales totaled $ 29 billion in 2010 and while the overall U.S. food sales grew by less than 1 % in 2010, organic food sales grew by 7.7 % (Willer and Kilcher 2012). With continuing consumer concerns about the environment and the chemicals used in food production, and the growing availability of certified organic production, the outlook for the continued growth of organic production is bright (Dimitri and Greene 2002).

Since 1981, the Rodale Institute Farming Systems Trial® has compared organic and conventional grain-based farming systems. This is a 22-year update of these farming systems based on environmental impacts, economic feasibility, energetic efficiency, soil quality, and other performance criteria. The information from these trials can be a tool for developing agricultural policies more in tune with the environment, while increasing economic returns to producers and increasing energy efficiency.

6.2 Methods and Materials

From 1981 through 2002, field investigations were conducted at The Rodale Institute Farming Systems Trial® in Kutztown, Pennsylvania on 6.1 ha. The soil is a Comly silt loam, which is moderately well drained. The land slopes ranged between 1 and 5 %. The growing season has 180 frost-free days, average temperature is 12.4 °C and average rainfall is 1,105 mm per year.

The main plots were 18 × 92 m, and these were split into three 6 × 92 m subplots, which allowed for the same crop comparisons in any 1 year. The main plots were separated with a 1.5 m grass strip to minimize cross movement of soil, fertilizers and pesticides. The subplots were large enough so that farm-scale equipment could be used in harvesting the crops.

The experimental design included three cropping systems (main plots) each replicated 8 times (see Figs. 6.1a and 6.1b):

6.2.1 Conventional (Synthetic Fertilizer and Herbicide-Based)

This system represented a typical cash grain, row-crop farming unit and used a simple 5-year crop rotation (See Figs. 6.1a and 6.1b) of corn, corn, soybeans,

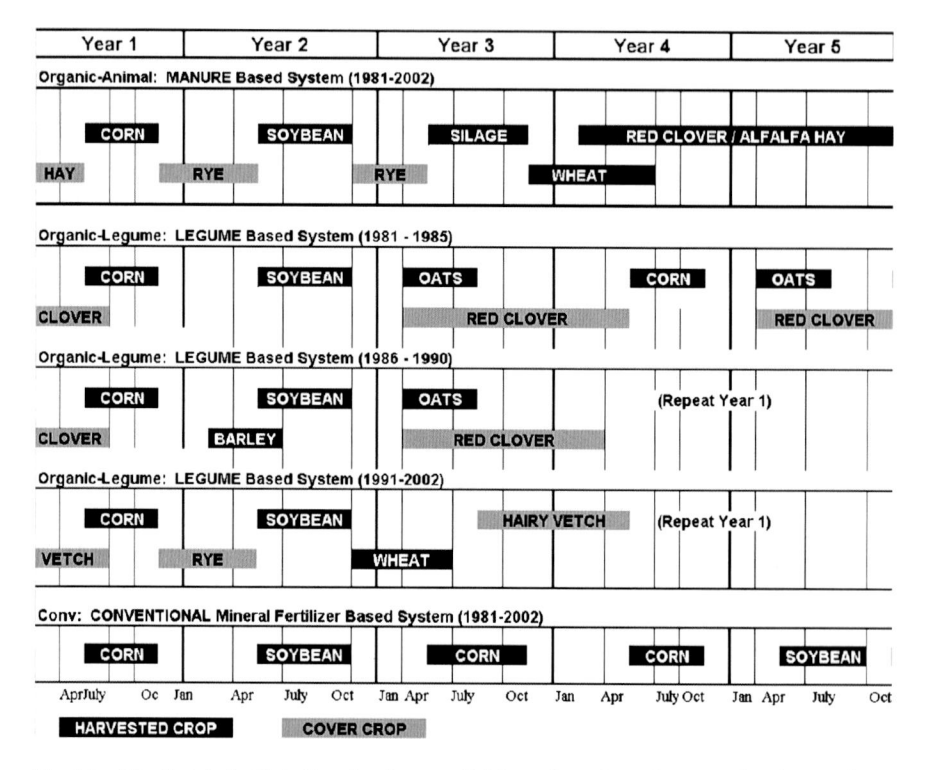

Fig. 6.1a The Rodale Institute Farming Systems Trial rotations. In each system the nitrogen input is added for the corn crop: Steer manure and legume plow-down in the organic-animal system; legume plow-down (red clover or hairy vetch) in the organic-legume system and mineral fertilizer in the conventional system. The rye cover crop was added as a catch crop to the animal system in 1992 and to the legume system in 1993

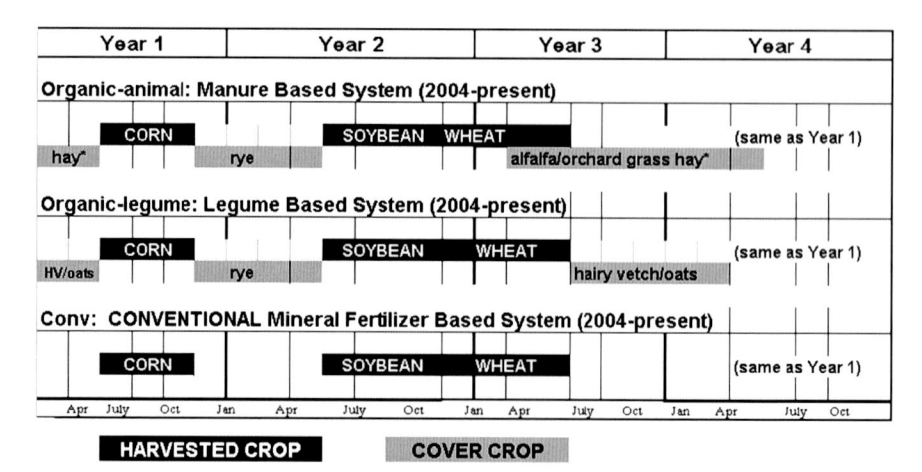

Fig. 6.1b The Rodale Institute Farming Systems Trial rotations. Each system has the same cash crops (corn, soybeans, wheat). In the two organic systems, nitrogen is only added for the corn crop: dairy manure-leaf compost and alfalfa-orchard grass plow-down in the organic-animal system; hairy vetch-oats plow-down in the organic-legume system. The conventional system receives mineral nitrogen fertilizer for both the corn and wheat crop

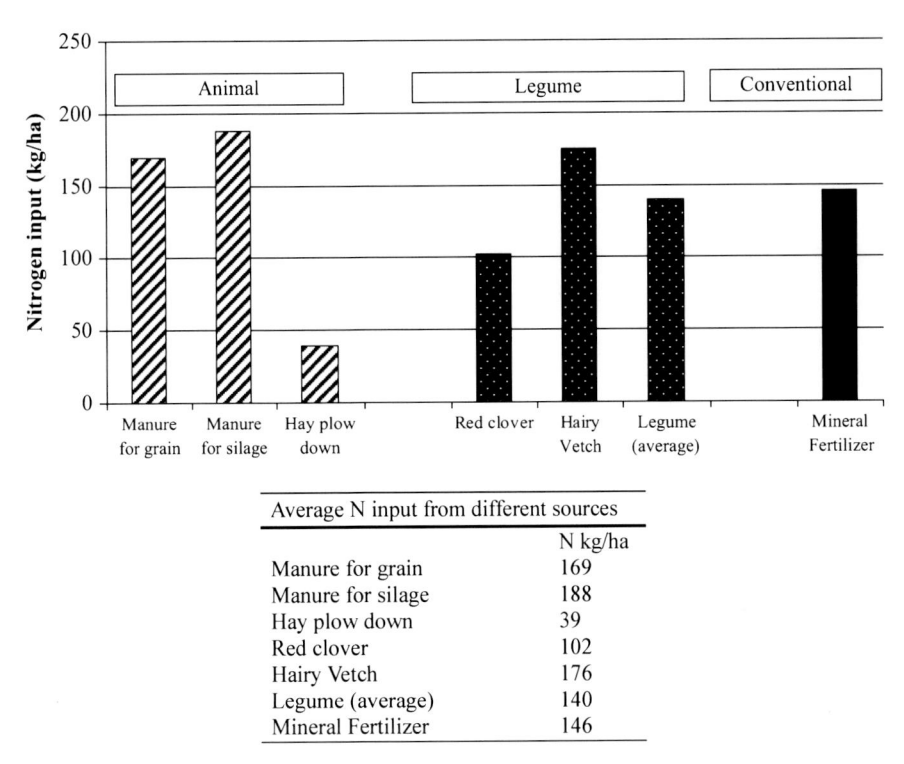

Average N input from different sources	
	N kg/ha
Manure for grain	169
Manure for silage	188
Hay plow down	39
Red clover	102
Hairy Vetch	176
Legume (average)	140
Mineral Fertilizer	146

Fig. 6.2 Average nitrogen inputs from different sources (mean values throughout the years, depending on the rotation). The Rodale Institute Farming Systems Trial 1981–2002 (ANIMAL = organic animal; LEGUME = organic legume)

corn, and soybeans, reflective of commercial conventional operations in the region and throughout the Midwest (over 40 million ha are in this production system in North America). Fertilizer and herbicide applications for corn and soybeans followed Pennsylvania State University Cooperative Extension recommendations (see Fig. 6.2). Crop residues were left on the surface of the land to conserve soil and water resources. The conventional system had no more exposed soil than in either the organic-animal or the organic-legume based systems during the growing season. However, it did not have cover crops during the non-growing season.

6.2.2 Organic, Animal Manure and Legume-Based

6.2.2.1 Organic, Animal Manure

This system represented a typical livestock operation in which grain crops were grown for animal feed, not cash sale. This Mid-Atlantic grain-rotation system included corn, soybeans, corn silage, wheat and red-clover-alfalfa hay plus a rye cover crop before corn silage and soybeans. This rotation (see Figs. 6.1a and 6.1b) was more complex than the rotation used in the conventional system.

Aged cattle manure served as the nitrogen source (see Fig. 6.2) and was applied at a rate of 5.6 t/ha (dry), 2 years out of every 5, immediately before plowing the soil for corn. Additional nitrogen was supplied by the plow-down of legume-hay crops. The system used no herbicides, relying instead on mechanical cultivation, weed-suppressing crop rotations, and relay cropping, in which one crop acted as a living mulch for another, for weed control.

6.2.2.2 Organic, Legume-Based

This system represented a cash grain operation, without livestock. Like the conventional system, it produced a cash grain crop every year, but used no commercial synthetic fertilizers, relying instead on nitrogen-fixing green manure crops as the primary source of nitrogen.

The final rotation included hairy vetch (winter cover crop), corn, rye (winter cover crop), soybeans, and winter wheat (see Figs. 6.1a and 6.1b). The hairy vetch winter cover crop was incorporated before corn planting as a green manure. The initial 5-year crop rotation (see Figs. 6.1a and 6.1b) in the legume-based system was modified twice to improve the rotation. Both organic systems (animal- and legume-based) included a small grain, such as wheat, grown alone or inter-seeded with a legume. Weed control practices were similar in both organic systems with no herbicide applied in either organic system.

6.3 Measurements Recorded in the Experimental Treatments

6.3.1 *Data Collection*

Cover crop biomass, crop biomass, weed biomass, grain yields, nitrate leaching, herbicide leaching, percolated water volumes, soil carbon, soil nitrogen, as well as soil water content were measured in all systems. In addition, seasonal total rainfall, energy inputs and returns, and economic inputs and returns were determined.

Plant biomass was determined by taking two to five 0.5 m^2 cuts in each plot. Corn grain yields were assayed by mechanically harvesting the center four rows of each plot. Soybean and wheat yields were obtained by mechanically harvesting a 2.4 m swath in the center of each plot.

A 76 cm long by 76 cm d steel cylinder (lysimeter) was installed in the fall of 1990 in four of the eight replications in each cropping system to enable the collec-

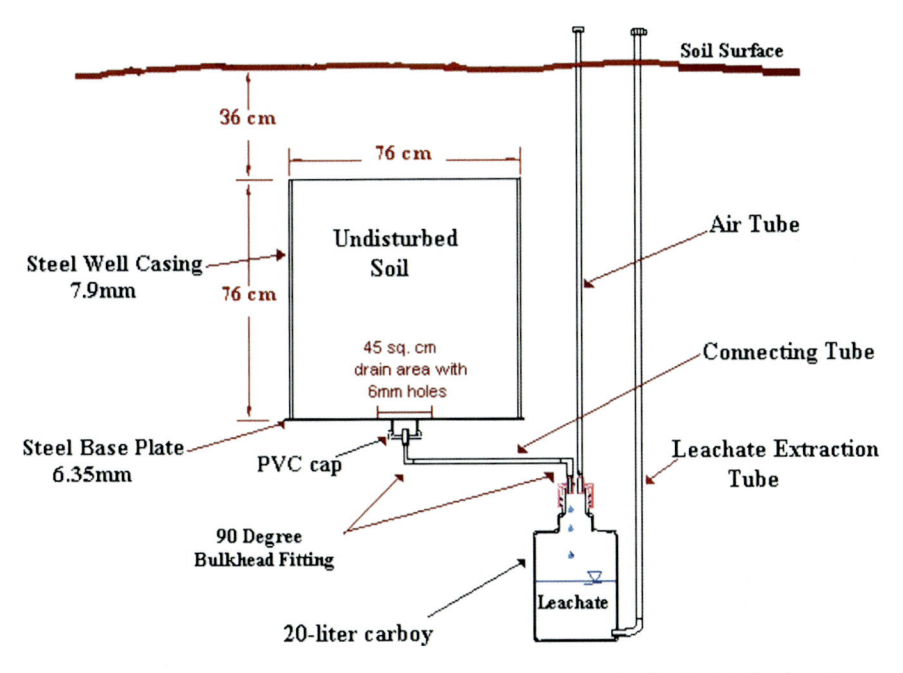

Fig. 6.3 Lysimeter used to collect percolated water in each system in The Rodale Institute Farming Systems Trial

tion of percolated water (Fig. 6.3). The top of each lysimeter was approximately 36 cm below the soil surface to allow field operations to be carried out in a normal fashion directly over the lysimeters. Approximately 20 holes were drilled in the center of the base plate to allow for unrestricted flow of percolate from the cylinder into the flexible tube leading to the collection vessel, a 20-liter polyethylene carboy. Two more tubes were connected to the carboy: the air tube, that ran from the cap of the carboy to the soil surface and the extraction tube that ran from the base of the carboy to the soil surface. The carboy was positioned below and offset to one side of the steel cylinder to enable gravitational flow of liquid to the collection vessel. Any percolate that flowed from the cylinder into the carboy was recovered using a marine utility pump connected to the extraction tube (Moyer et al. 1996). Water could not escape from the lysimeter system. Leachate samples were collected throughout the year.

6.3.2 *Analytical Methods*

Nitrate-nitrogen in leachate samples was determined by the cadmium reduction method using a Flow Injection Analysis (FIA) system from Lachat Instruments by the Soil and Plant Nutrient Laboratory, Michigan State University, East Lansing, MI.

Herbicides in leachate samples were analyzed using EPA 525.2 determination of organic compounds in water sample by liquid solid extraction and capillary column gas chromatography mass spectrometry by M.J. Reider Associates, Reading, PA.

Total soil carbon and nitrogen were determined by combustion using a Fisons NA1500 Elemental Analyzer by The Agricultural Analytical Services Laboratory, The Pennsylvania State University, University Park, PA.

Soil water content was determined gravimetrically on sieved soil (2 mm).

Statistical analyses were carried out using SPSS Version 10.1.3 General Linear Model Univariate Analysis of Variance (SPSS, Inc., Chicago, IL).

6.4 Results

6.4.1 Crop Yields under Normal Rainfall

From 1986 to 2001, corn grain yields averaged 6,700, 6,900, and 7,200 kg/ha for the conventional system, the organic-legume system, and for the organic-animal system, respectively (Pimentel et al. 2005). Corn yields in the animal system were essentially the same as for the conventional system (see Fig. 6.4). Soybean yields were 2,800, 2,400, and 2,500 kg/ha for the conventional system, for the organic-legume system, and for the organic-animal system, respectively (See Fig. 6.5) (Pimentel et al. 2005). In the conventional system, the soybean yield was not significantly higher than yields in either the organic-legume and organic-animal systems.

The 10-year period from 1988 to 1998 had 5 years in which the total rainfall from April to August was less than 350 mm (versus 500 mm in average years). Corn yields in those 5 dry years were significantly higher (28–34%) in the two organic systems: 6,947 and 7,234 kg/ha in the organic-animal and the organic-legume system, respectively, compared with 5,409 kg/ha in the conventional system. The two organic systems were not statistically different in terms of corn yields during the dry years but the corn yields in both organic systems were significantly different from the yields for the conventional system.

During the extreme drought of 1999 (total rainfall between April and August was only 224 mm compared with the normal average of 500 mm), the organic-animal system had significantly higher corn yields (1,511 kg/ha) than both the organic-legume (421 kg/ha) and the conventional system (1,100 kg/ha) (See Fig. 6.6). Crop yields in the organic-legume system were much lower in 1999 because of the high biomass of the hairy vetch winter cover crop used up a large amount of the soil water (Lotter et al. 2003).

Soybean yields responded differently than the corn during the 1999 drought. Specifically, soybean yields were about 1,800, 1,400, and 91 kg/ha for the organic-legume, the organic-animal, and the conventional systems, respectively (See Fig. 6.6). These treatments were statistically significant ($p=0.05$) from each other (Pimentel et al. 2005).

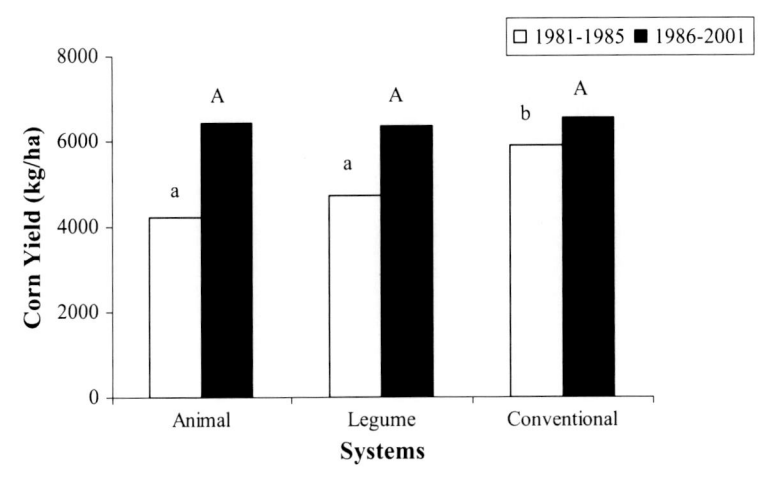

Fig. 6.4 Long-term average corn yields, The Rodale Institute Farming Systems Trial 1981–2001, (ANIMAL = organic animal; LEGUME = organic legume). Different letters above bars denote statistical differences at the 0.05 level for the same time period, according to Duncan's multiple range test

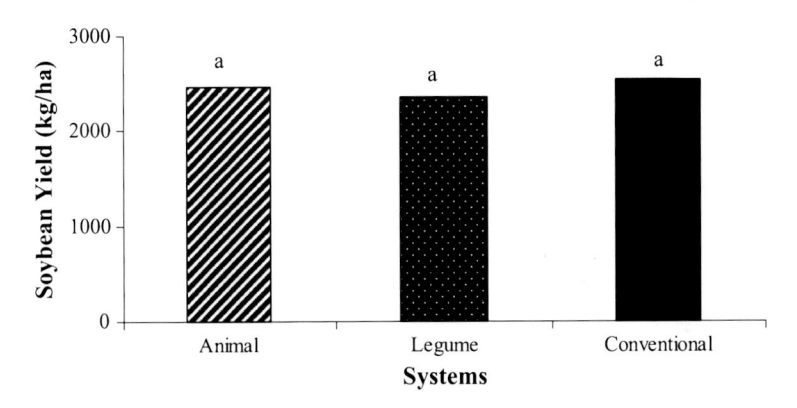

Fig. 6.5 Long-term average soybean yields, The Rodale Institute Farming Systems Trial 1981–2001, excluding 1988 (ANIMAL = organic animal; LEGUME = organic legume). Same letters above bars denote no statistical differences at the 0.05 level, according to Duncan's multiple range test

Over a 12-year period, water volumes percolating through each system (collected in lysimeters), were 15 and 20 % higher in the organic-legume and organic-animal systems, respectively, than in the conventional system. This indicated an increased groundwater recharge and reduced runoff in the organic systems compared to the conventional system (See Fig. 6.7). During the growing seasons of 1995, 1996, 1998 and 1999, soil water content was measured for the organic-legume and conventional systems. The measurements showed significantly more water in the organic-legume soil than in the conventional system (Fig. 6.7) (Pimentel et al. 2005). This accounted for the higher soybean yields in the organic-legume system in 1999 (Pimentel et al. 2005).

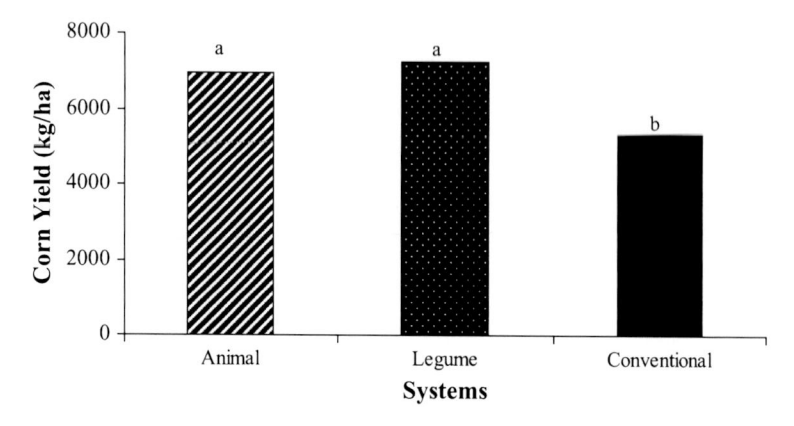

Fig. 6.6 Average corn yields in drought years (1988, 1994, 1995, 1997, 1998), The Rodale Institute Farming Systems Trial, (ANIMAL = organic animal; LEGUME = organic legume). Different letters above bars denote statistical differences at the 0.05 level, according to Duncan's multiple range test

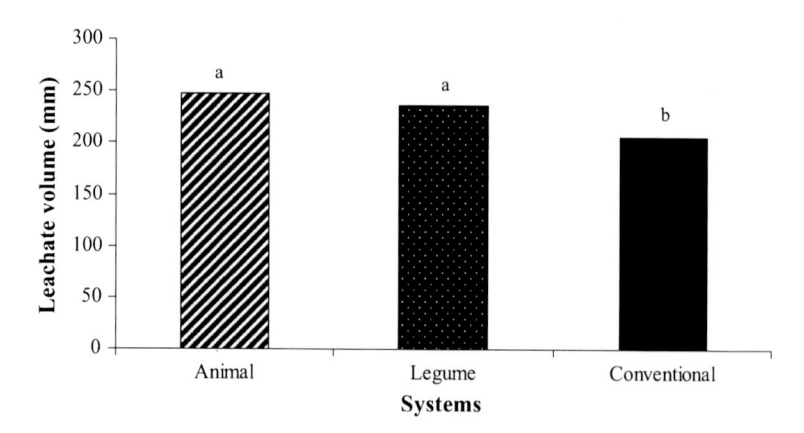

Fig. 6.7 Average amount of leachate volume per year, The Rodale Institute Farming Systems Trial 1991–2002 (ANIMAL = organic animal; LEGUME = organic legume). Different letters above bars denote statistical differences at the 0.05 level, according to Duncan's multiple range test

6.4.2 Energy Inputs

The energy inputs in the conventional, organic-legume, and organic-animal corn production systems were assessed. The inputs included fossil fuels for farm machinery, fuel, fertilizers, seeds, and herbicides. About 7.6 million kcal of energy per ha were invested in the production of corn in the conventional system (Fig. 6.8). The energy inputs for the organic-legume and organic-animal systems were about half that of the conventional system (3.8 million and 3.4 million kcal per ha, respectively) (Fig. 6.8). Commercial fertilizers for the conventional system were produced

Fig. 6.8 Farming systems trial from 1985 to 2000, average energy inputs for corn and soybeans in the three systems (CONV = conventional; LEGUME = organic legume; ANIMAL = organic animal)

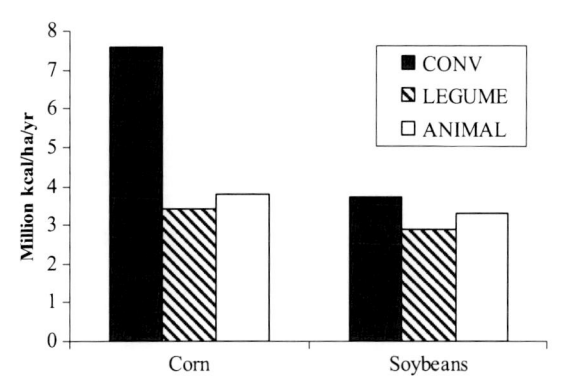

using energy from fossil fuel, but the nitrogen nutrients for the organic systems were obtained from legumes and/or cattle manure. Fossil energy inputs were required to transport and apply the manure to the field.

The energy inputs for soybean production in the conventional and organic-animal systems were similar, 3.7 million kcal and 3.3 million kcal per ha, respectively. The inputs for the organic-legume system of 2.9 million kcal per ha were somewhat lower than both the conventional and organic-animal based systems (Fig. 6.8).

6.4.3 Economics

Two economic studies were completed of the FST (Farm Systems Trial) evaluating the first 9 years and the first 15 years of operation (Hanson et al. 1990 and Hanson et al. 1997, respectively). As inclusive evaluations, these two studies captured the experiences of an organic farmer as s/he develops over time a rotation that best fits one's farm. With the development of the final rotation, however, a third evaluation was completed comparing this rotation with its conventional alternative (Hanson and Musser 2003). Many organic grain farmers in the Mid-Atlantic region have been adopting this 'Rodale rotation' on their farms and there was strong interest in an economic evaluation of only this rotation (i.e., without the transition period or learning curve).

The third economic comparison of the organic corn/soybean rotation and conventional corn/soybean systems covered the period 1991–2001. Without price premiums for the organic rotation, the net returns for both rotations were similar. The annual net return for the conventional system averaged about $ 184 per ha while the organic-legume system for cash grain production averaged $ 176 per ha. When the costs of the biological transition for the organic rotation (1982–1984) are included, then the net returns for the organic rotation are reduced to $ 162 per ha while the conventional net returns remain unchanged. Including the costs of family labor for both rotations reduces the net returns of conventional to $ 162 and organic to $ 127. However, even with the inclusion of the biological transition and family labor costs, the amount of an organic price premium required to equalize the organic and conventional returns is only 10%. Throughout the 1990s, the organic price premium for grains has exceeded this level and premiums now range between 65 and 140% (Pimentel et al. 2005).

The organic system requires 35% more labor, but since it is spread out over the growing season, the hired labor costs per ha are about equal between the two systems. Each system was allowed 250 h of "free" family labor per month. When labor requirements exceeded this level, labor was hired at $ 13.00/h. With the organic system, the farmer was busy throughout the summer with the wheat crop, hairy vetch cover crop, and mechanical weed control (but less than 250 h/month). In contrast, the conventional farmer had large labor requirements in the spring and fall, planting and harvesting, but very little in the summer months. This may have implications for the growing number of part-time farmers for whom the availability of family farm labor is severely limited. Other organic systems have been shown to require more labor per hectare than conventional crop production. On average, organic systems require about 15% more labor (Sorby 2002; Granatstein 2003), but the increased labor input may range from an increase of 7% (Brumfield et al. 2000) to a high of 75% (Nguyen and Haynes 1995; Karlen et al. 1995).

Over the 10-year period, organic corn (without price premiums) was 25% more profitable than conventional corn. This was possible because organic corn yields were only 3% less than conventional yields while costs were 15% less. This success is achieved by growing wheat with a high soil-investing value-crop in the previous year. More specifically, corn was grown 60% of the time in the conventional rotation, but only 33% of the time with the organic rotation. Stated in another way, the yields per ha between organic and conventional corn for grain may be similar within a given year; however, overall production of organic corn is diminished over a multiple-year period because it is grown less frequently.

6.4.4 Soil Carbon

Soil carbon, which correlates with soil organic matter levels, was measured in 1981 and 2002. Soil carbon values were statistically the same for all three systems at the start of the experiment in 1981 (Fig. 6.9). In 1981, soil carbon levels found in the soil of the three systems were not different ($p=0.05$). In 2002, however, soil carbon levels in the organic-legume and organic-animal systems were significantly higher than in the conventional system (Fig. 6.9). The soil carbon level in the conventional system from 1981 to 2002 did not differ statistically, whereas both organic systems had increased and were significantly higher in 2002 than in 1981. The higher level of soil organic matter (soil carbon) in both the organic-legume and organic-animal systems was associated with higher soil water content of the soils in these systems compared with the conventional system. Higher soil water content in the organic systems accounted for the higher corn and soybean yields during drought years in the organic-legume and organic-animal systems compared with the conventional system (Lotter et al. 2003).

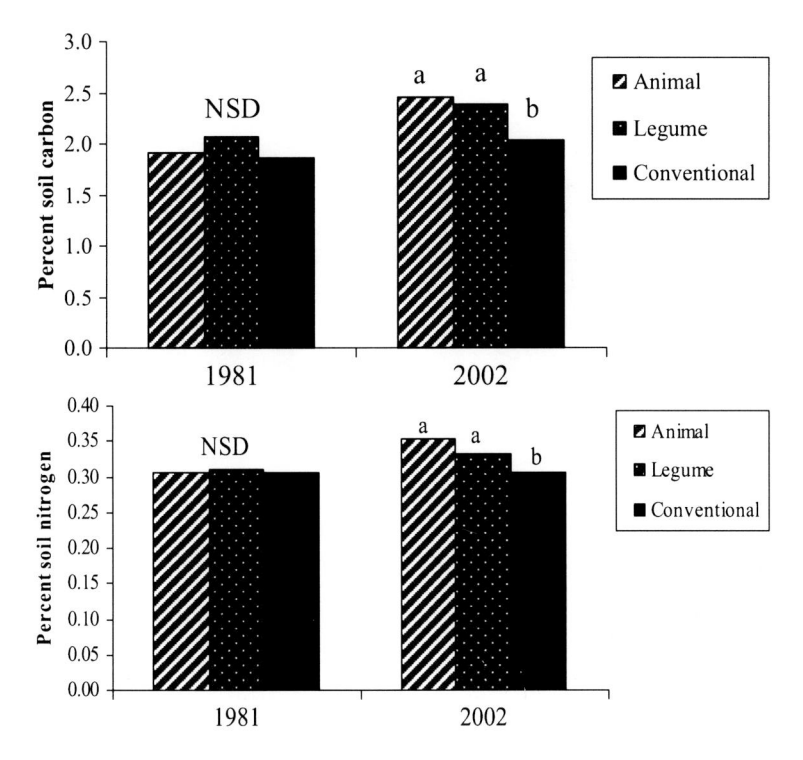

Fig. 6.9 Percent soil carbon and soil nitrogen for the three systems in 1981 and 2002, The Rodale Institute Farming Systems Trial, (ANIMAL = organic animal, LEGUME = organic legume). Different letters indicate statistically significant differences, NSD = not significantly different

6.4.5 Soil Nitrogen

Soil nitrogen levels were measured in 1981 and 2002 in the organic-legume, organic-animal, and conventional systems (Fig. 6.9). Initially the three systems had similar percentages of soil nitrogen or approximately 0.31%. By 2002, the conventional system remained unchanged at 0.31% while the organic-manure and organic-legume significantly increased to 0.35 and 0.33%, respectively. Thus, soil nitrogen was slowly increasing in both organic systems at a rate of 0.3–0.6% per year.

Harris et al. (1994) used N15 to demonstrate that 47, 38, and 17% of the nitrogen from the organic-animal, organic-legume, and conventional systems, respectively, were retained in the soil a year after application. The nitrogen losses were 53, 62, and 83% for the organic-animal, organic-legume, and conventional systems, respectively.

6.4.6 Nitrate Leaching

Overall, nitrate-nitrogen concentrations of leachates from the farming systems varied between 0 and 28 ppm throughout the year (Pimentel et al. 2005). Leachate

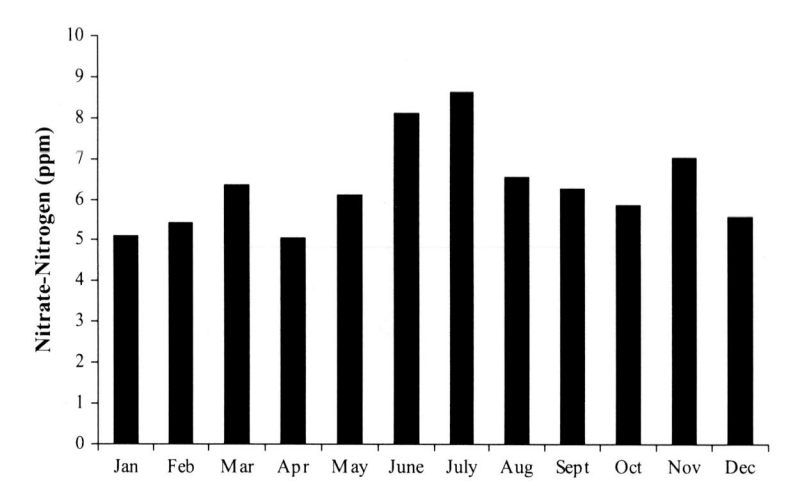

Fig. 6.10 Average monthly nitrate-nitrogen concentration in leachate across all systems, The Rodale Institute Farming Systems Trial, 1991–2001

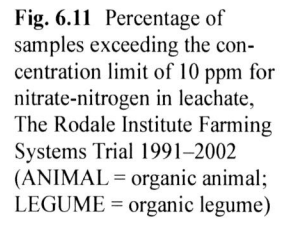

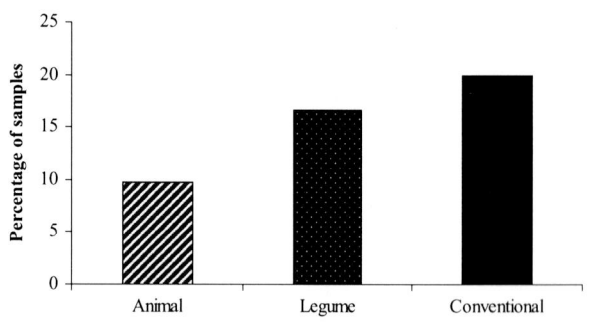

Fig. 6.11 Percentage of samples exceeding the concentration limit of 10 ppm for nitrate-nitrogen in leachate, The Rodale Institute Farming Systems Trial 1991–2002 (ANIMAL = organic animal; LEGUME = organic legume)

concentrations were usually highest in June and July, shortly after fertilizer application in the conventional systems or plow down of the animal manure and legume cover crop. In all systems, increased soil microbial activity during the growing season appears to have contributed to increased nitrate leaching (Fig. 6.10).

Water leachate samples from the conventional system most frequently exceeded the regulatory limit of 10 ppm for nitrate concentration in drinking water. A total of 20% of the conventional system samples were above the 10 ppm limit, while 10 and 16% of the samples from the organic-animal and organic-legume systems exceeded the limit, respectively (Fig. 6.11).

Over the 12-year period of monitoring (1991–2002), all three systems leached between 16 to 18 kg of nitrate-nitrogen per hectare per year (See Fig. 6.12). These rates were low compared to results from other similar experiments where nitrate-nitrogen leaching ranged from 30 to 146 kg/ha per year (Fox et al. 2001; Power et al. 2001). When measuring these nitrate-nitrogen losses as a percentage of

Fig. 6.12 Cumulative nitrate-nitrogen leached, The Rodale Institute Farming Systems Trial 1991–2002

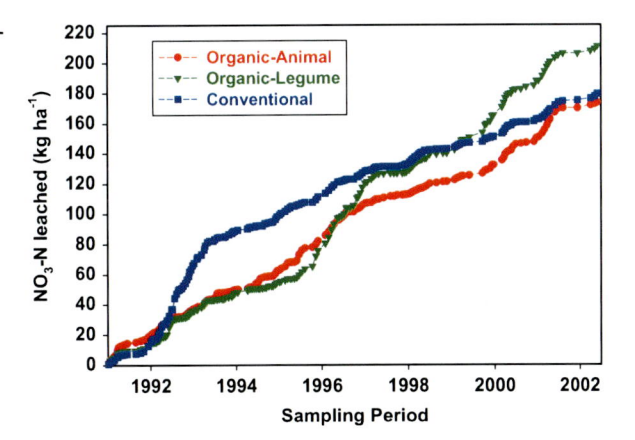

the nitrogen originally available to the crops in each system, the organic-legume, organic-animal, and the conventional systems lost about 32, 20, and 20 %, respectively, of the total nitrogen as nitrate.

The high nitrate leaching in the organic-legume system was not steady over the entire period of the study; instead, it occurred sporadically, especially during a few years of extreme weather. For example, in 1995 and 1999, the hairy vetch green manure supplied approximately twice as much nitrogen as needed for the corn crop that followed, contributing excess nitrogen in the soil and available for leaching. In 1999, the heavy nitrogen input was followed by a severe drought that stunted corn growth and reduced the corn's demand for nitrogen. In both years, these nitrogen-rich soils were also subjected to unusually heavy fall and winter rains that leached the excess nitrogen into the lower soil layers. Monitoring soil nitrogen and cover crop production are needed to manage excessive nitrate-nitrogen potential in all systems.

These data contrasts with experiments in Denmark that indicated that nitrogen leaching from the conventional treatments was twice that in the organic agricultural systems (Hansen et al. 2001). Overall nitrogen leaching levels were lower in this study than those reported by Hansen et al. (2001).

6.4.7 Herbicide Leaching

The following herbicides were applied to the conventional system: atrazine, metolachlor, and pendimethalin to corn and metolachlor and metribuzin to soybeans. From 2001 to 2003, atrazine and metolachlor were detected in water leachate samples collected only in the conventional system (Fig. 6.13). No metribuzin or pendimethalin were detected after application (Pimentel et al. 2005).

In all samples, in the conventional system atrazine concentrations exceeded the 0.1 ppb concentration known to produce deformities in frogs (Fig. 6.13) (Hayes

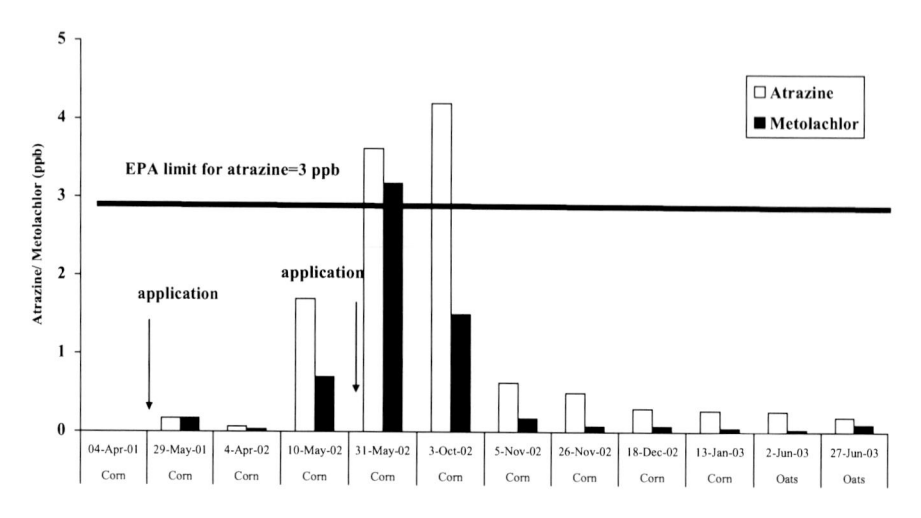

Fig. 6.13 Trends of atrazine and metolachlor concentrations in leachate found in corn after corn plots of the conventional system, The Rodale Institute Farming Systems Trial, 2001–2003

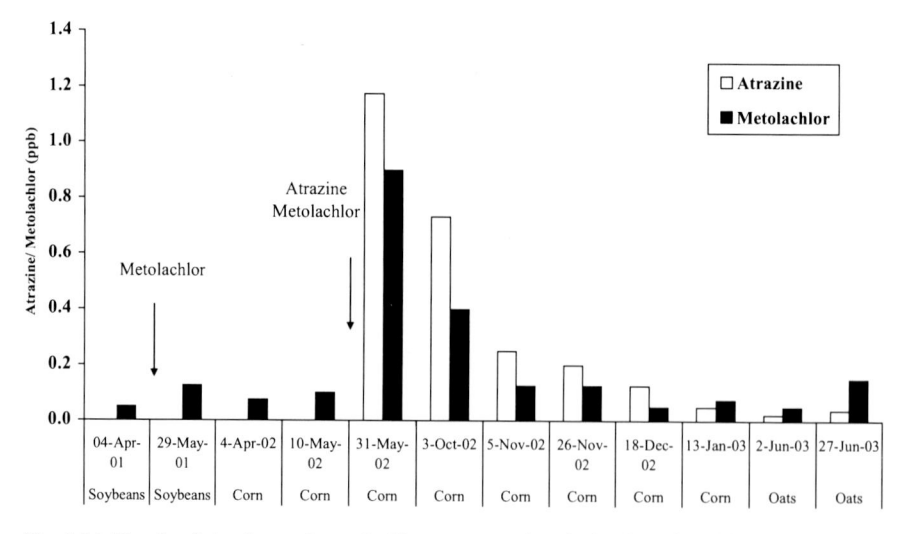

Fig. 6.14 Trends of atrazine and metolachlor concentrations in leachate found in corn after soybean plots of the conventional system, The Rodale Institute Farming Systems Trial, 2001–2003

et al. 2002). In the conventional plots where corn was planted after corn and atrazine was applied 2 years in a row, atrazine in the leachate sometimes exceeded 3 ppb (the MCL set by EPA for drinking water). These atrazine levels were higher than those in the corn-after-soybean treatment (Pimentel et al. 2005). In the conventional system, metolachlor was also detected at 0.2–0.6 ppb generally (Fig. 6.14).

When metolachlor was applied 2 years in a row in a corn-after-corn treatment, it peaked at 3 ppb (Pimentel et al. 2005). EPA has not yet established a MCL for metolachlor for drinking water.

6.5 Discussion

6.5.1 Soil Organic Matter and Biodiversity

Soil organic matter provides the base for productive organic farming and sustainable agriculture. Soil carbon (soil organic matter) was significantly higher in both the organic-animal and organic-legume systems than in the conventional system (Fig. 6.9). In 2002, the soil carbon was 2.5% in the organic-animal system, 2.4% in the organic-legume system, and 2.0% in the conventional system. Soil carbon increased in all three systems from 1981 to 2002. However, the soil carbon increased 27.9, 15.1, and 8.6% in the organic-animal, organic-legume, and the conventional systems, respectively (Fig. 6.9). The conventional system increase was not statistically significant ($p = 0.05$).

The amount of organic matter in the upper 15 cm of soil in the organic farming systems was approximately 110,000 kg/ha. The soil of this depth weighed about 2.2 million tons/ha. Approximately 41% of the volume of the organic matter in the organic systems consisted of water compared with only 35% in the conventional system (Sullivan 2002). The amount of water held in both of the organic systems is estimated at 816,000 liters/ha. The large amount of soil organic matter present in the organic systems aided in making these systems more drought tolerant in the 1999 drought and other drought years.

Large amounts of biomass (soil organic matter) significantly increase soil biodiversity (Pimentel et al. 1992; Troeh and Thompson 1993; Mader et al. 2002; Lavelle and Spain 2001). The arthropods per hectare can number from 2 to 5 million and earthworms from 1 to 5 million (Lavelle and Spain 2001; Gray 2003). The microarthropods and earthworms were reported to be twice as abundant in organic versus conventional agricultural systems in Denmark (Hansen et al. 2001). The weight of the earthworms per hectare in agricultural soils can range from 2,000 to 4,000 kg (Lavelle and Spain 2001). There can be as many as 1,000 earthworm and insect holes per square meter of land. Earthworms and insects are particularly helpful in constructing large holes in the soil that encourage the percolation of water into the soil and prevent excess water run off.

Biomass can help increase biodiversity which provides vital ecological services including crop protection (Altieri 1999). For example, adding compost and other organic matter reduces crop diseases (Cook 1988; Hoitink et al. 1991; Altieri 1999), and also increases the number of species of microbes in the agroecosystem (van Elsen 2000). In addition, in the organic systems, not using synthetic pesticides and commercial fertilizers minimizes the harmful effects of these chemicals upon non-target organisms (Pimentel 2005).

Among the natural biological processes upon which the organic rotations depend is symbiosis of arbuscular mycorrhizae (AB) and crop roots. Arbuscular mycorrhizal fungi are beneficial and indigenous to most soils. They colonize the roots of most crop plants, forming a mutualistic symbiosis (the "mycorrhiza"). The fungus receives sugars from the host-plant root and the plant benefits primarily from enhanced nutrient uptake from the fungus. The extraradical mycelium of the AM fungi act, in effect, as extensions of the root system, more thoroughly exploring the soil for immobile mineral nutrients such as phosphate (Smith and Read 1997). Arbuscular mycorrhizae have been shown to enhance crop disease resistance, improve water relations, and increase soil aggregation (Hooker et al. 1994; Miller and Jastrow 1990; Wright et al. 1999). Efficient utilization of this symbiosis contributes to the success of organic production systems.

Soils of The Rodale Institute Farming Systems Trial (FST) and other field trials at Rodale have been sampled to study the impact of conventional and organic agricultural management upon indigenous populations of AM fungi. Soils farmed with the two organic systems had both greater populations of spores of AM fungi and produced greater colonization of plant roots than in the conventional system (Douds et al. 1993). Most of this difference was ascribed to greater plant cover (70%) on the organic systems compared with the conventional corn-soybean rotation (40%). This was due to over-wintering cover crops in the organic rotation (Galvez et al. 1995). In addition to fixing or retaining soil nitrogen, these cover crops allow the AM fungi roots to colonize and maintain their viability during the interval from cash crop senescence to next year planting. Though levels of AM fungi were greater in the organically farmed soils, ecological species diversity indices were similar in the other farming system (Franke-Snyder et al. 2001).

Wander et al. (1994) demonstrated that soil respiration was 50% higher in the organic-animal system compared with the conventional system 10 years after initiation of The Rodale Institute Farming System Trial. Soil nitrogen and mineralized nitrogen in the organic-animal system increased 19 and 23%, respectively, compared with the conventional system.

Overall, environmental damage from agricultural chemicals was reduced in the organic systems. Overall, public health and ecological integrity could be improved through the adoption of practices that decrease the quantities of pesticides and commercial fertilizers applied in agriculture (NAS 2003; Pimentel 2005).

6.5.2 Oil and Natural Gas Inputs

Significantly less fossil energy was expended in The Rodale Institute's organic-legume and organic-animal systems compared with the conventional production system, especially with corn (Fig. 6.8). In the organic system, only small amounts of phosphorous (fertilizer) were applied once or twice. Other investigators have reported similar findings (Pimentel et al. 1983; Pimentel 1993; Smolik et al. 1995; Karlen et al. 1995; Dalgaard et al. 2001; Mader et al. 2002; Core 4 2003). In general, the utilization of less fossil energy and energy conservation by organic agriculture systems, reduces the amount of carbon dioxide released to the atmosphere, and therefore reduces the problem of global climate change (FAO 2002).

6.5.3 Crop Yields and Economics

Except for the 1999 drought year, the crop yields for corn and soybeans were similar in the organic-legume, organic-animal, and conventional farming systems. In contrast, Smolik et al. (1995) found corn yields in South Dakota were somewhat higher in the conventional system with average yields of 5,708 kg/ha compared with organic-legume system that averaged 4,767 kg/ha. However, the soybean yields in both systems were similar at 1,814 kg/ha. In a second study comparing wheat and soybean yields, the wheat yields were fairly similar averaging 2,600 kg/ha in the conventional and 2,822 kg/ha in the organic-legume system. Soybean yields were 1,949 and 2,016 kg/ha for the conventional and the organic-legume systems, respectively (Smolik et al. 1995). In The Rodale experiments, corn, soybeans, and wheat yields were considerably higher than those reported in South Dakota.

European field tests indicate that organic wheat and other cereal grain yields average from 30 to 50% lower than conventional cereal grain production (Mader et al. 2002). The lower yields for the organic system in their experiments compared with the conventional systems appear to be caused by lower nitrogen nutrient inputs in the organic systems. In New Zealand, organic wheat yields were reported to average 38% lower than those in the conventional system or similar to the results in Europe (Nguyen and Haynes 1995). In New Jersey, organically produced sweet corn yields were reported to be 7% lower than in a conventional system there (Brumfield et al. 2000). In The Rodale experiments, nitrogen levels in the organic systems have improved and have not been limiting the crop yields after the first 3 years. In the short term in organic systems, there may be nitrogen shortages that may reduce crop yields temporarily, but these can be eliminated by raising the soil nitrogen level through the use of animal manure and/or legume cropping.

In a subsequent field test in South Dakota, corn yields in the conventional system and the organic-alternative system were 7,652 and 7,276 kg/ha, respectively (Dobbs and Smolik 1996). Soybean yields were significantly higher in the conventional system averaging 2,486 kg/ha compared with only 1,919 kg/ha in the organic-alternative system.

The Rodale crop yields were similar to the results in the conventional and organic-legume farming system experiments conducted in Iowa (Delate et al. 2002). In the Iowa experiments, corn yields were 8,655 and 8,342 kg/ha for the conventional and organic-legume systems, respectively. Soybean yields averaged 2,890 and 2,957 kg/ha for the conventional and organic-legume systems, respectively.

Although the inputs for the organic-legume and conventional farming systems were quite different, the overall economic net returns were similar (Fig. 6.4, 6.5). Yet these net returns in The Rodale experiments differ from those of Dobbs and Smolik (1996) who reported a 38% higher gross income for the conventional than the organic-alternative system. In the latter comparison, however, the organic premiums were not calculated. Often in the market place, prices for organic corn and soybeans range from 20 to 140% higher than conventional corn, soybeans, and other grains (Dobbs 1998; Bertramsen and Dobbs 2002). Thus, when the market price differential was factored in, the differences between the organic-alternative and conventional would be relatively small and in most cases the returns on the organic produce would be higher, as in the results here for the FST.

In contrast to corn/soybeans, the economic returns (dollar return per unit) for organic sweet corn production in New Jersey were slightly higher (2%) than conventional sweet corn production (Brumfield et al. 2000). In the Netherlands, organic agricultural systems producing cereal grains, legume, and sugar beets reported a net return of 953 Euros/ha compared with conventional agricultural systems producing the same crops that reported 902 Euros/ha (Pacini et al. 2003).

In a California investigation of four crops (tomato, soybean, safflower, and corn) grown organically and conventionally, production costs for all four crops were 53% higher in the organic system compared with the conventional system (Clark et al. 1999). However, the profits for the four crops were only 25% higher in the conventional system compared with the organic system. If the 44% price advantage of the four organic-system crops were included, the organic crops would be slightly more profitable than the conventional (Clark et al. 1999).

6.5.4 Challenges for Organic Agriculture

Two primary problems identified with the organic system study in California were nitrogen deficiency and weed competition (Clark et al. 1999). This was also noted for the organic farming systems in the U.S. Midwest (Lockeretz et al. 1981). Nitrogen deficiencies and excessive weeds can be overcome with improved crop management, though "predicting the actual amount of nitrogen fixed is notoriously difficult as it depends on many factors including the legume species and cultivar management, weather conditions, and age of the ley" (Watson et al. 2002, p. 242). The *Trifolium pratense* L. cover crops used here supply an average of 138 kg N/ha, nearly the equivalent to the mean N supplied to corn as chemical fertilizer (141 kg N/ha; (Liebhardt et al. 1989).

Pest control can be a problem in organic crop production. Weed control is frequently a problem in organic crops because the farmer is limited to only mechanical and biological weed control, while under conventional production mechanical, biological, and chemical weed control options often are employed. Also weather conditions influence weed control. Mechanical weed control is usually more effective than chemical weed control under dry conditions, while the reverse holds under wet conditions. In the Rodale experiments, only the organic soybeans suffered negative impacts from weed competition.

Insect pests and plant pathogens can be effectively controlled in corn and soybean production by employing crop rotations (Pimentel et al. 1993). Some insect pests can be effectively controlled by an increase in parasitoids; reports on organic tomato production indicate nearly twice as many parasitoids in the organic compared with the conventional system (Letourneau and Goldstein 2001). However, increased plant diversity in tomato production was found to increase the incidence of plant diseases (Kotcon et al. 2001). With other crops, like potatoes and apples, dealing with pest insects and plant pathogens that adversely affect yields is a major problem in organic crop production (Pimentel et al. 1983).

6.5.5 Policy Needs

U.S. Government agricultural policies over time have resulted in increased use of pesticides, fertilizers, and reduced recycling of livestock wastes and reduced crop rotations (NAS 1989). For example, prior to the 1990s, farmers grew mostly program crops (in monocrop production) so as to protect their base hectares (and increase government payments). This reduced the diversity of crops grown and, in turn, livestock production was reduced. During the four decades from 1950 to 1990, pesticide and fertilizer use increased 10 times or more per hectare while soil erosion also increased significantly due to a reduction in crop rotation (Pimentel 1975; Pimentel 1993; Pimentel and Kounang 1998). Finally in 1990, new legislation was passed that encouraged farmers to rotate their crops and hopefully bring livestock back on the farm. If this could be accomplished, it would significantly improve the recycling of livestock wastes and improve the environment, plus reduce fossil energy inputs in crop production.

Some nations have already implemented programs to make their agriculture environmentally sound and sustainable (Kumm 2001; O'Riorda and Cobb 2001). As mentioned, Sweden has reduced pesticide use during the past decade by 68% without reducing crop yields. This major reduction in pesticide use has led to a 77% decrease in human poisonings from pesticides (Ekstrom and Bergkvist 2001). Studies have confirmed that U.S. pesticide use on average could be reduced by more than 50% without any reduction in crop yields (Pimentel et al. 1993).

6.6 Conclusion

Various organic agricultural technologies have been utilized for about 6,000 years to make agriculture sustainable while at the same time conserving soil, water, energy, and biological resources. Some of the benefits of organic technologies identified in this investigation are as follows:

- Soil organic matter (soil carbon) and nitrogen are higher in the organic farming systems providing many benefits to the overall sustainability of organic agriculture.
- Fossil energy inputs for organic crop production are from 30 to 50% lower than for conventionally produced crops.
- Depending on the crop, soil, and weather conditions, organically managed crop yields on a per hectare basis can equal those from conventional agriculture, but it is likely that organic cash crops cannot be grown as frequently over time because of the dependence on cultural practices to supply nutrients and control pests.
- Labor inputs average about 15% higher in organic farming systems and range from 7 to 75% higher.
- Because organic foods frequently bring higher prices in the market place, the net economic return per hectare is often equal or higher than conventionally produced crops.
- Crop rotations and cover cropping typical of organic agriculture, reduce soil erosion, pest problems, and the need for pesticides.
- High soil organic matter helps conserve soil and water resources and is proven beneficial during drought years.
- The recycling of livestock wastes reduces pollution and at the same time benefits organic agriculture.
- Abundant biomass both above and below ground (soil organic matter) also increases biodiversity which helps in the biological control of pests and increases crop pollination by insects.
- Traditional organic farming technologies may be adopted by conventional agriculture to make it more sustainable and ecologically sound.

Acknowledgement This research was supported in part by the Podell Emeriti Award at Cornell University.

References

Altieri, M. A. (1999). The ecological role of biodiversity in agroecosystems. *Agriculture Ecosystems & Environment, 74*, 19–31.
Bertramsen, S. K., & Dobbs, T. L. (2002). An update on prices of organic crops in comparison to conventional crops. *Economics Commentator. No. 426. February 22*. South Dakota State University, Brookings, South Dakota, USA.
Brumfield, R. G., Rimal, A., & Reiners, S. (2000). Comparative cost analyses of conventional, integrated crop management, and organic methods. *HortTechnology, 10*, 785–793.
Clark, S., Klonsky, K., Livingston, P., & Temple, S. (1999). Crop-yield and economic comparisons of organic, low-input, and conventional farming systems in California's Sacramento Valley. *American Journal of Alternative Agriculture, 14*, 109–121.
Colburn, T., Dumanoski, D., & Myers, J. P. (1997). *Our Stolen Future*. New York: Penguin.
Cook, R. J. (1988). Biological control and holistic plant-health care in agriculture. *American Journal of Alternative Agriculture, 3*, 51–62.
Core 4. (2003). Core 4: Conservation for agriculture's future. conservation technology information center, West Lafayette, Indiana. http://www.ctic.purdue.edu/Core4/Fact%20Sheets/. Accessed 18 July 2012.

Dalgaard, T., Halberg, N., & Porter, J. R. (2001). A model for fossil energy use in Danish agriculture used to compare organic and conventional farming. *Agriculture, Ecosystems & Environment, 87,* 51–65.

Delate, K., Duffy, M., Chase, C., Holste, A., Friedrich, H., & Wantate, N. (2002). An economic comparison of organic and conventional grain crops in a long-term agroecological research (LTAR) site in Iowa. *American Journal of Alternative Agriculture, 18,* 59–69.

Dimitri, C., & Greene, C. (2002). Recent growth patterns in the U.S. organic foods market. *Agriculture Information Bulletin No. 777.* http://webarchives.cdlib.org/sw1q52gj15/http://www.ers.usda.gov/publications/aib777/. Accessed 18 July 2012.

Dobbs, T., & Smolik, J. D. (1996). Productivity and profitability of conventional and alternative farming systems: A long-term on-farm paired comparison. *Journal of Sustainable Agriculture, 9,* 63–77.

Dobbs, T. L. (1998). Price premiums for organic crops. *Choices, (Second Quarter),* 39–41

Douds, D. D., Janke, R. R., & Peters, S. E. (1993). VAM fungus spore populations and colonization of roots of maize and soybean under conventional and low-input sustainable agriculture. *Agriculture Ecosystems & Environment, 43,* 325–335.

Doughty, R. S. (2003). *Use of wetlands to reduce nitrogen loads in the Mississippi Atchafalaya River Basin.* M.S. Thesis, Cornell University, Ithaca, USA.

Ekstrom, G., & Bergkvist, P. (2001). Persistence pays—lower risks from pesticides in Sweden. Pesticide Action Network International Website (Article first appeared in *Pesticides News No. 54,* December 2001, pages 10–11). http://www.pan-uk.org/pestnews/Issue/pn54/pn54p10.htm. Accessed 18 July 2012.

ERS. (2012). *Organic Production.* Washington, D.C.: Economic Research Service, U.S. Department of Agriculture. http://www.ers.usda.gov/data-products/organic-production.aspx. Accessed 18 July 2012.

FAO. (2002). Organic agriculture and climate change. In N. El-Hage Scialabba & C. Hattam (Eds.), *Organic Agriculture, Environment and Food Security.* Environment and Natural Resources Service, Sustainable Development Department. Series No. 4. Food and Agriculture Programme, United Nations, Rome. http://www.fao.org/docrep/005/y4137e/y4137e00.HTM. Accessed 10 Oct 2013.

Fox, R. H., Zhu, Y., Toth, J. D., Jemison, J. M. Jr., & Jabro, J. D. (2001). Nitrogen fertilizer rate and crop management effects on nitrate leaching from an agricultural field in central Pennsylvania. *Scientific World Journal, 1*(Suppl 2), 181–186.

Frankenberger, J., & Turco, R. (2003). Hypoxia in the Gulf of Mexico: A reason to improve nitrogen management. *Purdue Animal Issues.* Briefing, AI-6. Purdue Extension, Purdue University, West Lafayette, Indiana. http://www.ansc.purdue.edu/anissue/AI6.pdf. Accessed 26 Feb 2014.

Franke-Snyder, M., Douds, D. D., Galvez, L., Phillips, J. G., Wagoner, P., Drinkwater, L., & Morton, J. B. (2001). Diversity of communities of arbuscular mycorrhizal (AM) fungi present in conventional versus low-input agricultural sites in eastern Pennsylvania, USA. *Applied Soil Ecology, 16,* 35–48.

Galvez, L., Douds, D. D., Wagoner, P., Longnecker, L. R., Drinkwater, L. E., & Janke, R. R. (1995). An overwintering cover crop increases inoculum of VAM fungi in agricultural soil. *American Journal of Alternative Agriculture, 10,* 152–156.

Granatstein, D. (2003). *Tree Fruit Production with Organic Farming Methods.* Wenatchee: Center for sustaining agriculture and natural resources, Washington State University. http://modern-prepper.com/download/plants/Fruit%20tree%20Production%20with%20Organic%20Farming%20Methods.pdf. Accessed 18 July 2012.

Gray, M. (2003). Influence of agricultural practices on earthworm populations. *The Bulletin: Pest Management and Crop Development Information for Illinois,* April 24, 2003. University of Illinois Extension. http://bulletin.ipm.illinois.edu/pastpest/articles/200305d.html. Accessed 18 July 2012.

Hansen, B., Alroe, H. F., & Steen, K. E. (2001). Approaches to assess the environmental impact of organic farming with particular regard to Denmark. *Agriculture, Ecosystems & Environment, 83,* 11–26.

Hanson, J. C., Johnson, D. M., Peters, S. E., & Janke, R. R. (1990). The profitability of sustainable agriculture on a representative grain farm in the Mid-Atlantic region, 1981–1989. *Northeastern Journal of Agricultural and Resource Economics, 19*, 90–98.

Hanson, J. C., Lichenberg, E., & Peters, S. E. (1997). Organic versus conventional grain production in the mid-Atlantic: An economic and farming system overview. *American Journal of Alternative Agriculture, 12*, 2–9.

Hanson, J. C., & Musser, W. N. (2003). An economic evaluation of an organic grain rotation with regards to profit and risk. College of agriculture & natural resources, department of agricultural and resource economics, University of Maryland, *Working Papers 2003: No. 03–10.*

Harris, G. H., Hesterman, O. B., Paul, E. A., Peters, S. E., & Janke, R. R. (1994). Fate of legume and fertilizer nitrogen-15 in a long term cropping systems experiment. *Agronomy Journal, 86*, 910–915.

Hayes, T. B., Collins, A., Lee, M., Mendoza, M., Noriega, N., Stuart, A. A., & Vonk, A. (2002). Hermaphroditic, demasculinized frogs after exposure to the herbicide atrazine at low ecologically relevant doses. *Proceedings of the National Academy of Sciences of the United States of America, 99*, 5476–5480.

Hoitink, H. A. J., Inbar, Y., & Boehm, M. J. (1991). Status of compost-amended potting mixes naturally suppressive to soilborne diseases of floricultural crops. *Plant Disease, 75*, 869–873.

Hooker, J. E., Jaizme-Vega, M., & Atkinson, D. (1994). Biocontrol of plant pathogens using arbuscular mycorrhizal fungi. In S. Gianinazzi & H. Schüepp (Eds.)., *Impact of Arbuscular Mycorrhizas on Sustainable Agriculture and Natural Ecosystems* (pp.191–200). Basel: Birkhäuser Verlag. http://www.ansc.purdue.edu/anissue/AI6.pdf. Accessed 18 July 2012.

Karlen, D. L., Duffy, M. D., & Colvin, T. S. (1995). Nutrient, labor, energy, and economic evaluations of two farming systems in Iowa. *Journal of Production Agriculture, 8*, 540–546.

Kotcon, J. B., Collins, A., & Smith, L. J. (2001). Impact of plant biodiversity and management practices on disease in organic tomatoes. *Phytopathology, 91*(Suppl. 6), S50.

Kumm, K. I. (2001). Toward sustainable agriculture. *Journal of Sustainable Agriculture, 18*, 27–37.

Lavelle, P., & Spain, A. V. (2001). *Soil Ecology*. Dordrecht: Kluwer Academic Publishers.

Letourneau, D. K., & Goldstein, B. (2001). Pest damage and arthropod community structure in organic vs. conventional tomato production in California. *Journal of Applied Ecology, 38*, 557–570.

Liebhardt, W. C., Andrews, R. W., Culik, M. N., Harwood, R. R., Janke, R. R., Radke, J. R., & Rieger-Schwartz, S. L. (1989). Crop production during conversion from conventional to low-input methods. *Agronomy Journal, 81*, 150–159.

Lockeretz, W., Shearer, G., & Kohl, D. H. (1981). Organic farming in the corn belt. *Science, 211*, 540–547.

Lotter, D. W., Seidel, R., & Liebhardt, W. (2003). The performance of organic and conventional cropping systems in an extreme climate year. *American Journal of Alternative Agriculture, 18*, 146–154.

Mader, P., Fliessbach, A., Dubois, D., Gunst, L., Fried, P., & Niggli, U. (2002). Soil fertility and biodiversity in organic farming. *Science, 296*, 1694–1697.

Miller, R. M., & Jastrow, J. D. (1990). Hierarchy of root and mycorrhizal fungal interactions with soil aggregation. *Soil Biology and Biochemistry, 22*, 579–584.

Moyer, J. W., Saporito, L. S., & Janke, R. R. (1996). Design, construction, and installation of an intact soil core lysimeter. *Agronomy Journal, 88*, 253–256.

NAS. (1989). *Alternative Agriculture*. Washington, D.C.: National Academy of Sciences. http://www.nap.edu/openbook.php?record_id=1208&page=1. Accessed 19 July 2012.

NAS. (2003). *Frontiers in Agricultural Research: Food, Health, Environment, and Communities*. Washington, D.C.: National Academy of Sciences. http://www.nap.edu/openbook.php?record_id=10585&page=1. Accessed 19 July 2012.

Nguyen, M. L., & Haynes, R. J. (1995). Energy and labour efficiency for three pairs of conventional and alternative mixed cropping (pasture-arable) farms in Canterbury, New Zealand. *Agriculture, Ecosystems & Environment, 52*, 163–172.

O'Riorda, T., & Cobb, D. (2001). Assessing the consequences of converting to organic agriculture. *Journal of Agricultural Economics, 52*, 22–35.

Pacini, C., Wossink, A., Giesen, G., Vazzana, C., & Huirne, R. (2003). Evaluation of sustainability of organic, integrated and conventional farming systems: A farm and field-scale analysis. *Agriculture, Ecosystems & Environment, 95*, 273–288.

Pimentel, D. (1993). Economics and energetics of organic and conventional farming. *Journal of Agricultural and Environmental Ethics, 6*, 53–60.

Pimentel, D. (2005). Environmental and economic costs of the application of pesticides primarily in the United States. *Environment, Development and Sustainability, 7*, 229–252.

Pimentel, D. (Ed.). (1975). *Insects, Science and Society*. New York: Academic Press.

Pimentel, D., & Kounang, N. (1998). Ecology of soil erosion in ecosystems. *Ecosystems, 1*, 416–426.

Pimentel, D., & Pimentel, M. (1996). *Food, Energy and Society*. Niwot, Colorado, USA: Colorado University Press.

Pimentel, D., Berardi, G., & Fast, S. (1983). Energy efficiency of farming systems: Organic and conventional agriculture. *Agriculture, Ecosystems & Environment, 9*, 359–372.

Pimentel, D., Stachow, U., Takacs, D. A., Brubaker, H. W., Dumas, A. R., Meaney, J. J., O'Neil, J., Onsi, D. E., & Corzilius, D. B. (1992). Conserving biological diversity in agricultural/forestry systems. *Bioscience, 42*, 354–362.

Pimentel, D., McLaughlin, L., Zepp, A., Kakitan, B., Kraus, T., Kleinman, P., Vancini, F., Roach, W. J., Graap, E., Keeton, W. S., & Selig, G. (1993). Environmental and economic effects of reducing pesticide use in agriculture. *Agriculture, Ecosystems & Environment, 46*, 273–288.

Pimentel, D., Harvey, C., Resosudarmo, P., Sinclair, K., Kurz, D., McNair, M., Crist, S., Sphritz, L., Fitton, L., Saffouri, R., & Blair, R. (1995). Environmental and economic costs of soil erosion and conservation benefits. *Science, 267*, 1117–1123.

Pimentel, D., Hepperly, P., Hanson, J., Douds, D., & Seidel, R. (2005). Environmental, energetic, and economic comparisons of organic and conventional farming systems. *Bioscience, 55*, 573–582.

Power, J. F., Wiese, R., & Flowerday, D. (2001). Managing farming systems for nitrate control: A research review from management systems evaluation areas. *Journal of Environment Quality, 30*, 1866–1880.

Smith, S. E., & Read, D. J. (1997). *Mycorrhizal Symbiosis* (2nd ed.). London: Academic Press.

Smolik, J. D., Dobbs, T. L., & Rickert, D. H. (1995). The relative sustainability of alternative, conventional, and reduced-till farming systems. *American Journal of Alternative Agriculture, 16*, 25–35.

Sorby, K. (2002). Environmental benefits of sustainable coffee. Background paper to the World Bank Agricultural Technology Note 30, "Toward more sustainable coffee", published June 2002. http://www.wds.worldbank.org/external/default/WDSContentServer/WDSP/IB/2004/07/12/000090341_20040712112007/Rendered/PDF/295970Env0bene10also02453501public1.pdf. Accessed 19 July 2012.

Sullivan, P. (2002). Drought resistant soil. Agronomy technical note. Appropriate Technology Transfer for Rural Areas (ATTRA), National Center for Appropriate Technology (NCAT). http://www.plant-trees.org/resources/infomaterials/english/agroforestry_technologies/Drought%20Resistant%20Soils.pdf. Accessed 19 July 2012.

Surgeoner, G. A., & Roberts, W. (1993). Reducing pesticide use by 50% in the province of Ontario: Challenges and progress. In D. Pimentel & H. Lehman (Eds.)., *The Pesticide Question: Environment, Economic and Ethics* (pp. 206–222). New York: Chapman and Hall.

Troeh, F. R., & Thompson, L. M. (1993). *Soils and Soil Fertility*. New York: Oxford University Press.

USDA-AMS. (2002). National organic program. Final rule: 7 CFR Part 205. U.S. Department of Agriculture, Agricultural Marketing Service (USDA-AMS), Washington, D.C. http://www.ams.usda.gov/AMSv1.0/getfile?dDocName=STELPRDC5082652. Accessed 19 July 2012.

USGS. (2001). Selected findings and current perspectives on urban and agricultural water quality by National Water-Quality Assessment Program. U.S. Department of Interior, U.S. Geological Survey (USGS), Washington, D.C. http://pubs.usgs.gov/fs/fs-047–01/pdf/fs047–01.pdf. Accessed 19 July 2012.

van Elsen, T. (2000). Species diversity as a task for organic agriculture in Europe. *Agriculture, Ecosystems & Environment, 77*, 101–109.

Wander, M., Traina, S., Stinner, B. R., & Peters, S. E. (1994). Organic and conventional management effects on biologically active soil organic matter pools. *Soil Science Society of America Journal, 58*, 1130–1139.

Watson, C. A., Atkinson, D., Gosling, P., Jackson, L. R., & Rayns, F. W. (2002). Managing soil fertility in organic-farming systems. *Soil Use and Management, 18* (Supplement s1), 239–247.

Willer, H., & Kilcher, L. (2012). *The World of Organic Agriculture—Statistics and Emerging Trends 2012*. Research Institute of Organic Agriculture (FiBL), Frick, and International Federation of Organic Agriculture Movements (IFOAM), Bonn, Germany. http://www.organic-world.net/yearbook-2012.html. Accessed 14 March 2013.

Wright, S. F., Star, J. L., & Paltineau, I. C. (1999). Changes in aggregate stability and concentration of glomalin during tillage management transition. *Soil Science Society of America Journal, 63*, 1825–1829.

Chapter 7
Pesticides, Food Safety and Integrated Pest Management

Dharam P. Abrol and Uma Shankar

Contents

7.1	Introduction	168
	7.1.1 Food Supplies—Retrospect and Prospect	169
	7.1.2 Integrated Pest Management and Food Safety	170
	7.1.3 Impact of Plant Introductions/Invasions	171
	7.1.4 Biosecurity Concerns	171
	7.1.5 Nutrition, Food Quality and Food Safety	171
	7.1.6 Food Safety	172
	7.1.7 Increasing Production to Provide Food Security	172
	7.1.8 Current Trends in the Use of Agrochemicals and Food Safety	174
	7.1.9 The New Paradigms	175
7.2	Pesticides for Food Sovereignty	176
7.3	Externalities of Pesticides	177
	7.3.1 Pesticide Residues in Food	177
	7.3.2 Pesticide Residues in Milk and Livestock Poisonings Due to Pesticides	177
	7.3.3 Pesticide Residues in Honey	177
	7.3.4 Pesticide–Pollinator Conflict	178
	7.3.5 Natural Enemy/Beneficial Insects	179
	7.3.6 Wild Birds and Mammals	180
	7.3.7 Groundwater and Soil and Environment	180
7.4	Pesticide Resistance	181
7.5	Pesticide Poisoning	182
7.6	IPM for Food Safety through Eco-Friendly Pesticides	182
	7.6.1 Strategies for Minimizing Pesticide Use	182
	7.6.2 Role of Push–Pull Strategies	183
	7.6.3 Location-Specific IPM	184
7.7	Different Components of IPM for Sustainable Food Productivity	184
	7.7.1 Cultural Methods	184
	7.7.2 Behavioral Control	185
	7.7.3 Host-Plant Resistance	185

D. P. Abrol (✉) · U. Shankar
Division of Entomology, Sher-e-Kashmir University of Agricultural Sciences & Technology of Jammu, 180 009 Chatha, Jammu & Kashmir, India
E-mail: dharam_abrol@rediffmail.com

U. Shankar
E-mail: umashankar.ento@rediffmail.com

7.7.4 Biological Control 186
7.7.5 Biopesticides 186
7.7.6 Botanical Pesticides 187
7.7.7 Transgenics 187
7.7.8 Chemical Control 187
7.7.9 Biotechnological Approaches 188
7.8 Pesticides and Integrated Pest Management 189
7.9 Integrated Pest Management and Food Safety (Case Studies) 189
7.9.1 Adoption of IPM and Success Stories 189
7.9.2 The Impact of IPM Interventions on Crop Productivity 190
7.10 Ethical and Moral Issues 190
7.11 Conclusions 191
References 191

Abstract A rapidly growing human population has resulted in a demand for increased food production. In ancient times, traditional agriculture could meet human food needs when the size of human population was small. As human population grew, the demand for food increased, which was met by increasing the land area under cultivation. As the land resources became limited, efforts were made to increase productivity by fighting the losses inflicted by insects, weeds, and plant pathogens through the discovery of pesticides. At the same time, pesticides proved to be more dangerous due to their indiscriminate and excessive use, contaminating food (milk, honey, cereals, vegetables, and fruits) and the environment (ground water, soil, etc.), resulting in pest resistance, pest resurgence, and pest outbreaks. Consequently, the need arose for eco-friendly integrated pest management strategies to produce food safe from the negative impact of pesticide residues. The present chapter addresses information on the pesticide usage and their negative impacts on food safety leading to the development of integrated pest management (IPM). In this chapter, IPM for food safety through eco-friendly pesticides is discussed.

Keywords Pesticides · Food safety · IPM · Environmental problems · Residues · Food sovereignty

7.1 Introduction

The world today faces the challenge of food safety, food security, and environmental sustainability. In the near future, these challenges are expected to worsen further if measures are not taken to address them. Growing population pressure has hastened environmental degradation and depletion of essential natural resources (UNFPA 2008). Nearly 1 billion people in the world are undernourished or suffer from chronic diseases as a result of food insecurity due to population growth, climate change, and urban development (FAO 2010). In the next 50 years, the global population is expected to reach 9 billion, doubling the demand for food, feed, and crop (FAO 2009).

Pesticides have always been a matter of wonder and curiosity for human beings due to their miraculous overall impact on society both as a boon as well as a bane since 2,500 BC. Elemental sulfur was the first known pesticide used for dusting in Sumeria about 4,500 years ago. By the fifteenth century, toxic chemicals such as arsenic, mercury, and lead were being applied to crops to kill pests. In the seventeenth century, nicotine sulfate was extracted from tobacco leaves for use as an insecticide. The nineteenth century saw the introduction of two more natural pesticides—pyrethrum, which is derived from chrysanthemums, and rotenone, derived from the roots of tropical vegetables (Tyler 2002). In 1939, Paul Müller discovered that DDT (dichloro diphenyl trichloroethane) was a very effective insecticide. The discovery of DDT as an insecticide by Paul Müller in 1939 resulted in a great impact on the control of insect pests and soon was widely used all over the world. At that time, pesticides served as a boon due to the control of diseases like malaria transmitted by mosquitoes and bubonic plague transmitted by fleas, both of which had killed millions of people over time. Nevertheless, this opinion changed after the publication of the book *Silent Spring* by Rachel Carson in 1962 (Carson 1962) and research began to explore alternative control methods and safe use of pesticides. Pesticide use in general is recognized as having very detrimental long-term effects on environment. According to World Health Organization (WHO), more than 20,000 unintentional deaths and 3 million poisonings occur due to misuse of pesticides in the Third World every year (Lowel 1998). Available estimates reveal that global pesticide use has increased 50-fold since 1950, and 2.5 million tonnes of agricultural pesticides are now used each year worldwide (Tyler 2002). No doubt pesticides are credited to have saved millions of lives by controlling diseases, such as malaria and yellow fever which are insect borne; however, their use causes a variety of adverse health effects and environmental pollution as described above. Alternate pest control methods and the restricted use of pesticides can minimize the risk of pesticide usage (Soundararajan 2012). Pesticides can prove to be the most effective instruments in crop protection and if correctly used, their effect is fast and complete, which makes them applicable against nearly every pest (Oomen and Bouma 2003).

7.1.1 *Food Supplies—Retrospect and Prospect*

In ancient times, when the population levels were low, traditional agriculture could meet the human food demand. In the present time, a rapidly growing human population is posing a great challenge to produce more food and ensure food security (FAO 2005). Options to meet the growing demand include increasing the land area under cultivation, more efficient use of available natural resources, and mitigating the heavy losses inflicted by insect pests, weeds, and diseases. However, the introduction of high inputs of agro-chemicals during the Green Revolution era have proved to be more dangerous due to their overuse, which resulted in the deterioration of soil and plant health, pesticide contamination of food and the environment, and pesticide resistance and pest outbreaks. Consequently, the need arose for IPM strategies to produce safe food and reduce the negative externalities caused by pesticides (Kogan 1998; Wilson and Tisdell 2001; Abrol and Shankar 2012; Shankar and Abrol 2012a).

7.1.2 Integrated Pest Management and Food Safety

IPM is a constantly evolving and dynamic approach of managing crop pests to minimize ecological problems in agricultural ecosystems. Soil microbes play a crucial role in ecosystem functioning through regulating nutrient cycling, maintaining soil fertility and controlling or suppressing pests through their unique biological mechanisms. In view of the above, IPM can ensure higher productivity, maintain soil health and promote ecological sustainability.

The world's natural resources, especially soils and water, need to be improved and restored as they are already under great stress. Loss of ecosystem resilience, because of additional demands of the growing population and rising aspirations for increased standards of living, has a severe impact on ecosystem services. Sustainable management of the world's soil resources is essential to effectively address these issues. While scientific capacity to eradicate famines was achieved during the twentieth century (Devereux 2009) in developing countries, there still remain more than 1 billion food-insecure people in the world (FAO 2009a) for whom the food supply will have to be doubled between 2005 and 2050 (Borlaug 2009). The persistent problem of food deficit and famines in Sub-Saharan Africa and South Asia is also exacerbated by the increase in food prices (Pinstrup-Anderson 2009), soil degradation (Lal 2009), drought stress, and climate change (IPCC 2007).

The Green Revolution of the 1960s, which brought about a quantum jump in agricultural production in South Asia and elsewhere, was based on growing input-responsive varieties under irrigated conditions with use of chemical fertilizers and largely chemical-based pest-control measures. Despite the impressive gains, the Green Revolution has stalled in South Asia since mid-1980s, along with the threat of excessive water withdrawal in north India (Kerr 2009). Crop yields have been practically stagnant since the 1990s. Global average increase in crop yield was 4 % year^{-1} between 1960 and 1980, 2 % year^{-1} during 1990s, and <1 % year^{-1} during 2000s. An increase in mean global temperature by 2 °C may reduce agricultural output in the main grain-producing regions of the world by about a quarter. Chronically undernourished/food-insecure people in the world, estimated at 850 million around 2004 (Borlaug 2007), has increased to 1020 million in 2009 (FAO 2009a) with a severely adverse impact on children in the poorest nations (Dugger 2007). The per capita grain consumption peaked in 1985 at 335 kg, and decreased to 302 kg by 2000. Grain production per person in Sub-Saharan Africa has decreased from 150 kg in 1960 to <120 kg in 2005 (Brown 2004) and is projected to decrease drastically by 2030 (Funk and Brown 2009).

7.1.3 Impact of Plant Introductions/Invasions

Plant introductions have seriously affected agriculture resulting in more than 40 % of economic losses in crop production (Pimentel et al. 2001) and has become a major financial burden on the resources available to manage natural areas (Williams and Timmins 2002; Williamson 2002; Vila et al. 2010; Williams et al. 2010; Oreska and Aldridge 2011). Agricultural crops are facing the destructive activities of several pests and diseases which affect crop productivity and require heavy dosages of pesticides, which eventually result in health hazards, environmental degradation and economic losses. The excessive use of chemical pesticides has triggered several externalities such as degraded soils, groundwater pollution, pest/disease resistance and the resurgence of pests, food contamination, and residue problems. Evidently, robust tools are needed to identify which plant introductions are likely to cause harm (Groves et al. 2001) and their threat to species diversity in various habitats (Booth et al. 2003; Hulme 2003, 2012) around the world (Pimentel et al. 2005; Dogra et al. 2010).

7.1.4 Biosecurity Concerns

Biosecurity measures relevant to plants and plant products are required to prevent the entry and establishment of plant pests, including plant-disease-causing organisms (van der Graaff and Khoury 2010). The WTO SPS (Sanitary and Phytosanitary) agreement recognizes standards, guidelines and recommendations developed by the FAO/WHO Codex Alimentarius for food safety, the International Plant Protection Convention (IPPC) for Plant Health and the World Organization for Animal Health (OIE), and WTO SPS agreement strongly encourages the parties to base their SPS measures on these. On the other hand, food safety concerns due to pesticide residues in food are not often raised, but when raised, these concerns are more far reaching, addressing the principles of pesticide registration, such as maximum residue levels (MRLs) for pesticides not registered in the importing country and MRLs lower than the Codex Alimentarius standards. In view of the limited resources at the national level, considerable thought has to go into how to make best use of all biosecurity infrastructures in a country. While a substantial number of countries are concentrating on food safety, animal and plant health in single agencies, other countries seek other ways to make best use of synergies among those services.

7.1.5 Nutrition, Food Quality and Food Safety

In food production systems, plant biosecurity is a state of preparedness that ensures a safe, affordable, and available supply of food and feed. Food protests and riots in at least 30 nations are evidence of the significant linkage between food security and national security (Shelburne 2008). Without effective plant biosecurity programs to protect the world's staple crops, food safety and security will decline in the developing world (Shelburne 2008). The future of national governments has largely been determined by their food security (Shankar and Abrol 2012a). The failure of food security will further compromise global economic development and international programs to reduce hunger and improve health. Without effective plant biosecurity programs to protect the world's natural plant systems, the ecosystem services that they provide to support humans will decline, thus compromising the development of sustainable societies.

7.1.6 Food Safety

Food quality and safety have become serious issues, and a major concern to governments, industry, and consumers. Food safety issues focus most of the time on microorganisms, molds, or toxins which occur during storage. Stored-product insects can cause serious post-harvest losses, estimated to be from 9 % in developed countries to 20 % or more in developing countries (Pimentel et al. 1991). The most serious cause of concern is the contamination of food products by the presence of live insects, insect products such as chemical excretions or silk, and dead insects and insect body fragments. Furthermore, accumulation of chemical insecticide residues in food, as well as human exposure to dangerous chemical pesticides as a result of pest-control efforts is another cause of concern. Worldwide, an annual loss of 8–10 % (13 million t of grains lost due to insects and 100 million t due to failure to store grain properly) is estimated in stored-food commodities. Most storage losses are due to inadequate and poor storage facilities, which allow attacks by insect pests and diseases, causing enormous losses annually (Shankar and Abrol 2012b). Evidently, reliable pesticides and storage techniques are required to limit the damage to food grains in storage in most developing countries. For the last three decades, food-borne and water-borne viral infections have been increasingly recognized as causes of illness in humans because of the improved diagnostic methods that have enhanced detection of some virus groups and the increased marketing of fresh and frozen foods (Norrung 2000). Evidently, the judicious application of food processing and storage and preservation methods will help prevent outbreaks of food-borne diseases resulting from the consumption of contaminated food (Prokopov and Tanchev 2007).

7.1.7 Increasing Production to Provide Food Security

Natural resources such as agricultural land and water are being depleting at a rapid rate and making the food security situation horrible in all regions of the globe. Better management of water resources is needed in many regions, and use of plant varieties better adapted to regional weather conditions would increase water-use efficiency. In the present time, probably the immediate response to the need for increasing production of food is a more intensive use of agrochemicals. Agrochemicals include two large groups of compounds: chemical fertilizers and pesticides. The use of chemical fertilizers has tremendously increased worldwide since the 1960s and was largely responsible for the "Green Revolution"; that is, the massive increase in production obtained from the same area of land with massive inputs of mineral fertilizers (nitrogen, phosphorus, and potassium) and intensive irrigation. This has largely constituted the success story of the increased production of rice, corn, and wheat worldwide (Borlaug and Dowswell 1993). This revolution was assisted also with the introduction of more productive varieties of rice and dwarf wheat that more directly responded to increased fertilizer and irrigation inputs.

The use of pesticides, including insecticides, fungicides, herbicides, rodenticides, etc., to protect crops from pests, significantly reduced crop losses and improved yields of crops such as corn, maize, vegetables (Fig. 7.1), potatoes, cotton, as well as to protect cattle from diseases, insects, and ticks and to protect humans from malaria vectors. The world has known a continuous growth of pesticide usage, both in number of chemicals and quantities, utilized in agriculture. Pesticides are poisons intentionally dispersed onto croplands and pasturelands to control pests, but they also cause serious side effects on non-target species. Residues of pesticides contaminate soils and water, remain in the crops, enter the food chain, and finally are ingested by humans through foodstuffs and water (Barcelo and Hennion 1997; Taylor et al. 2003). Insect pests develop resistance to insecticides and, as a consequence, chemical companies continuously synthesize new chemicals. The European Union (EU) has registered more than 800 chemicals as pesticides. However, we know very little about the environmental behavior of these chemicals and about their effect upon human health (Sharpe 1999; EEA 2005).

The application of different agrochemicals varies with the region. For instance, in North America and Western Europe, due to high costs of labor, the chemical control of weeds is mostly done with herbicides, in contrast to East Asia and Latin America where herbicides are much less used. In the tropical regions, where insect pests and plant diseases are more frequent, pesticides are generally applied in massive amounts, both on small farms as well as on cash crops, that is, industrial plantations growing bananas, coffee, maize, and cotton. The residues of pesticides, especially the organochlorine and organophosphorous compounds, are found in soils, the atmosphere and in the aquatic and marine environments in relatively high concentrations (Carvalho et al. 1997). Studies of people living in rural areas of some countries, such as Costa Rica and Nicaragua, indicate direct exposure of many workers to pesticides and acute poisoning and chronic exposure with effects on human reproductive and central nervous systems (Munoz-de-Toro et al. 2006; Bretveld et al. 2006). The human population at large is exposed to pesticide residues that are dispersed in the environment. Recent studies, carried out in coastal areas of Mexico, Nicaragua, and Vietnam, show that marine species, such as clams and oysters, which are important components of the diet of riverine populations, may contain relatively high concentrations of DDT, lindane, hexachlorocyclohexane (HCHs), endosulfan, toxaphene, and chlorpyrifos, among other crop protection chemicals (Nhan et al. 1999; Carvalho et al. 2002; Taylor et al. 2003; Carvalho 2005).

Fig. 7.1 a) Exhibiting a spraying of crops using proper mask and clothing, and **b)** wrong practice without protective mask and clothing risking health hazards and pesticidal poisoning

7.1.8 Current Trends in the Use of Agrochemicals and Food Safety

Agrochemicals are an obvious part of current agriculture production systems. Regarding their use, in the present, there are two opposite trends, each one related to a geographic region. Developed countries, including EU, United States and Canada, approved new laws restraining the use of agrochemicals. This legislation aims at protecting consumers through a more thorough toxicological testing of compounds and the enforcement of lower pesticide residue limits for food and water (Harris 2002). For example, the maximum permitted concentration of pesticides in drinking water set by the EU is 0.1 mg/liter (Directive 80/778/EEC), challenging even the detection limits of current analytical methods (Barcelo and Hennion 1997).

This move is driven by health concerns of the public and consumer associations that perceive the presence of pesticide residues in the environment as detrimental to the quality of life. Results of scientific research support this point of view. Actually, it has been shown that even in low concentrations, the combined effect of xenobiotic chemicals causes a suppression of the immune response and a hypersensitivity to chemical agents. In many cases, a relationship between organochlorine residues and breast cancer, and between polychlorinated biphenyls (PCBs) and reduced sperm count and male sterility has been documented (Uri 1997; Rivas et al. 1997; Sharpe 1999; EEA 2005). Developed countries are moving, therefore, in the direction of fewer chemicals and more "green products." Furthermore, new pesticides are less persistent in the environment (more environmentally friendly) than classic pesticides. These new pesticides, however, are more costly compared with older chemicals, and developing countries generally cannot afford them.

Developing countries are moving in a different direction in this matter. They need to increase agricultural production, and the use of crop protection chemicals seems a simple way for obtaining higher crop yields. Therefore, they use chemicals that are cheap, such as DDT, HCH, and BCH, because either their patents have expired and are easy to synthesize or they are sold by developed countries. The increased use of cheap pesticides results in the contamination of environment, the exposure of the public, and increased residues in food. Risks to public health are higher too. Countries in the tropical belt and with industrial capability to produce pesticides, such as India, invest in cheap pesticides such as DDT and HCHs. The sale of these pesticides to Bangladesh, Philippines, and Latin America results in a massive use of organochlorine compounds in tropical agriculture.

7.1.9 The New Paradigms

Climate change, food insecurity, and energy demand are major concerns for modern agriculture and their impact is increasing rapidly. The last decade has seen new developments in food production: the genetic engineering of organisms and the organic chemical-free agriculture. Biotechnology and release of genetically modified organisms (GMOs), such as engineered soybean, colza, maize, and tomatoes, did promise a solution to food security needs and nutritional problems (Khush 2002). Interestingly, the development of genomics and patented GMOs is in the hands of private research companies that have largely surpassed the public research efforts (Pingali and Traxler 2002). According to the main private biotechnology companies (Aventis, Monsanto, Novartis, Zeneca, etc.), these GMOs may be resistant to insect pests, molds, frost, dry conditions, etc., and could revolutionize agriculture (Pingali and Traxler 2002). For example, soybean and other plants were modified to be tolerant to glyphosate, a common herbicide used to fight weeds allowing for much higher crop yields. However, because the weeds become increasingly resistant to this herbicide, the use of these genetically modified (GM) plants renders the farmers dependent on the use of more and more glyphosate. Interestingly, glyphosate is produced by the same company that produces GM herbicide-resistant plants (Sharpe 1999).

Concerns have been expressed also about the spread of GMOs and their impact on the genetic variability of wild plants and the (unknown) risk of health disorders they may cause in consumers. Because precaution has not been observed by the private biotechnology companies concerning possible health hazards of GMOs, many countries, including the EU, are reluctant to license GMOs in spite of optimism and self-confidence of these companies in the stated economic, social, and ecological qualities of their products (Nickson 1999). Moreover, it has not been demonstrated that GMOs would contribute to or solve the need for increased production of food for a growing world population (Falcon and Fowler 2002; Tripp 2002).

Another paradigm has been the development of organic agriculture. Although started in the 1920s, it has grown so much worldwide in the last 20 years that it has already been used on a few million hectares (Tamm 2001). Organic agriculture respects the normal functioning of ecosystems, avoids the use of agrochemicals, and leads to more healthy food, "free" of synthetic chemical residues. Notwithstanding the health value of better quality agriculture products, organic agriculture does not appear to have the potential for mass production of the amount of calories needed to feed humanity. The development of organic agriculture may, therefore, contribute to improved food safety but does not cope with food security. Agrichemicals will be needed to increase agricultural production further.

As discussed earlier, pesticides continue to play the key role in arthropod pest management. However, some key species (e.g., *Heliothis armigera* Hubner, *Plutella xylostella* L.) are developing high levels of insecticide resistance and with the human and environmental health concerns related to pesticide use, cultural techniques are increasingly being used. The combination of genetic resistance, hygiene, and monitoring of crops for threshold levels of infestation, allows the most economic and effective use of chemical controls with the result that economic yields can be maximized.

7.2 Pesticides for Food Sovereignty

Worldwide, about 4 million tons of pesticides per year are used although their distribution is uneven in different countries (FAOSTAT 2010). In the USA, approximately 500 million kg of more than 600 different pesticides are applied annually at a cost of US$ 10 billion (Pimentel and Greiner 1997). Despite the widespread application of pesticides in the United States at recommended dosages, pests (insects, plant pathogens, and weeds) destroy 37% of all potential crops (Pimentel 1997)—insects destroy 13%, plant pathogens 12%, and weeds 12%. In general, each dollar invested in pesticide control returns about US$ 4 in protected crops (Pimentel 1997). Although pesticides are generally profitable in agriculture, their use does not always decrease crop losses. For example, despite the more than 10-fold increase in insecticide (organochlorines, organophosphates, and carbamates) use in the United States from 1945 to 2000, total crop losses from insect damage have nearly doubled from 7 to 13% (Pimentel et al. 1991).

7.3 Externalities of Pesticides

7.3.1 Pesticide Residues in Food

The majority of foods purchased in supermarkets have detectable levels of pesticide residues. For instance, of several thousand samples of food, the overall assessment in 8 fruits and 12 vegetables is that 73 % have pesticide residues (Baker et al. 2002). In five crops (apples, peaches, pears, strawberries, and celery) pesticide residues were found in 90 % of the samples taken. Of interest is the fact that 37 different pesticides were detected in apples as residues (Groth et al. 1999). Up to 5 % of the foods tested in 1997 contained pesticide residues that were above the Food and Drug Administration (FDA) tolerance levels. Although these foods violated the U.S. tolerance of pesticide residues in foods, these foods were consumed by the public because the food samples were analyzed after the foods were sold in the supermarkets.

7.3.2 Pesticide Residues in Milk and Livestock Poisonings Due to Pesticides

In addition to pesticide problems that affect humans, several thousand domestic animals are accidentally poisoned by pesticides each year. In addition to animal carcasses, pesticide-contaminated milk cannot be sold and must be disposed of. In some instances, these losses are substantial. In Oahu, Hawaii in 1982, 80 % of the milk supply, worth more than US$ 8.5 million, was condemned by the public health officials because it had been contaminated with the insecticide heptachlor (Baker et al. 2002). This incident had immediate and far-reaching effects on the entire milk industry on the island. The best estimates indicate that about 20 % of the total monetary value of animal production, or about US$ 4.2 billion, is lost to all animal illnesses, including pesticide poisonings. It is reported that 0.5 % of animal illnesses and 0.04 % of all animal deaths reported to a veterinary diagnostic laboratory were due to pesticide toxicosis. Thus, US$ 21.3 and US$ 8.8 million, respectively, are lost to pesticide poisonings (FAO 1986).

7.3.3 Pesticide Residues in Honey

Wild bees are one of the most important groups of pollinators in the temperate zone (Kevan 1999). Therefore, population declines have potentially negative impacts for both crop and wildflower pollination. Bees provide key ecosystem services essential to maintaining wild plant diversity (Ashman et al. 2004; Aguilar et al. 2006; Potts et al. 2010) and agricultural productivity (Klein et al. 2007; Gallai et al. 2009; Lenda et al. 2010). Many plant species that are directly dependent on insect pollination for fruit and seed production (Wilkaniec et al. 2004; Morandin and Winston 2005; Velthuis and van Door 2006) might experience pollination limitation when pollinator species are scarce (Ashman et al. 2004). Therefore, the decline in wild bee populations reported throughout Europe and North America (Steffan-Dewenter et al. 2005; Biesmeijer et al. 2006; Potts et al. 2010) is alarming. One of the major factors causing declines in bee diversity and abundance is habitat loss and fragmentation, driven mostly by intensification of agriculture (Banaszak 1995; Steffan-Dewenter 2003; Le Feon et al. 2010). Other factors include pesticide use (Alston et al. 2007; Brittain et al. 2010), the impact of non-native invasive species (Moron et al. 2009), competition with managed populations of *Apis mellifera* (Walther-Hellwig et al. 2006) or *Bombus terrestris* (Kenta et al. 2007), the spread of bee pathogens (Colla et al. 2006) and genetic introgression (Kraus et al. 2011).

7.3.4 Pesticide–Pollinator Conflict

Wild pollinating bees play an important role in the farming industry and in maintaining ecosystem balance. For food supplies in the world, 35 % of crop plants are animal pollinated and many wild plants need bee pollination. However, due to the intensification of farming practices, indiscriminate use of agrochemicals, atmospheric pollution, pests and diseases, crop monoculture, and so on, the abundance of wild bees is declining throughout the world, which will pose threats to the integrity of biodiversity, global food webs, and human health. In the last 20 years, the exploitation and use of wild bees have rapidly advanced throughout the world.

Honeybees and wild bees are vital for pollination of fruits, vegetables, and other crops (Abrol et al. 2012). Bees are essential in the production of about one-third of crops in the USA and around the world; their benefits to US agriculture are estimated to be about US$ 40 billion year^{-1} (Pimentel et al. 1997). Because most insecticides used in agriculture are toxic to bees, pesticides have a major impact on both honeybee and wild bee populations. According to Mayer (cited by Pimentel and Greiner 1997), the yearly estimated loss from partial honeybee kills, reduced honey production plus the cost of moving colonies totals about US$ 25.3 million year^{-1}. Also, as a result of heavy pesticide use on certain crops, beekeepers are excluded from 4–6 million hectares of otherwise suitable apiary locations. Mayer estimates the yearly loss in potential honey production in these regions is about US$ 27 million each year. In addition to these direct losses caused by the damage to honeybees and honey production, many crops are lost because of the lack of pollination. Estimates of annual agricultural losses due to the reduction in pollination caused by pesticides may be as high as US$ 4 billion year.

The role of integrated pest management (IPM) in pollinator conservation is essential as it discourages the use of pesticides which pose negative effects on non-target organisms such as pollinators. Pesticide-reduction strategies in IPM may have multiple benefits, not just for crop pest predators, but also for pollinators (Johansen and Mayer 1990). A number of factors may be detrimental and limit the efficiency of bees as pollinators such as the effects of pesticides on pollinators (Incerti et al.

2003; Desneux et al. 2007; Chauzat et al. 2009; Krupke et al. 2012). Bee death is not the only outcome of pesticide contamination. An amount of pesticide too small to kill a bee may disrupt cognitive abilities, communication, various behaviors, and physiology. Exposure to chemicals that compromise the ability of worker bees to forage and communicate with other bees may negatively affect colony health (Desneux et al. 2007). Neonicotinoids have been found in pollen loads brought to hives by honey bees (Chauzat et al. 2006), in pollen stored within honey bee hives (Mullin et al. 2010; Krupke et al. 2012), and in honey stored within hives (Chauzat et al. 2009). In a separate study, bumble bees were exposed in a flight cage to blooming cucumbers treated with a foliar spray of imidacloprid applied "at field dose" (Incerti et al. 2003); a third of the bumble bees died within 48 hours of exposure. Krupke et al. (2012) reported residue levels of 3.9 ppb of clothianidin in corn pollen resulting from seed treatment at label rates. While residues in nectar resulting from thiamethoxam (e.g., Crusier) seed treatments remain unknown, residues in corn pollen after treatment to corn seed at label rates resulted in 1.7 ppb of thiamethoxam (Krupke et al. 2012). Residues are also found in contaminated dust released from seed-planting equipment (Greatti et al. 2006; Krupke et al. 2012; Tapparo et al. 2012) and in weeds growing within or adjacent to treated fields (Krupke et al. 2012). Based on the analysis of honeybee and related pollination losses from wild bees caused by pesticides, pollination losses attributed to pesticides are estimated to represent about 10 % of pollinated crops and have a yearly cost of about US$ 210 million year^{-1}.

7.3.5 Natural Enemy/Beneficial Insects

In both natural and agricultural ecosystems, many species, especially predators and parasites, control or help control plant-feeding arthropod populations. Indeed, these natural beneficial species make it possible for ecosystems to remain "green." With the parasites and predators keeping plant-feeding populations at low levels, only a relatively small amount of plant biomass is removed each growing season by arthropods (Hairston et al. 1960; Pimentel 1988). Like pest populations, beneficial natural enemies and biodiversity (predators and parasites) are adversely affected by pesticides (Pimentel et al. 1993). For example, the following pests have reached outbreak levels in cotton and apple crops after the natural enemies were destroyed by pesticides: cotton bollworm, tobacco budworm, cotton aphid, spider mites, and cotton loopers; European red mite on apples as well as red-banded leaf roller, San Jose scale, oyster shell scale, rosy apple aphid, wooly apple aphid, white apple aphid, two-spotted spider mite, and apple rust mite. Major pest outbreaks have also occurred in other crops. Also, because parasitic and predaceous insects often have complex searching and attack behaviors, sub-lethal insecticide dosages may alter this behavior and in this way disrupt effective biological controls. For example, from 1980 to 1985, insecticide use in rice production in Indonesia drastically increased (Oka 1991). This caused the destruction of beneficial natural enemies of the brown plant hopper and this pest population exploded. Rice yield decreased to the extent that rice had to be imported to Indonesia. The estimated cost of rice loss in just a 2-year period was US$ 1.5 billion (FAO 1988). Fungicides also can contribute to pest outbreaks when they reduce fungal pathogens that are naturally parasitic on many insects. For example, the use of benomyl reduces populations of entomopathogenic fungi, resulting in increased survival of velvet bean caterpillars and cabbage loopers in soybeans. This eventually leads to reduced soybean yields. When outbreaks of secondary pests occur because their natural enemies are destroyed by pesticides, additional and sometimes more expensive pesticide treatments have to be made in efforts to sustain crop yields. This raises the overall production costs and contributes to pesticide-related problems. An estimated US$ 520 million can be attributed to costs of additional pesticide application and increased crop losses, both of which follow the destruction of natural enemies by various pesticides applied to crops (Pimentel et al. 1991).

Parasites, predators, and host-plant resistance are estimated to account for about 80% of the non-chemical control of pest arthropods and plant pathogens in crops (Pimentel et al. 1991). Many cultural controls such as crop rotations, soil and water management, fertilizer management, planting time, crop-plant density, trap crops, polyculture, and others provide additional pest control. Together, these non-pesticide controls can be used to effectively reduce US pesticide use by more than 50% without any reduction in crop yields or cosmetic standards (Pimentel et al. 1993).

7.3.6 *Wild Birds and Mammals*

Wild birds and mammals are damaged and destroyed by pesticides and these animals make excellent "indicator species." Deleterious effects on wildlife include death from the direct exposure to pesticides or secondary poisonings from consuming contaminated food; reduced survival, growth, and reproductive rates from exposure to sub-lethal dosages; and habitat reduction occurs through the elimination of food resources and refuges.

7.3.7 *Groundwater and Soil and Environment*

Certain pesticides applied at recommended dosages to crops eventually end up in ground and surface waters. The three most common pesticides found in groundwater are aldicarb, alachlor, and atrazine (USEPA 1992; Barbash 1997). Estimates are that nearly one-half of the groundwater and well water in the United States is or has the potential to be contaminated (Holmes et al. 1988; USGS 1996).

Two major concerns about groundwater contamination with pesticides are that about one-half of the human population obtains its water from wells and once groundwater is contaminated, the pesticide residues remain for long periods of time. Not only are there extremely few microbes present in groundwater to degrade the pesticides, but the groundwater recharge rate is also less than 1% year^{-1} (CEQ 1980).

7.4 Pesticide Resistance

About 520 insect and mite species, a total of nearly 150 plant pathogen species, and about 273 weeds species are now resistant to pesticides (Stuart 2003). Increased pesticide resistance in pest populations frequently results in the need for several additional applications of the commonly used pesticides to maintain crop yields. These additional pesticide applications compound the problem by increasing environmental selection for resistance. Despite efforts to deal with the pesticide resistance problem, pesticide resistance continues to increase and spread to other species. A striking example of pesticide resistance occurred in northeastern Mexico and the Lower Rio Grande of Texas (NAS 1975). Over time, extremely high pesticide resistance had developed in the tobacco budworm population on cotton. Finally, approximately 285,000 ha of cotton had to be abandoned because the insecticides were totally ineffective due to the extreme pesticide resistance in the budworm. The economic and social impact of this on Texan and Mexican farmers dependent on cotton was devastating. The study by Carrasco-Tauber (1989) indicates the extent of costs associated with pesticide resistance; they reported a yearly loss of US$ 45–120 ha^{-1} to pesticide resistance in California cotton. A total of 4.2 million hectares of cotton were harvested in 1984; thus, assuming a loss of US$ 82.50 ha^{-1}, approximately US$ 348 million of the California cotton crop was lost to resistance. Since US$ 3.6 billion of US cotton was harvested in 1984 (USBC 1990), the loss due to resistance for that year was approximately 10%. Assuming a 10% loss in other major crops that receive heavy pesticide treatments in the United States, crop losses due to pesticide resistance are estimated to be about US$ 1.5 billion year^{-1}. Furthermore, efforts to control resistant *Heliothis* spp. (corn ear worm) exact a further cost when large, uncontrolled populations of Heliothis and other pests disperse onto other crops. In addition, populations of the cotton aphid and the whitefly exploded as secondary cotton pests because of their pesticide resistance and their natural enemies' exposure to high concentrations of insecticides. The total external cost attributed to the development of pesticide resistance is estimated to range between 10 and 25% of current pesticide treatment costs (Harper and Zilberman 1990), or more than US$ 1.5 billion each year in the United States. In other words, at least 10% of pesticide used in the USA is applied just to combat the increased resistance that has developed in several pest species.

Resistant plant varieties can be used as the primary method of insect control, or as a component of an integrated pest management program (Wiseman 1994). Insect-resistant varieties have been developed for corn (Wiseman et al. 1996) and for rice and soybean (Carozzi and Koziel 1997). For common beans, varieties resistant to pre- and post-harvest damage by beetles (Beebe et al. 1993; Ishimoto 1999; Kornegay and Cardona 1991) and varieties showing multiple resistance to insect attack (Bueno et al. 1999) are being selected.

7.5 Pesticide Poisoning

Human pesticide poisonings and illnesses are clearly the highest price paid for all pesticide use. The total number of pesticide poisonings in the United States is estimated to be 300,000 year^{-1} (EPA 1992). Worldwide, the application of 3 million metric tons of pesticides results in more than 26 million cases of non-fatal pesticide poisonings (Richter 2002). Of all the pesticide poisonings, about 3 million cases require hospitalization and there are approximately 220,000 fatalities and about 750,000 chronic illnesses every year (Hart and Pimentel 2002). Ample evidence exists concerning the carcinogenic threat related to the use of pesticides. These major types of chronic health effects of pesticides include neurological effects, respiratory and reproductive effects, and cancer. There is some evidence that pesticides can cause sensory disturbances as well as cognitive effects such as memory loss, language problems, and learning impairment (Hart and Pimentel 2002). Studies have also linked pesticides with reproductive effects. For example, some pesticides have been found to cause testicular dysfunction or sterility (Colborn et al. 1996). Sperm counts in males in Europe and the United States, for example, declined by about 50% between 1938 and 1990 (Carlsen et al. 1992). Currently, there is evidence that human sperm counts continue to decrease by about 2% year^{-1} (Pimentel and Hart 2001).

Certain pesticides have been shown to induce tumors in laboratory animals and there is some evidence that suggest similar effects occur in humans (Colborn et al. 1996). Many pesticides are also estrogenic—they mimic or interact with the hormone estrogen—linking them to an increase in breast cancer among some women; the breast cancer rate rose from 1 in 20 in 1960 to 1 in 8 in 1995 (Colborn et al. 1996). The negative health effects of pesticides can be far more significant in children than adults, for several reasons. First, children have higher metabolic rates than adults, and their ability to activate, detoxify, and excrete toxic pesticides differs from adults. Also, children consume more food than adults and thus can consume more pesticides per unit weight than adults. This problem is particularly significant for children because their brains are more than five times larger in proportion to their body weight than adult brains, making cholinesterase even more vital. According to the Environmental Protection Agency (EPA), babies and toddlers are 10 times more at risk for cancer than adults (Hebert 2003; USEPA 2005). Available estimates suggest that human pesticide poisonings and related illnesses in the United States cost about US$ 1 billion year^{-1} (Pimentel and Greiner 1997).

7.6 IPM for Food Safety through Eco-Friendly Pesticides

7.6.1 Strategies for Minimizing Pesticide Use

Different pest management strategies or practices are being used by farmers to reduce pest infestations. The most important tool for evaluating an insect problem is its economic threshold. The economic threshold is the pest density at which control measures need to be resorted to in order to prevent pesticide damage from reaching economic levels. Cultural practices can help protect the crop and reduce pesticide damage. Some of the most important cultural practices include tillage practices that disrupt the insect's life cycle and destroy crop residues, changing planting dates to minimize insect impact, and crop rotations that include non-susceptible crops. Some cultural practices also help increase or decrease the population of natural enemies of pests. The utilization of resistant crop varieties is another method for controlling pests. The use of resistant varieties also helps considerably to reduce pesticide selection pressure. In addition, biocontrol agents hold great promise to control pests and minimize pesticide use but often require a long time to establish. The IPM principle does not preclude chemical pesticide use, but rather uses it as one of the tools in the management package, to be used prudently and integrated with other tools. Thus, the concept of IPM of insects contains three basic elements: (i) maintaining insect populations below levels that cause economic damage; (ii) the use of multiple tactics to manage insect populations; and (iii) the conservation of environment quality. Excessive and indiscriminate use of pesticides endangers the health of farm workers and consumers of agricultural products worldwide (Goodell 1984). People do not want to rely on chemicals and look for alternative strategies for pest control such as cultural, biological, and bio-rational methods. According to Rola and Pingali (1993), this has been necessitated as a result of the negative impact of pesticides on biodiversity and food and water quality, and the high cost of pesticides and the development of resistance in pests. The Government of India adopted IPM as a cardinal principle of plant protection in 1985. Programs for training both extension workers and farmers in IPM were started throughout the country. As agricultural pests cause substantial crop losses throughout the world, in the past farmers had to manage this problem in order to secure their basic subsistence needs, and so they practiced and developed cultural and mechanical pest control based on trial and error. Over a period of time, these practices have become a part of their production management system.

7.6.2 Role of Push–Pull Strategies

Lepidopteran pests such as stem borers (*C. partellus* Swinhoe, *Eldana saccharina* Walker, *Busseola fusca* Fuller and *Sesamia calamistis* Hampson) are serious pests of maize (*Zea mays*) and sorghum (*Sorghum bicolor*) throughout eastern and southern Africa causing yield losses of 10–50%. The push–pull strategy has been a blessing for millions of people in South Africa to control these pests (Kfir et al. 2002; Khan and Pickett 2004). This strategy involves the combined use of intercrops and trap crops, using plants that are appropriate for the farmers and that also exploit natural enemies. Khan and Pickett (2004) report that this strategy has contributed to increased crop yields and livestock production, resulting in a significant impact on food security and livelihood of farmers in the region.

7.6.3 Location-Specific IPM

Many countries of the world have tropical or subtropical climates where the losses caused by pests are most serious, thereby requiring protection of the crop for any significant yield. The management of weeds, insect pests, and pathogens is one of the most challenging jobs in these areas, and consequently the use of chemicals for controlling these pests is increasing continuously. The increased use of insecticides not only results in health hazards but also increases the cost of production. The use of such chemicals can be minimized by the adoption of location-specific IPM techniques, which include the use of natural enemies, varieties with multiple resistance and less toxic chemicals such as biopesticides. Site-specific pest management utilizes spatial information about pest distribution to apply control tactics only where pest density is economically high within a field (Park et al. 2007). In Indonesia, IPM has proved a great success in rice cultivation through a combination of appropriate technology and government support (Roling and van de Fliert 1994).

7.7 Different Components of IPM for Sustainable Food Productivity

7.7.1 Cultural Methods

Since ancient times, farmers have been relying on cultural or physical practices for the management of pests. Cultural practices include making the cropping environment unfavorable or less suitable for pests and more suitable for natural enemies. The best example is the push–pull strategy (Cook et al. 2007) where the crop is made unattractive (push) while another food source is made attractive (pull). In eastern and southern Africa, stem borers in maize and sorghum were repelled by non-host intercrops (*Molasses minutiflora, Desmodium uncinatum*, and *Desmodium intortum*) (push) and concentrated on attractive trap plants (*Pennisetum purpureum* and *Sorghum vulgare* sudanense) (pull). This was due to the fact that *Melinis minutiflora* increased parasitism by *Cotesia flavipes*, and *Desmodium* suppressed the parasitic weed *Striga hermonthica*. Similar success was achieved in the management of melon bug (*Aspongopus viduatus*) in the Sudan where handpicking by women and children collected more than 200 t of bugs, which were then burned (Bashir et al. 2003). Picking and burning of bolls infested with the pink bollworm (*Pectinophora gossypiella*) at the end of the growing season also proved very successful for the management of this pest (Brader 1979).

Agro-ecosystem analysis In modern agriculture, the determination of economic threshold level can be replaced with agro-ecosystem analysis due to the complexity in fixing an arbitrary mean for major insect pests. Pest monitoring is one of the most important components of IPM to make proper decisions to manage the pest problem in the long term. Many agro-ecosystems are unfavorable environments for natural enemies because of the high levels of disturbance. Therefore, understanding biotic interactions in agro-ecosystems and how they can be utilized to support crop productivity and environmental health is one of the fundamental principles underlying IPM (Shennan 2008). Important elements for understanding biotic interactions include consideration of the effects of diversity; species composition and food-web structure on ecosystem processes; the impacts of timing, frequency and intensity of disturbance; and the importance of multitrophic interactions.

7.7.2 Behavioral Control

Utilization of behavioral attributes of insect pests is one of the best options for their management. Behavioral control can be achieved by variety of behavior-modifying chemicals such as pheromones, which are efficient even at low population densities without adversely affecting natural enemies. Pest management is becoming increasingly difficult because of the changing climatic patterns where insecticides are of little help compared to the pheromones and other semiochemicals, which hold great promise (Witzgall et al. 2010). The most important attribute of semiochemicals is their specificity, as most of them are bioactive only towards certain species or groups of insect pests, and as such some efforts have been directed towards the development of reliable controlled-release technologies for semiochemicals (Clarke 2001).

7.7.3 Host-Plant Resistance

Host-plant resistance, natural plant products, biopesticides, natural enemies, and agronomic practices offer potentially viable options for IPM. They are relatively safe for non-target organisms and humans. Host-plant resistance to insect pests is a key component in IPM as it is the most economical method and is compatible with other methods of pest control. Five diseases (blast, bacterial blight, sheath blight, tungro, and grassy stunt) and four insects (brown planthopper, green leafhopper, stem borer, and gall midge) are of major importance for rice in tropical and subtropical Asia. Most of the modern varieties of rice contain moderate resistance to one or more of these major diseases and insect pests. Resistance to bacterial blight has been achieved by marker-assisted breeding (Singh et al. 2001), and resistance to bacterial blight, sheath blight and stem borer has been achieved by transgene pyramiding (Datta et al. 2002).

7.7.4 Biological Control

Like other natural enemies, insect pathogens can exert considerable control over target populations. Spectacular crashes of insect pest populations have been reported to be caused by epizootics (Evans 1986; McCoy et al. 1988). The natural epizootics produced by nuclear polyhedrosis viruses (NPVs) of sawflies (*Gilpinia hercyniae* Hartig and *Neodiprion* spp.), gypsy moth (*Lymantria dispar L.*), Split NPV of Spodoptera litura and several other insects are often credited with eliminating the need for further interventions (Kaya 1976; 1986; Monobrullah and Shankar 2008). Management of these pests still relies heavily on the use of pesticides with their associated limitations. If appropriately applied, biological control offers one of the most promising, environmentally sound and sustainable tools for control of arthropod pests and weeds (van Lenteren et al. 2006; van Driesche et al. 2007). Management of non-indigenous and indigenous pests in many countries of the world has been achieved through public support for the biological control options. There exist significant opportunities for increasing the use and effectiveness of biological control agents.

7.7.5 Biopesticides

The term biopesticide encompasses many aspects of pest control such as entomophagous nematodes, plant-derived pesticides and secondary metabolites from microorganisms, pheromones, and genes used to transform crops to express resistance to pests. India has a rich biodiversity of flora and fauna with the potential for development into commercial technologies. Nevertheless, the adoption of biopesticides and bioagents remains extremely low because of a number of factors relating to technology, socio-economics, and institutional factors and policies. Some success stories of successfully utilizing biopesticides and biocontrol agents in Indian agriculture include control of the diamondback moth by *Bacillus thuringiensis,* mango hoppers, mealybugs, and coffee pod borer by *Beauveria bassiana, Helicoverpa armigera* on gram by *H. armigera* NPV, white fly on cotton by neem products, sugarcane borers by *Trichogramma* sp. and various types of rots and wilts in different crops by Trichoderma-based products (Kalra and Khanuja 2007). In view of consumers' awareness and perception of vegetables without chemical residues, the use of plant products can be an eco-friendly, effective and economical method for producing vegetables that are preferred in local and export markets (Gahukar 2007). Entomopathogens have become the most preferred method for managing a variety of invertebrate pests in greenhouses, row crops, orchards, ornamentals, stored products, and forestry, and for pests and vector insects of medical and veterinary importance (Burges 1981; Tanada and Kaya 1993; Lacey and Kaya 2000; Lacey et al. 2001).

7.7.6 Botanical Pesticides

Botanical pesticides are naturally occurring plant substances used for managing pests (Thacker 2002). Botanicals are endowed with a spectrum of properties such as insecticidal activity, repellence of pests, insect behavior modifiers, antifeedant activity and toxicity to mites, snails, slugs, nematodes, and other agricultural pests (Duke 1990; Narwal et al. 1997). The growing concern about the use of pesticides has resulted in renewed interest in the use of botanicals in IPM (Crazywacz et al. 2005; Isman 2006).

7.7.7 Transgenics

The area of land under transgenics or GM crops is continuing to increase throughout the world. However, production of most of the dominant crops such as soybean, maize, canola, and cotton remains concentrated in the USA, Canada, and Argentina, followed by Brazil, China, Paraguay, India, and South Africa (James 2004). The coexistence of GM crops and non-GM crops is a myth because the movement of transgenes beyond their intended destinations is a certainty, and this leads to genetic contamination of organic farms and other systems. However, organic agriculture is practiced in almost all countries of the world, and its share of agricultural land and farms is growing; in Europe, organic agriculture is increasing rapidly. It is unlikely that transgenes can be retracted once they have escaped and thus the damage to the purity of non-GM seeds is permanent. The dominant GM crops have the potential to reduce biodiversity further by increasing agricultural intensification. There are also potential risks to biodiversity arising from gene flow and toxicity to non-target organisms from herbicide-resistant and insect-resistant (Bt) crops (Altieri 2005).

7.7.8 Chemical Control

The use of pesticides is unavoidable once the pest population has built up on the crop (Dhawan 2001). However, judicious use can overcome the negative impacts of pesticides such as resurgence of pests and development of resistance in pests, with management of pesticide residues and conservation of natural enemy complexes and biodiversity in crop ecosystems. Pesticides provide a dependable, rapid, effective and economical means of controlling whole complexes of crop pests. The basis of using pesticides as pest management options and the consequences of misusing them need to be carefully analyzed in order to obtain maximum benefits from their application, while at the same time preventing and minimizing their possible hazardous effects on non-target organisms and the environment. In most developed countries, the bulk of the pesticides used are herbicides, which are less toxic compared with the insecticides used in developing countries (WRI 1999). Furthermore, the insecticides used in developing countries are generally obsolete types belonging to the organophosphates and carbamates, which are noted for their acute toxicity. Pesticides that are generally highly toxic and are known to have toxic residual effects are not recommended. To get more profit, farmers often do not wait for the correct period of time after use of the pesticide and harvest the crop to market. This leads to pesticide poisoning, chronic toxic effects and in some cases even death. Thus, more care and caution is required in applying pest-control practices in field crops.

7.7.9 Biotechnological Approaches

A shrinking natural resource base coupled with the burgeoning population demands a quantum jump in our productivity levels to meet food security requirements. Biotechnology offers unique opportunities to solve environmental problems, some of which derive from unsustainable agricultural and industrial practices, and has emerged as an important tool in IPM, providing new ways of manipulating plant resistance to pests. Using plant biotechnology, several herbicide-tolerant crops have been developed and commercialized that allow the use of herbicides that are effective, economical and have favorable environmental characteristics. Biotechnology provides the tools to modify the performance of important biological elements of pest control, such as natural enemies and plant varieties. New crop cultivars with resistance to insect pests and diseases combined with bio-control agents should lead to reduced reliance on pesticides, thereby reducing farmers' crop protection costs, while benefiting both the environment and public health (Sharma et al. 2002). Transgenic resistance to insects has been demonstrated in plants expressing insecticidal genes such as alfa-endotoxins from *Bacillus thuringiensis* (Bt), protease inhibitors, enzymes, secondary plant metabolites, and plant lectins. The protease inhibitor and lectin genes largely affect insect growth and development and, in most instances, do not result in insect mortality. The effective concentrations of these proteins are much greater than the Bt endotoxin proteins. Therefore, the potential of some of the alternative genes can only be realized by deploying them in combination with conventional host-plant resistance and *Bt* genes (Hilder and Boulter 1999).

Genes conferring resistance to insects can also be deployed as multiline or synthetic varieties. Impressive results in the control of Bt-susceptible pests have been obtained in the laboratory and in the field, and the first commercial Bt-transgenic crops are now in use. The application of biotechnology techniques in the agriculture sector can potentially improve food security by raising crop tolerance to adverse biotic and abiotic conditions by enhancing adaptability of crops to different climates and by improving yields, pest resistance and nutrition, particularly of staple food crops.

7.8 Pesticides and Integrated Pest Management

Basically, pesticides are applied to protect crops from pests in order to increase yields, but sometimes the crops are damaged by pesticide treatments when (1) the recommended dosages suppress crop growth, development and yield; (2) pesticides drift from the targeted crop to damage adjacent crops; (3) residual herbicides either prevent chemical-sensitive crops from being planted; and/or (4) excessive pesticide residue accumulates on crops, necessitating the destruction of the harvest. Crop losses translate into financial losses for growers, distributors, wholesalers, transporters, retailers, food processors and others.

7.9 Integrated Pest Management and Food Safety (Case Studies)

7.9.1 Adoption of IPM and Success Stories

FFSs (farmer field schools) on IPM in south India have proved very successful in strengthening the agricultural knowledge and skills of poor farmers to alleviate their hardship with respect to food and financial security. Jiggins and Mancini (2009) reported that FFSs comprised four principles: conservation of natural enemies, production of a healthy crop, performing regular field observations and improving the expertise of farmers to do this in their own fields. FFSs involve a group-based learning process to promote IPM, which is utilized by a number of governments and non-governmental organizations and international agencies. In 1989, the Food and Agricultural Organization (FAO) organized such FFSs in Indonesia and since then, more than 2 million farmers across Asia have participated in such schools (Bartlett 2005). Khan and Pickett (2004) successfully documented the application push–pull strategies in eastern Africa for the management of maize pests. According to Khan et al. (2006), more than 160,000 farmers are now using push–pull strategies to protect their maize and sorghum against stem borer (*Chilo partellus*), based on the combined use of intercrops such as molasses grass (*Melinis minutiflora*) and silverleaf desmodium (*Desmodium uncinatum*), and trap crops such as Napier grass (*Pennisetum purpureum*) or Sudan grass (*Sorghum vulgare sudanense*) that are locally available and exploit natural enemies. The rapid spread of FFSs and Junior Life Schools (for school-aged children) throughout the world are helping carry the strategy to an increasing number of farmers. The adoption of the push–pull strategy for stem borers has led to increased crop yields and livestock production, with a significant impact on food security throughout the region. A success story of the establishment of an exotic parasitoid was that of *Diglyphus isaea* (Hymenoptera: Eulophidae). The larval form of this ectoparasitoid of the leaf miner *Liriomyza huidobrensis* was released into the fields in 1997–1998. A post-evaluation survey was carried out in 2000 in six locations of Nuwara Eliya in the district of Sri Lanka where the parasitoid was released. The percentage of parasitism ranged from 1.3 to 65%, and in locations where there was a high level of parasitism, the farmers did not need to use highly toxic insecticides to control these vegetable pests (Nugaliyadde et al. 2001).

7.9.2 The Impact of IPM Interventions on Crop Productivity

IPM interventions have been found to minimize pest losses and increase productivity. For instance, IPM intervention in rice reduced pesticide usage by 67% and increased productivity by 25%. Interestingly, in sugarcane, pesticides have been eliminated altogether, resulting in increased income for farmers. Similarly, in other crops, superior yields have been obtained following IPM interventions. IPM interventions in South Africa are reported to have decreased crop losses by 90% in the case of cassava afflicted by cassava mealybugs. In other crops, losses dropped by 5% and yields increased by as much as 100%. Pesticide use was reduced considerably from 68% to only 11% for control of rice leaf feeders in the Philippines following IPM interventions.

7.10 Ethical and Moral Issues

Although pesticides provide about US$ 40 billion year^{-1} in saved US crops, the data of this analysis suggest that the environmental and social costs of pesticides to the nation totaled approximately US$ 10 billion. From a strictly cost/benefit approach, pesticide use is beneficial. However, the nature of the environmental and public health costs of pesticides has other trade-offs involving environmental quality and public health. One of these issues concerns the importance of public health versus pest control. For example, assuming that pesticide-induced cancers numbered more than 10,000 cases year^{-1} and that pesticides returned a net agricultural benefit of US$ 32 billion year^{-1}, each case of cancer is "worth" US$ 3.2 million in pest control. In other words, for every US$ 3.2 million in pesticide benefits, one person falls victim to cancer. Social mechanisms and market economics provide these ratios, but they ignore basic ethics and values.

In addition, pesticide pollution of the global environment raises numerous other ethical questions. The environmental insult of pesticides has the potential to demonstrably disrupt entire ecosystems. All through history, humans have felt justified in removing forests, draining wetlands and constructing highways and housing in various habitats. White (1967) has blamed the environmental crisis on religious teachings of mastery over nature. Whatever the origin, pesticides exemplify this attempt at mastery, and even a non-economic analysis would question its justification. There is a clear need for a careful and comprehensive assessment of the environmental impacts of pesticides on agriculture and natural ecosystems. Public concern over pesticide pollution confirms an international trend towards environmental values. Media emphasis on the issues and problems caused by pesticides has contributed to a heightened public awareness of ecological concerns. This awareness is encouraging research in sustainable agriculture and in non-chemical pest management.

7.11 Conclusions

World agriculture needs to keep pace with the continuous growth of the world population in a sustainable manner to reduce the number of undernourished people and promote health and welfare. The increase in the availability of food per capita for the world population might be obtained through an 'evergreen revolution' keeping in view the concerns of environment and food security through biotechnological innovations. Unwise application of technological tools may further deteriorate human health and environmental quality and compromise future development of human societies. The size of the tasks to be implemented as well as the complexity of the problems to be solved requires better coordination than ever among nations. It would be highly desirable that international organizations excel in their capacities, better coordinate efforts and take leadership.

References

Abrol, D. P., & Shankar, U. (2012). History, overview and principles of ecologically based pest management. In D. P. Abrol & U. Shankar (Eds.)., *Integrated Pest Management: Principles and Practice* (pp. 1–26). Wallingford, United Kingdom: CABI.

Abrol, D. P., Shankar, U., Chatterjee, D., & Ramamurthy, V. V. (2012). Exploratory studies on diversity of bees with special emphasis on non-*Apis* pollinators in some natural and agricultural plants of Jammu division, India. *Current Science, 103*(7),780–783.

Aguilar, R., Ashworth, L., Galetto, L., & Aizen, M. A. (2006). Plant reproductive susceptibility to habitat fragmentation: Review and synthesis through a meta-analysis. *Ecology Letters, 9*, 968–980.

Alston, D. G., Tepedino, V. J., Bradley, B. A., Toler, T. R., Griswold, T. L., & Messinger, S. M. (2007). Effects of the insecticide phosmet on solitary bee foraging and nesting in orchards of Capitol Reef National Park, Utah. *Environmental Entomology, 36*, 811–816.

Altieri, M. A. (2005). The myth of coexistence: Why transgenic crops are not compatible with agroecologically based systems of production. *Bulletin of Science, Technology and Society, 25*, 361–371.

Ashman, T. L., Knight, T. M., Steets, J. A., Amarasekare, P., Burd, M., Campbell, D. R., Dudash, M. R., Johnston, M. O., Mazer, S. J., Mitchell, R. J., Morgan, M. T., & Wilson, W. G. (2004). Pollen limitation of plant reproduction: Ecological and evolutionary causes and consequences. *Ecology, 85*, 2408–2421.

Baker, B. P., Benbrook, C. M., Groth, G., & Benbrook, K. L. (2002). Pesticide residues in conventional, integrated pest management (IPM)-grown and organic foods: Insights from three US data sets. *Food Additives and Contaminants, 19*(5), 427–446. http://www.consumersunion.org/food/orgnicsumm.htm. Accessed 12 July 2012.

Banaszak, J. (1995). *Changes in Fauna of Wild Bees in Europe*. Bydgoszez, Poland: Pedagogical University Press.

Barbash, J. E., & Resek, E. A. (1997). *Pesticides in Ground Water: Distribution, Trends and Governing Factors*. Boca Raton, Florida, USA: CRC Press.

Barcelo, D., & Hennion, M. C. (Eds.)., (1997). *Trace Determination of Pesticides and Their Degradation Products in Water*. Vol. 19 in the series, Techniques and Instrumentation in Analytical Chemistry. Amsterdam: Elsevier.

Bartlett, A. (2005). Farmer field schools to promote integrated pest management in Asia: The FAO experience. Case study presented to the Workshop on Scaling Up Case Studies in Agriculture, International Rice Research Institute, 16–18 August 2005, Bangkok, Thailand. http://www.share4dev.info/kb/output_view.asp?outputID=3371. Accessed 27 March 2013.

Bashir, Y. G. A., Elamin, E. M., & Elamin, E. M. (2003). Development and implementation of integrated pest management in the Sudan. In K. M. Maredia, D. Dakouo & D. Mota-Sanchez (Eds.)., *Integrated Pest Management in the Global Arena* (pp. 131–143). London: CAB International.

Beebe, S., Cardona, C., Diaz, O., Rodriguez, F., Mancia, E., & Ajquejay, S. (1993). Development of common bean (*Phaseolus vulgaris* L.) lines resistant to the bean pod weevil, *Apion godmani* Wagner, in Central America. *Euphytica, 69*(1–2), 83–88.

Biesmeijer J. C., Roberts S. P. M., Reemer M., Ohlemûller R., Edwards M., Peeters T., Schaffers A. P., Potts S. G., Kleukers R., Thomas C. D., Settele J., & Kunin W. E. (2006). Parallel declines in pollinators and insect-pollinated plants in Britain and the Netherlands. *Science, 313*, 351–354.

Booth, B. D., Murphy, S. P., & Swanton, C. J. (2003). *Weed Ecology in Natural and Agricultural Systems*. Willingford, United Kingdom: CABI Publishing.

Borlaug, N. E. (2007). Feeding a hungry world. *Science, 318*, 359–359.

Borlaug, N. E. (2009). Foreword [Editorial]. *Food Security, 1*, 1. doi:10.1007/s12571–009–0012–4.

Borlaug, N., & Dowswell, C. R. (1993). Fertilizer: To nourish infertile soil that feeds a fertile population that crowds a fragile world. Proceeding of the 61st International Fertilizer Association Annual Conference. *Fertiliser News*[Hyderabad, India], *38*(7),11–20.

Brader, L. (1979). Integrated pest control in the developing world. *Annual Review of Entomology, 24*, 225–254.

Bretveld, R. W., Thomas, C. M. G., Scheepers, P. T. J., Zielhuis, G. A., & Roeleveld, N. (2006). Pesticide exposure: The hormonal function of the female reproductive system disrupted? *Reproductive Biology and Endocrinology, 4*. doi:10.1186/1477–7827-4–30.

Brittain, C. A., Vighi, M., Bommarco, R., Settele, J., & Potts, S. G. (2010). Impacts of a pesticide on pollinator species richness at different spatial scales. *Basic and Applied Ecology, 11*, 106–115.

Brown, L. R. (2004). *Outgrowing the Earth: The Food Security Challenge in an Age of Falling Water Tables and Rising Temperatures*. New York: W.W. Norton.

Bueno, J. M., Cardona, C., & Quintero, C. M. (1999). Comparison between two improvement methods to develop multiple insect resistance in common bean (*Phaseolus vulgaris* L.). *Revista Colombiana de Entomologia, 25*, 73–78.

Burges, H. D. (Ed.) (1981). *Microbial Control of Pests and Plant Diseases 1970–1980*. London: Academic Press.

Carlsen, E. A., Giwercman, A., Kielding, N., & Skakkebaek, N. E. (1992). Evidence for decreasing quality of semen during the past 15 years. *British Medical Journal, 305*, 609–613.

Carozzi, N., & Koziel, M. (Eds.)., (1997). *Advances in Insect Control: The Role of Transgenic Plants*. London: Taylor and Francis.

Carrasco-Tauber, C. (1989). *Pesticide productivity revisited*. Masters dissertation, University of Massachusetts, USA.

Carson, R. (1962). *Silent Spring*. Boston: Houghton Mifflin.

Carvalho, F. P. (2005). Residues of persistent organic pollutants in coastal environments-a review. In F. V. Gomes, F. T. Pinto, L. Neves, O. Sena, & O. Ferreira (Eds.)., *Proceedings of the First International Conference on Coastal Conservation and Management in the Atlantic and Mediterranean (ICCCM'05)* (pp. 423–431). Tavira, Portugal, April 17–20. FEUP, Universidade do Porto.

Carvalho, F. P., Fowler, S. W., Villeneuve, J. P., & Horvat, M. (1997). Pesticide residues in the marine environment and analytical quality assurance of the results. In International Atomic Energy Agency (IAEA); Food and Agriculture Organization of the United Nations (Eds.)., *Environmental Behavior of Crop Protection Chemicals* (pp. 35–57): Proceedings of an International FAO/IAEA Symposium on the Environmental Behaviour of Crop Protection Chemicals. IAEA, Vienna.

Carvalho, F. P., Villeneuve, J. P., Cattini, C., Tolosa, I., Guillen, S. M., Lacayo, M., & Cruz, A. (2002). Ecological risk assessment of pesticide residues in coastal lagoons of Nicaragua. *Journal of Environmental Monitoring*, 4, 1778–1787.

CEQ. (1980). *The Global 2000 Report to the President of the U.S. Entering the 21st Century*, New York: Pergamon Press.

Chauzat, M., Carpentier, P., Martel, A. C., Bougeard, S., Cougoule, N., Porta, P., Lachaize, J., Madec, F., Aubert, M., & Faucon, J. P. (2009). Influence of pesticide residues on honey bee (Hymenoptera: Apidae) colony health in France. *Environmental Entomology*, 38, 514–523.

Chauzat, M. P., Faucon, J. P., Martel, A. C., Lachaize, J., Cougoule, N., & Aubert, M. (2006). A survey on pesticide residues in pollen loads collected by honey bees (*Apis mellifera*) in France. *Journal of Economic Entomology*, 99, 253–262.

Clarke, S. (2001). Review of the operational IPM programme for the southern pine beetle. *IPM Review*, 6, 293–301.

Colborn, T., Myers, J. P., & Dumanoski, D. (1996). *Our Stolen Future: How We are Threatening Our Fertility, Intelligence, and Survival: A Scientific Detective Story*. New York: Dutton.

Colla, S. R., Otterstatter, M. C., Gegear, R. J., & Thomason, J. D. (2006). Plight of the bumble bee: Pathogen spillover from commercial to wild populations. *Biological Conservation*, 129, 461–467.

Cook, S. M., Khan, Z. R., & Pickett, J. A. (2007). The use of push-pull strategies in integrated pest management. *Annual Review of Entomology*, 52, 375–400.

Crazywacz, D., Richards, A., Rabindra, R. J., Saxena, H., & Rupela, O. P. (2005). Efficacy of biopesticides and natural products for *Heliothis/Helicoverpa* control. In H. C. Sharma (Ed.)., *Heliothis Management: Emerging Trends and Strategies for Future Research* (pp. 371–389). New Delhi: Oxford & IBH Publishing Co.

Datta, K., Baisakh, N., Maung Thet, K., Tu, J., & Datta, S. K. (2002). Pyramiding transgenes for multiple resistance in rice against bacterial blight, yellow stem borer and sheath blight. *Theoretical and Applied Genetics*, 106, 1–8.

Desneux, N., Decourtye, A., & Delpuech, J. M. (2007). The sublethal effects of pesticides on beneficial arthropods. *Annual Review of Entomology*, 52, 81–106.

Devereux, S. (2009). Why does famine persist in Africa. *Food Security*, 1, 25–35.

Dhawan, A. K. (2001). Cotton pest scenario in India and its control. In *Proceedings of the 3rd Asia Pacific Crop Protection Conference* (pp. 115–127), 6–7 September 2001, New Delhi, India.

Dogra, K. S., Sood, S. K., Dobhal, P. K., & Sharma, S. (2010). Alien plant invasion and their impact on indigenous species diversity at global scale: A review. *Journal of Ecology and the Natural Environment*, 2(9), 175–186.

Dugger, C. W. (2007). Report on child deaths finds some hope in poorest nations. *The New York Times*, May 8, 2007. http://www.nytimes.com/2007/05/08/world/08children.html?_r=0. Accessed 26 Feb 2014.

Duke, S. O. (1990). Natural pesticides from plants. In J. Jaick & J. E. Simon (Eds.)., *Advances in New Crops* (pp. 511–517). Portland, Oregon, USA: Timber Press.

EEA. (2005). Environment and health. European Environment Agency, EEA Report No. 10.

EPA. (1992). Hired farm workers health and well-being at risk, United States General Accounting Office Report to Congressional Requesters. http://www.gao.gov/products/HRD-92-46. Accessed 26 Feb 2014.

Evans, H. F. (1986). Ecology and epizootiology of baculoviruses. In R. R. Granados & B. A. Federici (Eds.)., *The Biology of Baculoviruses, Vol. II, Practical Application for Insect Control* (pp. 89–132). Boca Raton, Florida, USA: CRC Press.

Falcon, W. P., & Fowler, C. (2002). Carving up the commons- emergence of a new international regime for germplasm development and transfer. *Food Policy*, 27, 197–222.

FAO. (1986). *Production Yearbook* (Vol. 40). Rome: Food and Agriclture Organization of the United Nations.

FAO. (1988). *Integrated Pest Management in Rice in Indonesia*. Jakarta: Food and Agriculture Organization of the United Nations.

FAO. (2005). *The State of Food Insecurity in the World 2005*. Economic and Social Department, Food and Agriculture Organization Of the United Nations, Rome, Italy. http://www.fao.org/docrep/008/a0200e/a0200e00.htm. Accessed 30 May 2013.

FAO. (2009). *High Level Expert Forum—How to Feed the World in 2050*. Rome: Food and Agriculture Organization of the United Nations. http://www.fao.org/fileadmin/templates/wsfs/docs/expert_paper/How_to_Feed_the_World_in_2050.pdf. Accessed 30 May 2013

FAO. (2009a). 1.02 billion people hungry. FAO Newsroom. http://www.fao.org/new/story/on/item/20568/icode/. Accessed 7 Nov 2013.

FAO. (2010). *The State of Food Insecurity in the World -Addressing Food Insecurity in Protracted Crises*. Food and Agriculture Organization of the United Nations, Rome. http://www.unscn.org/en/resource_portal/index.php?&themes=77&resource=684. Accessed 30 May 2013.

Food and Agriculture Organisation of the United Nations. FAOSTAT. (2010). http://faostat.fao.org/site/424/default.aspx#%23ancor. Accssesed 15 May 2013.

Funk, C. F., & Brown, M. E. (2009). Declining global per capita agricultural production and warming oceans threaten food security. *Food Security, 1,* 271–289.

Gahukar, R. T. (2007). Botanicals for use against vegetable pests and disease: A review. *International Journal of Vegetable Science, 13,* 41–60.

Gallai, N., Salles, J. M., Settele, J., & Vaissiere, B. E. (2009). Economic valuation of the vulnerability of world agriculture confronted with pollinator decline. *Ecological Economics, 68,* 810–821.

Goodell, G. E. (1984). Challenges to integrated pest management research and extension in the Third World: Do we really want IPM to work? *Bulletin of the Entomological Society of America, 30,* 18–26.

Greatti, M., Barbattini, R., Stravisi, A., Sabatini, A. G., & Rossi, S. (2006). Presence of the a.i. imidacloprid on vegetation near corn fields sown with Gaucho dressed seeds. *Bulletin of Insectology, 59,* 99–103.

Groth, E., Benbrook, C. M., & Lutx, K. (1999). *Do You Know What You're Eating?* An analysis of U.S. Government data on pesticide residues in foods. http://www/consumers union.org/food/do you know2.htm. Accessed 12 July 2012

Groves, R. H., Panetta, F. D., & Virtue, J. G. (2001). *Weed Risk Assessment*. Collingwood: CSIRO Publishing.

Hairston, N. G., Smith, F. E., & Slobodkin, L. B. (1960). Community structure, population control and competition. *The American Naturalist, 94,* 421–425.

Harper, C. R., & Zilberman, D. (1990). Pesticide regulation: Problems in trading off economic benefits against health risks. In D. Zilberman & J. B. Siebert (Eds.)., *Economic Perspectives on Pesticide Use in California: A Collection of Research Papers*, Berkeley: Department of Agricultural and Resource Economics, University of California.

Harris, C. A. (2002). The regulation of pesticides in Europe- Directive 91/414. *Journal of Environmental Monitoring, 4,* 28–31.

Hart, K., & Pimentel, D. (2002). Public health and costs of pesticides. In D. Pimentel (Ed.), *Encyclopedia of Pest Management*. New York: Marcel Dekker.

Hebert, H. J. (2003). *EPA Guidelines Address Kids, Cancer Risks*. Detroit Free Press. http://www.freep.-com/news/childrenfirst/risk4_20030304.htm. Accessed on 25 Feb 2013.

Hilder, V. A., & Boulter, D. (1999). Genetic engineering of crop plants for insect resistance—a critical review. *Crop Protection, 18,* 177–191.

Holmes, T., Nielsen, E., & Lee, L. (1988). Managing groundwater contamination in rural areas: Rural development perspectives. *Rural Development Perspectives*, October 1988. US Department of Agriculture, Economic Research Series, Washington, D.C. http://naldc.nal.usda.gov/download/IND89047077/PDF. Accessed 28 March 2013

Hulme, P. E. (2003). Biological invasions: Winning the science battles but losing the conservation war? *Oryx, 37*(2), 178–193.

Hulme, P. E. (2012). Weed risk assessment: A way forward or a waste of time? *Journal of Applied Ecology, 49,* 10–19.

Incerti, F., Bortolotti, L., Porrini, C., Micciarelli Sbrenna, A., & Sbrenna, G. (2003). An extended laboratory test to evaluate the effects of pesticides on bumblebees, preliminary results. *Bulletin of Insectology, 56*, 159–164.

IPCC. (2007). *Climate Change 2007: The Physical Science Basis*. Intergovernmental Panel on Climate Change (IPCC). Cambridge: Cambridge University Press.

Ishimoto, M. (1999). Evaluation and use of wild Phaseolus species in breeding. In K. Oono, D. Vaughan, N. Tomooka, A. Kaga, & S. Miyazaki (Eds.), *The Seventh Ministry of Agriculture, Forestry and Fisheries (MAFF), Japan, International Workshop on Genetic Resources Part 1: Wild Legumes.* (pp.183–190). October 13–15, Ibaraki, Japan.

Isman, M. B. (2006). Botanical insecticides, deterrents, and repellents in modern agriculture and an increasingly regulated world. *Annual Review of Entomology, 51*, 45–66.

James, C. (2004). *Global Review of Commercialized Transgenic Crops: 2004*. ISAAA Briefs No. 23. Ithaca, NY: International Service for the Acquisition of Agri-Biotech Applications (ISAAA).

Jiggins, J., & Mancini, F. (2009). Moving on: Farmer education in integrated insect pest and disease management. In R. Peshin & A. K. Dhawan (Eds.), *Integrated Pest Management: Dissemination and Impact, Vol.2* (pp. 307–332). Berlin: Springer-Verlag.

Johansen, C. A., & Mayer, D. F. (1990). *Pollinator Protection: A Bee and Pesticide Handbook*. Cheshire, Connecticut, USA: Wicwas Press.

Kalra, A., & Khanuja, S. P. S. (2007). Research and development priorities for biopesticide and biofertiliser products for sustainable agriculture in India. In P. S. Teng (Ed.)., *Business Potential for Agricultural Biotechnology* (pp. 96–102). Tokyo: Asian Productivity Organisation.

Kaya, H. K. (1976). Insect pathogens in natural and microbial control of forest defoliators. In J. F. Anderson & H. K. Kaya (Eds.)., *Perspectives in Forest Entomology* (pp. 251–263) New York: Academic.

Kaya, H. K. (1986). Steinernema carpocapsae: Use against foliage feeding insects and effect on nontarget insects. In R. A. Samson, J. M. Vlak, & D. Peters, (Eds.)., *Fundamental and Applied Aspects of Invertebrate Pathology* (pp. 268–270). Proceedings of the Fourth International Colloquium of Invertebrate Pathology, Veldhoven, The Netherlands. Wageningen, Netherlands: Foundation of the Fourth International Colloquium of Invertebrate Pathology.

Kenta, T., Inari, N., Nagamitsu, T., Goka, K., & Hiura, T. (2007). Commercialized European bumblebee can cause pollination disturbance: An experiment on seven native plant species in Japan. *Biological Conservation, 134*, 298–309.

Kerr, R. A. (2009). Northern India's ground water is going, going, going. *Science, 325*, 798–798.

Kevan, P. G. (1999). Pollinators as bioindicators of the state of the environment: Species, activity and diversity. *Agriculture, Ecosystems and Environment, 74*, 373–393.

Kfir, R., Overholt, W. A., Khan, Z. R., & Polaszek, A. (2002). Biology and management of economically important lepidopteran cereal stem borers in Africa. *Annual Review of Entomology, 47*, 701–731.

Khan, S., Tariq, R., Yuanlai, C., & Blackwell, J. (2006). Can irrigation be sustainable? *Agricultural Water Management, 80*, 87–99.

Khan, Z. R., & Pickett, J. A. (2004). The 'push-pull' strategy for stemborer management: A case study in exploiting biodiversity and chemical ecology. In G. M. Gurr, S. D. Wratten, & M. A. Altieri (Eds.)., *Ecological Engineering for Pest Management: Advances in Habitat Manipulation for Arthropods* (pp. 155–164). Wallingford, United Kingdom: CAB International.

Khush, G. S. (2002). The promise of biotechnology in addressing current nutritional problems in developing countries. *Food and Nutrition Bulletin, 23*(4), 354–357.

Klein, A. M., Vaissiere, B. E., Cane, J. H., Steffan-Dewenter, I., Cunningham, S. A., Kremen, C., & Tscharntke, T. (2007). Importance of pollinators in changing landscapes for world crops. *Proceedings of the Royal Society of London Series B, Biological Sciences, 274*, 303–313.

Kogan, M. (1998). Integrated pest management: Historical perspectives and contemporary developments. *Annual Review of Entomology, 43*(1), 243–270.

Kornegay, J., & Cardona, C. (1991). Breeding for insect resistance in beans. In A. van Schoonhoven & O. Voysest (Eds.)., *Common Beans: Research for Crop Improvement* (pp. 619–648). Wallingford, United Kingdom: CAB International.

Kraus, F. B., Szentgyorgyi, H., Rozej, E., Rhode, M., Moron, D., & Woyciezchowski, M. (2011). Greenhouse bumblebees (*Bombus terrestris*) spread their genes into the wild. *Conservation Genetics, 12,* 187–192.

Krupke, C., Hunt, G. J., Eitzer, B. D., Andino, G., & Given, K. (2012). Multiple routes of pesticide exposure for honey bees living near agricultural fields. *PLoS ONE, 7*(1). doi:10.1371/journal.pone.0029268.

Lacey, L. A., Frutos, R., Kaya, H. K., & Vail, P. (2001). Insect pathogens as biological control agents: Do they have a future? *Biological Control, 21,* 230–248.

Lacey, L. A., & Kaya, H. K. (Eds.)., (2000). *Field Manual of Techniques in Invertebrate Pathology: Application and Evaluation of Pathogens for Control of Insects and Other Invertebrate Pests.* Dordrecht, Netherlands: Kluwer Academic.

Lal, R. (2009). Soil carbon sequestration impacts on global climate change and food security. *Science, 304,* 1623–1627.

Le Feon, V., Schermann-Legionnet, A., Delettre, Y., Aviron, S., Billeter, R., Bugter, R., Hendrickx, F., & Burel, F. (2010). Intensification of agriculture, landscape composition and wild bee communities: A large study in four European countries. *Agriculture, Ecosystems and Environment, 137,* 143–150.

Lenda, M., Skorka, P., & Moron, D. (2010). Invasive alien plant species a threat or a chance for pollinating insects in agricultural landscapes? In T. H. Lee (Ed.)., *Agricultural Economics: New Research* (pp. 67–87). New York: Nova Science Publishers.

Lowel, J. F. (1998). *Producing Food Without Pesticides: Local Solutions to Crop Pest Control in West Africa.* Church World Service, Dakar Senegal: Technical Centre for Agricultural and Rural Cooperation (Ede, Netherlands).

McCoy, C. W., Samson, R. A., & Boucias, D. G. (1988). Entomogenous fungi. In C. M. Ignoffo & N. B. Mandava (Eds.)., *CRC Handbook of Natural Pesticides, Vol. 5, Microbial Pesticides, Part A: Entomogenous Protozoa and Fungi* (pp. 151–234). Boca Raton, Florida, USA: CRC Press.

Monobrullah, M., & Shankar, U. (2008). Sub-lethal effects of *Splt*MNPV infection on developmental stages of *Spodoptera litura* (Lepidoptera: Noctuidae). *Biocontrol Science and Technology, 18,* 431–437.

Morandin, L. A., & Winston, M. L. (2005). Wild bee abundance and seed production in conventional, organic, and genetically modified canola. *Ecological Applications, 15,* 871–881.

Moron, D., Lenda, M., Skorka, P., Szentgyorgyi, H., Settele, J., & Woyciechowski, M. (2009). Wild pollinator communities are negatively affcted by invasion of alien goldenrods in grassland landscapes. *Biological Conservation, 142,* 1322–1332.

Mullin, C. A., Frazier, M., Frazier, J. L., Ashcraft, S., Simonds, R., Vanengelsdorp, D., & Pettis, J. S. (2010). High levels of miticides and agrochemicals in North American apiaries: Implications for honey bee health. *PLoS One, 5*(3):e9754. doi: 10.1371/journal.pone.0009754.

Munoz-de-Toro, M., Durando, M., Beldomenico, P. M., Beldome´nico, H. R., Kass, L., Garcıa, S. R., & Luque, E. H. (2006). Estrogenic microenvironment generated by organochlorine residues in adipose mammary tissue modulates biomarker expression in ERa-positive breast carcinomas. *Breast Cancer Research, 8,* R47.

Narwal, S. S., Tauro, P., & Bisla, S. S. (1997). *Neem in Sustainable Agriculture.* Jodhpur, India: Scientific Publishers.

NAS. (1975). *Pest Control: An Assessment of Present and Alternative Technologies* (4 volumes). Washington, D.C.: National Academy of Sciences of the USA.

Nhan, D. D., Manh, A. N., Carvalho, F. P., Villeneuve, J. P., & Cattini, C. (1999). Organochlorine pesticides and PCBs along the coast of North Vietnam. *Science of the Total Environment, 237/238,* 363–371.

Nickson, T. E. (1999). Environmental monitoring of genetically modified crops. *Journal of Environmental Monitoring, 1,* 101–105.

Norrung, B. (2000). Foodborne viruses—Introduction. *International Journal of Food Microbiology, 59*(1–2), 79–80.

Nugaliyadde, M. M., Joshep, J. E., & Ahangama, D. (2001). Natural enemies of the vegetable leaf miner *Liriomyza huidobrensis* in up country of Sri Lanka. *Annals of the Sri Lanka Department of Agriculture, 3,* 377–380.

Oka, I. N. (1991). Success and challenges of the Indonesian national integrated pest management pro- gram in the rice based cropping system. *Crop Protection, 10*, 163–165.

Oomen, P. A., & Bouma, E. (2003). *Role of Pesticides in IPM*. Wageningen, Netherlands: Plant Protection Service.

Oreska, M. P. J., & Aldridge, D. C. (2011). Estimating the financial costs of freshwater invasive species in Great Britain: A standardized approach to invasive species costing. *Biological Invasions, 13*, 305–319.

Park, Y. L., Krell, R. K., & Carroll, M. (2007). Theory, technology and practice of site-specific insect pest management. *Journal of Asia-Pacific Entomology, 10*(2), 89–101.

Pimentel, D. (1988). Herbivore population feeding pressure on plant host: Feedback evolution and host conservation. *Oikos, 53*, 289–302.

Pimentel, D. (1997). *Techniques for Reducing pesticides: Environmental and Economic Benefits*. Chichester, United Kingdom: Wiley.

Pimentel, D., Acquay, H., Biltonen, M., Rice, P., Silva, M., Nelson, J., Lipner, V., Giordana, S., Horowitz, A., & D'Amore, M. (1993). Assessment of environmental and economic impacts of pesticide use. In D. Pimentel & H. Lehman (Eds.)., *The Pesticide Question: Environment, Economics and Ethics* (pp. 47–84). New York: Chapman & Hall.

Pimentel, D., & Greiner, A. (1997). Environmental and socio-economic costs of pesticide use. In D. Pimentel (Ed.)., *Techniques for Reducing pesticides: Environmental and Economic Benefits* (pp. 51–78). Chichester, United Kingdom: Wiley.

Pimentel, D., & Hart, K. (2001). Pesticide use: Ethical, environmental, and public health implications. In W. Galston & E. Shurr (Eds.)., *New Dimensions in Bioethics: Science, Ethics and the Formulation of Public Policy* (pp. 79–108). Boston: Kluwer Academic Publishers.

Pimentel, D., McLaughlin, L., Zepp, A., Latikan, B., Kraus, T., Kleinman, P., Vancini, F., Roach, W. J., Graap, E., Keeton, W. S., & Selig, G. (1991). Environmental and economic impacts of reducing US agricultural pesticide use. In D. Pimentel (Ed.)., *Handbook on Pest Management in Agriculture*, (Vol. 1, 2nd ed., pp. 679–718). Boca Raton, Florida, USA: CRC Press.

Pimentel, D., McNair, S., Janecka, J., Wightman, J., Simmonds, C., O'Connell, C., Wong, E., Russel, L., Zern, J., Aquino, T., & Tsomondo, T. (2001). Economic and environmental threats of alien plant, animal, and microbe invasions. *Agriculture, Ecosystem and Environment, 84*, 1–20.

Pimentel, D., Wilson, C., McCullum, C., Huang, R., Dwen, P., Flack, J., Tran, Q., Saltman, T., & Cliff, B. (1997). Economic and environmental benefits of biodiversity. *Bioscience, 47*, 747–757.

Pimentel, D., Zuniga, R., & Morrison, D. (2005). Update on the environmental and economic costs associated with alien-invasive species in the United States. *Ecological Economics, 52*(3), 273–288.

Pingali, P. L., & Traxler, G. (2002). Changing locus of agricultural research: Will the poor benefit from biotechnology and privatization trends? *Food Policy, 27*, 223–238.

Pinstrup-Anderson, P. (2009). Food security: Definition and measurement. *Food Security, 1*, 5–7.

Potts, S. G., Biesmeijer, J. C., Kremen, C., Neumann, P., Schweiger, O., & Kunin, W. E. (2010). Global pollinator declines: Trends, impacts and drivers. *Trends in Ecology & Evolution, 25*(6), 345–353.

Prokopov, T., & Tanchev, S. (2007). Methods of food preservation. In A. McElhatton & R. J. Marshall (Eds.)., *Food Safety: A Practical and Case Study Approach* (pp. 3–25). New York: Springer Science.

Richter, E. D. (2002). Acute human pesticide poisonings. In D. Pimentel (Ed.)., *Encyclopedia of Pest Management* (pp. 3–6). New York: Dekker.

Rivas, A., Nicolas, O., & Olea-Serrano, F. (1997). Human exposure to endocrine-disrupting chemicals: Assessing the total estrogenic xenobiotic effect. *Trends in Analytical Chemistry, 16*, 613–619.

Rola, A. C., & Pingali, P. L. (1993). *Pesticides, Rice Productivity, and Farmers' Health*. Manila: International Rice Research Institute. (Washington, D.C.: Word Resources Institute).

Roling, N. G., & van de Fliert, E. (1994). Transforming extension for sustainable agriculture: The case of integrated pest management in rice in Indonesia. *Agriculture and Human Values, 11*, 96–108.

Shankar, U., & Abrol, D. P. (2012a). Role of integrated pest management in food and nutritional security. In D. P. Abrol & U. Shankar (Eds.)., *Integrated Pest Management: Principles and Practice* (pp. 408–432). Wallingford, United Kingdom: CABI.

Shankar, U., & Abrol, D. P. (2012b). Integrated pest management in stored grains. In D. P. Abrol & U. Shankar (Eds.)., *Integrated Pest Management: Principles and Practice* (pp. 386–407). Wallingford, United Kingdom: CABI.

Sharma, H. C., Crouch, J. H., Sharma, K. K., Seetharama, N., & Hash, C. T. (2002). Applications of biotechnology for crop improvement: prospects and constraints. *Plant Science, 163*, 381–395.

Sharpe, M. (1999). Towards sustainable pesticides. *Journal of Environmental Monitoring, 1*, 33–36.

Shelburne, E. C. (2008). The world in numbers: The great disruption: How scarcity, affluence, and biofuel production are wreaking havoc on food prices. *The Atlantic Monthly*, September. http://www.theatlantic.com/magazine/archive/2008/09/the-great-disruption/306930/. Accessed 27 March 2013.

Shennan, C. (2008). Biotic interactions, ecological knowledge and agriculture. *Philosophical Transactions of the Royal Society B: Biological Sciences, 363*, 717–739.

Singh, S., Sidhu, J. S., Huang, N., Vikal, Y., Li, Z., Brar, D. S., Dhaliwal, H. S., & Khush, G. S. (2001). Pyramiding three bacterial blight resistance genes (*xa5, xa13* and *xa21*) using marker-assisted selection into Indica rice cultivar PR106. *Theoretical and Applied Genetics, 102*, 1011–1015.

Soundararajan, R. P. (Ed.). (2012). *Pesticides—Advances in Chemical and Botanical Pesticides*. In Tech Janeza Trdine, Rijeka, Croatia. http://dx.doi.org/10.5772/2609. Accessed 27 March 2013.

Steffan-Dewenter, I. (2003). Seed set of male–sterile an male–fertile oilseed rape (*Brassica napus*) in relation to pollinator density. *Apidologie, 34*, 227–235.

Steffan-Dewenter, I., Potts, S. G., & Packer, L. (2005). Pollinator diversity and crop pollination services are at risk. *Trends in Ecology and Evolution, 20*, 651–652.

Stuart, C. (2003). Section 2: Development of resistance in pest populations. *Report on Genetically Modified Food Crops*. (Chemistry and Public Policy Fall 1999, Notre Dame University, South Bend, Indiana, USA/Professor Marya Lieberman) Subcommittee E. Effects of genetically modified crops on the environment (not related to gene transfer). http://www3.nd.edu/~chem191/e2.html. Accessed 25 Feb 2013.

Tamm, L. (2001). Organic agriculture: Development and state of the art. *Journal of Environmental Monitoring, 3*, 92N–96N.

Tanada, Y., & Kaya, H. K. (1993). *Insect Pathology*. San Diego, California, USA: Academic Press.

Tapparo, A., Marton, D., Giorio, C., Zanella, A., Soldà, L., Marzaro, M., Vivan, L., & Girolami, V. (2012). Assessment of the environmental exposure of honeybees to particulate matter containing neonicotinoid insecticides coming from corn coated seeds. *Environmental Science & Technology, 46*, 2592–2599.

Taylor, M. D., Klaine, S. J., Carvalho, F. P., Barcelo, D., & Everaarts, J. (Eds.)., (2003). *Pesticide Residues in Coastal Tropical Ecosystems: Distribution, Fate and Effects*. London: Taylor and Francis.

Thacker, J. R. M. (2002). *An Introduction to Arthropod Pest Control*. Cambridge, United Kingdom: Cambridge University Press.

Tripp, R. (2002). Can the public sector meet the challenge of private research? Commentary on "Falcon and Fowler" and "Pingali and Traxler". *Food Policy, 27*, 239–246.

Tyler, M. G. Jr. (2002). *Living in the Environment: Principles, Connections and Solutions* (12th Edn.). Belmont, California, USA: Wadsworth/Thomson Learning.

UNFPA. (2008). Population and climate change- framework of UNFPA's agenda. United Nations Population Fund (UNFPA). http://www.unfpa.org/pds/climate/docs/climate_change_unfpa.pdf. Accessed on 25 Feb 2013.

Uri, N. D. (1997). A note on the development and use of pesticides. *Science of the Total Environment, 204*, 57–74.

USBC. (1990). *Statistical Abstract of the United States, 1990*. Washington, D.C.: US Bureau of the Census, US Government Printing Office.

USEPA. (1992). *Pesticides in Ground Water Database: A Compilation of Monitoring Studies, 1971–1991*, National summary (EPA 734/12–92-001). U.S. Environmental Protection Agency. Washington, D.C.: U.S. Government Printing Office.

USEPA. (2005). *Supplemental Guidance for Assessing Susceptibility from Early-Life Exposure to Carcinogens*. Office of the Science Advisor (OSA), U.S. Environmental Protection Agency. http://www.epa.gov/cancerguidelines/guidelines-carcinogen-supplement.htm. Accessed 28 March 2013.

USGS. (1996). *Pesticides in Public Supply Wells of Washington State*. USGS (US Geological Survey) Fact Sheet 122–96. http://wa.water.usgs.gov/pubs/fs/fs122–96/. Accessed 25 Feb 2013.

van der Graaff, N. A., & Khoury, W. (2010). Biosecurity in the movement of commodities as a component of global food security. In R. N. Strange & M. L. Gullino (Eds.)., *The role of Plant Pathology in Food Safety and Food Security, Plant Pathology in the 21st Century* (pp. 25–39). Dordrecht Netherlands: Springer. doi:10.1007/978–1-4020–8932-93.

van Driesche, R., Hoddle, M., & Center, T. (2007). Weed biocontrol agent diversity and ecology. In A. E. Hajek (Ed.)., *Control of Pests and Weeds by Natural Enemies—An Introduction to Biological Control* (pp. 45–55). Malden, Massachusetts: Blackwell Publishing.

van Lenteren, J. C., Bale, J., Bigler, F., Hokkanen, H. M. T., & Loomans, A. J. M. (2006). Assessing risks of releasing exotic biological control agents of arthropod pests. *Annual Review of Entomology, 51,* 609–634.

Velthuis, H. H. W., & van Door, A. (2006). A century of advances in bumblebee domestication and the economic and environmental aspects of its commercialization for pollination. *Apidologie, 37,* 421–451.

Vila, M., Basnou, C., Pysek, P., Josefsson, M., Genovesi, P., Gollasch, S., Nentwig, W., Olenin, S., Roques, A., Roy, D., & Hulme, P. E. (2010). How well do we understand the impacts of alien species on ecosystem services? A pan-European cross-taxa assessment. *Frontiers in Ecology and the Environment, 8,* 135–144.

Walther-Hellwig, K., Fokul, G., Buchler, R., Ekschmitt, K., & Walters, V. (2006). Increased density of honeybee colonies affects foraging bumblebees. *Apidologie, 37,* 517–532.

White, L. (1967). The historical roots of our ecological crisis. *Science, 155,* 1203–1207.

Wilkaniec, Z., Giejdasz, K., & Proszynski, G. (2004). Effect of pollination of onion seeds under isolation by red mason bee (*Osmia rufa* L.) (Apoidea: Megachilidae) on the setting and quality of obtained seeds. *Journal of Apicultural Science, 49,* 35–41.

Williams, F., Eschen, R., Harris, A., Djeddour, D., Pratt, C., Shaw, R. S., Varia, S., Lamontagne-Godwin, J., Thomas, S. E., & Murphy, S. T. (2010). *The Economic Cost of Invasive Non-native Species on Great Britain.* Wallingford, United Kingdom: CABI Publishing.

Williams, P., & Timmins, S. (2002). Economic impacts of weeds in New Zealand. In D. Pimentel (Ed.)., *Biological Invasions, Economic and Environmental Costs of Alien Plant, Animal, and Microbe Species* (pp. 175–184). London: CRC Press.

Williamson, M. (2002). Alien plants in the British Isles. In D. Pimentel (Ed.)., *Biological Invasions, Economic and Environmental Costs of Alien Plant, Animal, and Microbe Species* (pp. 91–112). London: CRC Press.

Wilson, C., & Tisdell, C. (2001). Why farmers continue to use pesticides despite environmental, health and sustainability costs. *Ecological Economics, 39,* 449–462.

Wiseman, B. R. (1994). Plant resistance to insects in integrated pest management. *Plant Disease, 78,* 927–932.

Wiseman, B. R., Davis, F. M., & Williams, W. P. (1996). Resistance of a maize genotype, Fawcc (C5), to fall armyworm larvae. *Florida Entomologist, 79,* 329–336.

Witzgall, P., Krisch, P., & Cork, A. (2010). Sex pheromones and their impact on pest management. *Journal of Chemical Ecology, 36,* 80–100.

WRI. (1999). *World Resources, 1998/1999: Environmental change and human health*. New York: World Resources Institute, Oxford University Press. http://www.wri.org/publication/world-resources-1998–99-environmental-change-and-human-health. Accessed 27 March 2013.

Chapter 8
Crop Losses to Arthropods

Thomas W. Culliney

Contents

8.1	Introduction	202
8.2	Crop Loss Assessment	203
	8.2.1 Elements of Crop Loss Assessment	204
8.3	Crop Losses Due to Pests	205
	8.3.1 Losses to Arthropod Pests	206
8.4	Addressing the Crop Loss Challenge	208
	8.4.1 Roots of Pest Problems in Modern Agriculture	208
	8.4.2 Alleviating Pest Problems to Reduce Crop Losses in Modern Agriculture	214
	8.4.3 Economic Benefits	215
8.5	Concluding Remarks	216
	References	217

Abstract Arthropods destroy an estimated 18–26 % of annual crop production worldwide, at a value of more than $ 470 billion. The greater proportion of losses (13–16 %) occurs in the field, before harvest, and losses have been heaviest in developing countries. With the Earth's human population expected to reach 10 billion by the end of the current century, raising agricultural productivity through the prevention of crop losses to pests has assumed considerable urgency. The techniques employed in crop loss assessment provide a framework useful in identifying the causes and magnitude of crop losses, and a basis for the evaluation of control options. Crop losses to arthropods have been reported to be lower in traditional than in modern, industrial agriculture, and are thought to result from the more environmentally sound and sustainable practices employed in traditional agriculture. Many of the pest problems in modern agriculture have arisen through an over-reliance on synthetic chemicals for pest control. More environmentally sound pest control practices not only are more sustainable, but may provide greater economic benefits

T. W. Culliney (✉)
Plant Protection and Quarantine Programs, Center for Plant Health Science and Technology,
Plant Epidemiology and Risk Analysis Laboratory, USDA-APHIS, 1730 Varsity Dr., Ste. 300,
Raleigh, NC 27606, USA.
e-mail: thomas.w.culliney@aphis.usda.gov

as well. The return per dollar invested in ecologically-based biological and cultural pest controls has been estimated to range from $ 30 to $ 300, significantly higher than the $ 4 estimated for control based on synthetic pesticides. Crop losses to pests must be reduced in ways that are compatible with sustainable production, which requires pest control to be approached in a holistic manner with a focus on the entire agroecosystem. Key to averting or minimizing crop losses to pests is a commitment, by government or other entities, to collect the data, on which reliable estimates of losses are based.

Keywords Agroecosystems · Alternative pest control · Crop loss assessment · Economic damages · Herbivory

> ...little attention has, comparatively, been paid to those noxious animals which annually consume an amount of produce that sets calculation at defiance; and, indeed, if an approximation could be made to the quantity thus destroyed, the world would remain sceptical of the result obtained, considering it too marvellous to be received as truth
>
> J. Curtis (1860)

> The struggle between man and insects began long before the dawn of civilization, has continued without cessation to the present time, and will continue, no doubt, as long as the human race endures. It is due to the fact that both men and certain insect species constantly want the same things at the same time...Here and there a truce has been declared, a treaty made, and even a partnership established, advantageous to both parties to the contract—as with the bees and silkworms, for example; but wherever their interests and ours are diametrically opposed, the war still goes on and neither side can claim a final victory. If they want our crops they still help themselves to them...Not only is it true that we have not really won the fight with the world of insects, but we may go farther and say that by our agricultural methods, by the extension of our commerce, and by other means connected with the development of our civilization, we often actually aid them most effectively in their competition with ourselves
>
> S.A. Forbes (1915)

8.1 Introduction

As of October 31, 2011, the world's human population was estimated to total 7 billion (UN 2011b). Based on current trends in fertility and population growth rates, that figure is projected to increase to 10 billion by the end of the century (UN 2011a). The demand for food will rise two to four times over the next few decades if anticipated increases in population and living standards occur (Hall 1995). Raising agricultural productivity and expanding food and fiber supplies have therefore become priorities for agricultural policy worldwide. At the beginning of the 21st century, a host of new and increasingly urgent challenges to maintaining or expanding crop production face mankind, including loss of land productivity due to soil erosion and conversion of cropland to nonfarm uses, falling water tables and increasing competition for water for industrial and domestic uses, and rising global temperatures (Brown 2004). One challenge, however, is as old as agriculture itself: crop losses caused by pest organisms (primarily weeds, arthropods, and plant pathogens) in their competition with humans for food and other plant products.

The historical record of man's struggle with pests for the products of cultivation extends far back into antiquity. The chronology compiled by Mayer (1959), for example, begins with a reference to locust attack on grain crops, dating from the 6th Dynasty in Egypt (2625-2475 B.C.E.). Other early accounts of agricultural pests appear throughout the Old Testament of the Bible and in texts from ancient Greece and Rome, and continue through the Middle Ages and Renaissance period (Ordish 1976). Throughout history, the ravages of agricultural pests have resulted in periods of famine and, occasionally, of economic and social upheaval. They have continued to plague mankind into the modern era (Anonymous 1958b; Cox and Large 1960; Ullstrup 1972).

Despite recent increases in production, world food supplies are barely keeping up with demand (FAO 2011a). In order to feed the burgeoning global population, overall food production will have to increase at least 70% over the next few decades; production in developing countries will need almost to double (FAO 2011b). Ensuring the success of this endeavor will require improvements in crop health monitoring and crop protection. Given such imperatives, it is vitally important that the causes of crop losses to pests be identified and their magnitude assessed.

8.2 Crop Loss Assessment

The need for a framework to facilitate identifying the causes and magnitude of crop losses to pests is clearly evident if such losses are to be averted as far as possible. This need has largely been met by the development of a methodology known in the aggregate by the term *crop loss assessment* (Teng 1987).

Crop loss assessment is a necessary prerequisite for a pest management program. By comparing the value of a reduction in crop production or quality with the cost of limiting or preventing the loss, an informed decision may be made by the grower as to when, where, how, or whether to apply control measures. For the policymaker, such information permits limited resources to be concentrated on those problems of significant economic importance. In particular, accurate data on losses to pests are essential to establish the *economic injury level*, the threshold of pest density or disease intensity, above which control becomes necessary to prevent economic damage, and to provide a basis for the evaluation of costs versus benefits in the application of control measures (Chiarappa et al. 1972).

8.2.1 Elements of Crop Loss Assessment

Crop loss assessment has three general components: *detection* of the harmful organism; *measurement* of the extent of the infestation or infection; and *evaluation* of the response of the crop to infection or infestation (Strickland 1971). Detection involves both careful sampling and accurate identification of the collected specimens. Also, as more than one kind of harmful organism may be present, a *crop loss profile* may be determined, through field experiment, to elucidate the relative importance of each (Chiarappa 1981; Zadoks 1981).

Methods for measuring the extent or intensity of pest attack may be divided into direct counts of pests or assessments of indirect proxies, such as injury or damage to crops (Walker 1981). In the employ of direct methods, the absolute, total number of individual pests, such as arthropods, can be counted, if practical, or a sample taken to estimate this number. Direct counts on plants may be made, for example, by: cutting open (for stem borers, larvae in fruit or pods); crushing or imprinting on gloss or ninhydrin paper (for aphids or mites); x-rays of seeds, stems, or galls for internal feeders; beating onto sheets; brushing onto a sticky surface by hand or machine (for small insects or mites); washing, using detergent or solvent, to estimate pest density by volume (for small insects or mites); sweep-net sampling on the basis of number of sweeps or total time sweeping; knockdown by application of a non-persistent pesticide, specimens falling onto a sheet; or suction, to collect all individuals on a plant. Direct counts in the wider environment may be made by: soil and debris sampling, separating specimens by sieving, Berlese or Tullgren funnels, or flotation; and various forms of trapping, employing, for example, attractant, color, water, sticky, suction, light, emergence, and pitfall traps.

In the indirect estimation of pest populations, various measures of pest impact or presence may be employed, such as: damage to the whole plant (number of plants killed, wilted, dying back—for stem borers, aphids, soil larvae); damage to roots (reduced weight, volume, length); damage to stems (cuts, as by cutworms or sawflies, dead-hearts produced by borers, exit holes, internode attacks); damage to leaves (mining, uniform or irregular areas eaten, skeletonization); damage to fruit and seeds by grazing, boring, or oviposition; and quantity of honeydew (for Homoptera).

The ultimate response of the crop to attack by pest organisms commonly is expressed as the effect on yield, the quantity of harvestable economic product, typically given as mass or weight of product per unit area, such as kilograms or tonnes per hectare. Several ways of categorizing yield have been proposed (Nutter et al. 1993). *Maximal obtainable yield* is the theoretical yield that could be achieved if the crop were grown under optimal environmental conditions, in the absence of pests, and is influenced primarily by the genetics of the crop species. *Attainable yield* is the yield obtained at a specific location when all available crop protection measures are used to alleviate the stresses caused by pests, although other, abiotic factors, such as soil fertility, water availability, and growing degree days, may still be limiting. It is the yield achieved typically only under experimental conditions. *Actual yield* is the real-world level of production characteristically realized within modern cropping systems employing recommended pest controls, but which is still subject to a variety of environmental constraints. Finally, *primitive yield* is defined as the yield achievable without the employment of modern methods of pest control, particularly synthetic pesticides, and is characteristic of subsistence agriculture. Crop loss to pests commonly is expressed as the difference between the actual and attainable yields (Walker 1987a; Nutter et al. 1993), a value, which has been termed the *avoidable loss* (Walker 1983). The aim of pest management programs is to narrow the gap between actual and attainable yield, and thus minimize the avoidable loss.

Crop losses may be quantitative or qualitative (Oerke 2006). Quantitative losses result from reduced productivity, leading to a smaller yield. Qualitative losses from pests may result from reduced nutritional content, reduced market quality, reduced storage characteristics, or contamination of the harvested product with pests, parts of pests, or the toxic products of pests (e.g., mycotoxins from plant-parasitic fungi). From a different perspective, Ordish (1952) viewed losses in terms of the resources (land, labor, energy, money) wasted in the production of crop harvests that were never realized—the "untaken" harvests. These represent a significant opportunity cost, for the wasted resources could have been diverted to other productive use or uses in the absence of pests.

8.3 Crop Losses Due to Pests

Arriving at reliable figures for crop losses to pest organisms is fraught with difficulty, and the resulting estimates come burdened with considerable uncertainty. Accordingly, reports on crop losses, in nearly all cases, have been compiled from indirect data with recognized weaknesses (Van der Graaff 1981). Crop yield is affected by a multitude of variables, many interacting, and the response of yield to one of these, pests, is itself almost infinitely variable. The distribution of pests in space and time, the response of plants and pests to different climates and soils, multiple cropping, interactions between pests and pathogens, and the diseases they produce, are just some of the complications in measuring yield loss due to pests (Walker 1987b). Consequently, there is a shortage of data concerning the extent and value of the losses caused by pests, particularly for developing countries (Reed 1983; Yudelman et al. 1998). Most of the data that do exist come from North America, western Europe, Australia, and India; estimates of losses on a global scale are necessarily extrapolated from the available data (Oerke et al. 1994).

Despite the challenges involved, a few creditable attempts have been made to assess the magnitude of crop losses to pests, the higher of which occur pre-harvest. The foci of these exercises have been the major classes of agricultural pests: arthropods, plant pathogens (including nematodes), and weeds, as well as molluscs and some vertebrates, and losses occurring in storage, as well as in the field. Estimates of losses in monetary terms for major crops in the United States were compiled by USDA (1965). Production losses in various field crops, fruits, and vegetables in Great Britain were assessed by Ordish (1952) and Strickland (1965), and were estimated for Canadian agriculture by Jaques et al. (1994). The most comprehensive account of crop losses to pests on a global scale was provided by Cramer (1967), and was probably the first based on well-founded analysis. That author estimated overall annual losses in major crops (including cereals, potato, vegetables, fruits, oil crops, fiber crops, and natural rubber) to be about 34%. More recent estimates of overall crop losses include those of Pimentel (1997) (35–42%) and Oerke (2006) (69%). Despite the application of pesticides and other prophylactic or control measures, pests destroy a substantial proportion of annual production in individual crops worldwide: as much as 50% in rice, 41% in potato, 40% in coffee, 39% in maize, 38% in cotton, 34% in wheat, 32% in soybean, 30% in barley, and 26% in sugar beet (Oerke et al. 1994; Oerke and Dehne 2004). Losses have been heaviest in developing countries.

8.3.1 Losses to Arthropod Pests

Chief among pests responsible for the unacceptably high losses in crops are arthropods, and these losses have been increasing over recent decades (Pimentel 1997). Insects and mites account for the majority of the damage to crops, with a much smaller proportion attributable to other groups, such as the Collembola, Symphyla, and Oniscidea. Among the mites, the worst pests are found in the orders Trombidiformes (families Tydeidae, Phytoptidae, Diptilomiopidae, Eriophyidae, Tetranychidae, Tenuipalpidae, Tuckerellidae, and Tarsonemidae) and Sarcoptiformes (Acaridae) (Jeppson et al. 1975; Krantz and Walter 2009). The Orthoptera, Hemiptera (Heteroptera and Homoptera), Thysanoptera, Coleoptera, Lepidoptera, and Diptera are considered the most important insect orders containing agricultural pests (Metcalf and Metcalf 1993; Triplehorn and Johnson 2005). Recent years have seen a growing recognition of the importance, as pests of stored grains and other food products, of the Psocoptera (book and bark lice), a group hitherto considered of no significance in that context (Ahmedani et al. 2010). However, although they are the major destroyers of crops, probably fewer than 1% of all insect species may be considered pests in any way, and, of these, perhaps 3500 species require regular attention (Pedigo and Rice 2006).

Arthropods have been consuming the tissues of living green plants since the Late Silurian Period, about 416 million years ago (Labandeira 2007). In natural ecosystems, this herbivory tends to be minor, amounting to 0.5–15% of net plant (primary) production (Soholt 1973), and usually does not impair overall plant productivity (Mattson and Addy 1975). However, the conversion of natural systems to arable agriculture brings arthropod feeding into conflict with human interests, and puts crop production at risk. In their simplified structures, open mineral cycles, and high rates of biomass accumulation (high yields), agroecosystems resemble the early stages of ecological succession (Odum 1969; Vitousek and Reiners 1975). That description applies to modern, industrialized agriculture, in particular, with its tendency towards single-stand cropping and extensive dependence on external inputs.

Crop losses due to insect pests have been reported to be greater under modern than under traditional agricultural practices (Andow and Hidaka 1989; Reddy and Zehr 2004; Dhaliwal et al. 2007). This is thought to result from the greater spatial and temporal complexity of traditional agriculture, which fosters conditions inconducive to increases in pest populations (Brown and Marten 1986). Odum (1984) pointed out that the power density level (rate of energy flow per unit area) of traditional agriculture, due to low energy and chemical subsidies, is little different from that of natural ecosystems at a mature stage, which tend towards complexity and stability. In contrast, modern cropping practices tend to simplify agroecosystems and encourage pest problems through monoculture and the planting of more susceptible crop varieties, destruction of beneficial arthropods (predators and parasites), promotion of pesticide resistance, decreased crop rotations, reduced crop sanitation, reduced tillage, and increased cosmetic standards for crop quality (Pimentel et al. 1978; Metcalf 1980; NRC, 1989; Oerke 2006). The reduction in plant species richness brings about changes in the composition of the resident pest community, and typically results in greater crop losses from a pest complex that is less diverse, but more abundant, and dominated by specialist herbivores with a narrow host range (Matson et al. 1997). Also, the myriad stresses and imbalances, to which crop plants are subjected under modern production systems, are thought to lower their resistance to pests (Hodges and Scofield 1983). In particular, the increased crop productivity characteristic of modern agriculture, an advance made possible by the "green revolution," often is associated with higher vulnerability to pest attack (Oerke 2006).

Arthropods damage crops in a variety of ways (Metcalf and Metcalf 1993). They attack growing plants by: chewing leaves, buds, stems, bark, or fruit; sucking the sap from leaves, buds, stems, or fruit; boring into or tunneling through the bark, stems, or twigs, into fruit, nuts, or seeds, or between the surfaces of leaves; causing cancer-like growths on plants (galls), within which they live; attacking roots or underground stems in any of the above ways; laying eggs in plant parts; taking plant parts for the construction of nests or shelters; and transmitting plant pathogens. They destroy or depreciate the value of stored plant products by: consuming the items as food; contaminating them with their secretions, fecal material, eggs, or their own bodies; and increasing the labor and expense of sorting, packing, and preserving foods.

Fairly detailed figures on crop losses to insect and mite pests, in monetary terms, were compiled and published by the U.S. Department of Agriculture during the previous century (e.g., Haeussler 1952; USDA 1965), several compilations appearing in the Department's *Cooperative Economic Insect Report* and the succeeding *Cooperative Plant Pest Report* series (Anonymous 1958a, 1961, 1966, 1967, 1968, 1969, 1971, 1973, 1974, 1976, 1978). In an early report, overall losses in the United States were estimated to range from 10–20% (Marlatt 1905). Recent estimates of the amount of pre-harvest crop production destroyed globally each year by arthropods, incorporating the earlier-published data, similarly are on the order of 13–16% (e.g., Pimentel 1997; Yudelman et al. 1998). Oerke (2006) found "animal pests" (the most important of which are arthropods; Pimentel and Pimentel 1978) to account for losses of about 18%. A further 5–10% of crop production is lost to arthropods post-harvest (Boxall 2001). Assuming the worst case (percentage losses at the higher end), and based on current figures for production values of the most important cereals and other crops (FAO 2012a), economic losses due to crop-destroying arthropods exceed $ 470 billion annually, undoubtedly an underestimate, given the paucity of data from the developing world. For individual crops, information is more readily available for pre-harvest losses than for losses in storage, and a large volume of the data appears to have been gathered in India (Tables 8.1 and 8.2).

8.4 Addressing the Crop Loss Challenge

8.4.1 Roots of Pest Problems in Modern Agriculture

Modern, industrialized agriculture creates its own problems by providing arthropod pests with vast tracts of densely planted, genetically uniform monocultures of high-yielding, highly palatable, and nutritious crops, on which to dine and multiply. In natural ecosystems, arthropod population outbreaks arise in response to stress or high density in host plant populations. Under such circumstances, they tend to be self-limiting, as the most stressed, and therefore vulnerable, plants are targeted, leading to a decrease in their density, which mitigates competition for resources, a major source of stress. A reduction in the food supply, in turn (and in conjunction with an increase in the action of natural enemies), results in a decline in arthropod densities, restoring the system to equilibrium. Particularly vulnerable to arthropod outbreaks are agroecosystems, with their high crop densities and stress-inducing plant-plant competition and cultivation practices. However, in these, the natural equilibrium-restoring process, which would lead to a loss in yield, is short-circuited through artificial control of the arthropod population. Artificially maintaining production in this way represents a destabilization of the ecosystem, and requires the addition of energy, nutrients, and other inputs (Schowalter 2007).

Bountiful crop harvests are achieved only with the aid of modern pest control technologies, which have tended to emphasize a single approach, the liberal and frequent use of chemical pesticides. This reliance on pesticides as the sole or major means of control has had unintended and unforeseen negative consequences by selecting for resistant populations of the targeted, primary pests and destroying natural enemies, which has resulted in outbreaks of other, secondary pests and, in turn, in the need for ever-increasing applications of pesticides to control them. With regard to the latter outcome, it has been estimated that, worldwide, about 50% of arthropod pests now controlled with pesticides became pests as a direct result of pesticide use (Thacker 2002). Before the widespread use of pesticides, these species were not considered important enough to warrant control measures, and undoubtedly were kept in check by their natural enemies. Paradoxically, as pesticide use has steadily increased over the past few decades, crop losses to arthropods also have increased, nearly doubling, from about 7% to 13% (Pimentel 1993).

In recognition of this "pesticide treadmill," efforts to address the problem of arthropod pest control in agriculture have become organized into a program, integrated pest management (IPM) (Vandermeer 1995). The philosophy of IPM is simple, and espouses two basic principles: (1) use pesticides only when necessary,

Table 8.1 A sampling of reported estimates of pre-harvest crop loss caused by arthropod pests[a]

Crop	Pest species	Taxonomic group	Country	Loss (%)[b, c]	Reference
Abelmoschus esculentus	*Tetranychus urticae* Koch	Acari: Tetranychidae	India	32–36	Prasad and Singh (2009)
Allium cepa	*Delia antiqua* (Meigen)	Diptera: Anthomyiidae	Romania	20–30	Becherescu (2009)
Apium graveolens	*Oribius destructor* Marshall, *O. inimicus* Marshall	Coleoptera: Curculionidae	Papua New Guinea	36	Wesis et al. (2010)
Beta vulgaris	*Pemphigus betae* Doane	Homoptera: Aphididae	United States	31–34	Hutchison and Campbell (1994)
Brassica spp.	*Brevicoryne brassicae* (L.), *Lipaphis erysimi* (Kaltenbach)	Homoptera: Aphididae	Pakistan	70–80	Razaq et al. (2011)
Brassica campestris, B. juncea	*Lipaphis erysimi* (Kaltenbach)	Homoptera: Aphididae	India	10–90	Rana (2005)
Brassica napus	*Crocidolomia binotalis* (Zeller)	Lepidoptera: Pyralidae	India	48	Maity et al. (2001)
Brassica oleracea	*Oribius destructor* Marshall, *O. inimicus* Marshall	Coleoptera: Curculionidae	Papua New Guinea	18	Wesis et al. (2010)
	Plutella xylostella (L.)	Lepidoptera: Plutellidae	Ethiopia	50–91	Ayalew (2006)
Capsicum annuum	*Helicoverpa armigera* (Hübner)	Lepidoptera: Noctuidae	India	77	Katagihallimath (1963)
Carum carvi	*Aceria carvi* (Nalepa)	Acari: Eriophyidae	Czech Republic	44	Zemek et al. (2005)
Cicer arietinum	*Helicoverpa armigera* (Hübner)	Lepidoptera: Noctuidae	India	41	Deshmukh et al. (2010)
Citrus unshiu	*Ceratitis capitata* (Wiedemann)	Diptera: Tephritidae	Italy	58	D'Aquino et al. (2011)
Cocos nucifera	*Aceria guerreronis* Keifer	Acari: Eriophyidae	Colombia	60	Navia et al. (2005)
			India	25	Jagadeesh and Nanjappa (2009)
			Tanzania	10–100	Navia et al. (2005)
			Venezuela	70	Navia et al. (2005)
	Opisina arenosella Walker	Lepidoptera: Oecophoridae	India	45	Mohan et al. (2010)
Coffea arabica	*Leucoptera coffeella* (Guérin-Méneville and Perrottet)	Lepidoptera: Lyonetiidae	Brazil	80	Domingos Scalon et al. (2011)
Cucumis sativus	*Bactrocera cucurbitae* (Coquillet)	Diptera: Tephritidae	India	39	Kate et al. (2009)

Table 8.1 (continued)

Crop	Pest species	Taxonomic group	Country	Loss (%)[b,c]	Reference
Fragaria × ananassa	Tetranychus urticae Koch	Acari: Tetranychidae	United States	25	Walsh et al. (1998)
Glycine max	Etiella zinckenella (Treitschke)	Lepidoptera: Pyralidae	Indonesia	1–52	Van den Berg et al. (1998)
Gossypium hirsutum	Earias insulana (Boisduval)	Lepidoptera: Nolidae	Egypt	51	Abd El-Rahman et al. (2008)
	Phenacoccus solenopsis Tinsley	Homoptera: Pseudococcidae	India	50	Joshi et al. (2010)
Helianthus annuus	Cochylis hospes Walsingham	Lepidoptera: Tortricidae	United States	0.5–79	Charlet et al. (2009)
Lactuca sativa	Oribius destructor Marshall, O. inimicus Marshall	Coleoptera: Curculionidae	Papua New Guinea	31–53	Wesis et al. (2010)
Lens culinaris	Melanoplus bivittatus (Say)	Orthoptera: Acrididae	Canada	70	Olfert and Slinkard (1999)
Mangifera indica	Bactrocera invadens Drew, Tsura and White, Ceratitis cosyra (Walker)	Diptera: Tephritidae	Benin	>75	Vayssieres et al. (2008)
Manihot esculenta	Mononychellus tanajoa (Bondar)	Acari: Tetranychidae	Colombia	21–53	Herrera Campo et al. (2011)
	Polyphagotarsonemus latus (Banks)	Acari: Tarsonemidae	India	50	Rajalakshmi et al. (2009)
Morus sp.	Aleurodicus dispersus Russell	Homoptera: Aleyrodidae	India	28	Qadri et al. (2010)
Morus alba	Cosmopolites sordidus (Germar)	Coleoptera: Curculionidae	India	85	Shukla (2010)
Musa sp.	Odoiporus longicollis (Olivier)	Coleoptera: Curculionidae	India	10–90	Shukla (2010)
Nicotiana tabacum	Acherontia lachesis (F.)	Lepidoptera: Sphingidae	Japan	<1	Imai et al. (2011)
Olea europaea	Prays oleae (Bernard)	Lepidoptera: Yponomeutidae	Spain	40	Ramos et al. (1998)
Oryza sativa	Dicladispa armigera (Olivier)	Coleoptera: Chrysomelidae	Bangladesh	40–50	Murphy (2005)
	Lissorhoptrus oryzophilus Kuschel	Coleoptera: Curculionidae	United States	20	Zou et al. (2004)
	Orseolia oryzivora Harris and Gagné	Diptera: Cecidomyiidae	Burkina Faso	68–87	Nacro et al. (2006)
Pennisetum glaucum	Heliocheilus albipunctella (de Joannis)	Lepidoptera: Noctuidae	Niger	12–42	Youm and Owusu (1998)

Table 8.1 (continued)

Crop	Pest species	Taxonomic group	Country	Loss (%)[b, c]	Reference
Phaseolus vulgaris	Ophiomyia phaseoli (Tryon)	Diptera: Agromyzidae	Kenya	3–69	Ojwang et al. (2010)
	Oribius destructor Marshall, O. inimicus Marshall	Coleoptera: Curculionidae	Papua New Guinea	40	Wesis et al. (2010)
Prunus persica	Ceratitis capitata (Wiedemann)	Diptera: Tephritidae	Italy	32–45	D'Aquino et al. (2011)
Prunus persica var. nucipersica	Ceratitis capitata (Wiedemann)	Diptera: Tephritidae	Italy	63–93	D'Aquino et al. (2011)
Saccharum officinarum	Ceratovacuna lanigera Zehntner	Homoptera: Aphididae	India	30	Galande et al. (2005)
	Eldana saccharina Walker	Lepidoptera: Pyralidae	South Africa	22	Berry et al. (2010)
	Sesamia grisescens (Walker)	Lepidoptera: Noctuidae	Papua New Guinea	5–18	Kuniata and Sweet (1994)
Sesamum indicum	Elasmolomus sordidus (F.)	Heteroptera: Lygaeidae	India	78	Kalaiyarasan and Palanisamy (2005)
Solanum melongena	Aculus lycopersici (Wolffenstein)	Acari: Eriophyidae	India	16–20	Prasad and Singh (2009)
	Leucinodes orbonalis (Guenée)	Lepidoptera: Pyralidae	India	25	Haseeb et al. (2009)
Solanum tuberosum	Oribius destructor Marshall, O. inimicus Marshall	Coleoptera: Curculionidae	Papua New Guinea	22–42	Wesis et al. (2010)
Theobroma cacao	Conopomorpha cramerella (Snellen)	Lepidoptera: Gracillariidae	Malaysia	22–54	Day (1989)
	Helopeltis theivora Waterhouse	Heteroptera: Miridae	Malaysia	95	Muhamad and Way (1995)
	Planococcus citri (Risso)	Homoptera: Pseudococcidae	Brazil	4–9	Delabie and Cazorla (1991)
Theobroma grandiflorum	Conotrachelus humeropictus Fielder	Coleoptera: Curculionidae	Brazil	58	Lopes and Silva (1998)
Triticum aestivum	Calamobius filum (Rossi)	Coleoptera: Cerambycidae	France	20	Dordolo (2008)
	Diuraphis noxia (Mordvilko)	Homoptera: Aphididae	United States	50–83	Mirik et al. (2009)
Vicia faba	Sitona lineatus (L.)	Coleoptera: Curculionidae	Denmark	28	Nielsen (1990)
Vigna mungo	Riptortus pedestris (F.)	Heteroptera: Alydidae	India	27	Borah and Sharma (2009)
Vigna radiata	Apion amplum (Faust)	Coleoptera: Apionidae	India	63	Deshmukh et al. (2007)

Table 8.1 (continued)

Crop	Pest species	Taxonomic group	Country	Loss (%)[b,c]	Reference
Vigna unguiculata	Megalurothrips sjostedti (Trybom)	Thysanoptera: Thripidae	Uganda	60–65	Nabirye et al. (2003)
Vitis vinifera	Lobesia botrana (Denis and Schiffermüller)	Lepidoptera: Tortricidae	Greece	13–27	Moschos (2006)
	Taedia scrupeus (Say)	Heteroptera: Miridae	United States	26	Martinson et al. (1998)
Withania somnifera	Coccidohystrix insolita (Green)	Homoptera: Pseudococcidae	India	8	Murali Baskaran et al. (2009)
	Epilachna vigintioctopunctata (F.)	Coleoptera: Coccinellidae	India	19	Murali Baskaran et al. (2009)
Zea mays	Chilo partellus (Swinhoe)	Lepidoptera: Pyralidae	India	24–35	Singh and Sharma (2009)
	Ostrinia furnacalis (Guenée)	Lepidoptera: Pyralidae	Philippines	9–49	Litsinger et al. (2007)
	Rhopalosiphum maidis (Fitch)	Homoptera: Aphididae	United States	10–53	Everly (1960)
	Spodoptera frugiperda (Smith)	Lepidoptera: Noctuidae	Brazil	20	Gianluppi et al. (2002)
Ziziphus mauritiana	Larvacarus transitans (Ewing)	Acari: Tenuipalpidae	India	31	Sharma and Naqvi (1993)

[a] Experimental studies, in which pest populations were artificially manipulated, are omitted
[b] Under standard cultivation practices or in the absence of control measures
[c] Ranges express differences in variables, such as growing region, crop variety, growth stage, and sample date

Table 8.2 A sampling of reported estimates of post-harvest crop loss caused by arthropod pests[a]

Crop	Pest species	Taxonomic group	Country	Loss (%)[b]	Reference
Arachis hypogaea	Caryedon serratus (Olivier)	Coleoptera: Bruchidae	India	25–90	Nandagopal et al. (2007)
Brassica juncea	Bagrada cruciferarum Kirkaldy	Heteroptera: Pentatomidae	India	30	Singh and Malik (1993)
Cicer arietinum	Callosobruchus chinensis (L.)	Coleoptera: Bruchidae	Eritrea	28	Haile (2006)
Coffea sp.	Hypothenemus hampei (Ferrari)	Coleoptera: Scolytidae	Jamaica	21	Reid and Mansingh (1985)
Hordeum vulgare	Rhizopertha dominica (F.)	Coleoptera: Bostrichidae	India	1–3	Chander and Bhargava (2005)
Ipomoea batatas	Cylas formicarius (F.)	Coleoptera: Curculionidae	India	58	Ray et al. (1994)
Lens culinaris	Bruchus lentis Frölich	Coleoptera: Bruchidae	Turkey	0.30–0.86	Dörtbudak et al. (1999)
Manihot esculenta	Prostephanus truncatus (Horn)	Coleoptera: Bostrichidae	Ghana	8–21	Stumpf (1998)
Oryza sativa	Corcyra cephalonica (Stainton)	Lepidoptera: Pyralidae	India	4	Lal et al. (2000)
	Oryzaephilus surinamensis (L.)	Coleoptera: Cucujidae	India	5	Lal et al. (2000)
	Rhizopertha dominica (F.)	Coleoptera: Bostrichidae	India	1	Lal et al. (2000)
	Sitophilus oryzae (L.)	Coleoptera: Curculionidae	India	66	Lal et al. (2000)
	Sitotroga cerealella (Olivier)	Lepidoptera: Gelechiidae	Bangladesh	3–12	Shahjahan (1974)
	Tribolium castaneum (Herbst)	Coleoptera: Tenebrionidae	India	24	Lal et al. (2000)
Solanum tuberosum	Phthorimaea operculella (Zeller)	Lepidoptera: Gelechiidae	India	70	Rondon (2010)
Sorghum bicolor	Sitophilus oryzae (L.)	Coleoptera: Curculionidae	Cameroon	1–13	Ladang et al. (2008)
Triticum sp.	Trogoderma granarium Everts	Coleoptera: Dermestidae	India	9–14	Bains et al. (1976)
Vigna radiata	Callosobruchus maculatus (F.)	Coleoptera: Bruchidae	India	78	Singh and Jakhmola (2011)
Vigna unguiculata	Bruchidius atrolineatus (Pic), Callosobruchus maculatus (F.)	Coleoptera: Bruchidae	Niger	19–86	Van Alebeek (1996)
Zea mays	Prostephanus truncatus (Horn)	Coleoptera: Bostrichidae	Tanzania	30	Farrell and Schulten (2002)
	Sitophilus zeamais (Motschulsky)	Coleoptera: Curculionidae	Kenya	10–20	Bett and Nguyo (2007)
	Sitotroga cerealella (Olivier), Sitophilus spp.	Lepidoptera: Gelechiidae, Coleoptera: Curculionidae	Malawi	15–90	Schulten (1975)

[a] Experimental studies, in which pest populations were artificially manipulated, are omitted
[b] Ranges express differences in variables, such as crop variety, sample source locality, and storage conditions

and (2) manage the agroecosystem in such a way that they do not become necessary or their use is kept to a minimum. Alternatively, put in the form of a question, these become: what agroecosystem conditions lead to unacceptable pest numbers, and how best can these be mitigated (Schowalter 2007)? The methodology of IPM integrates ecological information into the pest control process, and is based on a thorough understanding of the entire crop production system, including the role played by pests. One goal is for ecological theory to be used as a guide to predict how changes in production practices and inputs may affect pest problems (Nicholls and Altieri 2007). IPM relies largely on alternative means of pest control, particularly cultural and biological techniques, to produce a crop environment unfavorable to pest establishment or increase, or in which crop susceptibility to damage is reduced. Many of its precepts have long been a part of traditional agriculture. Rather than complete pest annihilation, the goal of IPM is to *manage* pests, to maintain their numbers below levels at which economically significant crop damage would occur.

8.4.2 Alleviating Pest Problems to Reduce Crop Losses in Modern Agriculture

Research has suggested that one practical means by which yield losses to pests may be reduced is through diversification of the crop environment in space or time. The conceptual basis of such work is the view, arising from mathematical theory as well as observation in nature, that ecological diversity begets stability in biological populations, particularly those of herbivorous arthropods (Goodman 1975). Diverse cropping systems, in the form of polycultures, may experience lower yield losses than monocultures (Andow 1991b). The greater vegetational complexity in mixed cropping systems is thought to prevent pest increase by fostering populations of natural enemies and disrupting the ability of pests to find and colonize hosts (Root 1973; Andow 1991a). Diversification of the agroecosystem need not be wholesale to reap pest control benefits, which may make such manipulation of the crop environment more acceptable to growers. Rather, refuges for beneficial predators and parasitoids may be created in field borders or other areas of low value for crop production (Wratten and van Emden 1995). This could be achieved, for example, by maintenance of some weeds within the cropping system (Altieri 1995).

Crop rotation can prevent pest population buildup over several cropping cycles by disrupting pest life cycles through the substitution of unfavorable hosts for favorable ones (Bullock 1992). The improved pest control and other benefits of alternating crops over time contribute to a well-known "rotation effect," whereby a crop following almost any other crop does better than when grown in continuous monoculture (Magdoff 1995). Other cultural techniques, such as varying planting dates, crop sanitation, tillage, and increasing plant density, also limit the ability of pests to find or colonize a crop.

8.4.3 Economic Benefits

Such alterations of the crop environment, which mitigate against pest outbreaks, can produce immediate economic benefits, particularly an increase in marketable crop yields with reduced losses to pests. For example, in a comparison of conventional maize production with maize grown under conditions incorporating environmentally sound practices (no pesticides, tillage for weed control, organic soil amendments, crop rotation), yield was 16% higher (8100 vs. 7000 kg per ha) and crop loss to arthropod pests 71% lower (3.5 vs. 12%) in the latter system, at a total cost of production 36% lower (Pimentel 1993). Further, the increased yields are achieved without the monetary cost, health risks (to farm workers, consumers), and environmental impacts (groundwater contamination, nontarget species kills) resulting from precautionary, fixed-schedule applications of pesticides (Letourneau 1997). The return per dollar invested in ecologically-based biological and cultural pest controls has been estimated to range from $ 30 to $ 300, compared to $ 4 for control based on synthetic pesticides (Pimentel 1986). When costs arising from the negative impacts of pesticides on human health and the environment are factored in, the ratio of benefits to costs of chemical control is barely greater than unity (Pimentel 2005).

Unfortunately, recent economic trends have resulted in a reduction of regional crop diversity in the United States. An increase in hectarage planted to maize monoculture to satisfy the surge in demand for maize as a feedstock for ethanol production has effected a significant restructuring of the agricultural landscape in the midwestern part of the country, with negative consequences for production in other crops. For example, for soybean producers employing IPM, natural biological control of soybean aphid, *Aphis glycines* Matsumura, has an estimated value of $ 33 per ha (Landis et al. 2008). These authors found that the reduced landscape diversity resulting from increased maize production disrupted the activity of natural enemies, diminishing the level of biological control by 24%. This loss of biocontrol services costs soybean producers an estimated $ 58 million per year in reduced yield and increased insecticide use. For those, such as organic producers, who rely solely or largely on biological control, costs would be considerably higher.

Beyond suppression of pest populations, habitat diversification may provide additional economic and other benefits to a sustainable agriculture through continuous vegetational cover for soil protection, constant food production for subsistence and marketing, closing nutrient cycles and effective use of local resources, increased organic matter and soil fertility, improved soil physical and biological properties, soil and water conservation through mulching and windbreaks, increased activity of pollinators, increased multiple-use capacity of the landscape, and sustained crop production without relying on environmentally degrading chemical inputs (Bullock 1992; Pimentel et al. 2005; Nicholls and Altieri 2007).

The conditions, under which crops are lost to pests, are complex and unpredictable, influenced by numerous physical and biological variables. Some, such as the extremes of weather that induce stress in crop plants and make them susceptible to pest outbreaks, may be impossible to mitigate. Others, however, are amenable to at least some measure of control. Through the application of ecological principles to agricultural production, a return to the tried-and-true cultural methods traditionally employed, and less reliance on synthetic chemical pesticides for pest control, crop losses to pests may be alleviated and agriculture put on a more sustainable footing.

8.5 Concluding Remarks

The continuing onslaught of pests, particularly arthropods, on the world's plant resources underscores the crucial need for effective and viable means of crop protection to ensure the food supplies for a growing human population. An estimated 4.9 billion hectares, 38 % of the Earth's land area, are managed for agriculture (FAO 2012b), and recent years have seen no increase in arable capacity (Gerard 1995). Most of the land optimal for agriculture already is in production, and this, increasingly, is being converted to other uses or otherwise lost (e.g., through erosion, salinization, desertification). As cultivable land is a diminishing resource, there is no longer the opportunity of expanding crop production into new areas, as once might have been done, to meet increased demand. Thus, the only remaining option for increasing the food supply is to raise crop productivity on existing cultivated land. A significant contributor to this increase in productivity would be the reduction of crop losses achieved through application of the methods of integrated pest management.

Crop losses to pests must be reduced in ways that are compatible with sustainable production, which requires pest control to be approached in a holistic manner with a focus on the entire agroecosystem. This in turn requires knowledge of the factors that promote a healthy crop environment, enabling crops to ward off or withstand pest attack. Worth more consideration are the potential benefits of diversifying the crop environment. Cropping regimes, such as those making use of intercropping or strip cropping, selective weeding, and crop rotation, simulate natural systems, and reduce pest colonization or population increase by disrupting the host-finding process and fostering populations of natural enemies.

From past experience, it is fair to conclude that pests are continuing to take an unacceptable toll on agricultural production. However, the true extent of the problem at the present time is difficult to gauge. Unfortunately, there do not appear to be any current or recent, comprehensive, and readily accessible data on crop losses in the United States or elsewhere in the world. Neither the U.S. Department of Agriculture nor the Food and Agriculture Organization of the United Nations currently compiles statistics on crop losses to pests. Inquiries, made by the author, of the departments of agriculture of the top 10 U.S. agricultural states (California, Iowa, Texas, Nebraska, Illinois, Minnesota, Kansas, North Carolina, Indiana, and Missouri; ERS 2010) revealed that such data also are not routinely collected at the state level. Thus, an updating of the magnitude of overall, present-day crop losses to arthropods or other pests cannot here be provided. These data would be particularly valuable in establishing a baseline of information, from which to assess the influence of emerging environmental trends, such as global climate change, on agricultural production losses to pests.

The monumental work of Cramer (1967), who had access to concurrent sources of global crop production and regional loss data, from which to make extrapolations, was the last thorough analysis of crop loss to pests, but was published decades ago, and the estimates derived therein undoubtedly have little if any remaining applicability to inform the situation as it exists today. If crop losses to pests are to be averted or minimized, reliable estimates of the losses that are now occurring are essential, and collecting the data on which they are based should be a priority, especially for governmental agencies and international organizations, which, in the past, provided the funds to support this important work.

References

Abd El-Rahman, A. G., Abd El-Hafez, A. M., El-Sawar, B. M., Refaie, B. M., & Imam, A. I. (2008). Efficacy of the egg parasitoid, *Trichogramma evanescens* West. in suppressing spiny bollworm, *Earias insulana* (Boisd.) infestation in El-Farafra cotton fields, New Valley Governorate, Egypt. *Egyptian Journal of Biological Pest Control, 18*(2), 265–269.

Ahmedani, M. S., Shagufta, N., Aslam, M., & Hussnain, S. A. (2010). Psocid: A new risk for global food security and safety. *Applied Entomology and Zoology, 45*(1), 89–100.

Altieri, M. A. (1995). *Agroecology: The Science of Sustainable Agriculture* (2nd ed.). Boulder: Westview.

Andow, D. A. (1991a). Vegetational diversity and arthropod population response. *Annual Review of Entomology, 36*, 561–586.

Andow, D. A. (1991b). Yield loss to arthropods in vegetationally diverse agroecosystems. *Environmental Entomology, 20*(5), 1228–1235.

Andow, D. A., & Hidaka, K. (1989). Experimental natural history of sustainable agriculture: Syndromes of production. *Agriculture, Ecosystems and Environment, 27*(1–4), 447–462.

Anonymous. (1958a). Some insect loss estimates for 1957. *Cooperative Economic Insect Report, 8*(11), 205–206.

Anonymous. (1958b). Crop pests make history. *World Crops, 10*(2), 52–57.

Anonymous. (1961). Losses and production costs attributable to insects and related arthropods—1960. *Cooperative Economic Insect Report, 11*(47), 1062–1081.

Anonymous. (1966). Estimated losses and production costs attributed to insects and related arthropods—1965. *Cooperative Economic Insect Report, 16*(42), 997–1007.

Anonymous. (1967). Estimated losses and production costs attributed to insects and related arthropods—1966. *Cooperative Economic Insect Report, 17*(45), 991–1007.

Anonymous. (1968). Estimated losses and production costs attributed to insects and related arthropods—1967. *Cooperative Economic Insect Report, 18*(43), 1012–1028.

Anonymous. (1969). Estimated losses and production cost attributed to insects and related arthropods—1968. *Cooperative Economic Insect Report, 19*(50), 878–893.

Anonymous. (1971). Estimated losses and production cost attributed to insects and related arthropods—1970. *Cooperative Economic Insect Report, 21*(45–48), 759–772.

Anonymous. (1973). Estimated losses and production costs attributed to insects and related arthropods—1972. *Cooperative Economic Insect Report, 23*(49–52), 783–796.

Anonymous. (1974). Estimated losses and production costs attributed to insects and related arthropods—1973. *Cooperative Economic Insect Report, 24*(49–52), 881–903.

Anonymous. (1976). Estimated losses and production costs attributed to insects and related arthropods—1975. *Cooperative Plant Pest Report, 1*(48–52), 875–893.

Anonymous. (1978). Estimated losses and production costs attributed to insects and related arthropods—1976. *Cooperative Plant Pest Report, 3*(11), 91–117.

Ayalew, G. (2006). Comparison of yield loss on cabbage from diamondback moth, *Plutella xylostella* L. (Lepidoptera: Plutellidae) using two insecticides. *Crop Protection, 25*(9), 915–919.

Bains, S. S., Battu, G. S., & Atwal, A. S. (1976). Distribution of *Trogoderma granarium* Everts and other stored grain insect pests in Punjab and losses caused by them. *Bulletin of Grain Technology, 14*(1), 18–29.

Becherescu, A. (2009). Efficacy of integrated protection complexes in fighting against the pest *Delia antiqua* Meig. *Journal of Horticulture, Forestry and Biotechnology, 13,* 151–154.

Berry, S. D., Leslie, G. W., Spaull, V. W., & Cadet, P. (2010). Within-field damage and distribution patterns of the stalk borer, *Eldana saccharina* (Lepidoptera: Pyralidae), in sugarcane and a comparison with nematode damage. *Bulletin of Entomological Research, 100*(4), 373–385.

Bett, C., & Nguyo, R. (2007). Post-harvest storage practices and techniques used by farmers in semi-arid Eastern and Central Kenya. *African Crop Science Conference Proceedings, 8,* 1023–1027.

Borah, B. K., & Sharma, K. K. (2009). Assessment of losses in yield of black gram caused by *Riptortus pedestris* Fab. *Insect Environment, 14*(4), 156–157.

Boxall, R. A. (2001). Post-harvest losses to insects—a world overview. *International Biodeterioration and Biodegradation, 48*(1–4), 137–152.

Brown, B. J., & Marten, G. G. (1986). The ecology of traditional pest management in Southeast Asia. In G. G. Marten (Ed.), *Traditional Agriculture in Southeast Asia: A Human Ecology Perspective* (pp. 241–272). Boulder Colorado, USA: Westview.

Brown, L. R. (2004). *Outgrowing the Earth: The Food Security Challenge in an Age of Falling Water Tables and Rising Temperatures.* New York: W.W. Norton.

Bullock, D. G. (1992). Crop rotation. *Critical Reviews in Plant Sciences, 11*(4), 309–326.

Chander, R., & Bhargava, M. C. (2005). Effect of different storage containers on incidence of lesser grain borer, *Rhizopertha dominica* (Fabr.) in stored barley. *Journal of Applied Zoological Researches, 16*(1), 102–103.

Charlet, L. D., Seiler, G. J., Miller, J. F., Hulke, B. S., & Knodel, J. J. (2009). Resistance among cultivated sunflower germplasm to the banded sunflower moth (Lepidoptera: Tortricidae) in the Northern Great Plains. *Helia, 32*(51), 1–10.

Chiarappa, L. (1981). Establishing the crop loss profile. In L. Chiarappa (Ed.), *Crop Loss Assessment Methods—Supplement 3* (pp. 21–24). Farnham Royal, Slough, United Kingdom: Commonwealth Agricultural Bureaux.

Chiarappa, L., Chiang, H. C., & Smith, R. F. (1972). Plant pests and diseases: Assessment of crop losses. *Science, 176,* 769–773.

Cox, A. E., & Large, E. C. (1960). *Potato Blight Epidemics Throughout the World. USDA Agriculture Handbook No. 174.* Washington, D.C.: Government Printing Office.

Cramer, H. H. (1967). Plant protection and world crop production. *Pflanzenschutz-Nachrichten, 20*(1), 1–524.

Curtis, J. (1860). *Farm Insects: Being the Natural History and Economy of the Insects Injurious to the Field Crops of Great Britain and Ireland.* Glasgow, United Kingdom: Blackie and Son.

D'Aquino, S., Cocco, A., Ortu, S., & Schirra, M. (2011). Effects of kaolin-based particle film to control *Ceratitis capitata* (Diptera: Tephritidae) infestations and postharvest decay in citrus and stone fruit. *Crop Protection, 30*(8), 1079–1086.

Day, R. K. (1989). Effect of cocoa pod borer, *Conopomorpha cramerella*, on cocoa yield and quality in Sabah, Malaysia. *Crop Protection, 8*(5), 332–339.

Delabie, J. H. C., & Cazorla, I. M. (1991). Danos causados por *Planococcus citri* Risso (Hemiptera: Pseudococcidae) na produção do cacaueiro. *Agrotrópica, 3*(1), 53–57.

Deshmukh, S. S., Goud, K. B., & Giraddi, R. S. (2007). Seasonal incidence and crop loss estimation of pod weevil, *Apion amplum* (Faust) on greengram, *Vigna radiata* (L.) Wilczek. *Karnataka Journal of Agricultural Sciences, 20*(4), 855–856.

Deshmukh, S. G., Sureja, B. V., Jethva, D. M., & Chatar, V. P. (2010). Estimation of yield losses by pod borer *Helicoverpa armigera* (Hubner) on chickpea. *Legume Research, 33*(1), 67–69.

Dhaliwal, G. S., Dhawan, A. K., & Singh, R. (2007). Biodiversity and ecological agriculture: Issues and perspectives. *Indian Journal of Ecology, 34*(2), 100–109.

Domingos Scalon, J., Lopes Avelar, M. B., de Freitas Alves, G., & Sérgio Zacarias, M. (2011). Spatial and temporal dynamics of coffee-leaf-miner and predatory wasps in organic coffee field in formation. *Ciência Rural, 41*(4), 646–652.

Dordolo, M. (2008). Recrudescence des degats d'aiguillonnier des cereales dans le sud-ouest suivi biologique et gestion du risque en interculture. In *AFPP—8ème Conférence Internationale sur les Ravageurs en Agriculture* (p. 665), Montpellier SupAgro, France, 22–23 October 2008. Association Française de Protection des Plantes (AFPP). Alfortville.

Dörtbudak, N., Erdoğan, P., & Aydemir, M. (1999). Orta Anadolu Bölgesi'nde depolanan mercimek ve fasulyede zararlı olan baklagil tohum böceklerinin yayılışı, bulaşma oranı, yoğunlukları ve meydana getirdikleri ürün kayıpları üzerinde araştırmalar. *Bitki Koruma Bülteni, 39*(1/2), 57–75.

ERS. (2010). *Which are the top 10 Agricultural Producing States?* USDA Economic Research Service. Available via Common Questions About ERS Subject Areas. http://www.ers.usda.gov/AboutERS/FAQs.htm. Accessed Feb 2012.

Everly, R. T. (1960). Loss in corn yield associated with the abundance of the corn leaf aphid, *Rhopalosiphum maidis*, in Indiana. *Journal of Economic Entomology, 53*(5), 924–932.

FAO. (2011a). *Food Outlook: Global Market Analysis (November 2011)* No. 4, November. Rome: Food and Agriculture Organization of the United Nations.

FAO. (2011b). *FAO FAO in the 21st Century: Ensuring Food Security in a Changing World*. Rome: Food and Agriculture Organization of the United Nations.

FAO. (2012a). Gross production value (constant 2004–2006 million US$) (USD). Statistics Division, Food and Agriculture Organization of the United Nations. http://faostat.fao.org/site/613/DesktopDefault.aspx?PageID=613#ancor. Accessed Feb 2012.

FAO. (2012b). *FAO Statistical Yearbook 2012: World Food and Agriculture*. Rome: Food and Agriculture Organization of the United Nations.

Farrell, G., & Schulten, G. G. M. (2002). Larger grain borer in Africa: A history of efforts to limit its impact. *Integrated Pest Management Reviews, 7*(2), 67–84.

Forbes, S. A. (1915). *The Insect, the Farmer, the Teacher, the Citizen, and the State*. Urbana, Illinois, USA: Illinois State Laboratory of Natural History.

Galande, S. M., Ankali, S. M., & Bhoi, P. G. (2005). Effect of sugarcane woolly aphid, *Ceratovacuna lanigera* Zehntner incidence on cane yield and juice quality. *Journal of Applied Zoological Researches, 16*(2), 190–191.

Gerard, P. W. (1995). *Agricultural practices, farm policy, and the conservation of biological diversity. USDI National Biological Service Biological Science Report No. 4*. Washington, D.C.: U.S. Fish and Wildlife Service.

Gianluppi, D., Gianluppi, V., & Smiderle, O. J. (2002). *Recomendações técnicas para o cultivo do milho nos cerrados de Roraima. Ministério da Agricultura, Pecuária e Abastecimento, Brasil Circular Técnica No. 5*. Rodovia: Embrapa Roraima.

Goodman, D. (1975). The theory of diversity-stability relationships in ecology. *Quarterly Review of Biology, 50*(3), 237–266.

Haeussler, G. J. (1952). Losses caused by insects. In A. Stefferud (Ed.), *Insects: The Yearbook of Agriculture 1952* (pp. 141–146). Washington, D.C.: U.S. Government Printing Office.

Haile, A. (2006). On-farm storage studies on sorghum and chickpea in Eritrea. *African Journal of Biotechnology, 5*(17), 1537–1544.

Hall, R. (1995). Challenges and prospects of integrated pest management. In R. Reuveni (Ed.)., *Novel Approaches to Integrated Pest Management* (pp. 1–19). Boca Raton Florida, USA: Lewis.

Haseeb, M., Sharma, D. K., & Qamar, M. (2009). Estimation of the losses caused by shoot and fruit borer, *Leucinodes orbonalis* Guen. (Lepidoptera: Pyralidae) in brinjal. *Trends in Biosciences, 2*(1), 68–69.

Herrera Campo, B. V., Hyman, G., & Bellotti, A. (2011). Threats to cassava production: Known and potential geographic distribution of four key biotic constraints. *Food Security, 3*(3), 329–345.

Hodges, R. D., & Scofield, A. M. (1983). Effect of agricultural practices on the health of plants and animals produced: A review. In W. Lockeretz (Ed.)., *Environmentally Sound Agriculture* (pp. 3–34). New York: Praeger.

Hutchison, W. D., & Campbell, C. D. (1994). Economic impact of sugarbeet root aphid (Homoptera: Aphididae) on sugarbeet yield and quality in southern Minnesota. *Journal of Economic Entomology, 87*(2), 465–475.

Imai, T., Kasaishi, Y., Harada, H., Takahashi, R., & Kuramochi, K. (2011). The first report of *Acherontia lachesis* (F.) (Lepidoptera: Sphingidae) infestation on tobacco in Oyama, Tochigi Prefecture. *Japanese Journal of Applied Entomology and Zoology, 55*(2), 65–67.

Jagadeesh, H. V., & Nanjappa, D. (2009). Crisis encountered by the coconut growers of Hassan district in Karnataka. *Mysore Journal of Agricultural Sciences, 43*(4), 823–826.

Jaques, R. P., Jarvis, W. R., Seaman, W. L., Howard, R. J., Vrain, T. C., Ebsary, B. A., & Garland, J. A. (1994). Crop losses and their causes. In R. J. Howard, J. A. Garland & W. L. Seaman (Eds.)., *Diseases and Pests of Vegetable Crops in Canada: An Illustrated Compendium* (pp. 11–21). Ottawa, Canada: Canadian Phytopathological Society/Entomological Society of Canada.

Jeppson, L. R., Keifer, H. H., & Baker, E. W. (1975). *Mites Injurious to Economic Plants*. Berkeley: University of California Press.

Joshi, M. D., Butani, P. G., Patel, V. N., & Jeyakumar, P. (2010). Cotton mealy bug, *Phenacoccus solenopsis* Tinsley—a review. *Agricultural Reviews, 31*(2), 113–119.

Kalaiyarasan, S., & Palanisamy, S. (2005). Estimation of yield loss caused by sesame podbug *Elasmolomus sordidus* Fabricius under field condition. *Journal of Plant Protection and Environment, 2*(2), 108–109.

Katagihallimath, S. S. (1963). Chilli (*Capsicum annuum*)—A new host plant of *Heliothis armigera*, Hb. *Current Science, 32*(10), 464–465.

Kate, A. O., Bharodia, R. K., Joshi, M. D., Pardeshi, A. M., & Makadia, R. R. (2009). Estimation on yield losses in cucumber due to fruit fly, *Bactrocera cucurbitae* (Coquillet). *International Journal of Plant Protection, 2*(2), 276–277.

Krantz, G. W., & Walter, D. E. (Eds.). (2009). *A Manual of Acarology* (3rd ed.). Lubbock, Texas, USA: Texas Tech University Press.

Kuniata, L. S., & Sweet, C. P. M. (1994). Management of *Sesamia grisescens* Walker (Lep.: Noctuidae), a sugar-cane borer in Papua New Guinea. *Crop Protection, 13*(7), 488–493.

Labandeira, C. (2007). The origin of herbivory on land: Initial patterns of plant tissue consumption by arthropods. *Insect Science, 14*(4), 259–275.

Ladang, Y. D., Ngamo, L. T. S., Ngassoum, M. B., Mapongmestsem, P. M., & Hance, T. (2008). Effect of sorghum cultivars on population growth and grain damages by the rice weevil, *Sitophilus oryzae* L. (Coleoptera: Curculionidae). *African Journal of Agricultural Research, 3*(4), 255–258.

Lal, M., Vaidya, D. N., & Mehta, P. K. (2000). Relative abundance and extent of losses in unhusked rice due to stored grain insect pests in Kangra district of Himachal Pradesh. *Pest Management and Economic Zoology, 8*(2), 129–132.

Landis, D. A., Gardiner, M. M., van der Werf, W., & Swinton, S. M. (2008). Increasing corn for biofuel production reduces biocontrol services in agricultural landscapes. *Proceedings of the National Academy of Sciences U S A, 105*(51), 20552–20557.

Letourneau, D. K. (1997). Plant–arthropod interactions in agroecosystems. In L. E. Jackson (Ed.)., *Ecology in Agriculture* (pp. 239–290). San Diego, California, USA: Academic Press.

Litsinger, J. A., dela Cruz, C. G., Canapi, B. L., & Barrion, A. T. (2007). Maize planting time and arthropod abundance in southern Mindanao, Philippines. I. Population dynamics of insect pests. *International Journal of Pest Management, 53*(2), 147–159.

Lopes, C. M. D. A., & Silva, N. M. (1998). Impacto econômico da broca do cupuaçu, *Conotrachelus humeropictus* Field (Coleoptera: Curculionidae) nos estados do Amazonas e Rondônia. *Anais da Sociedade Entomológica do Brasil, 27*(3), 481–483.

Magdoff, F. (1995). Soil quality and management. *Agroecology: The Science of Sustainable Agriculture* (2nd ed., pp. 349–364). Boulder, Colorado, USA: Westview.

Maity, B. K., Tripathy, M. K., & Panda, S. K. (2001). Estimation of crop loss due to *Crocidolomia binotalis* Zell in Indian rapeseed and determination of its economic threshold level. *Indian Journal of Agricultural Research, 35*(1), 52–55.

Marlatt, C. L. (1905). The annual loss occasioned by destructive insects in the United States. In G. W. Hill (Ed.)., *Yearbook of the United States Department of Agriculture, 1904* (pp. 461–474). Washington, D.C.: Government Printing Office.

Martinson, T., Bernard, D., English-Loeb, G., & Taft, T., Jr. (1998). Impact of *Taedia scrupeus* (Hemiptera: Miridae) feeding on cluster development in Concord grapes. *Journal of Economic Entomology, 91*(2), 507–511.

Matson, P. A., Parton, W. J., Power, A. G., & Swift, M. J. (1997). Agricultural intensification and ecosystem properties. *Science, 277*, 504–509.

Mattson, W. J., & Addy, N. D. (1975). Phytophagous insects as regulators of forest primary production. *Science, 190*, 515–522.

Mayer, K. (1959). *4500 Jahre Pflanzenschutz: Zeittafel zur Geschichte des Pflanzenschutzes und der Schädlingsbekämpfung unter besonderer Berücksichtigung der Verhältnisse in Deutschland*. Stuttgart, Germany: Verlag Eugen Ulmer.

Metcalf, R. L. (1980). Changing role of insecticides in crop protection. *Annual Review of Entomology, 25*, 219–256.

Metcalf, R. L., & Metcalf, R. A. (1993). *Destructive and Useful Insects: Their Habits and Control* (5th ed.). New York: McGraw-Hill, Inc.

Mirik, M., Ansley, J., Michels, J., Jr., & Elliott, N. (2009). Grain and vegetative biomass reduction by the Russian wheat aphid in winter wheat. *Southwestern Entomologist, 34*(2), 131–139.

Mohan, C., Nair, C. P. R., Nampoothiri, C. K., & Rajan, P. (2010). Leaf-eating caterpillar (*Opisina arenosella*)-induced yield loss in coconut palm. *International Journal of Tropical Insect Science, 30*(3), 132–137.

Moschos, T. (2006). Yield loss quantification and economic injury level estimation for the carpophagous generations of the European grapevine moth *Lobesia botrana* Den. et Schiff. (Lepidoptera: Tortricidae). *International Journal of Pest Management, 52*(2), 141–147.

Muhamad, R., & Way, M. J. (1995). Damage and crop loss relationships of *Helopeltis theivora*, Hemiptera, Miridae and cocoa in Malaysia. *Crop Protection, 14*(2), 117–121.

Murali Baskaran, R. K., Rajavel, D. S., & Suresh, K. (2009). Yield loss by major insect pests in ashwagandha. *Insect Environment, 14*(4), 149–151.

Murphy, S. T. (2005). Ecology and management of rice hispa (*Dicladispa armigera*) in Bangladesh. *Final Technical Report No. 7891 (ZA 0445)*. U.K. Department for International Development, Crop Protection Programme, [s.l.].

Nabirye, J., Nampala, P., Kyamanywa, S., Ogenga-Latigo, M. W., Wilson, H., & Adipala, E. (2003). Determination of damage-yield loss relationships and economic injury levels of flower thrips on cowpea in eastern Uganda. *Crop Protection, 22*(7), 911–915.

Nacro, S., Barro, S. A., Sawadogo, L., Gnamou, A., & Tankoano, H. (2006). The effect of planting date on the African rice gall midge *Orseolia oryzivora* (Diptera: Cecidomyiidae) damage under irrigated conditions in Boulbi, central Burkina Faso. *International Journal of Tropical Insect Science, 26*(4), 227–232.

Nandagopal, V., Prasad, T. V., Gedia, M. V., Prakash, A., & Rao, J. (2007). Life history, distribution and management of groundnut beetle, *Caryedon serratus* (Olivier): A review. *Journal of Applied Zoological Researches, 18*(2), 93–107.

Navia, D., de Moraes, G. J., Roderick, G., & Navajas, M. (2005). The invasive coconut mite *Aceria guerreronis* (Acari: Eriophyidae): Origin and invasion sources inferred from mitochondrial (16S) and nuclear (ITS) sequences. *Bulletin of Entomological Research, 95*(6), 505–516.

Nicholls, C. I., & Altieri, M. A. (2007). Agroecology: Contributions towards a renewed ecological foundation for pest management. In M. Kogan & P. Jepson (Eds.), *Perspectives in Ecological Theory and Integrated Pest Management* (pp. 431–468). Cambridge, United Kingdom: Cambridge University Press.

Nielsen, B. S. (1990). Yield responses of *Vicia faba* in relation to infestation levels of *Sitona lineatus* L. (Col., Curculionidae). *Journal of Applied Entomology, 110*(4), 398–407.

NRC (National Research Council). (1989). *Alternative Agriculture*. Washington, D.C.: National Academy Press.

Nutter, F. W., Jr., Teng, P. S., & Royer, M. H. (1993). Terms and concepts for yield, crop loss, and disease thresholds. *Plant Disease, 77*(2), 211–215.

Odum, E. P. (1969). The strategy of ecosystem development. *Science, 164,* 262–270.

Odum, E. P. (1984). Properties of agroecosystems. In R. Lowrance, B. R. Stinner, & G. J. House (Eds.), *Agricultural Ecosystems: Unifying Concepts* (pp. 5–11). New York: Wiley.

Oerke, E.-C. (2006). Crop losses to pests. *Journal of Agricultural Science, 144*(1), 31–43.

Oerke, E.-C., & Dehne, H.-W. (2004). Safeguarding production—losses in major crops and the role of crop protection. *Crop Protection, 23*(4), 275–285.

Oerke, E.-C., Dehne, H.-W., Schönbeck, F., & Weber, A. (1994). *Agricultural Ecosystems: Unifying Concepts*. Amsterdam: Elsevier Science B.V.

Ojwang, P. P. O., Melis, R., Songa, J. M., & Githiri, M. (2010). Genotypic response of common bean to natural field populations of bean fly (*Ophiomyia phaseoli*) under diverse environmental conditions. *Field Crops Research, 117*(1), 139–145.

Olfert, O., & Slinkard, A. (1999). Grasshopper (Orthoptera: Acrididae) damage to flowers and pods of lentil (*Lens culinaris* L.). *Crop Protection, 18*(8), 527–530.

Ordish, G. (1952). *Untaken Harvest: Man's Loss of Crops from Pest, Weed and Disease: An Introductory Study*. London: Constable and Company Ltd.

Ordish, G. (1976). *The Constant Pest: A Short History of Pests and Their Control*. New York: Charles Scribner's Sons.

Pedigo, L. P., & Rice, M. E. (2006). *Entomology and Pest Management* (5th ed.). Upper Saddle River, New Jersey, USA: Pearson Prentice Hall.

Pimentel, D. (1986). Acroecology [sic] and economics. In M. Kogan (Ed.)., *Ecological Theory and Integrated Pest Management Practice* (pp. 299–319). New York: Wiley.

Pimentel, D. (1993). Environmental and economic benefits of sustainable agriculture. In M. G. Paoletti, T. Napier, O. Ferro, B. R. Stinner & D. Stinner (Eds.), *Socio-economic and Policy Issues for Sustainable Farming Systems* (pp. 5–20). Padova, Italy: Cooperativa Amicizia S.r.l.

Pimentel, D. (1997). Pest management in agriculture. In D. Pimentel (Ed.)., *Techniques for Reducing Pesticide Use: Economic and Environmental Benefits* (pp. 1–11). Chichester, United Kingdom: Wiley.

Pimentel, D. (2005). Environmental and economic costs of the application of pesticides primarily in the United States. *Environment, Development and Sustainability, 7*(2), 229–252.

Pimentel, D., & Pimentel, M. (1978). Dimensions of the world food problem and losses to pests. In D. Pimentel (Ed.)., *World Food, Pest losses, and the Environment. AAAS Selected Symposium* (Vol. 13) (pp. 1–16). Boulder, Colorado, USA: Westview.

Pimentel, D., Krummel, J., Gallahan, D., Hough, J., Merrill, A., Schreiner, I., Vittum, P., Koziol, F., Back, E., Yen, D., & Fiance, S. (1978). Benefits and costs of pesticide use in U.S. food production. *Bioscience, 28*(12), 772, 778–784.

Pimentel, D., Acquay, H., Biltonen, M., Rice, P., Silva, M., Nelson, J., Lipner, V., Giordano, S., Horowitz, A., & D'Amore, M. (1993). Assessment of environmental and economic impacts of pesticide use. In D. Pimentel & H. Lehman (Eds.)., *The Pesticide Question: Environment, Economics, and Ethics* (pp. 47–84). New York: Chapman & Hall.

Pimentel, D., Hepperly, P., Hanson, J., Douds, D., & Seidel, R. (2005). Environmental, energetic, and economic comparisons of organic and conventional farming systems. *Bioscience, 55*(7), 573–582.

Prasad, R., & Singh, J. (2009). Yield loss estimation in okra and brinjal caused by phytophagous mites during summer. *Journal of Plant Protection and Environment, 6*(1), 125–130.

Qadri, S. M. H., Sakthivel, N., & Punithavathy, G. (2010). Estimation of mulberry crop loss due to spiralling whitefly, *Aleurodicus dispersus* Russell (Homoptera: Aleyrodidae) and its impact on silkworm productivity. *Indian Journal of Sericulture, 49*(2), 106–109.

Rajalakshmi, E., Sankaranarayanan, P., & Pandya, R. K. (2009). The yellow mite, *Polyphagotarsonemus latus* (Banks)—a serious pest of mulberry under Nilgiris hill conditions. *Indian Journal of Sericulture, 48*(2), 187–190.

Ramos, P., Campos, M., & Ramos, J. M. (1998). Long-term study on the evaluation of yield and economic losses caused by *Prays oleae* Bern. in the olive crop of Granada (southern Spain). *Crop Protection, 17*(8), 645–647.

Rana, J. S. (2005). Performance of *Lipaphis erysimi* (Homoptera: Aphididae) on different *Brassica* species in a tropical environment. *Journal of Pest Science, 78*(3), 155–160.

Ray, R. C., Chowdhury, S. R., & Balagopalan, C. (1994). Minimizing weight loss and microbial rotting of sweet potatoes (*Ipomoea batatas* L.) in storage under tropical ambient conditions. *Advances in Horticultural Science, 8*(3), 159–163.

Razaq, M., Mehmood, A., Aslam, M., Ismail, M., Afzal, M., & Ali Shad, S. (2011). Losses in yield and yield components caused by aphids to late sown *Brassica napus* L., *Brassica juncea* L. and *Brassica carrinata* A. Braun at Multan, Punjab (Pakistan). *Pakistan Journal of Botany, 43*(1), 319–324.

Reddy, K. V. S., & Zehr, U. B. (2004). Novel strategies for overcoming pests and diseases in India (symposia papers 3.7). In T. Fischer, N. Turner, & J. Angus, et al. (Eds.)., *New Directions for a Diverse Planet: Proceedings of the 4th International Crop Science Congress* (pp. 1–8). Gosford, NSW, Australia: The Regional Institute Ltd.

Reed, W. (1983). Crop losses caused by insect pests in the developing world. In *Plant Protection for Human Welfare: 10th International Congress of Plant Protection 1983* (pp. 74–80). Croydon, England: British Crop Protection Council.

Reid, J. C., & Mansingh, A. (1985). Economic losses due to *Hypothenemus hampei* Ferr. during processing of coffee berries in Jamaica. *Tropical Pest Management, 31*(1), 55–59.

Rondon, S. I. (2010). The potato tuberworm: A literature review of its biology, ecology, and control. *American Journal of Potato Research, 87*(2), 149–166.

Root, R. B. (1973). Organization of a plant-arthropod association in simple and diverse habitats: The fauna of collards (*Brassica oleracea*). *Ecological Monographs, 43*(1), 95–124.

Schowalter, T. D. (2007). Ecosystems: Concepts, analyses and practical implications in IPM. In M. Kogan & P. Jepson (Eds.)., *Perspectives in Ecological Theory and Integrated Pest Management* (pp. 411–430). Cambridge, United Kingdom: Cambridge University Press.

Schulten, G. G. M. (1975). Losses in stored maize in Malawi (C. Africa) and work undertaken to prevent them. *EPPO Bulletin, 5*(2), 113–120.

Shahjahan, M. (1974). Extent of damage of unhusked stored rice by *Sitotroga cerealella* Oliv. (Lepidoptera, Gelechiidae) in Bangladesh. *Journal of Stored Products Research, 10*(1), 23–26.

Sharma, A., & Naqvi, A. R. (1993). Assessment of losses in ber from the attack of *Larvacarus transitans* (Ewing). *Indian Journal of Entomology, 55*(2), 220–222.

Shukla, A. (2010). Insect pests of banana with special reference to weevil borers. *International Journal of Plant Protection, 3*(2), 387–393.

Singh, H., & Malik, V. S. (1993). Biology of painted bug (*Bagrada cruciferarum*). *Indian Journal of Agricultural Sciences, 63*(10), 672–674.

Singh, P., & Sharma, R. K. (2009). Effect of insecticides for the control of maize stem borer *Chilo partellus* (Swinhoe). *Mysore Journal of Agricultural Sciences, 43*(3), 577–578.

Singh, P., & Jakhmola, S. S. (2011). Evaluation of plant extracts on the losses caused by *Callosobruchus maculatus* (Fab.) in green gram. *Journal of Applied Bioscience, 37*(1), 31–34.

Soholt, L. F. (1973). Consumption of primary production by a population of kangaroo rats (*Dipodomys merriami*) in the Mojave Desert. *Ecological Monographs, 43*(3), 357–376.

Strickland, A. H. (1965). Pest control and productivity in British agriculture. *Journal of the Royal Society of Arts, 113,* 62–81.

Strickland, A. H. (1971). The actual status of crop loss assessment. *EPPO Bulletin, 1*(1), 39–51.

Stumpf, E. (1998). *Post-harvest loss due to pests in dried cassava chips and comparative methods for its assessment: A case study on small-scale farm households in Ghana*. Dissertation. Berlin: Humboldt-Universität.

Teng, P. S. (Ed.). (1987). *Crop Loss Assessment and Pest Management*. St. Paul, Minnesota, USA: APS.

Thacker, J. R. M. (2002). *An Introduction to Arthropod Pest Control*. Cambridge, United Kingdom: Cambridge University Press.

Triplehorn, C. A., & Johnson, N. F. (2005). *Borror and DeLong's Introduction to the Study of Insects* (7th ed.). Belmont, California, USA: Thomson Brooks.

Ullstrup, A. J. (1972). The impacts of the southern corn leaf blight epidemics of 1970–1971. *Annual Review of Phytopathology, 10*, 37–50.

UN. (2011a). World population to reach 10 billion by 2100 if fertility in all countries converges to replacement level. *Press Release, May 3, 2011*. United Nations Department of Economic and Social Affairs, Population Division, New York.

UN. (2011b). *The State of World Population 2011*. New York: United Nations Population Fund.

USDA (United States Department of Agriculture). (1965). *Losses in Agriculture. USDA Agriculture Handbook No. 291*. Washington, D.C.: U.S. Government Printing Office.

Van Alebeek, F. A. N. (1996). Natural suppression of bruchid pests in stored cowpea (*Vigna unguiculata* (L.) Walp.) in West Africa. *International Journal of Pest Management, 42*(1), 55–60.

Van den Berg, H., Shepard, B. M., & Nasikin (1998). Damage incidence by *Etiella zinckenella* in soybean in East Java, Indonesia. *International Journal of Pest Management, 44*(3), 153–159.

Van der Graaff, N. A. (1981). Increasing reliability of crop loss information: The use of "indirect" data. In L. Chiarappa (Ed.), *Crop Loss Assessment Methods—Supplement 3* (pp. 65–69). Farnham Royal, Slough, United Kingdom: Commonwealth Agricultural Bureaux.

Vandermeer, J. (1995). The ecological basis of alternative agriculture. *Annual Review of Ecology and Systematics, 26*, 201–224.

Vayssieres, J. F., Korie, S., Coulibaly, O., Temple, L., & Boueyi, S. P. (2008). The mango tree in central and northern Benin: Cultivar inventory, yield assessment, infested stages and loss due to fruit flies (Diptera Tephritidae). *Fruits, 63*(6), 335–348.

Vitousek, P. M., & Reiners, W. A. (1975). Ecosystem succession and nutrient retention: A hypothesis. *Bioscience, 25*(6), 376–381.

Walker, P. T. (1981). The measurement of insect populations, their distribution and damage. In L. Chiarappa (Ed.), *Crop Loss Assessment Methods—Supplement 3* (pp. 43–49). Farnham Royal, Slough, United Kingdom: Commonwealth Agricultural Bureaux.

Walker, P. T. (1983). Crop losses: The need to quantify the effects of pests, diseases and weeds on agricultural production. *Agriculture, Ecosystems and Environment, 9*(2), 119–158.

Walker, P. T. (1987a). Quantifying the relationship between insect populations, damage, yield and economic thresholds. In P. S. Teng (Ed.), *Crop Loss Assessment and Pest Management* (pp. 114–125). St. Paul, Minnesota, USA: APS.

Walker, P. T. (1987b). Losses in yield due to pests in tropical crops and their value in policy decision-making. *Insect Science and Its Application, 8*(4–6), 665–671.

Walsh, D. B., Zalom, F. G., & Shaw, D. V. (1998). Interaction of the twospotted spider mite (Acari: Tetranychidae) with yield of day-neutral strawberries in California. *Journal of Economic Entomology, 91*(3), 678–685.

Wesis, P., Niangu, B., Ero, M., Masamdu, R., Autai, M., Elmouttie, D., & Clarke, A. R. (2010). Host use and crop impacts of *Oribius* Marshall species (Coleoptera: Curculionidae) in Eastern Highlands Province, Papua New Guinea. *Bulletin of Entomological Research, 100*(2), 133–143.

Wratten, S. D., & van Emden, H. F. (1995). Habitat management for enhanced activity of natural enemies of insect pests. In D. M. Glen, M. P. Greaves, & H. M. Anderson (Eds.), *Ecology and Integrated Farming Systems* (pp. 117–145). Proceedings of the 13th Long Ashton International Symposium. Chichester, United Kingdom: Wiley.

Youm, O., & Owusu, E. O. (1998). Assessment of yield loss due to the millet head miner, *Heliocheilus albipunctella* (Lepidoptera: Noctuidae) using a damage rating scale and regression analysis in Niger. *International Journal of Pest Management, 44*(2), 119–121.

Yudelman, M., Ratta, A., & Nygaard, D. (1998). Pest management and food production: Looking to the future. *Food, Agriculture, and the Environment Discussion Paper No. 25*. International Food Policy Research Institute, Washington, D.C.

Zadoks, J. C. (1981). Crop loss today, profit tomorrow: An approach to quantifying production constraints and to measuring progress. In L. Chiarappa (Ed.)., *Crop Loss Assessment Methods—Supplement 3* (pp. 5–11). Farnham Royal, Slough, United Kingdom: Commonwealth Agricultural Bureaux.

Zemek, R., Kurowská, M., Kameníková, L., Rovenská, G. Z., Havel, J., & Reindl, F. (2005). Studies on phenology and harmfulness of *Aceria carvi* Nal. (Acari: Eriophyidae) on caraway, *Carum carvi* L., in the Czech Republic. *Journal of Pest Science, 78*(2), 115–116.

Zou, L., Stout, M. J., & Dunand, R. T. (2004). The effects of feeding by the rice water weevil, *Lissorhoptrus oryzophilus* Kuschel, on the growth and yield components of rice, *Oryza sativa*. *Agricultural and Forest Entomology, 6*(1), 47–53.

Chapter 9
Crop Loss Assessment in India- Past Experiences and Future Strategies

T. V. K. Singh, J. Satyanarayana and Rajinder Peshin

Contents

9.1	Introduction	228
9.2	Losses in Various Crops	230
	9.2.1 Rice	230
	9.2.2 Wheat	231
	9.2.3 Sorghum	232
	9.2.4 Maize	232
	9.2.5 Pulses	233
	9.2.6 Commercial Crops	233
	9.2.7 Oilseed Crops	235
	9.2.8 Vegetables	237
9.3	Current Status	237
9.4	Conclusions	239
	References	240

Abstract India is basically an agrarian country. The Green Revolution in India led to a quantum jump in agricultural production thereby allowing domestic food availability to comfortably meet domestic food demand. In the Indian sub-continent, insect pest problems in agriculture have shown a considerable shift from the Green Revolution era to the first decade of the twenty-first century due to agro-ecosystem and technological changes. Global losses due to insect pests have declined from 13.6% in post-Green Revolution era to 10.8% towards the beginning of this century. In India, the crop losses have declined from 23.3% in post-Green Revolution

---

T. V. K. Singh (✉) · J. Satyanarayana
Department of Entomology, Acharya N. G. Ranga Agricultural University,
Rajendranagr, Hyderabad 500 030, India
e-mail: tvksingh@yahoo.com

J. Satyanarayana
e-mail: snjella@gmail.com

R. Peshin
Sher-e-Kashmir University of Agricultural Sciences and Technology of Jammu, Main Campus,
Chatha, Jammu 180009, India
e-mail: rpeshin@rediffmail.com

D. Pimentel, R. Peshin (eds.), *Integrated Pest Management*,
DOI 10.1007/978-94-007-7796-5_9,
© Springer Science+Business Media Dordrecht 2014

era to 17.5% at present. In terms of monetary value, Indian agriculture suffers an annual loss of about US$ 42.66 millions due to insect pests. With the intensive cultivation, commiserative improvement and intensification measures to protect the crops have to be taken. During the last decade, visible progress has been made in the development of biological control strategies, insect resistant crops plants and genetically engineered crops. All these measures if suitably employed in integrated pest management (IPM) along with improvement in crop protection services should lead to a substantial reduction in crop losses due to pests.

Keywords India · Crop losses · Insect pests · Monetary losses

9.1 Introduction

Food plants of the world are damaged by more than 10,000 species of insects, 30,000 species of weeds, diseases (caused by fungi, viruses, bacteria and other microorganisms) and 1,000 species of nematodes (Dhaliwal et al. 2007). However, less than 10% of the total identified pest species are generally considered as major pests. The severity of pest problems has been changing with the developments in agricultural technology and modifications of farming practices.

India is basically an agricultural country and has highly variable climatic regions owing to its geographic features. Total arable land area is 168 million ha and a major part of it falls under tropical climate, and allows for the cultivation of a variety of cereals, oil seeds, pulses, vegetable and horticultural crops. India has achieved self-sufficiency in food grains but there is an urgent need to improve the productivity in all crops to meet future challenges. India needs to produce an additional 5–6 million tons of food grains every year to keep pace with the growth of our population (Paroda 1999). To realize this goal, one of the important stumbling blocks seems to be the yield losses due to insect pests. There is an urgent need to assess such losses, and frame strategies to reduce it.

To assess the yield losses in many crops, studies have to be carried out systematically, but still the losses caused by individual pests are not distinguished from the whole pest complex. Yield loss estimates vary depending on type of the cultivar, density of pest population, time of pest attack in relation to crop phenology and cultural practices followed. Another problem is that most of the studies are conducted in small experimental plots in research stations rather than in farmers' fields, which may not give an exact picture of the losses caused by insects (Table 9.1).

Extensive surveys carried out by various agencies in India during the 1950s revealed that fruits, cotton, rice and sugarcane suffered 25, 18, 10 and 10% yield losses, respectively, due to ravages of insect pests (Pradhan 1964) (Table 9.1)

The introduction of high yielding varieties (HYVs) along with application of agrochemicals including fertilizers increased the productivity of cropland with a concomitant increase in the proportion of crop yield lost to insect pests. Unfortunately, very little information is available regarding the extent of losses due to insect

Table 9.1 Losses in field crops due to insect pests in traditional agriculture. (Source: Pradhan 1964)

Crop	Loss in yield (%)
Rice	10
Wheat	3
Maize	5
Sorghum & millets	5
Cotton	18
Sugarcane	10
Fruits	25

Table 9.2 Worldwide crop losses (%) due to insect and mite pests during pre- and post-Green Revolution era. (Source: Benedict 2003)

Crop	Pre-Green Revolution (1965) (A)	Post-Green Revolution (1988–90) (B)	Changes in loss (B-A)
Barley	3.9	8.8	+4.9
Cotton	16.0	15.4	−0.6
Maize	13.0	14.5	+1.5
Potatoes	5.9	16.1	+10.2
Rice	27.5	20.7	−6.8
Soybean	4.4	10.4	+6.0
Wheat	5.1	9.3	+4.2
Average	10.8	13.6	+2.8

pests in different crops in India and other developing Asian countries (Dhaliwal and Arora 1994). However, even the limited available information reveals that crop losses due to pests are higher for India than for other parts of the world (APO 1993).

The overall losses due to insect pests were estimated to be US$ 2.9 million in 1983 (Krishnamurthy Rao and Murthy 1983), US$ 2.9 million in 1986 (Atwal 1986), US$ 9.80 million in 1993 (Jayaraj 1993) and US$ 14.4 million in 1996 (Dhaliwal and Arora 1996). Raheja and Tewari (1997) had indicated that losses due to *H. armigera* alone may be US$ 18,225,000 annually. Studies have revealed that the losses caused by key insect pests varied from 10 to 50%. Apart from inflicting direct losses, insect pests also act as vectors for transmission of several viral diseases, of which aphids alone transmit about 160 viruses and leafhoppers about 35 viruses (Puri 2000).

Crop protection aims to avoid or prevent crop losses or to reduce them to economically acceptable losses; the availability of quantitative data on the effect of different categories of pests is very limited. The first attempt to estimate crop losses due to various pests on a global scale was made by Cramer (1967). Subsequently, Oerke et al. (1994) carried out an extensive study to estimate losses in principal food and cash crops. In spite of the widespread use of synthetic pesticides and other control measures, the losses due to insect and mite pests increased in the post-Green Revolution era (Oerke et al. 1994) over the pre-Green Revolution era (Cramer 1967). Worldwide total pre-harvest losses for post-Green Revolution era (1988 through 1990) period was estimated to value at US$ 801,900,000 for eight principal food and cash crops (barley, coffee, cotton, maize, potato, rice, soybean and wheat) (Benedict 2003) (Table 9.2).

Table 9.3 Crop losses (%) due to insect pests during pre- and post-Green Revolution in India.

Crop	Pre-Green Revolution (early 1960s)	Post-Green Revolution (early 2000s)
Cotton	18.0	50.0
Groundnut	5.0	15.0
Other oilseeds	5.0	25.0
Pulses	5.0	15.0
Rice	10.0	25.0
Maize	5.0	25.0
Sorghum and millets	3.5	30.0
Wheat	3.0	5.0
Sugarcane	10.0	20.0
Average	7.2	23.3

Losses due to insect pests in Indian agriculture have been estimated from time to time (Pradhan 1964; Krishnamurthy and Murthy 1983; Atwal 1986, Jayaraj 1993; Lal 1997; Dhaliwal and Arora 1996; 2002; Dhaliwal et al. 2003, 2004). The actual losses due to various pests have been estimated as 26–29% for soybean, wheat and cotton, and 31, 37 and 40% for maize, rice and potatoes, respectively. In general, the losses in the post-Green Revolution era (Dhaliwal et al. 2004) have been increasing compared to losses during the pre-Green Revolution era (Pradhan 1964). Overall, the losses increased from 7.2% in the early 1960s to 23.3% in the early 2000s (Table 9.3). The maximum increase in loss occurred in cotton (18.0–50.0%), followed by other crops like sorghum and millets (3.5–30.0%), maize (5.0 to 25.0%) and oilseeds (other than groundnut) (5.0–25.0%).

There has been a paradigm shift in crop management in Indian agriculture since the beginning of this century. *Bacillus thuringiensis* (Bt) cotton was released in the country in 2002 and the area under Bt cotton increased from 50,000 ha in 2002 to 8.4 million ha in 2009. Of the estimated 9.6 million ha of cotton in India, 87% was under Bt cotton in 2009 (James 2009). Second, concerted efforts were made to implement integrated pest management programs in principal food and cash crops. As a result of these developments, losses due to insect pests in several agricultural crops are declining. However, in terms of monetary value, the decline in losses does not appear to be significant due to both the increase in production levels and the increase in prices of different commodities.

9.2 Losses in Various Crops

9.2.1 Rice

Rice, *Oryza sativa* (L.), is the staple food in India for 65% of the population. Rice is the most important source for meeting the caloric and dietary protein needs of the people as well as for generating employment and income, particularly for low income groups in rural areas. The crop accounts for about 22% (44.6 million ha) of the total cropped area with an output of 87 million t, which forms approximately 46 % of the cereals production and 42 % of the total food grains.

More than hundred species of insects have been recorded as pests of rice, of which about a dozen are of economic significance in India (Gururaj Katti et al. 2010).

Assessment of losses through analysis of 135 multi-location trials under the All India Coordinated Rice Improvement Project (AICRIP) revealed that the avoidable losses due to insect pests averaged 28.8 % in rice. Worldwide, Oerke et al. (1994) reported losses of 51.4 % of the attainable yields in rice of which 20.7 % was due to animal pests including insect pests, 15.6 % due to weeds and 15.1 % due to pathogens. The equivalent monetary loss in yield for rice amounted to $ 113 billion.

The estimated loss from stem borer damage varies from 3 to 95 %. The relation between injury and yield is considered to be generally non-linear and plants exhibited a high degree of tolerance to initial injury. Plants with as high as 30 % dead hearts from stem borer attack may have no significant yield losses and as much as 10 % white earheads can be tolerated (Teng et al. 1993). A higher degree of tolerance has been recorded under higher doses of fertilizer applications. Insect damage is thus speculated to depend on the crop age and nutritional status when the crop was infected, in addition to factors such as insect densities and feeding duration.

Brown plant hopper, *Nilaparvata lugens* Stal, causes 10–70 % losses, whereas related species, white backed plant hopper, *Sogatella furcifera* Horvath causes 10–80 % loss. Gall midge, *Orseolia oryzae* (Wood-Mason), an endemic pest in certain parts of the sub-continent is known to cause 15–60 % loss (Puri et al. 1999)

Leaf folder, *Cnaphalocrocis medinalis* (Guen.), once considered a minor pest, has now attained a major pest status with the spread of high-yielding rice varieties, continuous availability of rice crop in double cropping areas and the accompanying changes in cultural practices like increased application of fertilizers and use of broad-spectrum insecticides. Rice yield losses due to leaf folders range from 63–80 %, with high-yielding or hybrid rice varieties being more susceptible (Teng et al. 1993). Shanmugam et al. (2006) identified leaf folder as one of the most serious productivity constraints responsible for yield gap of rice in Tamil Nadu, India, accounting for an 11.18 % loss. However, the field larval stage and density as well as stage of the crop are the major factors that determine the quantum of larval feeding and yield reduction (Heong 1990).

The overall field losses due to insect damage in rice were estimated at 35–44 % (Pathak and Dhaliwal 1981). The average yield loss due to insect pests was estimated to vary from 21 to 51 % (Singh and Dhaliwal 1994). Damage during vegetative phase (50 %) contributed more to yield reduction than during the reproductive (30 %) or ripening phase (20 %) (Litsinger et al. 1987).

9.2.2 Wheat

Wheat, *Triticum aestivum* L., is a major rabi (spring harvest) cereal crop and insect pests are usually not a limiting factor in wheat production. The annual monetary loss of US$ 75.3 million has been reported due to insect pests in wheat in India

(Dhaliwal and Koul 2010). It has been further reported that the losses due to insect pests in wheat have increased from 3 to 5 % after the Green Revolution. This could be ascribed to change in the pest scenario in wheat. Till late 1960s, barring Gujhia weevil, *Tanymecus indicus* Faust and grasshopper there were hardly any serious pests of wheat, but with the introduction of high yielding semi-dwarf varieties, changed cropping systems, and use of new agro techniques, some new insects attained pest status. The termites, aphids, armyworm, American pod borer and brown mite are now major pests of wheat (Deol 2002). Severe losses may result from termites (up to 91.4 %), shoot fly (30.7 %), armyworms (42.2 %), aphids (36.4 %) and mites (15 %) under favorable environmental conditions (Deol 1990).

9.2.3 Sorghum

Sorghum, *Sorghum bicolor* (L.) Moench is an important cereal crop in Asia, Africa, North and South America and Australia. Grain yield for farmers' fields in the semi-arid tropics are generally low (500–800 kg ha^{-1}) mainly due to insect pest damage. Nearly 150 insect species have been reported as pests on sorghum, of which shoot fly, stem borer, shoot bug, midge, ear head bug and head caterpillars are the major pests.

Shoot fly, *Atherigona soccata* (Rondani), and stem borer, *Chilo zonellus* Swinhoe, are reported causing 2,534 and 860 kg ha^{-1} loss in grain and 2,511 and 4,100 kg ha^{-1} loss in fodder yield of the crop grown during rabi (winter) and *kharif* (rainy season), respectively. Shoot bug, *Peregrinus madis* (Ashmead), which was previously a minor pest, has now assumed major status in certain parts of the states of Karnataka, Maharashtra, Andhra Pradesh and Tamil Nadu. In India, it is reported to cause crop loss up to 41 % (Manisegaran and Soundarajan 2008).

The National Council of Applied Economic Research, on the basis of multi-location trials over several years before the advent of high yielding varieties, estimated a yield loss of 12.1 % due to insect pest damage to sorghum crop (NCAER 1967). Total avoidable and real losses due to the insect pest complex in sorghum hybrid CSH-5 were estimated to be 32.2 and 53 %, respectively. During recent years, yield losses varying from 55 to 100 % have been recorded in northern India due to stem borer damage in sorghum. Shoot fly has been estimated to cause 22–80 % loss to crop in Maharashtra.

9.2.4 Maize

Maize, *Zea mays* (L.), originated in Central America and is now a principal cereal crop in the tropics and subtropics. Stem borer, *Chilo partellus* Swinhoe is a limiting factor in the successful cultivation of this crop, and is reported to cause 24–83 % crop loss in India. The losses due to *C. partellus* ranged from 27.6 to 80.4 % and the combined avoidable losses in maize yield due to stem borer and shoot fly has been estimated to be 20–87 % and the shoot fly, *Atherigona* spp, alone is known to cause damage ranging from 69 to 97 % (Mathur 1992) and up to a 20 % grain yield loss (Pathak et al. 1971).

Pink stem borer, *Sesamia inferens* Walker, causes serious damage to maize in the winter season particularly in peninsular India where the average yield loss in maize may vary from 25.7 to 78.9% (Sarup et al. 1971).

9.2.5 Pulses

Various pulse crops are attacked by around 250 insect species. Pigeon pea, *Cajanus cajan* (L.), is one of the major pulse crops grown in India. It covers about 16.5% of the total area under pulses in India and contributes about 18.5% towards the total pulse production in the country.

In a survey conducted by International Crop Research Institute for Semi-Arid Tropics (ICRISAT), *Melanagromyza obtusa* Malloch was reported to damage 22.5% pigeon pea pods in North India, 21% of pods in Central India and 13.2% of pods in South India, whereas the pod borer damage was reported to be 29.7% in North-West region of India, 13.2% in North India, 24.3% in Central India and 36.4% in South India. The annual loss of pigeon pea production due to pod fly alone has been estimated at 25–30%. The total grain loss due to pod sucking bugs damage has been worked out to be 50,000 t. In case of pigeon pea, pod borer, *Helicoverpa armigera* (Hub), has been reported to cause yield losses varying from 40 to 50% in major pigeon pea growing states.

In case of pulses, pod borer, *H. armigera*, has been reported to cause yield losses varying from 4.2 to 39.7% in 12 major chickpea growing states. The average loss due to insect pest damage in chickpea has been estimated at 29.2% at the national level. In case of pigeon pea, losses often exceed 50%, while in blackgram, avoidable losses were estimated at 34.7 and 28.7% in Bihar and Andhra Pradesh, respectively (Krishnamurthy Rao and Murthy 1983; Rao et al. 1990). In Haryana, *H. armigera* was found to damage 13.7% pods and 5.3% of the grain of pigeon pea, whereas *M. obtusa* damaged 9.4–10.1% pods and 3.1–3.5% of the grain (Yadav and Choudhary 1993). On an average, a single larva per plant of pigeon pea reduces the yield by 138.5 kg/ha of the expected yield (Reddy et al. 2001). On average 30–80% crop losses occur in pulses due to ravages of insect pests valued at US$ 72,900,000 (Asthana et al. 1997). The average loss due to the insect pests in urdbean, *Vigna radiate* (L.), and mung bean, *Vigna mungo* (L.), was estimated to be 34.7 and 28.7% in different states of India, respectively.

9.2.6 Commercial Crops

9.2.6.1 Cotton

The losses due to animal pests in cotton, *Gossypium hirsutum* L., have been estimated at 30% in India and 20% in Pakistan (Table 9.4). Bollworms and sucking pests have been causing various degrees of loss in different states of India (Table 9.5). The National Council of Applied Economic Research, on the basis of experiments

Table 9.4 Actual and potential crop losses in cotton due to various pests in Asia. (Source: Oerke et al. 1994)

Region	Crop losses (%) due to					
	Animal pests		Diseases		Weeds	
	Actual	Potential	Actual	Potential	Actual	Potential
Near east south Asia	17	55–60	10	15–20	12	50–55
India, Myanmar, Sri Lanka	30	55–60	15	15–20	25	55–60
Bangladesh, Pakistan	20	55–60	12	15–20	15	55–60
Southeast Asia	22	55–60	12	15–20	IS	57–62
East Asia	15	55–60	10	15–20	10	55–60

Table 9.5 Potential crop losses in cotton due to insect pests in various states of India. (Source: Oerke et al. 1994)

State	Pests	Potential loss %
Punjab	*Bemisiatabaci* (Gennadius)	8.31
	Bollworms, jassids, *B. tabaci*	63
	Bollworms and jassids	52
	Bollworms (on *G. arboreum)*	37
	Bollworms (on *G. hirsutum)*	51
	Bollworms	18–66
Delhi	Bollworms	79
	Bollworms	25–75
	Bollworms	48
	All	52
Haryana	Bollworms	39
Gujarat	Bollworms	69
	Jassids, bollworms	39–44
	Jassids, bollworms	67–81
Andhra Pradesh	*B. tabaci*	11–49
Maharashtra	Bollworms	63
	Bollworms	59
	Bollworms	46
	Bollworms	51
	Aphids	5
Tamil Nadu	Jassids, bollworms, aphid	At least 55
	Jassids, thrips, aphids	81–82

undertaken in seven states between 1950–1951 and 1965–1966, estimated an average loss of 40.3% in seed cotton yield due to insect pests. In Maharashtra, still higher losses of 16.3, 71.7 and 94.2% were attributed to sucking pests, bollworms and all insect pests, respectively (Taley et al. 1988). The Central Institute of Cotton Research (CICR), Nagpur, attributed an average loss of 50–60% in seed cotton yield to infestation by insect pests. *H. armigera* and whitefly, *Bemisia tabaci* (Gennadius), attained the status of most severe pests, especially in northern India. Before the introduction of Bt cotton hybridization in India, Satpute et al. (1988) reported that minimum losses in cotton were caused by sucking pests (4.6%) while maximum losses were caused by bollworm in cotton (51.3%).

9.2.6.2 Sun Hemp

The fiber loss in sun hemp, *Crotalaria juncea* L., due to top shoot borer, *Cydia* (*Laspeyresia*) *tricentrata* Meyr, has been estimated to be 16.67–20.63% (Prakash 1990).

9.2.6.3 Sugarcane

In sugarcane, *Saccharum officinale* (Sacch.), insect pests cause enormous losses both in tonnage and recovery of sugar in mills. Losses due to different insect pests vary greatly for different locales. However, by conservative estimate, growers lose about 20% in sugarcane yield and sugar factories suffer a loss of about 15% sugar recovery due to the ravages by insect pests (Avasthy 1983).

9.2.7 Oilseed Crops

The 'Yellow Revolution' in India during the early 1990s made the country self-sufficient in oilseeds production, but this self-sufficiency lasted only for a short period. At present the demand for edible oils has outstripped the supply. About 40% of the country's edible oil requirements is met by imports. Oil seeds crops are mainly grown under energy deprived conditions with low inputs. According to one estimate, insect pests are known to cause 15–30% loss in yield in different oilseed crops (Dhaliwal et al. 2004), thus devouring up to one third of the agricultural produce.

9.2.7.1 Groundnut

Among the various pests infesting groundnut, *Arachis hypogaea* L., white grubs, leaf miners and sucking pests have been reported to cause heavy damage. The losses due to insect pests varied from 5 to 10% in Maharashtra to 4 to 70% in Gujarat. At the national level losses due to the insect pest complex in groundnut have been estimated at 48% (Krishnamurthy Rao and Murthy 1983). The loss in yield due to leafhopper and thrips in groundnut has been estimated to be 48.5% in Tamil Nadu (Shivalingaswamy and Palanisamy 1986). Pod yield losses of 49–56% due to the leaf miner, *Aproaerema modicella* Deventer, have been reported (Wightman and Amin 1988).

9.2.7.2 Mustard

A number of insect pests are known to attack rapeseed-mustard, *Brassica* spp. in India. Of these, the mustard aphid, *Lipaphis erysimi* Kalt, is the key pest and causes colossal yield losses year after year. Continuous feeding by large aphid colonies debilitate plants by sucking sap, Debilitated plants produce less seeds. Due to high fecundity, short generation time and prolific breeding the mustard aphid is a major constraint in the successful cultivation of rapeseed-mustard. Yield losses estimated due to mustard aphid varied from 4 to 81% during different years at various locales in the country. The losses also varied among different Brassica species. Higher losses were reported in *B. campestris* and *B. mapus*, lower in *B. carinola* and highly variable in *B. juncea* (Arora 1999).

In India, the yield losses due to mustard aphid was estimated to be 35–73 % and in addition, there was a 6–10% reduction in oil content, seed size and seed viability in unprotected crop (Arora 1999). In addition to aphid, mustard sawfly, *Athalia lugens* Klug, and leaf miner, *Chromatomyia horticola* Goureau, have also been reported to cause yield losses of 5–18 % and 15.2%, respectively (Arora 1999).

9.2.7.3 Sesamum and Niger

The avoidable yield losses due to insect pests were estimated to be 18.4, 24.0 and 25.7% in multi-location trials under the All India Coordinated Research Project on Oilseeds (AICORPO) in sesamum, *Sesamum indicum* L., sunflower and linseed, respectively. Saxena and Jakhmola (1993) reported 10–60% loss in yield due to shoot webber and pod borer, *Antigastra catalaunalis* (Dup.), in sesame. The seed damage in infested pods due to *Penthicoides seriatoporus* and *Anisolabis annulipes* (Lucas) in groundnut, *A. catalaunalis* in sesame and *Conogethes punctiferalis* (Guenee) in castor was 63.5, 73.4 and 42.3%, respectively and the corresponding weight loss of damaged seeds was 59.2, 100.0 and 63.0% (Kapadia 1996). Gupta et al. (2000) reported that avoidable losses due to *A. catalaunalis* varied from 6.2 to 43.1% in different genotypes of sesamum. The mean yield loss due to insect pests in Niger was estimated to be 36.2% at Hyderabad (Basappa 2000).

9.2.7.4 Sunflower

Crop losses due to insect pests in sunflower *Helianthus annuus* L. vary from region to region. Defoliators attack before flower initiation, affecting the source partitioning between stem, leaves and roots and in the later stages affecting the growth of both the vegetative parts and inflorescence. The plant stand of sunflower crop can be reduced by more than 30% (Basappa and Bhat 1998).

Leaf hoppers alone have reportedly caused crop loss ranging from 18.5 to 46.3% in Maharashtra (AICRP 1979). Capitulum borer, *Helicoverpa armigera* (Hub), is the key pest that has worldwide distribution and has been observed in most of the agro-ecological regions of India. Capitulum borer alone causes up to 50% yield loss by directly inflicting damage to flower buds, ovaries and developing seeds. Crop loss due to capitulum borer is more if the star bud and bloom stage of the crop coincides with peak activity of the pest. The loss in seed yield due to defoliators in a rain fed kharif crop was up to 268 kg/ha at Bangalore. Panchabhavi and Krishnamurthy (1978) reported yield loss of 120 kg/ha due to *H. armigera* damage in Karnataka.

9.2.7.5 Castor

The defoliation of castor *Ricinus communis* L. by semilooper, *Achaea janata* Linn. and *Spodoptera* is common on the Deccan plateau; defoliators are quite unimportant in Gujarat except in isolated locations in some years. Heavy infestations of semilooper often result in the abandonment the affected fields. *Spodoptera* and hairy caterpillars lay eggs in groups and the gregarious larvae feed on the same leaf for 3–4 days before dispersing to other plants.

The capsule borer which was a minor pest in the past has become more serious in recent years causing 20–50% capsule damage in southern India. The damage is high when the larvae bore at the base of the young spike, which results in the withering and death of the whole spike.

9.2.8 Vegetables

Crop loss in vegetables is more important because even a little deterioration in quality of the produce results in complete loss in marketability. Thus, besides the quantitative loss, qualitative loss is more magnified in case of vegetable products.

In vegetables, the pest status in a particular crop varies from place to place depending on different agroclimatic zones of India. The yield loss caused by these pests not only depends on the prevalence of the pest but also the type of cultivars in vogue in a given region. The biotic factors are regulated by climate and these biotic factors ultimately influence the pests. The yield loss caused by important insect pests of vegetables is given in Table 9.6.

9.3 Current Status

From the available data, an effort has been made to quantify the monetary loss for all the major crops based on the latest figures for national production and minimum support prices for these crops. Rice alone suffers a loss of US $ 2679.95 million closely followed by cotton with US $ 2230.59 million annually due to insect pest damage. All other major crops including sugarcane, maize, other coarse cereals, pulses, rapeseed-mustard, groundnut, other oilseeds and wheat suffer losses well in excess of US $ 161 million each. The monetary value of annual yield losses in all important agricultural crops, based on 2000-2001 minimum support prices fixed by the Ministry of Agriculture, Government of India, is estimated to be US $ 7214.2 million (Dhaliwal and Arora 2002).

Table 9.6 Yield loss due to insect pests in vegetables (by state)

Crop/Pest	Yield loss (%)	State	Source
Cabbage			
Diamond back moth, *Plutella xylostella* (L.)	9.8–16.8	Karnataka	Viraktamath et al. 1994
Cabbage whitefly, *Peiris brassicae* (L.)	68.5	Meghalaya	Thakur 1996
Cabbage caterpillar, *Crocidolomia binotalis* Zeller	28.9–50.8	Karnataka	Peter et al. 1988
Cabbage borer, *Hellula undalis* Fab	30–58	Karnataka	Shivalingaswamy et al. 2002
Chinese cabbage sawfly, *Athalia proxima* Klug	36.5	Uttar Pradesh	Ram et al. 1987
Aphid, *Lipaphis erysimi* Kalt	36.5	Uttar Pradesh	Ram et al. 1987
	44–54	Karnataka	Shivalingaswamy et al. 2002
Chillies			
Thrips, *Scirtothrips dorsalis* Hood	11.8	Assam	Borah and Langthasa 1995
	50	Tamil Nadu	Nelson and Natarajan 1994
	>90	Karnataka	Kumar 1995
Egg plant			
Fruit and shoot borer, *Leucinodes orbonalis* Guenée	50	Tamil Nadu	Srinivasan and Gowder 1969
	48	Maharashtra	Mote 1981
	11.1–47.18	Punjab	Shivalingaswamy et al. 2002
	54–66	Karnataka	Krishnaiah 1980
	25.82–92.50	Rajasthan	Kumar and Shukla 2002
	20.54	Uttar Pradesh	Mall et al. 1992
Tomato			
Fruit borer, *Helicoverpa armigera* (Hub.)	22.9–37.7	Karnataka	Tewari and Moorthy 1984
Okra			
Jassids, *Amrasca biguttula biguttula* Ishida	54–66	Karnataka	Krishnaiah 1980
Whitefly, *Bemisia tabaci* (Gennadius)	54.04	Rajasthan	Shivalingaswamy et al. 2002
Shoot and fruit borer, *Earias vittella* (Fab.)	54.04	Rajasthan	Shivalingaswamy et al. 2002
	38.43	Uttar Pradesh	Satpathy and Rai 1998
	22.9–50.5	Punjab	Brar et al. 1994
Cucurbits (only fruit fly damage) Fruit fly, *Bactrocera cucurbitae* (Coquillett)			
Cucumber	20–39	Assam	Borah 1996
	80	Himachal Pradesh	Gupta and Verma 1992
Little gourd	63	Gujarat	Patel 1994
Muskmelon	76–100	Rajasthan	Pareek and Kavadia 1994
Snake gourd	63	Assam	Borah and Dutta 1997
Sponge gourd	50	Andhra Pradesh	Gupta and Verma 1992
Potato			
Potato tuber moth, *Phthorimaea operculella* Zeller	14.4–55.4	Karnataka	Trivedi et al. 1994
Agrotisi psilon Hufnagel	2.78–7.39	Himachal Pradesh	Misra and Sharma 1988
Holotrichia sp.	98.3	Himachal Pradesh	Misra and Sharma 1988

Table 9.7 Estimation of crop losses caused by insect pests to major agricultural crops in India. (Source: Dhaliwal et al. 2010)

Crop	Actual production[a] (million tons)	Approximate estimated loss in yield		Hypothetical production in the absence of losses (million tons)	Monetary value of estimated losses in millions of US $
		Percentage	Total (million tons)	–	
Cotton	44.03	30	18.9	62.9	6,877.80
Rice	96.7	25	32.2	128.9	4,862.57
Maize	19	20	4.8	23.8	596.33
Sugarcane	348.2	20	87.1	435.3	1,430.94
Rapeseed-mustard	5.8	20	1.5	7.3	528.50
Groundnut	9.2	15	1.6	10.8	509.57
Other oilseeds	14.7	15	2.6	17.3	752.95
Pulses	14.8	15	2.6	17.4	881.87
Coarse cereals	17.9	10	2.0	19.9	241.63
Wheat	78.6	5	4.1	82.7	837.66
Total/Average	–	17.5	–	–	17,492.84

[a] Production and minimum support price (MSP) fixed by Government of India for 2007–2008. Adapted from Anonymous (2010)

There has been a paradigm shift in crop management in Indian agriculture since the beginning of this century. Bt cotton was released in the country in 2002 and the area under Bt cotton increased from 50,000 ha in 2002 to 8.4 million ha in 2009. Of the estimated 9.6 million ha of cotton in India, 87% was under Bt cotton in 2009 (James, 2009). Second, concerted efforts were made to implement integrated pest management programs in principal food and cash crops. As a result of these developments, losses due to insect pests in several agricultural crops are declining (Table 9.7). However, in terms of monetary value, the decline in losses does not appear to be significant. This is due to both the increase in production levels and the increase in prices of different commodities.

It is imperative to contain these colossal losses, year after year, in order to meet the rising demand for food grains and other agricultural commodities. Moreover, this has to be done in such a way that environmental quality is maintained and long term sustainability of the agro ecosystem does.

9.4 Conclusions

The global losses due to insect pests, diseases and weeds have increased tremendously during the last 2–3 decades, despite the fact that high priority has been given to crop protection measures. Although the Green Revolution technology substantially increased the yields per unit area, in many regions the crop protection technology failed to keep pace with the increase in intensity of cultivation. The sole dependence on pesticides led to rapid development of pesticide resistance among insect populations and has resulted in outbreaks of pests in many crops. The developments of other methods of pest control did not keep pace with the need for pest control. As cultivation becomes more intensive, there should be a commensurate improvement and intensification of the measures taken to protect the crops. During the last decade, spectacular progress has been achieved in the development of biological control strategies, insect-resistant crop plants and genetically engineered crops. All these measures suitably employed in the integrated pest management (IPM) system along with improvements in crop protection extension services should lead to a substantial reduction in crop losses due to pests.

References

AICRP. (1979). *Annual Progress Report: Sunflower, 1978–1979. All India Coordinated Research Project on Oilseeds* (pp. 181–184). Hyderabad, India: DOR.
Anonymous. (2010). *Economic Survey 2009–2010, Ministry of Finance and Company Affairs*. New Delhi: Government of India.
APO. (1993). *Pest Control in Asia and the Pacific.* Report of a Seminar, September 24–October 4, 1991. Tokyo: Asian Productivity Organization.
Arora, R. (1999). Insect pests of rapeseed mustard and their management. In R. K. Upadhyay, K. G. Mukeiji & R. L. Rajak (Eds.)., *IPM System in Agriculture Oilseeds* (Vol. 5, pp. 35–75). New Delhi: Aditya Books Pvt. Ltd.
Asthana, A. N., Lal, S. S., & Dhar, V. (1997). Current problems in pulse crops and future needs. In *National Seminar on Plant Protection Towards Sustainability* (p. L-3). Hyderabad, India: Plant Protection Association of India.
Atwal, A. S. (1986).Future of pesticides in plant protection. *Proceeding of Indian National Science Academy Allahabad*, B52, 77–90.
Avasthy, P. N. (1983). Insect pest management for sugarcane in India. In M. Balasubramanian & A. R. Solayappan (Eds.)., *Sugarcane Pest Management in India* (pp. 71–77). Madras: Tamil Nadu Cooperative Sugar Federation.
Basappa, H. (2000). Integrated insect pest management in sunflower and other oilseeds. In *National Training on Integrated Pest Management in Oilseeds.* Directorate of Oilseeds Research, Rajendranagar, Hyderabad, India, October 15–22, 2000.
Basappa, H., & Bhat, N. S. (1998). Pest management in sunflower seed production, In K. Virupakshappa, A. R. G. Ranganath, & B. N. Reddy (Eds.)., *Hybrid Sunflower Seed Production Technology* (pp. 62–66). Hyderabad, India: Directorate of Oilseeds Research.
Benedict, J. H. (2003). Strategies for controlling insect, mite and nematodes pests. In M. J. Chrispeels & D. E. Sadava (Eds.)., *Plants, Genes, and Crop Biotechnology* (pp. 414–442). Sudbury, Massachusetts, USA: Jones and Bartlett Publishers.
Borah, R. K. (1996). Influence of sowing season and varieties on the infestation of fruitfly, *Bactrocera cucurbitae* in cucumber in the hill zone of Assam. *Indian Journal of Entomology, 58*, 382–383.
Borah, S, R., & Dutta, S.K. (1997). Infestation of fruit fly in some cucurbitaceous vegetables. *Journal of the Agricultural Science Society of North East India, 10*, 128–131.
Borah, R. K., & Langthasa (1995). Incidence thrip *(Scirtothrips dorsalis* Hood) in relation to date of planting on chilli in the Hill Zone of Assam. *PKV Research Journal, 19*, 92.
Brar, K. S., Arora, S. K., & Ghai, T. R. (1994). Losses in fruit yield of okra due to *Earias* spp as influenced by dates of sowing and varieties. *Journal of Insect Science, 7*, 33–135.
Cramer, H. H. (1967). Plant protection and world crop production. *Bayer Pflanzenschutz-Nachrichten, 20*, 1–24.

Deol, G. S. (1990). Key pests of wheat and barley and their management. *Summer Institute on Key Insect Pests of India, Their Bioecology with Special Reference to Integrated Pest Management*, June 6–15, 1990 (pp. 226–235). Ludhiana, Punjab, India: Punjab Agricultural University.

Deol, G. S. (2002). Latest trends for insect management in wheat. *Paper presented in Specialized Workshop on Identification and Management of Weeds, Insect Pests and Diseases in Wheat*, held at CETWPT, P.A.U., Ludhiana, Punjab, India.

Dhaliwal, G. S., & Arora, R. (Eds.). (1994). *Trends in Agricultural Insect Pest Management*. New Delhi: Commonwealth Publishers.

Dhaliwal, G. S., & Arora, R. (1996). An estimate of yield losses due to insect pests in Indian agriculture. *Indian Journal of Ecology, 23*(1), 70–73.

Dhaliwal G. S., & Arora, R. (2002). Estimation of losses due to insect pests in field crops In B. Sarath Babu, K. S. Varaprasad, K. Anitha, R. D. V. J. Prasada Rao, S. K. Chakrabarty, & P. S. Chandukar (Eds.)., *Resources Management in Plant Protection* (Vol. 1, pp. 11–23). Hyderabad, India: Plant Protection Association of India.

Dhaliwal, G. S., & Koul, O. (2010). *Quest for Pest Management: From Green Revolution to Gene Revolution*. New Delhi: Kalyani Publishers.

Dhaliwal, G. S., Arora, R., & Dhawan, A. K. (2003). Crop losses due to insect pests and determination of economic threshold levels. In A. Singh, T. P. Trivedi, H. R. Sardana, O. P. Sharma, & N. Sabir (Eds.)., *Recent Advances in Integrated Pest Management* (p. 12–20). New Delhi: National Centre for Integrated Pest Management.

Dhaliwal, G. S., Arora, R., & Dhawan, A. K. (2004). Crop losses due to insect pests in Indian agriculture: An update. *Indian Journal of Ecology, 31*(1), 1–7.

Dhaliwal, G. S., Dhawan, A. K., & Singh, R. (2007). Biodiversity and ecological agriculture: Issues and perspectives. *Indian Journal of Ecology, 4*(2), 100–109.

Dhaliwal, G. S., Jindal, V., & Dhawan, A. K. (2010). Insect pest problems and crop losses: Changing trends. *Indian Journal of Ecology,* 37,1–7.

Gupta, M. P., Rai, H. S., Jakhmola, S. S. (2000). Assessment of avoidable losses in grain yield due to the incidence of leaf roller and capsule borer *Antigastra catalaunolis* Dup. in sesame varieties. *Extended Summaries. National Seminar on Oilseeds and Oils Research and Development Needs in the Millennium* (p. 226). Feb, 2–4, 2000, Directorate of Oilseeds Research, Hyderabad, India.

Gupta, D., & Verma, A. K. (1992). Population fluctuations of the maggots of fruit flies (*Daucus cucurbitae* Coquillett and *Dacus tau* Walker) infesting cucurbitaceous crops. *Advances in Plant Sciences,* 5(2), 518–523.

Gururaj Katti, A. P., Padmakumari, Ch. P., Shankar, C. (2010). Integrated pest management in rice: A bio-intensive approach. In H. R. Sardana, O. M. Bambawale, & D. Prasad, (Eds.)., *Sustainable Crop Protection Strategies* (Vol. I, pp. 385–396). Delhi: Daya Publishing house.

Heong, K. L. (1990). Feeding rates of the rice leaf-folder, *cnaphalocrocis medinalis* (Lepidoptera: Pyralidae) on different plant stages. *Journal of Agricultural Entomology, 7,* 81–90.

James, C. (2009). *Global Status of Commercialized Biotech/GM Crops: 2009. ISAAA Brief No. 41.* Ithaca, New York, USA: International Service of the Acquisition of Agro-Biotech Applications.

Jayaraj, S. (1993). Biopesticides and integrated pest management for sustainable crop production. In N. K. Roy (Ed.)., *Agrochemicals and Sustainable Agriculture* (pp. 65–81). New Delhi: APC Publications Pvt. Ltd.

Kapadia, M. N. (1996). Estimation of losses due to pod borers in oilseed crops. *Oilseeds Research, 13*(1), 189–190.

Krishnaiah, K. (1980). Assessment of crop losses due to pests and diseases. In H. C. Govindu (Ed.)., *University of Agricultural Sciences Technology Series* (Vol. 33, pp. 259–267). Bangalore Karnataka, India.

Krishnamurthy Rao, B. H., & Murthy, K. S. R. K., Entomological Society of India. (Eds.)., (1983). *Proceedings National Seminar on Crop Losses Due to Insect Pests* (Vols. I–II). Hyderabad, India: Entomological Society of India.

Kumar, N. K. K. (1995). Yield loss in chilli and sweet pepper due to *Scirtothrips dorsalis* Hood (Thysanoptera: Thripidae). *Pest Management in Horticultural Ecosystems, 1,* 61–69.

Kumar, A., & Shukla, A. (2002). Varietal preference of fruit and shoot borer, *Leucinodes orbonalis* Guen. on brinjal. *Insect Environment, 8,* 44–45.

Lal, O. P. (Ed.). (1997). *Recent Advances in Entomology.* New Delhi: APC Publications Pvt. Ltd.

Litsinger, J. A., Canapi, B. L., Bandong, J. P., Dela- Cruz, C. G, Apostol, R. F., Pantua, P. O., Lumaban, M. D., & Alviola, A. L. (1987). Rice crop losses from insect pests in wetland and dryland environments of Aisa with emphasis on Philippines. *Insect Science Applications, 8,* 677–692.

Mall, N. P., Pandey, R. S., Singh, S. V., & Singh, S. K. (1992). Seasonal incidence of insect-pests and estimation of the losses caused by shoot and fruit borer on brinjal. *Indian Journal of Entomology, 54,* 241–247.

Manisegaran, S., & Soundararajan, R. P. (2008). *Pest Management in Field Crops: Principles and Practices.* Jodhpur, India: Agribios.

Mathur, L. M. L. (1992). Insect pest management and its future in Indian maize programme. *Proceedings of the XI International Congress of Entomology,* June 21 to July 4, 1992, Beijing, China.

Misra, S. S., & Sharma, H. C. (1988). Assessment of crop losses due to major insect pests (White Grubs And Cutworms) of potato crop (pp. 142–149). *Annual Scientific Report 1987.* India: Central Potato Research Institute.

Mote, U. N. (1981). Varietal resistance in eggplant to *Leucinodes orbonalis* Guen. screening under field conditions. *Indian Journal of Entomology, 43,* 112–115.

NCAER. (1967). *Pesticides in Indian Agriculture.* New Delhi: National Council of Applied Economic Research.

Nelson, S. J., & Natarajan, S. (1994). Economic threshold level of thrips in semi dry chilli. *South Indian Horticulture, 42,* 336–338.

Oerke, E. C., Dehne, H. W., Schoobeck, F., & Weber, A. (1994). *Crop Production and Crop Protection.* Amsterdam: Elsevier Science.

Panchabhavi, K. S., & Krishnamoorthy, P. N. (1978). Estimation of avoidable loss by insect pests on sunflower at Bangalore. *Indian Journal of Agriculture Science, 48,* 264–265.

Pareek, B. L., & Kavadia, V. S. (1994). Relative preference of fruit fly *Dacus cucurbitae* Coquillett on different cucurbits. *Indian Journal of Entomology, 56,* 72–75.

Paroda, R. S. (1999). For a food secure future. *Hindu Survey of Indian Agriculture, 17*(19), 21–23.

Patel, Ravindra Kumar K. (1994). *Bionomics and control of fruit fly, Dacus cillatus Loew (Tephritidae, Diptera) infesting bitter gourd, Coccinia indica.* Ph.D. Thesis, Gujarat Agricultural University, Navasari, Gujarat, India.

Pathak, M. D., & Dhaliwal, G.S . (1981). Trends and strategies for rice insect problems in tropical Asia. *IRRI Research Paper Series No. 64.* Los Banos, Philippines: International Rice Research Institute.

Pathak, P. K., Sharma, V. K, & Singh, J. M. (1971). Effect of date of planting of spring sown maize on the incidence of Shoot fly, *Atheriona* sp. and loss in yield due to its attack. *Annual Report 1970–71.* Experimental Station, Uttar Pradesh Agricultural University, Pantnagar, India.

Peter, C., Singh, I., ChannaBasavanna, G. P., Suman, C. L., Krishnalah, K., & Singh, L. (1988). Loss estimation in cabbage due to leaf webber *Crocldolomia* (Lepidbptera: Pyralidae). *Journal of the Bombay Natural History Society, 85*(3), 642–644.

Pradhan, S. (1964). Assessment of losses caused by insect pests of crops and estimation of insect population. In *Entomology in India, 1938–1963.* New Delhi: Entomological Society of India.

Prakash, S. (1990). Estimation of loss in sun hemp (*Crotalaria juncea*) fiber yield due to attack of top shoot borer *Cydia* (*Laspeyresia*) *tricehtrata* Meyr. *Jute Development Journal, 20*(1), 20–24.

Puri, S. N., Murthy, K. S., & Shanna, O. P. (1999). Pest problems in India- current status. *Indian Journal of Plant Proection, 27*(1 & 2), 20–31.

Puri, S. N. (2000). India (1). In Asian Productivity Organization (Ed.)., *Farmer-led Integrated Pest Management in India and the Pacific* (pp. 83–108). Tokyo: Asian Productivity Organization.

Raheja, A. K., & Tewari, G. C. (1997). *Awareness Program on Pesticides and Sustainable Agriculture* (p. 69). New Delhi: Indian Council of Agricultural Research.

Ram, S., Gupta, M. P., & Patil, B. D. (1987). Incidence and avoidable losses diie to insect-pests in green-fodder yield of Chinese cabbage. *Indian Journal of Agricultural Sciences, 57*, 955–956.

Rao, K. T., Rao, N. V., Abbaiah, K., & Satyanarayana, A. (1990). Yield losses due to pests and diseases of black gram *(Vigna mungo)* in Andhra Pradesh. *Indian Journal of Plant Protection, 18*(2), 287–289.

Reddy, C. N., Singh, Y., & Singh, V. S. (2001). Economic injury level of gram pod borer (*Helicoverpa armigera*) on pigeonpea. *Indian Journal of Entomology, 63*(4), 381–387.

Sarup, P., Sircar, P., Sharma, D. N., Singh, D. S., Amarpuri, S. R. S., & Lai, R. (1971). Effect of formulations on the toxicity on pesticidal granules to some important pests of mustard. *Indian Journal of Entomology, 33*, 82–89.

Satpathy, S., & Rai, S. (1998). Influence of sowing date and crop phenology on pest infestation in okra. *Vegetable Science, 25*, 175–180.

Satpute, U. S., Sarnaik, D. N., & Bhalerao, P. D. (1988). Assessment of avoidable field tosses in cotton yield due to sucking pests and bollworms. *Indian Journal of Plant Protection, 16*, 37–39.

Saxena, A. K., & Jakhmola, S. S. (1993). Effect of spray time and number on sesame leaf webber and capsule borer, *Antigastra catalaunalis* Dup. *Agriculture Science Digest, 13*, 131–133.

Shanmugam, T. R., Sendhil, R., Thirumalavalavam. (2006). Quantification and prioritization of constraints causing yield losses in rice (*Oryza sativa* L.) in India. *Agricultura Tropicaet Subtropica, 39*(6), 194–201.

Shivalingaswamy, T. M., Satpathy, S., & Banerjee, M. K. (2002). Estimation of crop losses due to insect pests in vegetables. In B. Sarath Babu, K. S. Varaprasad, K. Anitha, D. V. J. Prasada Rao, S. K. Chakrabarthy, & P. S. Chandurkar (Eds.)., *Resource Management in Plant Protection* (Vol. I, pp. 24–31). New Delhi: PPAI.

Singh, J., & Dhaliwal, G. S. (1994). Insect pest management in rice: A perspective. In G. S. Dhaliwal, & R. Arora (Eds.)., *Trends in Agricultural Insect Pest Management* (pp. 56–112). New Delhi: Commonwealth Publishers.

Shivalingaswamy, P., & Palanisamy, G. A. (1986). Losses in yield due to leafhopper and thrips in groundnut. *Madras Agriculture Journal, 73*, 530–531.

Srinivasas, P. N., & Gowder, R. B. (1969). A note on the control of brinjal shoot and fruit borer *(Leucinodes orbonalis)*. *Indian Journal of Agricultural Sciences, 29*, 71–78.

Taley, Y. M., Thote, R. L., & Nimbalkar, S. A. (1988). Assessment of crop losses due to insect pests of cotton and cost benefit of protection schedule. *PKV Research Journal, 12*(2), 126–128.

Teng, P. S., Heong, K. L., & Moody, K. (1993). Advances in tropical rice integrated pest management research. In K. Muralidharan, & E. A. Siddiq (Eds.)., *New Frontiers in Rice Research* (pp. 241–255). Hyderabad, India: Directorate of Rice Research.

Tewari, G. C., & Moorthy, P. N. K. (1984). Yield loss in tomato caused by fruit borer. *Indian Journal of Agricultural Science, 54*, 341–343.

Thakur, N. S. A. (1996). Relationship of cabbage butterfly larvae (*Pieris brassicae* L.). Population on the marketable yield of cabbage. *Journal of Hill Research, 9*, 356–358.

Trivedi, T. P., Rajagopal, D., & Tandon, P. L. (1994). Assessment of losses due to potato tuber moth. *Journal of the Indian Potato Association, 21*, 207–210.

Viraktamath, S., Sbekarappa, Reddy, B. S., Patil, M. G., & Satyanarayana Reddy, B. (1994). Effect of date of planting on the extent of damage by the diamondback moth, *Plutella xylostella* on cabbage. *Karnataka Journal of Agricultural Sciences, 7*, 238–239.

Wightman, J. A., & Amin, P. W. (1988). Groundnut pests and their control in the semi-arid tropics. *Tropical Pest Management, 34*, 218–226.

Yadav, L. S., & Chaudhary, J. P. (1993). Estimation of losses due to pod borers in pigeon pea. *Indian Journal of Entomology, 55*(4), 375–379.

Chapter 10
Review of Potato Biotic Constraints and Experiences with Integrated Pest Management Interventions

Peter Kromann, Thomas Miethbauer, Oscar Ortiz and Gregory A. Forbes

Contents

10.1	Importance of Potato in Poverty Alleviation, Food Security and Culture	246
10.2	Socio-economic Impact of Biotic Constraints in Potato Production	247
	10.2.1 Pathogens and Impacts	247
	10.2.2 Insect Pests and Impacts	249
	10.2.3 Pesticide Costs	250
	10.2.4 Health Effects of Pesticides	252
10.3	Pathogen Management Interventions	253
	10.3.1 Resistant Potato Cultivars	253
	10.3.2 On-farm Management	254
10.4	Arthropod Management Interventions	256
	10.4.1 On-farm Management	256
	10.4.2 Technological Innovations	258
10.5	Sustainability of IPM Extension and Scaling-up	259
	10.5.1 Farmer Capacity Building for Pathogen Management	259
	10.5.2 Farmer Capacity Building for Arthropod Management	261
10.6	Risk Assessment for Insects and Late Blight Modelling	261
10.7	Future Research	262
10.8	Conclusion	263
References		264

Abstract Potato (*Solanum* spp.) ranks third in importance as a single food crop worldwide. Late blight, caused by *Phytophthora infestans*, is considered to be the most important single biotic constraint of potato, but degeneration of vegetative planting material, caused primarily by a complex of viruses, potentially causes even greater yield losses. Arthropod pests are also important, with the primary problems on a global scale being the potato tuber moth complex (*Phthorimaea operculella*, *Symmetrischema tangolias* and *Tecia solanivora*), leaf miner fly (*Liriomyza huidobrensis*), Colorado potato beetle (*Leptinotarsa decemlineata*), and Andean

---

P. Kromann (✉) · T. Miethbauer · O. Ortiz · G. A. Forbes
The International Potato Center, Avenida La Molina 1895,
12, Lima, Peru
e-mail: p.kromann@cgiar.org

D. Pimentel, R. Peshin (eds.), *Integrated Pest Management*,
DOI 10.1007/978-94-007-7796-5_10,
© Springer Science+Business Media Dordrecht 2014

potato weevil (*Premnotrypes* spp.). Potato is one of the most pesticide-demanding agricultural crops and health risks related to pesticide use in potato production are high, especially in developing countries where protective clothing is generally not used. Experiences with potato integrated pest management (IPM) interventions have been multiple, but some of the most promising for disease management involve efforts to integrate the use of resistant cultivars, fungicides (for late blight) and capacity building of farmers. Interventions for arthropod pests rely less on host resistance and focus more on sustaining biodiversity and habitat management, as well as technological innovations to improve on-farm management, for example, cultural management practices and biological control. It is concluded that farmer capacity building is one of the most important elements needed to improve potato IPM in developing countries and that farmer acceptance of new technologies is best achieved through their understanding of the economic, ecological and practical benefits of the new technologies.

Keywords Late blight · Tuber moth · Leaf miner fly · Andean potato weevil · Socio-economic impact · Pesticides · Farmer capacity building · On-farm management · Biological control

10.1 Importance of Potato in Poverty Alleviation, Food Security and Culture

The potato crop was domesticated in the Andes about 8,000 years ago, and from the beginning it played a key role in food security. Initially, pre-Inca and Inca cultures relied on potatoes as one of the most important sources of food, particularly in the high Andes (Moseley 1992). Currently, a high diversity of potato varieties still plays a key role in the diet and culture of the Andean populations. Potato ranks third in importance as a food crop worldwide; its production has expanded globally in the last three decades, with about a billion people currently eating potato, and its flexibility and adaptability to a wide range of agro-ecologic conditions (altitudes, latitudes, etc.) make it an essential crop for food security (Birch et al. 2012). The flexibility and adaptability of the potato have been recognized in the development policies of some countries, such as China, the largest consumer of potatoes worldwide. Potatoes are also cultivated as a cash crop in several countries in Sub-Saharan Africa and Asia, representing an income-generation opportunity for farmers located in mountain regions or in the lowlands during the winter season. However, this increasing importance of potato for food security and poverty alleviation is threatened by a number of biotic constraints, the most important of which are dealt with in this review.

10.2 Socio-economic Impact of Biotic Constraints in Potato Production

The socio-economic impact of biotic constraints of potato is paradoxical because the importance of biotic constraints is well known to anyone working with this crop, yet there is little information that accurately quantifies these constraints, whether individually or as a composite. Generally it is difficult to attribute an observed productivity gap to a specific insect pest or disease problem when several problems occur simultaneously and abiotic stress factors are present as well. In this review, we attempt to compile what little is available in the literature and also identify critical information needed for a better understanding of the impact of constraints and for better management of constraints. To facilitate a review of the topic, we divide the constraints by the traditional academic disciplines of pathology (diseases and nematodes) and entomology (insects and other arthropods). Within each of those groupings, we focus on a few of the dominant constraints, caused by either individual organisms (e.g., potato late blight) or complexes (e.g., potato tuber moths).

10.2.1 Pathogens and Impacts

Late blight of potato, caused by *Phytophthora infestans*, is one of the world's most important food-crop diseases, with global annual losses estimated to be between US$ 3 and 15 billion (Judelson and Blanco 2005; Haldar et al. 2006; Haverkort et al. 2009). Late blight is a threat to potato farmers for several reasons: (i) it may affect food security in areas where the disease is severe and the crop is an important food source; (ii) it reduces household finances due to direct losses of commercial product and/or purchase of fungicides; and (iii) it threatens human health and environment because of the often excessive number of fungicide applications needed to control the disease, which are generally applied without protective clothing in developing countries (Orozco et al. 2009).

Potato late blight is possibly one of the most well-known examples of a plant disease causing human suffering because of the proximate role it played in the Irish Famine, which led to starvation and massive emigration from Ireland in the mid-nineteenth century (Bourke 1993). Late blight continues to cause much damage, and is considered by the International Potato Center (CIP) to be the single-most important disease of potato on a global scale. Fungicide use is probably the easiest loss indicator to estimate; it can be measured in volume or the monetary cost of the fungicide applications. Haverkort et al. (2009) estimate fungicide applications cost about US$ 500 per ha in the Netherlands, however these values are difficult to extrapolate globally for numerous reasons (e.g., location-specific costs of labor). Nonetheless, for example, if one assumes a conservative estimate of five sprays per season on the 19 million hectares of potatoes grown in the world, and a unit spray cost of US$ 30, this means that roughly US$ 2.8 billion are spent on fungicide applications per year on a global scale. Yield losses due to late blight are much more difficult to estimate. Haverkort et al (2009) estimated losses globally to be about 16%, but this figure is derived through vast extrapolation based on many assumptions. Even when fungicides are routinely used, yield losses may occur. Fungicides are readily used in Indonesia, however, severe late blight in 2010 lead to heavy importation of ware potato from abroad (Nikardi Gunadi, personal communication). Similarly, one research project in Ecuador lost about 10% of the on-farm sample fields because they had been abandoned due to severe late blight (Oyarzún et al. 2005). Estimated benefits accrued to farmers by adoption of late blight control technologies are discussed below in the section on insects (see Sect. 10.2.2.)

In spite of the global incidence and popular awareness of late blight, even greater yield losses are believed to be due to degeneration of planting material (tubers), caused primarily by viruses. For the vast majority of potato farmers, the crop is routinely propagated vegetatively,[1] and as with other vegetatively propagated crops, a potato crop can accumulate increasingly higher incidences of seed-borne diseases with each subsequent generation. Pathogens of many types are known to infect seed, including viruses, phytoplasmas, bacteria, fungi and oomycetes (Stevenson et al. 2001). Of these, viruses are probably the most damaging on a global scale. Seed-borne viral diseases caused by potato virus Y (PVY) and potato leaf role virus (PLRV) have been estimated to potentially cause more than 50% yield reduction and are considered the main causes of seed degeneration in tropical lowland potato production (Salazar 1996).

A recent study done in Ecuador indicated that up to 29% of the yield variability found in farmers' fields could be explained by seed health (Panchi et al. 2012); however, the average loss or amount of variability was not determined. Interestingly, this same study also showed that mixed infection by a combination of pathogens, mainly viruses, may affect tuber yield positively compared to the plants that harbor only one pathogen, and that the effect of pathogens in the seed tuber on the yield of daughter tubers is highly related to the potato genotype (Panchi et al. 2012).

Nonetheless, the problem of degenerated seed is routinely considered to be among the most extreme of constraints to productivity in potato in the developing world. A study done in Kenya showed that only 3% of seed tubers sold in markets was virus free (Gildemacher et al. 2009). The extent of yield losses due to degeneration can be gathered from some recent studies on the effects of positive selection. This simple procedure of farmers selecting symptomless plants as seed donors gave yield increases in the first season of about 20% on average in a study done in Kenya (Gildemacher et al. 2011), and similar results have been found in recent extensive studies in Ecuador and Peru (Unpublished data, the authors). These studies represent only a partial estimate of losses due to degeneration, because the process of positive selection is not perfect and the selected seed still harbor many pathogens.

In some highland tropical potato production areas, viral diseases are of less importance in degeneration. For example, in the Andes, PVY and PLRV, the viruses that cause the greatest yield reductions in potato globally, are generally present

---

[1] In very few locations sexually derived seed is used

at low incidence (Bertschinger et al. 1990; Fankhauser 2000; Panchi et al. 2012). However, bacteria such as *Ralstonia solanacearum* (Pradhanang et al. 1992) or fungi such as *Rhizoctonia solani* (Fankhauser 2000) may also be important in the degeneration of planting material.

Bacterial wilt of potato caused by *Ralstonia solanacearum* has a seed-borne phase and thus must be counted as one of the causes of seed degeneration described above. However, bacterial wilt also has a soil-borne phase and can cause losses due to primary infection (i.e., not coming from the seed) if potato is planted into contaminated soil. The incidence of the bacterial wilt pathogen declines rapidly in soil if no hosts are planted, so that crop rotation is an effective way of managing soil-borne inoculum. However, in areas where appropriate crop rotation is not practiced, primary infection of bacterial wilt can be very severe, although we do not know of studies where this has been quantified.

Early blight of potato, caused by *Alternaria solani*, is not a widespread disease problem like those already mentioned but can cause serious yield loss under certain conditions. Early blight and other foliage diseases are becoming a bigger problem because broad spectrum fungicides previously used for late blight control are now being replaced by late blight specific fungicides in many parts of the world; many of these fungicides are not effective against non-oomycete pathogens. As with bacterial wilt, quantitative estimates of losses due to early blight are not known.

10.2.2 Insect Pests and Impacts

Like pathogens, insect pests pose severe constraints to sustainable potato production and impact on farm households, consumers and the environment. Direct yield losses by tuber infestation in field and storage, and tuber yield losses resulting from reduced plant foliage cause economic or food security losses for farmers. Additionally, consumers at large are affected by a reduced food supply and higher prices. Farmers investing in the purchase of pesticides for loss abatement are prone to short-term financial risk and are often constrained by lack of liquidity. Alternative practices of insect pest control and prevention might not be accessible to farmers for various reasons. Dependence on pesticides may produce pest insect resistances, or pathogen resistance in the case of fungicides, and the necessity for new and often more expensive chemical products. A quantification of what might be an aggregate impact of insect pests is difficult to obtain, as is the case for pathogen constraints. Insect pests are estimated to cause 16 % of all potato crop losses globally (Oerke et al. (1994) cited in Kroschel and Schaub 2013, p. 165). And worldwide reductions in tuber yields and losses due to quality reduction caused by insect pests are estimated in a wide interval from 30 to 70 % (Raman and Radcliffe (1992) cited in Kroschel and Schaub 2013, p. 165). Scientists from CIP undertook a systematic effort to calculate the aggregate impact in a pragmatic but comprehensive and methodologically coherent way (Fuglie 2007). Integrated pest management (IPM), as one of the nine CIP potato technologies, was analyzed with regard to its possible contribution in alleviating worldwide crop losses and expenses for insecticide application, rather than current loss estimates directly. Based on expert surveys and classical cost-benefit analysis (using the aggregate-economic-surplus model), estimations were obtained on the global area affected by the insect pest constraint (but limited to CIP worldwide target regions). The analysis has been done for those insect pests considered as the most important ones: potato tuber moths (PTM) (*Phthorimaea operculella, Symmetrischema tangolias* and *Tecia solanivora*), leaf miner fly (LMF) (*Liriomyza huidobrenis*), Colorado potato beetle (CPB) (*Leptinotarsa decemlineata*) and Andean potato weevil (APW) (*Premnotrypes* spp.). Cost savings by substitution of pesticide use have been considered ranging from US$ 100 per ha (e.g., for PTM control in Indonesia, Kenya and Ethiopia) up to US$ 400 per ha (e.g., LMF control in coastal Peru). Yield increases and loss abatement that can be achieved through implementation of potato insect pest IPM are estimated to range between 0 and 40% in field, and in storage from 15 to 50%. The total area considered as affected by the analyzed pest species (within CIP's target regions in the developing world) is around 797,000 ha (in comparison: area affected by late blight is an estimated 5,652,000 ha). It is striking that the achievable positive impact of the insect-IPM technology (like for late blight control as well) is reduced by significant dissemination constraints. Of those 797,000 ha, only 129,000 ha are considered as likely IPM adoption area (693,000 ha for late blight control technology). The aggregated impact of the insect-IPM technology is an estimated US$ 28.4 million per year (late blight: US$ 319.4 million per year) which accrue as an economic benefit to farmers and consumers. Other literature sources offer good overviews of past country case studies on insect pests and IPM impact and on respective methodological issues (e.g., Norton et al. 2005; Peshin and Dhawan 2009, especially chap. 13 of Ortiz et al. 2009; Radcliffe, Hutchison, and Cancelado 2009).

10.2.3 Pesticide Costs

Potato has been grown for millennia without the use of pesticides. However, that is not the case today. Modern-day potato is one of the most pesticide demanding agricultural crops. The worldwide migration of the devastating potato late blight pathogen *P. infestans* in the 1800s (Fry and Goodwin 1997) and the development of effective fungicides in the mid-twentieth century have been the principal drivers for today's heavy pesticide use in potato production (Haverkort et al. 2009). Since the 1960s, intensification of potato production has been followed by increased use of pesticides practically everywhere in the world (Haverkort et al. 2009). The use of pesticides undoubtedly gives potato farmers a short-term gain in production yield. Farmers may invest more than 20% of their production expenditures on pesticide applications, which generally gives an immediate return on investment of more than 10%. The investment in pesticides is, nevertheless, principally an insurance against crop loss.

Table 10.1 Cost of potato pest and disease management by use of insecticides and fungicides, Peruvian highlands, 2010/11 season. (Miethbauer 2012)

Community	Farmers	Plots	Avg. insecticide costs—Andean potato weevil control	Avg. insecticide costs—Total	Avg. fungicide costs	Avg. total input costs—pest and disease control
	Number		US$/ha (% of total pest and disease input costs)			
Aymara	49	67	121 (*70%*)	127 (*74%*)	45 (*26%*)	172 (*100%*)
Chuquitambo	48	62	110 (*62%*)	122 (*69%*)	55 (*31%*)	177 (*100%*)
Miravalle	11	14	266 (*64%*)	382 (*92%*)	33 (*8%*)	415 (*100%*)
Ñuñunhuayo	14	16	107 (*63%*)	107(*63%*)	64 (*37%*)	171 (*100%*)
Yanamarca	53	61	191 (*69%*)	198 (*72%*)	79 (*28%*)	277 (*100%*)
Total	175	220	146 (*67%*)	160 (*73%*)	58 (*27%*)	218 (*100%*)

The costs of pesticide use in potato production and cost comparisons between countries have been the aim of limited research mainly because the socio-economic conditions are very different, even between neighboring countries and regions in the same country. Potato farmers in northern Ecuador commonly apply pesticides 7 times on average to the same crop using a mixture of at least 2 pesticide products in each spray (Kromann et al. 2011), and some farmers use fungicides at rates which result in a cost of more than US$ 600 per ha (Oyarzún et al. 2005). Similar data exists for some parts of Peru, but research suggests that household economies influence the amounts of pesticide used by Andean farmers, and at least one study from Bolivia indicated that many farmers do not spray enough for adequate disease control (Torrez et al. 1999). However, many potato farmers around the world apply fungicides more than 15 times, and in parts of Indonesia farmers are known to use up to 30 fungicide sprays to harvest a single crop, representing an estimated cost of more than US$ 400 per ha. As mentioned, pesticide costs often vary among countries; a recent study showed that the popular fungicide mancozeb was three times more expensive in Kenya than in Nepal (Kromann et al. 2012). In many developing countries, the most popular pesticides are the cheapest, and the cheapest pesticides available are often older, non-specific and sometimes highly hazardous. Pesticides that have been banned or subject to strict government regulations in the developed world are still traded frequently without restriction in many developing countries.

An analysis of recent household surveys in the Peruvian highlands reveals that even in the same region (Mantaro Valley, 3500–3900 meters above sea level) the intensity of pesticide use and costs differs considerably among different communities (Table 10.1).

The results from five communities show a range of around US$ 170–400 per ha input costs for insect and disease control with conventional pesticides. Control of APW is the most important cost component in these Peruvian communities. Control intensity and, thus, costs differ according to farmers' plot sizes (Table 10.2). An interesting implication is that the situation for farmers' decision-making and their incentives are different within the same communities concerning possible introduction of alternative control practices to substitute for pesticide use. In this case study,

Table 10.2 Estimation of possible cost savings by plastic barrier use for APW control. (Miethbauer 2012)

Plot size	Average cost for plastic barrier	Average cost for APW control by insecticides	Average estimated possible cost saving
	US$/ha		
≤500 m²	364	498	134
> 500 to < 2 ha	153	132	−21
≥2 ha	54	96	42

it became evident that only for small plot sizes (≤500 m²) or for larger plots (≥2 ha) are there potential cost savings when plastic barriers would be adopted for APW control (described below in sect. 10.4.2.1). Ideally, farmers should be encouraged to collectively do a common rotation system management of their individual and/or communal lands, and coordination should be encouraged to install common plastic barriers in larger areas.

10.2.4 Health Effects of Pesticides

In addition to the risk of crop loss and the cost of pesticides, the pesticide used in potato production also creates a considerable risk to human health. This is particularly true for applicators and their families, but also indirectly for consumers and others handling tubers contaminated with pesticide residues. In economic terms, increased morbidity and mortality also means reduced productivity and increased direct costs for healthcare (Antle et al. 1998). When health and environmental effects are included in productivity analyses, results can be very different from traditional financial analysis. The popular ethylene-bisdithio-carbamate fungicides (EBDCs), including mancozeb, are among the most common pesticide compounds used in potato production in developing countries and among the cheapest fungicides in most parts of the world. EBDCs break down into ethylene-thiourea, which is a carcinogen and an anti-thyroid compound (Panganiban et al. 2004). EBDCs are also known as skin irritants and high levels of dermatitis have been attributed to these products in Ecuador (Cole et al. 1997). Both mancozeb and chlorothalonil, a non-EBDC, were considered highly dangerous for low-input farmers in developing countries (Wesseling et al. 2005). Similar examples of disturbing health risks can be described for other old and cheap pesticides that have been widely used in the developing world (Orozco et al. 2009).

A study in Peru and Ecuador showed that the use of highly and moderately hazardous pesticides is common among Andean potato farmers (Orozco et al. 2009). This same study concluded that the worst indicators of pesticide abuse can be observed in places with lower education and greater poverty. Limited government enforcement capacity, social irresponsibility of the pesticide industry and lack of farmers' knowledge are all factors adding to adverse health effects of pesticide use in potato production (Orozco et al. 2009). A common scenario of inefficient use of pesticides is seen when farmers contract non-trained laborers to apply pesticides. Many farmers and their laborers regard the smell of a pesticide as the main indicator of toxicity and believe that pesticides enter through the nose and mouth and are unaware that pesticides mainly enter the body through the skin (Orozco et al. 2009). Research into adverse health effects of pesticide usage led by CIP revealed that many Andean farmers mix pesticides with their hands without the use of gloves; few use personal protective equipment when spraying for a variety of reasons, including social pressure, as well as the limited availability and high cost of protective equipment. Pesticide exposure and rates of intoxication are therefore predictably high among potato farmers in many developing countries as a result of the heavy use of pesticides and the low level of awareness of the risks involved with their use (Orozco et al. 2009).

10.3 Pathogen Management Interventions

10.3.1 Resistant Potato Cultivars

Potato late blight is an aggressive disease that can completely destroy a susceptible crop in a few weeks when the climatic conditions are favorable for disease development. Cultivars with resistance to the disease slow down the epidemic rate, reducing the risk of crop loss and facilitate management of the disease with fungicides. Fry (1978) found that a resistant cultivar used approximately half as much fungicide in weekly sprays as did the susceptible ones. Grünwald et al. (2002) compared resistant and susceptible cultivars in Mexico and found that host resistance allowed for disease control with one-third or one-half of sprays needed for a susceptible cultivar. These results are consistent with those of Kromann et al. (2009) who found that in Peru and Ecuador, sprays could be reduced to about one-half or one-third by using resistant cultivars. The reduction in fungicide use by half or more has obvious economic advantages for the farmer, and extrapolating this across a large area could result in very large savings at a regional or national level, but resistant cultivars also play an important role in reducing the risk of crop loss. This was indirectly demonstrated in the study of Kromann et al. (2009) who found that in about half of the cases studied, weekly calendar spraying with a contact fungicide did not provide adequate protection on susceptible cultivars. For this reason, in areas of high late blight pressure, farmers must resort to using more expensive systemic fungicides with susceptible cultivars (Kromann et al. 2008). Therefore, an important aspect of using a resistant cultivar is not just the economic benefit of reduced fungicide use, but also the reduced risk of losing part or even all of the crop's production. Recently, a new method for quantifying the level of late blight host resistance was developed, which assigns a relative value on a susceptibility scale to tested cultivars (Yuen and Forbes 2009). This tool could become important for developing cultivar and resistance specific management packages.

The level of host resistance to late blight actually in use in farmers' fields is not well known but it is generally considered that the most widely grown cultivars are susceptible (Forbes 2012). CIP and partners are currently mapping the level of resistance of cultivars in use in many developing countries where late blight is a major constraint. This task has been made possible by the existence of the susceptibility scale of Yuen and Forbes (2009) mentioned above.

Unfortunately, there is no robust scale for measuring resistance in potato cultivars to the major viruses causing seed degeneration described earlier. Furthermore, researchers at CIP and other research institutes have only recently realized the high potential impact of virus resistance in potato cultivars for the control of seed degeneration. Lack of attention to the importance of virus resistance can probably be attributed to a historical emphasis on formal disease-free seed systems as the primary means of managing degeneration caused primarily by seed-borne viruses (Forbes et al. 2009). However, as potato researchers studied patterns of cultivars which have been widely adopted in developing countries, it became evident that most have an appreciable level of virus resistance. As a result, CIP is now increasing efforts to ensure that all materials sent to national programs for evaluation have resistance to the main viruses causing degeneration.

CIP has also worked for decades on resistance to bacterial wilt, and as with late blight, results have been mixed. Unlike late blight, the phenotypic response of a plant to exposure to bacterial wilt is very irregular, undoubtedly due to inconsistent infection which has made progress in potato breeding difficult. Nonetheless, some cultivars known to be consistently produced in bacterial wilt-infested areas have been identified. Curiously, one of the potato cultivars with the highest level of field resistance to bacterial wilt is Cruza-148, widely grown in the Lake Kivu area of Africa; this cultivar also has a very high level of field resistance to late blight. One problem associated with resistance to bacterial wilt is that tubers of symptomless plants often harbor the latent pathogen (Priou et al. 2001).

10.3.2 On-farm Management

10.3.2.1 Fungicides

The use of fungicides continues to be the most common practice to manage late blight and its efficacy is arguably the reason for the continued use of susceptible cultivars in most parts of the world. The majority of recent research on late blight management for conventional farming has dealt either with fungicide dynamics or fungicide optimization, which reflects the common perception regarding the areas where late blight management can be improved. An important goal of IPM of potato late blight is an optimized integrated use of fungicides and host resistance; host resistance is still not widely used since as noted earlier, the most widely grown potato varieties are late blight susceptible. The global importance of potato late blight, tomato late blight and other oomycete diseases has provided incentive to the agro-chemical industry to find novel molecules for control of this class of diseases.

For example, the EuroblightNetwork (www.euroblight.net) lists 20 formulations involving over 15 different compounds for the control of potato blight in Europe. However, in any particular developed country, the profile of fungicides available for late blight control may be much restricted.

In many industrialized countries, sophisticated decision support systems have been developed to assist farmers in making decisions about fungicide application. These systems are generally based on detailed weather data, employ computers to manage the data and often involve large geographic coordination. In developing countries, these systems are not used because farmers do not have the necessary knowledge and/or equipment, and regional infrastructure does not exist. CIP and partners developed a simple decision support system based on accumulated rainfall thresholds that can be used by small-scale farmers for timing of contact fungicide applications (Kromann et al. 2009). Research shows that thresholds up to 50 mm resulted in significant reductions in fungicide applications compared to calendar spraying (Kromann et al. 2009). However, this technique has not been adopted by farmers, in part because it is most effective on cultivars with moderate to high levels of resistance to late blight, which are not available in the area where the tests were done. Currently, scientists at CIP are testing another simple decision support tool comprising overlapping discs that help farmers decide when to apply fungicide and when a systemic fungicide is needed.

Given the fact that farmers in developing countries seldom use protective equipment when spraying, evaluating promising late blight control products of low toxicity is important. Many biological control products have been evaluated, including biopesticides based on natural chemistries (Mizubuti et al. 2007). Products that have been evaluated include biological control with fungi (e.g., *Trichoderma harzianum*), bacterial control with, for example, *Bacillus* and *Pseudomonas* and different compost and plant extracts, but none have been adopted widely in conventional agriculture. A promising technology with low toxicity and reduced environmental impact gaining acceptance from farmers in different parts of the world is phosphonate, available in many formulations as phosphite salts (Kromann et al. 2012) that can compete with conventional fungicides for late blight control efficacy.

10.3.2.2 Cultural Management Practices

Many cultural management practices help reduce disease pressure in the field.

Farmers may avoid high disease pressure by planting at high altitudes where temperatures are low. Planting at high altitudes has been particularly useful for traditional farmers in the Andes (Thurston 1990). Altitude has also been used to manage degeneration and produce seed under conditions where virus pressure is low (Thiele 1999). Another important strategy for smallholder management of late blight is planting outside the rainy season when low humidity retards disease development. However, escaping disease in time and space often leads to lower yields as the potato crop is not planted at a time or site for optimal plant growth. Late blight evasion by planting outside of the rainy season in the Andes is also characterized by increased risk of crop loss due to frost or drought.

For many soil-borne diseases, such as bacterial wilt, rotating planting of potatoes with other crops reduces initial inoculum in the soil and number of volunteer plants that may harbor pathogens (Lemaga et al. 2001). High hilling may reduce disease and insect pressure by increasing the barrier of soil that can protect tubers. Management practices that reduce the leaf wetness period and the air humidity in the foliage can help reduce disease progress caused by many pathogens, especially late blight. Orienting rows with the prevailing winds may dry foliage faster. Similarly, row orientation, planting density and potato plots can be selected for optimal solarization, which can reduce humidity and spore survival (Mizubuti et al. 2000). Intercropping, for example potato and faba bean, has been used for centuries in the Andes; it is also very common in Asia and Africa and can be used in integrated disease and insect management, although there are relatively few published accounts (Bouws and Finckh 2008; Autrique and Potts 2008). Harvesting and storing tubers adequately is important to avoid both insect and disease problems in tubers after harvest. Although the requirement for cash investment is often low, few of cultural management practices are widely implemented by farmers who are not organic growers. The easy access to pesticides and farmers' limited knowledge of the benefits of these tactics are at least two reasons for their poor adoption.

10.4 Arthropod Management Interventions

10.4.1 On-farm Management

Successful management of potato arthropod pests with IPM technologies is highly related to farmers' knowledge of the on-average maximum of two to four economically significant arthropod pests that require control. A major part of CIP's efforts to help resource-poor farmers implement alternative strategies to toxic chemicals have been based on training farmers to better understand the pests' biology and specific life cycles, and on-farm management practices that reduce pest infestation sources (Ortiz et al. 1996). Many cultural practices have been introduced to farmers via a range of technology dissemination strategies with extension partners, including use of high-quality pest-free seed, on-farm seed management, adequate crop rotation, optimal planting and harvest dates and best practice of weeding and hilling. The adoption of many of these practices has, however, in many cases been low and insecticides continue to be the single preferred strategy used by many farmers to manage arthropods. The reason most often given by farmers is that IPM technologies are not as immediately effective as pesticides, they are difficult to apply and insecticides continue to be more cost effective. CIP scientists have therefore developed an approach based on a holistic framework for the development and adoption of on-farm IPM that focuses on practical, economic and ecological solutions to pest management, which are applicable to resource-poor farmers. The approach focuses on agro-ecosystems research to understand the relationships of pests and natural enemies, and aims to build fauna inventories and descriptions of functional diversity of natural antagonists; research which contributes to understanding the stability of agro-ecosystems and their resilience to counteract pests, and intends to identify biocontrol agents and biopesticides (entomopathogens). Thorough economic and ecological evaluations are conducted to verify and support the benefits of on-farm IPM interventions, and the IPM interventions are compared to farmers' practice to promote adoption by farmers, a global approach, but based on specific characteristics of the local agro-ecosystem.

10.4.1.1 Insecticides

The rational use of insecticides is an important practice in IPM for the control of many arthropods. The main research efforts for rational use of insecticides have been focused on selecting low-toxicity insecticides with minimal effects on non-target organisms, appropriate application according to pest life cycles and plant or field-part specific application opposed to field-wide applications with broad-spectrum insecticides. For example, abamectin and cyromazine are effective and selective insecticides that can control specific stages of the leaf miner fly (*Liriomyza huidobrensis*) and have no detrimental effects on populations of leaf miner fly parasitoids. Research in recent decades has also been strongly focused on finding alternative technologies to the use of extremely and highly hazardous pesticides (World Health Organization categories Ia and Ib), which are still commonly used by potato farmers in the Andes to control APW and other arthropods (Orozco et al. 2009).

10.4.1.2 Sustaining Biodiversity

Sustaining natural biodiversity is one of the most important components of IPM. IPM stabilizes agro-ecosystems and allows natural antagonists to help avoid extreme pest incidence increases. Under many circumstances, sustaining the resilience of the agro-ecosystems is sufficient to maintain the incidence of certain pest species below an economic damage threshold; this is also the case in potato. Although little is still known about existing beneficial organisms present in many of the wide range of agro-ecosystems where potato is produced worldwide, experiences have shown that potato agro-ecosystems can function in a self-regulating manner, keeping, for example, aphids (*Myzus persicae*, *Macrosiphum euphorbiae*) or the cutworm (*Agrotis ipsilon*) below the-need-to-control threshold (Kroschel 1995). Recently, it was reported that high numbers of carabids of different genera greatly affect APW in the central highlands of Peru (Kroschel et al. 2009). CIP scientists have started in-depth ecological studies in two agro-ecological zones, one in the coastal area and the other in the Andes of Peru, to develop practical recommendations for IPM. Potentially, this research will lead to recommendations for natural vegetation conservation, and how to establish vegetation-diverse field boundaries and biological corridors to improve habitat management for the promotion of beneficial organisms.

10.4.1.3 Biological Control

Here, we consider biological control as interventions concerning the release of natural enemies for classical biocontrol (parasitoids) and the use of biopesticides (entomopathogens). CIP scientists are studying the biology and ecology of natural enemies of potato pests. When efficacy has been determined at field scale, systems for a natural enemy, formulation and application are developed in collaboration with governmental programs and private companies. A cheaply formulated granulovirus-and-talcum-based biopesticide to control the potato tuber moth complex in tuber storage has been developed and currently is gaining important acceptance by farmers in Bolivia, Peru and Ecuador. Recent studies have shown that a commercial formulation of *Bacillus thuringiensis* subsp. *kurstaki* (Btk) reformulated in magnesium silicate (15 g/1 kg talcum) effectively protects tubers in storage against the potato tuber moth complex and is highly competitive compared to the granulovirus-based product and chemical pesticides. This experience has helped turn the trend of thinking that the formulation of biopesticides has not been economically viable.

Parasitoids are reared by CIP and are used in classical biocontrol; *Copidosoma koehleri*, *Orgilus lepidus* and *Apanteles subandinus* for potato tuber moth control and, for example, *Halticoptera arduine*, *Chrysocharis flacilla* and *Phaedrotoma scabriventris* for leaf miner fly control. However, maintaining parasitoid populations in the field has proven difficult and so far adoption of inoculative strategies has been limited due to the high cost of rearing control organisms. The establishment of a habitat close to potato fields that conserve biological control organisms and the continued effort to confirm the potential of unidentified efficacious natural antagonists are crucial for the successful implementation of biological control. Recent research identified 9 leaf miner fly species in 27 crops and 63 parasitoids in the coastal area of Peru. The high number of parasitoids suggests that leaf miner fly has its center of origin in the Americas. *Halticoptera arduine* was the most abundant and efficient parasitoid and was found to parasitize all leaf miner fly species in 25 crops, making it a potentially interesting species as a commercial biocontrol agent along with other abundant leaf miner fly parasitoids: *Chrysocharis flacilla*, and *C. caribea* and *Diglyphus websteri* (Mujica and Kroschel 2011).

10.4.2 Technological Innovations

CIP research has led to technological innovations that can be used by small-scale and resource-poor farmers. Recent technology innovation has concentrated on the development of one technology to control APW – physical barriers that prevent weevil migration to potato plots – and two technologies to control potato tuber moths – attracticides to control the species *Phthorimaea operculella* and *Symmetrischema tangolias* in the field and in storage, and a formulation of talc and *Bacillus thuringiensis* subsp. *kurstaki* (Btk) to protect potatoes in storage from the same two moth species.

10.4.2.1 Plastic Barriers for Andean Potato Weevil Management

The APW cannot fly, which restricts its movement to walking. New emerging adults migrate from fields having potato in the previous season to infest newly planted potato fields. Installing 30 cm high plastic barriers around potato fields at planting prevents adult weevils from migrating into these fields from other areas, and has proven to significantly reduce weevil tuber damage more than the common use of insecticide. The installation cost of the barriers can be less than that of insecticide application, although this depends on the size of the plot (Table 10.2) and insecticide use intensity. Ninety-five per cent of farmers who participated in one study stated that the plastic barriers were easy to install and did not interfere with the cultural practices of potato production. The main limitations to the use of plastic barriers include damage to the plastic due to inclement weather, and finding an ecologically sound system for disposing of or reusing the plastic (Kroschel et al. 2009).

10.4.2.2 Attract and Kill for Field and Storage Management of the Potato Tuber Moth (PTM) Complex

This strategy consists of an insecticide-pheromone co-formulation whereby the male moths are attracted by the pheromone and killed through contact with a low volume of beta-cyfluthrin, an insecticide of low human toxicity. This technology effectively controls *P. operculella* and *S. tangolias* under both field and storage conditions. A droplet size of 100 µl and 2,500 drops per ha of each of *AdiosMacho-Po* and *AdiosMacho-St* effectively reduces the male populations of *P. operculella* and *S. tangolias*, respectively. Consequently, foliage and tuber damage is significantly reduced up to 90% compared to untreated controls. The treatment costs are estimated between US$ 20 and 30 per ha, while in storage conditions of small-scale farmers costs are between US$ 1 to 2 (1 drop/m^2 of storage area). This low-cost method can easily be integrated in potato pest management programs for small-scale agriculture in the tropics (Kroschel and Zegarra 2013).

10.5 Sustainability of IPM Extension and Scaling-up

10.5.1 Farmer Capacity Building for Pathogen Management

One common characteristic of most resource-poor potato farmers in developing countries is that they know little about processes that cause plant disease. Farmers know about biological entities they can see, such as crops, animals and even to some extent insects, but they logically know almost nothing about invisible microorganisms (Ortiz et al. 2004). For example, surveys have repeatedly demonstrated that common answers to the question of 'what causes potato late blight' will be anything but correct: lightening, low temperature, rain, sun while it rains, stages of the moon, bad seed or even mystical explanations (Ortiz and Forbes 2003). In spite of having access to new technologies, particularly agro-chemicals, many rural people have not gained new knowledge from agricultural science.

The lack of knowledge about basic aspects of plant disease makes it difficult to simply and rapidly show farmers how to manage a disease with fungicides or other technologies. For that reason, extension workers in developing countries have been using knowledge-intensive, participatory techniques to help farmers increase their understanding of how disease occurs and how it can be managed. The most commonly used participatory approach is probably the farmer field school (FFS). CIP and partners initiated an FFS program in the late 1990s with support from the Food and Agriculture Organization of the United Nations (FAO) and the International Fund for Agricultural Development (IFAD)(Ortiz et al. 2004). This program evolved over the years and generated interesting lessons regarding implementation of integrated disease management programs. Basically, it confirmed that farmers needed special learning activities to be able to understand biophysical principles involved in disease management, so that new knowledge could be reflected in improved disease management and in improved yields (Godtland et al. 2004; Ortiz et al. 2004; Ortiz 2006). To provide FFS facilitators with materials related to potato, a guide was developed by CIP and partners in Peru with a strong focus on potato late blight. Initially, the FFSs were intended to focus primarily on late blight, but this rapidly evolved into a focus on potato IPM and potato production in general in response to needs expressed by farmers. To address the need for effective training materials, CIP and partners implemented a program to develop competency-based training, where a competency is a 'standardized requirement for an individual to properly perform a specific job, including a combination of knowledge, skills and behaviour' (Zapata Sánchez 2006, pp. xii to xiii). Competency analysis was used to develop a training guide for extension workers in Ecuador. A group of farmers, extension workers and plant pathologists identified six competencies needed to manage late blight efficiently: (i) identify the disease symptoms, (ii) know its causal agent and how it lives, (iii) identify the characteristics and benefits of using resistant potato cultivars, (iv) use fungicides appropriately, (v) visit the potato plot frequently and (vi) select control measures for late blight. Knowledge and behavior specific for each competency were identified and from those, learning objectives were defined. From the objectives, the contents, learning strategies, and evaluation questions for each training session were developed. A Spanish version of the training guide was developed and iteratively tested and improved in three FFSs in the central highlands of Ecuador. The guide was then published in Spanish and Ecuadorian Quechua. Spanish and English versions of the guides are available for local adaptation (Cáceres et al. 2008). A similar farmer-focused approach was used to develop materials for farmer capacity building to manage diseases causing seed degeneration in potato. These materials focus on improving farmers' understanding of the problem, particularly identifying the types of disease involved in degeneration. As a control tactic, farmers are trained to select plants that have no symptoms of infection (Gildemacher et al. 2011).

10.5.2 Farmer Capacity Building for Arthropod Management

IPM of arthropod pests is knowledge intensive. Information on the ecology of pests and appropriate management practices has been communicated to farmers and their families via field demonstrations, pamphlets, posters, pest collection competitions, and entomological kits. Nevertheless, similar to farmer capacity building for pathogen management, the most common participatory approach for capacity building for arthropod management has probably been the farmer field school (FFS). IPM of arthropod pests was included in the FFS program initiated in the late 1990s, mentioned above (see Sect. 10.5.1), which was initially developed for late blight IPM. Following farmers' requests, capacity-building modules were included in subsequent FFSs on APW and potato tuber moth to help farmers increase their understanding of how arthropod life cycles take place and how the pests can be managed. In the Andean countries, most emphasis was given to APW. The use of refuge traps, in which a small amount of potato foliage treated with insecticide placed in the field under a cover of cardboard or straw, is not uncommon in the Andes for monitoring and controlling APW; a technology principally introduced through FFSs. Other practices introduced to farmers for control of APW include: timely harvesting, stirring up soil in areas of high soil infestation combined with chickens that feed and eliminate larvae, nocturnal manual collection of weevil adults, vegetative barriers of plants, and other practices. Technologies promoted by CIP for control of potato tuber moth include timely planting and harvesting, irrigation up to harvest to keep soil humid and reduce cracking, improved high hilling, rational use of low toxicity insecticides and other practices. The majority of potato farmers, in areas where these practices have been included in extension programs, use one or more of these tactics to reduce tuber damage by potato tuber moth. Adequate management practices of harvested and stored tubers, including solarization and use of biopesticides, have been adopted by farmers for potato tuber moth control. However, in communities of small-scale farmers where farms are located close to each other, inadequate tuber management and pest management in general by one farmer can easily be a source of pest infestation for the neighboring farmers.

10.6 Risk Assessment for Insects and Late Blight Modelling

Modelling of insects and late blight disease, through the processes of computing multi-parameter mathematical models to simulate real insect and disease systems, is used currently to investigate insect and disease dynamics, and epidemiological principles. Pests' development and their population dynamics are to a great extent driven by climatic factors, principally temperature, influenced by some species-specific biotic factors and external factors such as cultural practices implemented by farmers. These factors are used in insect life-cycle modelling software to develop phenology models for important insect pests.

Late blight disease is principally driven by epidemic parameters related to host, pathogen and climate. A late-blight-simulation model developed at Cornell University has been modified and parameterized for the highland tropics and was qualified for a broad range of environments including the highland tropics (Andrade-Piedra et al. 2005; Andrade-Piedra et al. 2005a; Andrade-Piedra et al. 2005b; Blandón-Díaz et al. 2011).

Simulation software can help predict insect and disease population growth potentials under given agro-ecologies and extensive experimentation can be carried out with simulations that would be impossible or very expensive using the real systems in the laboratory or field. Using models of this type within a geographic information system (GIS), regional and global distribution maps of the potential intensities of pests and diseases can be assessed. Insect and disease risks can be predicted based on changes in climate and the occurrence of extreme climate events. The approach is increasingly being used in climate change research.

10.7 Future Research

As indicated above (see Sect. 10.2.3), potato production is highly dependent on pesticide use, although this dependency is rather recent in the history of this millenniums-old crop. Furthermore, in spite of this pesticide dependency, crop losses to both pests and diseases are substantial. Actual yields in developing countries are less than half of attainable yields, and much of this gap is probably due to pests and diseases (Fuglie 2007), although very few studies have actually quantified losses. Better quantification of the potato pest and disease problems will require intensified research, better coordination, and more holistic approaches to estimate impact.

Recently, CIP evaluated the utility in developing countries of one pesticide risk indicator, the EIQ, which combines the pesticide hazard posed to farm workers (applicator and picker effects), consumers (consumer effect and groundwater effect), and the local environment (aquatic and terrestrial effects) into a composite hazard indicator (Kromann et al. 2011). In this study, the environmental impact per hectare varied greatly among the different potato systems tested and ranged from 40 for an IPM system (resistant cultivar plus less hazardous pesticides) to 1,235 for a high-input conventional system (susceptible cultivar plus frequent use of hazardous pesticides), thus emphasizing the importance of pesticide risk studies related to the social impact of pesticide use and the potential benefits of IPM technologies. Further research on and the use of environmental and health risk indicators can improve our understanding of the importance of biotic constraints and the benefits of IPM technologies, and should in time increasingly complement the efficacy and economic assessments traditionally used to evaluate and contrast different technologies.

The direct effects of constraints on yield are more difficult to estimate than pesticide use and therefore the problem has often been approached with statistical models associating the degree of disease severity (e.g., late blight) with yield (Olofsson 1968; James et al. 1971) or mechanistic models that predict yield based on foliage loss (Shtienberg et al. 1990). The statistical models have the disadvantage of not being very applicable to areas outside of where they originated. None of the models were implemented in a spatially specific GIS that would allow for large-scale estimates taking into account variations for climate, soil or other factors that may affect disease severity.

CIP and partners have recently been working on a process of estimating the potential severity of late blight based on geo-referenced weather data, which give large-scale estimates, taking climatic variation into account (Sparks et al. 2011). While this can easily be interpreted in terms of fungicide use needs based on potential severity, it is not easy to estimate yield loss because of the uncertainty related to farmers' disease management capacity. CIP and partners are now implementing research to model the human component of disease severity and thereby get more accurate estimates of actual yield loss. While these efforts are ongoing for late blight, and geographical risk maps for potato tuber moth and leaf miner fly are being developed, little else has been done to estimate the effects of other diseases or pests on potato yield over large geographic areas. This is one of the objectives of new projects, for example, one project on degeneration of vegetatively propagated crops involving CIP and a number of partners in developing and industrialized countries. As part of this project, researchers will attempt to develop modelling tools to estimate yield losses due to degeneration and extrapolate these over different geographic scales. A similar geographic approach linking pest damage with crop models is being sought for potato tuber moth and leaf miner fly to estimate spatial yield losses by these insect pests.

10.8 Conclusion

The most important biotic constraint of potato caused by an individual organism is potato late blight, the single major driver for heavy pesticide use in potato production worldwide and one of the world's most important food-crop diseases. Research on late-blight management at CIP and other research institutes has to a large extent focused on the integration of host-plant resistance and reduced fungicide use. Research has shown that fungicidal sprays to control late blight can be reduced by two-thirds by using resistant cultivars, which serves to illustrate how important host plant resistance can be to manage biotic constraints. Degeneration of planting material (tubers), caused primarily by viruses, is another major biotic constraint of potato. Apart from expensive and generally unsuccessful efforts to clean up seed material for large geographical areas in the developing world, this disease complex has received little attention. Researchers at CIP estimate that degeneration of planting material causes more yield reductions than any other biotic constraint of potato. Thus, more research effort has recently been concentrated on breeding for virus resistance.

Potato producers are faced with about 20 arthropod pests, but on average each farmer needs to control two to four economically important insects. Insecticide use is still the most common strategy deployed by farmers, and pesticide exposure and health risks from pesticides are very high among potato farmers in developing countries, where protective equipment generally is not used. In the last decades, CIP research has led to technological innovations that can be used by small-scale and resource-poor farmers. Technologies that have been developed for arthropod problems include the use of attract and kill for managing potato tuber moths, the most important insect problem in potato in the developing world; the use of plastic barriers that inhibit migration of the APW; and the rational use of selective and low-toxicity insecticides. However, many of the technologies have not been widely adopted by farmers for different reasons, and the achievable positive impact of many IPM technologies has been reduced by major dissemination constraints. The reason most often given by farmers for not adopting IPM technologies is that they are not as immediately effective as pesticides, they are difficult to apply, and insecticides continue to be more cost effective.

The main conclusion that can be drawn from numerous potato IPM intervention projects is that the most important element of IPM in developing countries is the pest management capacity of farmers. To increase farmer acceptance of IPM technologies, it has proven essential to demonstrate and train farmers to understand the economic, ecological, and practical benefits of the new technologies compared to their traditional strategies (Ortiz 2006). To develop practical recommendations for IPM, CIP has developed a holistic approach considering all economically important pests at farm level, and which aims to develop technological innovations to replace farmers' pesticide applications with equal efficacy. The approach includes participatory on-farm research to adapt new technologies to local conditions and farmer training. Future research will increasingly focus on the use of environmental and health risk indicators to improve the understanding of the benefits of IPM technologies. We anticipate that systems for formulation and application of IPM technologies will increasingly be developed in collaboration between public research institutes and private companies. There is an urgent need for new technology dissemination strategies. Thus, we foresee a similar increase in the interest for public-private partnerships that will work towards supplying quality IPM extension services needed for sustainable potato production.

References

Andrade-Piedra, J. L., Forbes, G. A., Shtienberg, D., Grünwald, N. J., Taipe, M. V., & Fry, W. E. (2005a). Simulation of potato late blight in the Andes: II: Validation of the LATEBLIGHT Model. *Phytopathology, 95*(10), 1200–1208.

Andrade-Piedra, J. L., Forbes, G. A., Shtienberg, D., Grünwald, N. J., Taipe, M. V., Hijmans, R. J., & Fry, W. E. (2005b). Qualification of a plant disease simulation model: Performance of the LATEBLIGHT model across a broad range of environments. *Phytopathology, 95*(12), 1412–1422.

Andrade-Piedra, J. L., Hijmans, R. J., Forbes, G. A., Fry, W. E., & Nelson, R. J. (2005). Simulation of potato late blight in the Andes: I: Modification and parameterization of the LATEBLIGHT model. *Phytopathology, 95*(10), 1191–1199.

Antle, J. M., Cole, D. C., & Crissman, C. C. (1998). Further evidence on pesticides, productivity and farmer health: Potato production in Ecuador. *Agricultural Economics, 18*(2), 199–207.

Autrique, A., & Potts, M. J. (2008). The influence of mixed cropping on the control of potato bacterial wilt (Pseudomonas solanacearum). *Annals of Applied Biology, 111*(1), 125–133.

Bertschinger, L., Scheidegger, U. C., Luther, K., Pinillos, O., & Hidalgo, A. (1990). La incidencia de virus de papa en cultivaresnativos y mejorados en la sierra peruana. *Revista Latinoamericana De La Papa, 3*(1), 62–79.

Birch, P. R. J., Bryan, G., Fenton, B., Gilroy, E., Hein, I., Jones, J., Prashar, A., Taylor, M., Torrance, L., & Toth, I. (2012). Crops that feed the World 8: Potato: Are the trends of increased global production sustainable. *Food Security, 4*(4), 477–508.

Blandón-Díaz, J. U., Forbes, G. A., Andrade-Piedra, J. L., & Yuen, J. E. (2011). Assessing the adequacy of the simulation model LATEBLIGHT under Nicaraguan conditions. *Plant Disease, 95*(7), 839–846.

Bourke, A. (1993). *"The Visitation of God"? The Potato and the Great Irish Famine*. Dublin: Lilliput Press Ltd.

Bouws, H., & Finckh, M. R. (2008). Effects of strip intercropping of potatoes with non-hosts on late blight severity and tuber yield in organic production. *Plant Pathology, 57*(5), 916–927.

Cáceres, P. A., Pumisacho, M., Forbes, G. A., & Andrade-Piedra, J. L. (2008). *Learning to Control Potato Late Blight: A Facilitator's Guide*. International Potato Center (CIP), Instituto Nacional Autónomo de Investigaciones Agropecuariasdel Ecuador (INIAP), Secretaría Nacional de Ciencia y Tecnologíadel Ecuador (SENACYT). Quito, Ecuador. http://cipotato.org/publications/pdf/004358.pdf. Accessed 10 Oct 2012.

Cole, D. C., Carpio, F., Math, J. J. M., & Leon, N. (1997). Dermatitis in Ecuadorian farm workers. *Contact Dermatitis, 37*(1), 1–8.

Fankhauser, C. (2000). *Seed-transmitted diseases as constraints for potato production in the tropical highlands of Ecuador*. PhD Dissertation. Zurich: Swiss Federal Institute of Technology.

Forbes, G. A. (2012). Using host resistance to manage potato late blight with particular reference to developing countries. *Potato Research, 55*(3–4), 205–216.

Forbes, G. A., Shtienberg, D., & Mizubuti, E. (2009). Plant disease epidemiology and disease management—has theory had an impact on practice? In R. Peshin. & A. K. Dhawan (Eds.)., *Integrated Pest Management: Innovation—Development Process, Vol.1.* (pp. 351–368). Dordrecht: Netherlands: Springer.

Fry, W. E. (1978). Quantification of general resistance of potato cultivars and fungicide effects for integrated control of potato late blight. *Phytopathology, 68*(11), 1650–1655.

Fry, W. E., & Goodwin, S. B. (1997). Re-emergence of potato and tomato late blight in the United States. *Plant Disease, 81*(12), 1349–1357.

Fuglie, K. (2007). *Research Priority Assessment for the CIP 2005–2015 Strategic Plan: Projecting Impacts on Poverty, Employment, Health and Environment*. Lima, Peru: International Potato Center.

Gildemacher, P. R., Demo, P., Barker, I., Kaguongo, W., Woldegiorgis, G., Wagoire, W. W., Wakahiu, M., Leeuwis, C., & Struik, P. C. (2009). A description of seed potato systems in Kenya, Uganda and Ethiopia. *American Journal of Potato Research, 86*(5), 373–382.

Gildemacher, P. R., Schulte-Geldermann, E., Borus, D., Demo, P., Kinyae, P., Mundia, P., & Struik, P. C. (2011). Seed potato quality improvement through positive selection by smallholder farmers in Kenya. *Potato Research, 54*(3), 253–266.

Godtland, E., Sadoulet, E., de Janvry, A., Murgai, R., & Ortiz, O. (2004). The impact of farmer-field-schools on knowledge and productivity: A study of potato farmers in the Peruvian Andes. *Economic Development and Cultural Change, 53*(1), 63–92.

Grünwald, N. J., Romero Montes, G., LozoyaSaldaña, H., Rubio Covarrubias, O. A., & Fry, W. E. (2002). Potato late blight management in the Toluca Valley: Field validation of SimCast modified for cultivars with high field resistance. *Plant Disease, 86*(10), 1163–1168.

Haldar, K., Kamoun, S., Hiller, N. L., Bhattacharje, S., & van Ooij, C. (2006). Common infection strategies of pathogenic eukaryotes. *Nature Reviews Microbiology, 4*(12), 922–931.

Haverkort, A. J., Struik, P. C., Visser, R. G. F., & Jacobsen, E. (2009). Applied biotechnology to combat late blight in potato caused by Phytophthora Infestans. *Potato Research, 52*(3), 249–264.

James, W. C., Callate Blighteck, L. C., Hodgson, W. A., & Shih, C. S. (1971). Evaluation of a method used to estimate loss in yield of potatoes caused by late blight. *Phytopathology, 61*(12), 1471–1476.

Judelson, H. S., & Blanco, F. A. (2005). The spores of Phytophthora weapons of the plant destroyer. *Nature Reviews Microbiology, 3,* 47–58.

Kromann, P., Leon, D., Andrade-Piedra, J. L., & Forbes, G. A. (2008). Comparison of alternation with a contact fungicide and sequential use of the translaminar fungicide cymoxanil in the control of potato late blight in the highland tropics of Ecuador. *Crop Protection, 27,* 1098–1104.

Kromann, P., Pérez, W. G., Taipe, A., Schulte-Geldermann, E., Prakash Sharma, B., Andrade-Piedra, J. L., & Forbes, G. A. (2012). Use of phosphonate to manage foliar potato late blight in developing countries. *Plant Disease, 96*(7), 1008–1015.

Kromann, P., Pradel, W., Cole, D., Taipe, A., & Forbes, G. A. (2011). Use of the environmental impact quotient to estimate health and environmental impacts of pesticide usage in Peruvian and Ecuadorian potato production. *Journal of Environmental Protection, 2*(5), 581–591.

Kromann, P., Taipe, A., Pérez, W. G., & Forbes, G. A. (2009). Rainfall thresholds as support for timing fungicide applications in the control of potato late blight in Ecuador and Peru. *Plant Disease, 93*(2), 142–148.

Kroschel, J. (1995). *Integrated pest management in potato production in the Republic of Yemen with special reference to the integrated biological control of the potato tuber moth (Phthorimaea operculella Zeller)* (Doctoral Dissertation). Weikersheim, Germany: Margraf Verlag.

Kroschel, J., & Schaub, B. (2013). Biology and ecology of potato tuber moths as major pests of potato. In A. Alyokhin, P. Giordanengo, & C. Vincent. (Eds.)., *Insect Pests of Potato Global Perspective on Biology and Management. (pp.165–192).* Waltham, Massachusetts, USA: Elsevier.

Kroschel, J., Alcazar, J., & Pomar, P. (2009). Potential of plastic barriers to control Andean potato weevil Premnotrypes Suturicallus Kuschel. *Crop Protection, 28,* 466–476.

Kroschel, J., & Zegarra, O. (2013). Attract-and-kill as a new strategy for the management of the potato tuber moths *Phthorimaea operculella* (Zeller) and *Symmetrischema tangolias* (Gyen) in potato: Evaluation of its efficacy under potato field and storage conditions. *Pest Management Science, 69*(11), 1205–1215.

Lemaga, B., Kanzikwera, R., Kakuhenzire, R., Hakiza, J. J., & Manzi, G. (2001). The effect of crop rotation on bacterial wilt incidence and potato tuber yield. *African Crop Science Journal, 9*(1), 257–266.

Miethbauer, T. (2012). Collective action and on-farm benefits of pesticide substitution: A case study. Preliminary analysis of household survey results (survey 03/2012; CIP-GIZ Project GnC1070). Poster presentation at Tropentag Conference 2012. Book of Abstracts, p. 158. Göttingen, Germany.

Mizubuti, E. S. G., Aylor, D. E., & Fry, W. E. (2000). Survival of *Phytophthora infestans* sporangia exposed to solar radiation. *Phytopathology, 90*(1), 78–84.

Mizubuti, E. S. G., LourençoJúnior, V., & Forbes, G. A. (2007). Management of late blight with alternative products. *Pest Technology, 1*(2), 106–116.

Moseley, M. (1992). *The Incas and their Ancestors: The Archeology of Peru.* London: Thames and Hudson.

Mujica, N., & Kroschel, J. (2011). Leafminer fly (Diptera: Agromyzidae) occurrence, distribution and parasitoid associations in field and vegetable crops along the Peruvian coast. *Environmental Entomology, 40*(2), 217–230.

Norton, G. W., Heinrichs, E. A., Luther, G. C., & Irwin, M. E. (2005). *Globalizing Integrated Pest Management. A Participatory Research Process.* Oxford, United Kingdom: Blackwell Publishing.

Oerke, E. C., Dehne, H. W., Schönbeck, F., & Weber, A. (1994). *Crop Production and Crop Protection. Estimated Losses in Major Food and Cash Crops.* Amsterdam: Elsevier.

Olofsson, B. (1968). Determination of the critical injury threshold for potato late blight (Phytophthora infestans). *Statens Växtskyddsanstalt (Stockholm), 14,* 85–93.

Orozco, F. A., Donald, C. C., Forbes, G. A., Kroschel, J., Wanigaratne, S., & Arica, D. (2009). Monitoring adherence to the international FAO code of conduct on the distribution and use of pesticides: Highly hazardous pesticides in central Andean agriculture and farmers' rights to health. *International Journal of Occupational and Environmental Health, 15*(3), 255–268.

Ortiz, O. (2006). Evolution of agricultural extension and information dissemination in Peru: An historical perspective focusing on potato-related pest control. *Agriculture and Human Values, 23*(4), 477–489.

Ortiz, O., Alcazar, J., Catalan, W., Villano, W., Cerna, V., Fano, H., & Walker, T. (1996). Economic impact of IPM practices on the Andean potato weevil in Peru. In T. S. Walker., C. C. Crissman (Eds.)., *Case Studies of the Economic Impact of CIP-related Technologies* (pp. 157). Lima, Peru: International Potato Center.

Ortiz, O., & Forbes, G. A. (2003). Fighting a global problem: Managing potato late blight through partnership. Poster presented at: *Conference of the Global Forum for Agricultural Research* (GFAR), Senegal 22–24 May.

Ortiz, O., Garret, K. A., Heath, J. J., Orrego, R., & Nelson, R. J. (2004). Management of potato late blight in the Peruvian highlands: Evaluating the benefits of farmer field schools and farmer participatory research. *Plant Disease, 88*(5), 565–571.

Ortiz, O., Kroschel, J., Alcazar, J., Orrego, R., & Pradel, W. (2009). Evaluating dissemination and impact of IPM: Lessons from case studies of potato and sweetpotato IPM in Peru and other Latin American countries. In R. Peshin & A. K. Dhawan (Eds.)., *Integrated Pest Management: Dissemination and Impact* (Vol. 2, pp. 419–434). Dordrecht, Netherlands: Springer.

Oyarzún, P. J., Garzón, C. D., Leon, D., Andrade, I., & Forbes, G. A. (2005). Incidence of potato tuber blight in Ecuador. *American Journal of Potato Research, 82*(2), 117–122.

Panchi, N., Navarrete, I., Taipe, A., Orellana, H., Pallo, E., Yumisaca, F., Montesdeoca, F., Kromann, P., & Andrade-Piedra, J. L. (2012). Incidencia, severidad y pérdidas causadas por plagas de la semilla se papa en Ecuador.Poster presented at: *Congreso de la Asociación Latinoamericana de la Papa—ALAP*. Brasil: Uberlandia.17–20 Sep. 2012.

Panganiban, L., Cortes-Maramba, N., Dioquino, C., Suplido, M. L., Ho, H., Francisco-Rivera, A., & Manglicmot-Yabes, A. (2004). Correlation between blood ethylenethiourea and thyroid gland disorders among banana plantation workers in the Philippines. *Environmental Health Perspectives, 112*(1), 42–45.

Peshin, R., & Dhawan, A. (2009). *Integrated Pest Management: Dissemination and Impact* (Vol. 2). Dordrecht, Netherlands: Springer.

Pradhanang, P. M., Pandey, R. R., Ghimere, S. R., Dhital, B. K., & Subedi, A. (1992). *An Approach to Management of Bacterial Wilt of Potato Through Crop Rotation and Farmers' Participation. ACIAR Proceedings, No. 45*. Taiwan: Australian Centre for International Agricultural Research (ACIAR).

Priou, S., Salas, C., De Mendiburu, F., Aley, P., & Gutarra, L. (2001). Assessment of latent infection frequency in progeny tubers of advanced potato clones resistant to bacterial wilt: A new selection criterion. *Potato Research, 44*(4), 359–373.

Radcliffe, E. B., Hutchison, W. D., & Cancelado, R. E. (2009). *Integrated Pest Management: Concepts, Tactics, Strategies and Case Studies*. Cambridge, United Kingdom: Cambridge University Press.

Raman, K., & Radcliffe, E. B. (1992). Pest aspects of potato production, Part 2. Insect pests. In P. M. Harris (Ed.)., *The Potato Crop: The Scientific Basis for Improvement* (2nd edn.). London: Chapman and Hall.

Salazar, L. F. (1996). *Potato Viruses and Their Control*. Lima, Peru: International Potato Center.

Shtienberg, D., Bergeron, S. N., Nicholson, A. G., Fry, W. E., & Ewing, E. E. (1990). Development and evaluation of a general model for yield loss assessment in potatoes. *Phytopathology, 80*(5), 466–472.

Sparks, A. H., Forbes, G. A., Hijmans, R. J., & Garrett, K. A. (2011). A metamodeling framework for extending the application domain of process-based ecological models. *Ecosphere, 2*(8) (August): art90. doi:10.1890/ES11-00128.1.

Stevenson, W., Loria, R., Franc, G. D., & Weingartner, D. P. (2001). *Compendium of Potato Diseases*, 2nd edn. St. Paul, Minnesota, USA: American Phytopathological Society.

Thiele, G. (1999). Informal potato seed systems in the Andes: Why are they important and what should we do with them? *World Development, 27*(1), 83–99.

Thurston, H. D. (1990). Plant disease management practices of traditional farmers. *Plant Disease, 74*(2), 96–102.

Torrez, R., Tenorio, J., Valencia, C., Orrego, R., Ortiz, O., Nelson, R., & Thiele, G. (1999). Implementing IPM for Late Blight in the Andes. *In Impact on a Changing World: Program Report 1997–98* (pp. 91–99). Lima, Peru: International Potato Center.

Wesseling, C., Corriols, M., & Bravo, V. (2005). Acute pesticide poisoning and pesticide registration in Central America. *Toxicology and Applied Pharmacology, 207*(2), 697–705. doi:10.1016/j.taap.2005.03.033.

Yuen, J. E., & Forbes, G. A. (2009). Estimating the level of susceptibility to Phytophthora infestans in potato genotypes. *Phytopathology, 99,* 783–786.

Zapata Sánchez, V. (2006). *Manual Para La Formación De Gestores De Conocimiento* (Manual for the Formation of Knowledge Managers). Cali, Colombia: CIAT.

Chapter 11
Biological Control: Perspectives for Maintaining Provisioning Services in the Anthropocene

Timothy R. Seastedt

Contents

11.1	Introduction	270
11.2	Biological Control as Viewed from a Food Web and Ecosystem Perspective	271
11.3	Biological Protectors of Crops	273
11.4	Biological Control in Grasslands and Rangelands	274
11.5	Advances in Conservation Biology Control	275
11.6	Identifying Risks, Non-target Effects, and Unintended Consequences of Biological Control	276
11.7	Summary Thoughts	276
References		277

Abstract Biological control is an essential component of sustainable crop management that attempts to maximize one ecosystem service—production of food and fiber—while concurrently contributing in a positive manor to other ecosystem services required for human health and wellbeing. Biological control techniques are both plant species and site specific, so efficacy of controls and specific methodologies will vary among crop species or for a single crop species over resource and climatic gradients. However, techniques to enhance food web diversity within croplands via maximizing spatial and temporal heterogeneity of these local landscapes appear to be the appropriate framework with which to attempt specific biological control techniques. Such a framework also provides an agricultural system with potential resilience to climatic extremes, emergent diseases, and other factors deleterious to food security.

Keywords Biological control · Anthropocene · Crop protection · Agroecosystems

T. R. Seastedt (✉)
Department of Ecology and Evolutionary Biology,
University of Colorado, Boulder, Colorado 80309, USA
e-mail: Timothy.seastedt@colorado.edu

11.1 Introduction

Crop pests continue to reduce food production by about 40%, and that percentage has remained relatively constant through the 20th century and into the 21st century in spite of decades of pest management implementation (Foreward by David Pimentel, in Gurr et al. 2004). Optimization for one property of a crop such as high potential productivity can have negative effects on other plant characteristics such as ability to withstand consumer and pathogen attacks (Garratt et al. 2011). Karieva (1999, p. 10) summarized this succinctly when he stated, "…agriculture just may have to accept some substantial level of crop losses as unavoidable". That said, the challenge to feed 9 billion people in a sustainable manner requires that scientists and managers continue their focus on crop protection from pests.

In a world that wants to maximize crop productivity per unit area, the negative impacts associated with keeping crop losses to pests and pathogens at historical averages include both substantial economic as well as ecological penalties, and the latter appear to be amplified as a function of the amount of land under cultivation and used for livestock. These lands now account for about 50% of the terrestrial surface of the planet (Ellis et al. 2010). Holding current production per unit area constant but reducing pest losses can therefore have substantial win-win-win consequences for growers, consumers, and the environment.

Today's agricultural systems, particularly in developed nations, are often large-area monocultures that have largely forfeited their ability to defend themselves from predators and pathogens. Humans have taken on the role of protector, subsidizing the plant with defenses, often with chemical defenses. Genetic engineering makes this a more sophisticated activity, at least temporarily for selected crops, but the goal to maximize one ecosystem service (food production or provisioning) at the expense of all other ecosystem services (including supporting, regulating and cultural services, c.f. Chapin et al. 2009) may not be sustainable regardless of the technology used to maximize annual crop productivity. Given the amount of land required to feed 9 billion people in the coming decades, the goal of agroecosystems should be to contribute in a neutral to positive way to all ecological services. In doing so pest reduction becomes not only feasible but may be a logical consequence of smart management techniques. Biological control of pests is therefore viewed as one dimension of truly sustainable agriculture (e.g., Altieri 1999; Van Driesche et al. 2010).

The concern that herbicides, fungicides, and insecticides have failed to protect food supplies and human heath remains a common view among a diverse group of scientists. In addressing the problems with invasive weeds, McFadyen (1998) stated that "biocontrol offers the only safe, economic, and environmentally sustainable solution". A subset of biotechnological advances offers similar promises (Lord 2005). Both approaches use the observation of Matson et al. (1997, p. 508), "…by manipulating trophic levels above and below the pest, we can influence pest population dynamics and behavior in ways that reduce crop damage without the negative environmental consequences that often accompany pesticide use". This manipulation of food webs becomes 'biological control' in the broadest sense.

Eilenberg et al. (2001, p. 390) defined biological control as "The use of living organisms to suppress the population density or impact of a specific pest organism, making it less abundant or less damaging than it would otherwise be." Within this definition Eilenberg and Hokkanen (2006) identified four complimentary strategies used in biological control that have been used to organize the current chapter. These include: 1) classical biological control, which releases a control agent for long-term establishment, 2) inoculation biological control, to release an agent that will work for an anticipated length of time, but not be permanent, 3) inundation biological control, whereby the large number or amount of control agents applied by the land steward are themselves the agent of control, an intensive and short-term result, and 4) conservation biological control, whereby the environment or management practices are altered to enhance control by existing agents. In a perfect world, we would discover mechanisms whereby we could focus on methods 1 and 4 and create a perpetuating, sustainable control system that did not require continued economic investment. However, the reality is that all approaches are usually necessary, and these strategies can also be combined with management activities focused at supporting and regulating rather than provisioning services.

An intensified research effort in the last 20 years has resulted in an explosion of the literature documenting the four approaches to biological control described above. This work is spread out across a large spectrum of biological and biochemical disciplines, but in particular can be found in specialty journals such as *Biological Control* or *Agronomy for Sustainable Development*, and in the book series *Progress in Biological Control* (currently 12 volumes) and *Sustainable Agriculture Reviews* (currently 9 volumes). Early successes and failures of biological control have been documented in Debach and Rosen (1991). A brief history of pest management and some of the unresolved, broader issues that have been taken up by current research efforts were outlined by the National Research Council (NRC 1996). While that report is dated, the issues identified within that volume remain largely unchanged. Agriculture, when viewed as a reconstruction of a food web designed to benefit humans that is sustainable at decadal scales, requires a diverse and heterogeneous landscape. This view also emerges from contributions to edited volumes on pest management produced after the NRC report (Barbosa 1998; Gurr et al. 2004; Eilenberg and Hokkanen 2006). Here, I attempt to summarize a portion of the key points identified by those researchers, and put these findings into a sustainability framework suggested by Chapin et al. (2009, 2011).

11.2 Biological Control as Viewed from a Food Web and Ecosystem Perspective

The food web of a single crop system is shown in Fig. 11.1. The myopic view that agronomic systems should be a two trophic level system composed of the crop and humans is perhaps one reason that sustainable agroecosystems have been slow in developing. Here in Fig. 11.1, the human connection is omitted, yet the complexity

Fig. 11.1 The food web of a well-weeded crop. (modified from Swift et al. 2004). *Solid arrows* depict flow of materials from one component to another. *Dashed arrows* indicate strong indirect controls by one component on the other. Crops 'feed' service providers via production of dead organic matter

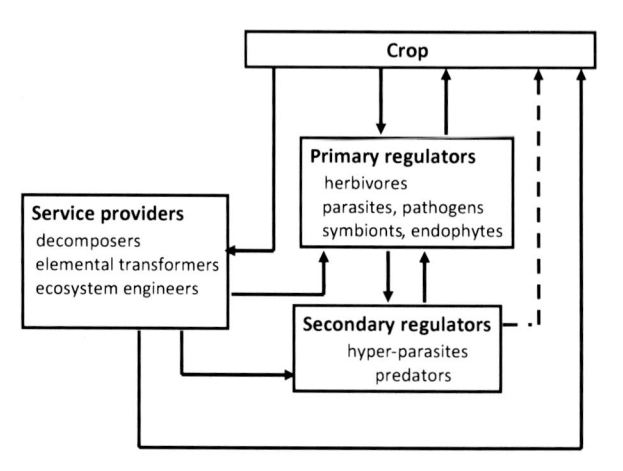

of the food web of a most simplified agroecosystem is apparent. The solid arrows in this figure indicate matter transfers from one biotic component to another, whereas the dashed lines indicate indirect controls or transfers. The goal of classical biological control of crop pests is to create 'a green world', one where the negative impacts to the crop from other biotic components are minimized, and positive effects maximized. From a food chain perspective, this is generated by designing the system to have a three component system which includes the crop, primary regulator (pest), and secondary regulator (biological control agent). However, the primary regulators of plants also include beneficial organisms (symbionts, whose contributions to the crop are represented by the arrow from the primary regulators in Fig. 11.1), including organisms that can directly or indirectly defend plants from herbivores and parasites. These symbionts include bacteria and fungi living entirely within the plants (endophytes) or extending outside of the plant, functioning as accessory roots (mycorrhizae). The service providers identified in Fig. 11.1 are often ignored in such models and include mostly decomposers and mineralizers which are fed by plant residues and themselves provide resources for both primary and secondary regulators. The feedbacks from the service providers in the form of plant nutrients and physical alterations to soils may mitigate the impacts of plant-pest interactions (Altieri et al. 2012). All agroecosystems possess this minimum level of complexity.

Managers know that biological control is very much crop species specific (sometimes cultivar specific) and site specific. Not shown in Fig. 11.1 are the abiotic conditions—the physical and chemical factors—that provide the environment for the food web. These abiotic conditions can strongly influence the abundance and activities of species within the web and thereby influence the strength of negative and positive interactions (e.g., Stirling 2011; Altieri et al. 2012). Further, the indirect effects generated by biotic-biotic or abiotic-biotic interactions have the ability to amplify, neutralize or even reverse direct effects and also can make the outcome of biological control a site-specific response (e.g., McEvoy and Coombs 2000; Whipps 2001).

Table 11.1 Organisms involved in classical biological control, targets, and recent volumes or reviews on topic

Biocontrol agent	Target	Recent reference
Arthropods	Crop Weeds	Muniappan et al. (2009)
Arthropods	Crop herbivores	Smagghe and Diaz (2012)
Arthropods	Orchards	Simon et al. (2010)
Conservation biological control	Arthropod pests of crops	Jonsson et al. (2008)
Endophytes	Pathogens, nematodes, insects	Backman and Sikora (2008)
Fungal derivatives (Mycocides)	Insects and mites	Faria and Wraight (2007)
Microbes	Parasitic nematodes	Davies and Spiegel (2011)
Microorganisms	Multiple crop components	Andrews et al. (2011)
Pathogens and nematodes	Crop herbivores	Hajek et al. (2007)
Pathogens and nematodes	Orchard, Forest arthropods	Hajek and Tobin (2010)
Symbiotic Fungi	Insect herbivores	Hartley and Gange (2009)

11.3 Biological Protectors of Crops

Biological control groups, their targets, and recent references or reviews of these groups are listed in Table 11.1. Classical biological control involves finding an enemy of the pest affecting crop production. In a perfect world one might search for the 'silver bullet' of biological control, whereby the addition of a single biological control species removes the single pest. But single controls rarely work under all environmental conditions, therefore multiple controls are often necessary, and these controls may vary in their effects across environmental gradients (McEvoy and Coombs 2000; Denoth et al. 2002). As noted by those authors and many others, species additions can sometimes have unintended consequences, and increased complexity of food webs may not necessarily lead to better control of pests. That said, the combination of multiple controls on pests, including using multiple phyla of controls, continues to provide greater insights—and in many cases greater biological control—of crop pests.

The emerging importance of primary regulator symbiotic organisms—endophytic bacteria and fungi, along with mycorrhizae fungi—may represent an opportunity for substantial advances in biotechnological innovations for biological control (e.g., Schulz et al. 2002, Lodewyckx et al. 2002, Guo et al. 2008). For example, mycorrhizal fungi assist plants not only in nutrient uptake, but can reduce attack rates of plant parasitic nematodes (Vos et al. 2012). Using both endophytic microbes and free-living microbes in a comprehensive biological control effort has the potential to improve plant productivity, substitute for hazardous chemicals and decrease production costs (Tikhonovich and Provorov 2011).

When weed control becomes the focus of pest management, the level of complexity increases (Fig. 11.2). Reducing the competition represented by one or more non-crop plants becomes the focus. Again, the food web can both directly and indirectly affect the outcome. Biological control efforts attempt to minimize competition from non-crop plants by constructing a food web that maximizes the negative feedbacks to the weed,

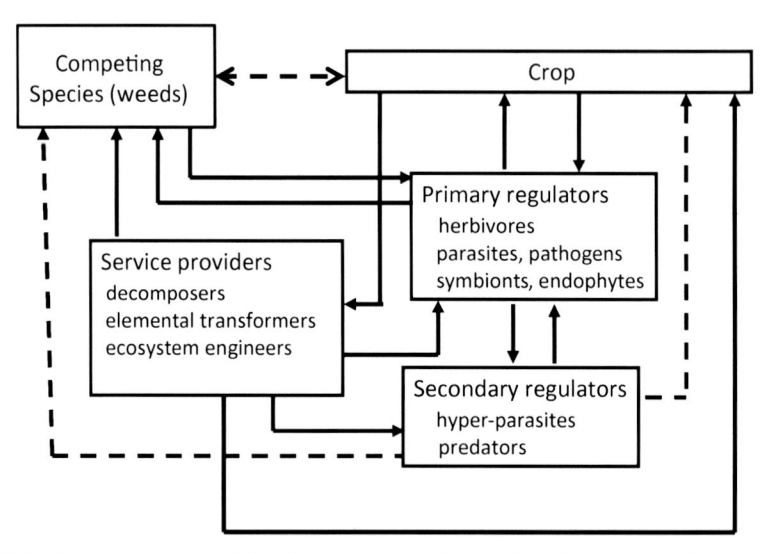

Fig. 11.2 An agroecosystem with at least two competing species. The presence of an additional species does not necessarily imply significant competition, and the presence of the food web may alter the strength of the interactions between plant species

while remaining neutral or positive towards the crop species. Usually, this means having a primary regulator that substantially harms the weed, but ignores or even benefits the crop, and secondary regulators that lack sufficient control on the primary regulators of the weeds. The selection of the crop itself becomes a key biological control strategy. If the species is a good competitor for light, water and nutrients, then the crop can constrain the competition. The use of reduced tillage and no-tillage agriculture as a necessary management tool to increase carbon storage and reduce erosion and nutrient losses (c.f., Montgomery 2007) has the unintended (but almost inevitable) consequence of increasing weeds in croplands. Thus, the need to develop sustainable weed control in these systems has intensified with soil conservation efforts.

11.4 Biological Control in Grasslands and Rangelands

Native grasslands and shrublands and areas with planted grasses for hay production or for direct consumption by livestock (cattle, sheep, goats) represents a variation on the conceptual model shown in Fig. 11.2. The 'crop' is now the desirable forage species (often multiple species), and the weeds are now those species not desired by livestock. By using a mix of livestock (some combination of cattle, goats, and sheep, either in some rotational scheme or together), such systems may often lack species considered to be weeds, and the livestock function as generalists, controlling all species on site. Thus, the goal is now to maximize productivity of a selected group of the primary regulators of the system. Ironically, social rather than scientific constraints seem to preclude this practice in some societies. In the western United

States for example, some cattleman appear to prefer to kill weeds themselves rather than grazing with sheep or goats that could control the weeds while producing a potentially marketable product. The large economic losses attributed to weeds in rangelands might disappear overnight if a cultural shift occurred. However, the presence of secondary regulators in these systems (coyotes and, in limited areas, wolves), would demand further cultural shifts in the way that livestock are allowed to utilize these areas and graze without protective 'symbionts'.

A substantial number of weed problems in rangelands have been addressed successfully using classical biological control procedures. Well-known examples include the control of prickly pear cactus (*Opuntia* spp.) in numerous parts of the world (e.g., Zimmermann and Moran 1991), control of St. John's wort or Klamath weed (*Hypericum perforatum*) in North America (e.g., Harris et al. 1969), and more recently control of diffuse knapweed (*Centaurea diffusa*; (Seastedt et al. 2005; Myers et al. 2009) or leafy spurge in United States (e.g., Kalischuk et al. 2004). These are termed 'successes' in that the target species has been reduced to levels below an established economic or ecological threshold in most habitats. Other rangeland weed species, such as spotted knapweed (*Centaurea stoebe*), in North America (termed "the wicked weed of the West", by Alpers 2004), and cheatgrass, an Asian annual grass species that is globally abundant, are works in progress, but both may have potential control organisms already in place (e.g., Knochel and Seastedt 2010; Dooley and Beckstead 2010). Most weeds are weeds because of their ability to grow rapidly under high resource conditions, but also because of the absence of their native pathogens in these introduced landscapes (e.g., Blumenthal et al. 2009). Advances in our knowledge of pathogen use on weeds have been substantial over the last decade (Barton 2012) and are contributing to sustainable solutions for our grazing systems.

11.5 Advances in Conservation Biology Control

The recognition that temporal and spatial patterns of crops, pests, and their biological control agents influence the outcome of crop production has become an integral part of the larger biological control effort (Barbosa 1998, Altieri 1999, Landis et al. 2000). Timing of crop growth can influence pest abundance and pest control (articles in *Annals of Applied Biology*, virtual issue 2010). Enhancing the abundance of generalist predators of pests, often a consequence of creating a more heterogeneous agricultural landscape is considered an undervalued benefit of these systems (Mensah and Sequeira 2004). Maintaining the heterogeneity of the agricultural landscape can reduce costs and amounts of chemical use required to maintain productivity (Meehan et al. 2011). Maintaining habitats that generate the desirable trophic interactions is not a straightforward exercise (Greenstone et al. 2010; Ratnadass et al. 2012) and requires an expert systems approach of knowledgeable managers, but clearly is part of a sustainable solution that can enhance several ecosystem services (e.g., Pfiffner and Wyss 2004) while addressing food security issues. Finally, an emerging paradigm is the importance of maximizing local biodiversity to sustainably control pests (Gurr et al. 2012).

11.6 Identifying Risks, Non-target Effects, and Unintended Consequences of Biological Control

Risks associated the release of any living organism include the possibility that non-target hosts will be attacked or that other, negative unintended consequences will emerge (Cook et al. 1996, Louda et al. 2003). Concerns about these unintended consequences have slowed the rate of release of biological control agents in recent decades (e.g., Hajek et al. 2007). There's strong evidence that food web alterations almost always have indirect effects, and these will range the spectrum of negative, neutral, and positive with respect to the crop and nontarget species. The problems associated with unintended biocontrol releases appear to be relatively few in comparison to the gains made by their use (Hokkanen and Pimentel 1989), and this appears to be particularly true for those biocontrol agents released during the last decade of the previous century to date (Van Driesche et al. 2008). Except for expected changes in food web dynamics, I am unaware of any serious unintended consequences resulting from biological control releases over the last decade. The history of human civilization—which arguably began with agriculture—is a history of food web alterations. One would have to say that the net effect of these alterations has been beneficial to human societies, and, with care, future alterations of food webs may continue to provide net positive effects.

11.7 Summary Thoughts

Classical biological control of weeds and herbivore pests of crops is a mature science, but with global transport of new pests in conjunction with changes in growing season temperatures and precipitation, (and concurrent changes in plant and biological control phenologies) this science needs to expand to address these new threats. Opportunities afforded by direct use of endophytes and symbionts, or by using the biochemicals produced by endophytes and symbionts, appears to be a rapidly emerging and promising approach to reduce crop losses to pest consumers (Rosenblueth and Martinez-Romero 2006). As an ecosystem ecologist, I'm poorly trained to make evaluations about relative threats and benefits of proactive genetic engineering in crops or in controls of crop pests. However, a biotechnical approach to expand upon traditional organic methods identified by Pimentel et al. (2005) seems prudent (Martin et al. 2010). The search for genetic traits in plants, endophytes, mycorrhizae, and other microbes should not be discouraged as these traits have the potential to reduce pathogen and herbivore damage.

A discussion of the risks and benefits of incorporating biological control into the plant itself via genetically modified organism-style techniques is outside the scope of this chapter. However, the argument against genetic engineering is one that paralleled the argument Steven Gould used against the use of non-native plant species in human societies. He stated, "Speaking biologically, the only general defense that I can concoct…—and I regard this as no mean thing—lies in protection thus afforded by our overweening arrogance" (Gould 1997, p. 9). Approaches that fail to consider non-target effects or do not explore possible unintended consequences to existing trophic systems must be avoided and such risks minimized. However, the search for sustainable solutions to pest problems affecting food security that do not concurrently reduce supporting, regulating and cultural services provided by agroecosystems must be pursued and all scientifically credible approaches be given their due consideration. Trophic manipulations employing the diversity of biological control techniques discussed here have the potential to optimize the ecosystem services of lands whose primary function is to provide food security, and provide additional ecosystem services as well.

Acknowledgments The author acknowledges a series of grants from the USDA invasive plant programs that led to the author's affirmation of the value of biological control efforts in agriculture. I thank Meredith Chedsey for her help in editing and formatting this chapter.

References

Alpers, J. (2004). The wicked weed of the West. *Smithsonian Magazine, 35*(7), 33–36.

Altieri, M. A. (1999). The ecological role of biodiversity in agroecosystems. *Agriculture, Ecosystems & Environment, 74,* 19–31.

Altieri, M. A., Ponti, L., & Nicholls, C. I. (2012). Soil fertility, biodiversity and pest management. In G. M. Gurr, S. D. Wratten, W. E. Snyder, & D. M. Y. Read (Eds.)., *Biodiversity and Insect Pests: Key Issues for Sustainable Management* (pp. 72–84). Boston: Wiley-Blackwell.

Andrews, M., Cripps, M. G., & Edwards, G. R. (2011). The potential of beneficial microorganisms in agricultural systems. *Annals of Applied Biology, 160*(1), 1–5.

Backman, P. A., & Sikora, R. A. (2008). Endophytes: An emerging tool for biological control. *Biology Control, 46,* 1–3.

Barbosa, P. (1998). *Conservation Biological Control*. San Diego, California, USA: Academic.

Barton, J. (2012). Predictability of pathogen host range in classical biological control of weeds: An update. *Biology Control, 57,* 289–305.

Blumenthal, D., Mitchell, C. E., Pyšek, P., & Jarošík, V. (2009). Synergy between pathogen release and resource availability in plant invasion. *Proceedings of the National Academy of Sciences U S A, 106*(19), 7899–7904.

Chapin, F. S. III, Kofinas, G. P., & Folke, C. (2009). *Principles of Ecosystem Stewardship: Resilience-based Natural Resource Management in a Changing World*. New York: Springer.

Chapin, F. S. III, Power, M. E., Pickett, S. T. A., Freitag, A., Reynolds, J. A., Jackson, R. B., Lodge, D. M., Duke, C., Collins, S. L., Power, A. G., & Bartuska, A. (2011). Earth stewardship: Science for action to sustain the human-earth system. *Ecosphere, 2*(8), art89. doi:10.1890/ES11-00166.1.

Cook, R. J., Bruckart, W. L., Coulson, J. R., Goettel, M. S., Humber, R. A., Lumsden, R. D., Maddox, J. V., McManus, M. L., Moore, L., Meyer, S. E., Quimby, P. C., Stack, J. P., & Vaughn, J. L. (1996). Safety of microorganisms intended for pest and plant disease control: A framework for scientific evaluation. *Biology Control, 7,* 333–351.

Davies, K. G., & Spiegel, Y. (2011). *Biological Control of Plant-parasitic Nematodes, Building Coherence Between Microbial Ecology and Molecular Mechanisms*. Dordrecht, Netherlands: Springer.

DeBach, P., & Rosen, D. (1991). *Biological Control by Natural Enemies* (2nd ed.). Cambridge: Cambridge University Press.

Denoth, M., Frid, L., & Myers, J. H. (2002). Multiple agents in biological control: Improving the odds? *Biological Conservation, 24*, 20–30.

Dooley, S. R., & Beckstead, J. (2010). Characterizing the interaction between a fungal seed pathogen and a deleterious rhizobacterium for biological control of cheatgrass. *Biology Control, 53*, 197–203.

Eilenberg, J., Hajek, A., & Lomer, C. (2001). Suggestions for unifying the terminology in biological control. *Biology Control, 46*, 387–400.

Eilenberg, J., & Hokkanen, H. M. T. (2006). *An Ecological and Societal Approach to Biological Control*. Dordrecht, Netherlands: Springer.

Ellis, E. C., Goldewijk, K. K., Siebert, S., Lightman, D., & Ramankutty, N. (2010). Anthropogenic transformation of the biomes, 1700 to 2000. *Global Ecology Biogeography, 19*, 589–606.

Faria, M. R. D., & Wraight, S. P. (2007). Mycoinsecticides and Mycoacaricides: A comprehensive list with worldwide coverage and international classification of formulation types. *Biology Control, 43*, 237–256.

Garratt, M. P. D., Wright, D. J., & Leather, S. R. (2011). The effects of farming system and fertilizers on pests and natural enemies: A synthesis of current research. *Agriculture Ecosystem Environment, 141*, 261–270.

Gould, S. J. (1997). An evolutionary perspective on strengths, fallacies, and confusions in the concept of native plants. In J. Wolschke-Bulmahn (Ed.)., *Nature and Ideology: Natural Garden Design in the Twentieth Century* (Vol. 18, pp. 11–19). Washington, D.C.: Dumbarton Oaks Research Library and Collection.

Greenstone, M. H., Szendrei, Z., Payton, M. E., Rowley, D. L., Coudron, T. C., & Weber, D. C. (2010). Choosing natural enemies for conservation biological control: Use of the prey detectability half-life to rank key predators of Colorado potato beetle. *Entomology Experiment Application, 136*, 97–107.

Guo, B., Wang, Y., Sun, X., & Tang, K. (2008). Bioactive natural products from Endophytes: A review. *Application Biochemical Micrology, 44*(2), 153–158.

Gurr, G. M., Wratten, S. D., & Altieri, M. A. (2004). *Ecological Engineering for Pest Management: Advances in Habitat Manipulation for Arthropods*. Collingwood, Victoria, Australia: CSIRO Publishing.

Gurr, G. M., Wratten, S. D., Snyder, W. W., & Read, M. Y. (2012). *Biodiversity and Insect Pests: Key Issues for Sustainable Management*. Boston: Wiley-Blackwell.

Hajek, A. E., McManus, M. L., & Delalibera, I. (2007). A review of introductions of pathogens and nematodes for classical biological control of insects and mites. *Biology Control, 41*, 1–13.

Hajek, A. E., & Tobin, P. C. (2010). Micro-managing arthropod invasions: Eradication and control of invasive arthropods with microbes. *Biology Invasions, 12*, 2895–2912.

Harris, P., Peschken, D., & Milroy, J. (1969). The status of biological control of the weed *Hypericum perforatum* in British Columbia. *Canadian Entomologist, 101*, 1–15.

Hartley, S. E., & Gange, A. C. (2009). Impacts of plant symbiotic fungi on insect herbivores: Mutualism in a multitrophic context. *Annual Review of Entomology, 54*, 323–342.

Hokkanen, H. M. T., & Pimentel, D. (1989). New associations in biological control: Theory and practice. *Canadian Entomologist, 95*, 785–792.

Jonsson, M., Wratten, S. D., Landis, D. A., & Gurr, G. M. (2008). Recent advances in conservation biological control of arthropods by arthropods. *Biological Control, 45*, 172–175.

Kalischuk, A. R., Bourchier, R. S., & McClay, A. S. (2004). Post hoc assessment of an operational biocontrol program: Efficacy of the flea beetle *Aphthona lacertosa* Rosenhauer (Chrysomelidae: Coleoptera), an introduced biocontrol agent for leafy spurge. *Bioloigy Control, 29*, 418–426.

Karieva, P. (1999). Coevolutionary arms races: Is victory possible? *Proceedings of the National Academy of Sciences U S A, 96*, 8–10.

Knochel, D. G., & Seastedt, T. R. (2010). Field measurements of herbivory, resource limitation, and plant competition on the growth and reproduction of spotted knapweed (*Centaurea stoebe*). *Ecology Application, 20*, 1903–1912.

Landis, D., Wratten, S., & Gurr, G. M. (2000). Habitat management to conserve natural enemies of arthropod pests in agriculture. *Annual Review of Entomology, 45*, 175–201.

Lodewyckx, C., Vangronsveld, J., Porteous, F., Moore, E. R. B., Taghavi, S., Mezgeay, M., & van der Lelie, D. (2002). Endophytic bacteria and their potential applications. *CRC Critical Review Plant Science, 21,* 583–606.

Lord, J. C. (2005). Mechnikoff to Monsanto and beyond: The path of microbial control. *The Journal of Invertebrate Pathology, 89,* 19–29.

Louda, S. M., Pemberton, R. W., Johnson, M. T., & Follett, P. A. (2003). Nontarget effects—The Achilles heel of biological control? Retrospective analyses to reduce risk associated with biocontrol introductions. *Annual Review of Entomology, 48,* 366–396.

Martin, H., Burgess, E. P. J., Masarik, M., Kramer, K. J., Beklova, M., Adam, V., & Kizek, R. (2010). Avidin and plant biotechnology to control pests. In E. Lichtfouse (Ed.)., *Genetic Engineering, Biofertilisation, Soil Quality and Organic Farming Sustainable Agriculture Reviews,* (*Vol.* 4, pp. 1–21). Dordrecht, Netherlands: Springer. http://appius.claudius.free.fr/Download/Agribio/divers/Genetic%20Engineering,%20Biofertilisation,%20Soil%20Quality%20 and%20Organic%20Farming.pdf. Accessed 9 Nov 2012.

Matson, P. A., Parton, W. J., Power, A. G., & Swift, M. J. (1997). Agricultural intensification and ecosystem properties. *Science, 277,* 504–509.

McEvoy, P. B., & Coombs, E. M. (2000). Why things bite back: Unintended consequences of biological weed control. In P. A. Follett & J. J. Duan (Eds.)., *Nontarget Effects of Biological Control* (pp. 167–194). Boston: Kluwer Academic.

McFadyen, R. E. (1998). Biological control of weeds. *Annual Review Entomology, 43,* 369–393.

Meehan, T. D., Werling, B. P., Landis, D. A., & Gratton, C. (2011). Agricultural landscape simplification and insecticide use in the Midwestern United States. *Proceedings of the National Academy of Sciences U S A, 108,* 11500–11505.

Mensah, R. K., & Sequeira, R. V. (2004). Habitat manipulation for insect pest management in cropping systems. In G. M. Gurr., S. D. Wratten, & M. A. Altieri (Eds.)., *Ecological Engineering for Pest Management* (pp. 187–197). Collingwood, Victoria, Australia: CSIRO Publishing.

Montgomery, D. R. (2007). Soil erosion and agricultural sustainability. *Proceedings of the National Academy of Sciences U S A, 104,* 13268–13272.

Muniappan, R., Reddy, G. V. P., & Raman, A. (2009). *Biological Control of Tropical Weeds Using Arthropods.* Cambridge: Cambridge Press.

Myers, J. H., Jackson, C., Quinn, H., White, S. R., & Cory, J. S. (2009). Successful biological control of diffuse knapweed, *Centaurea diffusa,* in British Columbia, Canada. *Biology Control, 50,* 66–72.

National Research Council (NRC) Board on Agriculture. (1996). *Ecologically Based Pest Management: New Solutions for a New Century.* Washington, D.C.: National Academy Press.

Pfiffner, L., & Wyss, E. (2004). Use of sown wildflower strips to enhance natural enemies of agricultural pests. In G. M. Gurr., S. D. Wratten, & M. A. Altieri (Eds.)., *Ecological Engineering for Pest Management: Advances in Habitat Manipulation for Arthropods* (pp. 167–188). Collingwood, Victoria, Australia: CSIRO Publishing.

Pimentel, D., Hepperly, P., Hanson, J., Douds, D., & Seidel, R. (2005). Environmental, energetic, and economic comparisons of organic and conventional farming systems. *Bioscience, 55,* 573–582.

Ratnadass, A., Fernades, P., Avelina, J., & Habib, R. (2012). Plant species diversity for sustainable management of crop pests and diseases in agroecosystems: A review. *Agronomy for Sustainable Development, 32,* 273–303.

Rosenblueth, M., & Martinez-Romero, E. (2006). Bacterial endophytes and their interactions with hosts. *Molecular Plant-Microbe Interactions, 19,* 827–837.

Schulz, B., Boyle, C., Draeger, S., Rommert, A. K., & Krohn, K. (2002). Endophytic fungi: A source of novel biologically active secondary metabolites. *Mycological Research, 106,* 996–1004.

Seastedt, T. R., Suding, K. N., & LeJeune, K. D. (2005). Understanding invasions: The rise and fall of diffuse knapweed (*Centaurea diffusa*) in North America. In S. Inderjit (Ed.), *Invasive Plants: Ecological and Agricultural Aspects* (pp. 129–139). Basal, Switzerland: Birkhauser-Verlag AG.

Simon, S., Bouvier, J. C., Debras, J. F., & Sauphanor, B. (2010). Biodiversity and pest management in orchard systems: A review. *Agronomy for Sustainable Development, 30,* 139–152. doi: 10.1051/agro/2009013.

Smagghe, G., & Diaz, I. (2012). *Arthropod-plant Interactions: Novel Insights and Approaches for IPM.* Dordrecht, Netherlands: Springer.

Stirling, G. R. (2011). Biological control of plant-parasitic nematodes: An ecological perspective, a review of progress and opportunities for further research. *Programm Biology Control, 11,* 1–38.

Swift, M. J., Izac, A. M. N., & van Noordwijk, M. (2004). Biodiversity and ecosystem services in agricultural landscapes—Are we asking the right questions? *Agriculture Ecosystem Environment, 104,* 113–134.

Tikhonovich, I. A., & Provorov, N. A. (2011). Microbiology is the basis of sustainable agriculture: An opinion. *Annals of Applied Biology, 159,* 155–168.

Van Driesche, R. G., Carruthers, R. I., Center, T., Hoddle, M. S., Hough-Goldstein, J., Morin, L., Smith, L., Wagner, D. L., Blossey, B., Brancatini, V., Casagrande, R., Causton, C. E., Coetzee, J. A., Cuda, J., Ding, J., Fowler, S. V., Frank, J. H., Fuester, R., Goolsby, J., Grodowitz, M. et al. (2010). Classical biological control for the protection of natural ecosystems. *Biological Control, 54,* S2–S33. http://ag.udel.edu/enwc/research/biocontrol/pdf/Van%20Driesche%20 et%20al.%202010.pdf. Accessed 9 Aug 2012.

Van Driesche, R. G., Hoddle, M., & Center, T. (2008). *Control of Pests and Weeds by Natural Enemies: An Introduction to Biological Control.* Malden, Massachusetts, USA: Blackwell Publishing.

Vos, C., Geerinckx, K., Mkandawire, R., Panis, B., DeWaele, D., & Elsen, A. (2012). Arbuscular mycorrhizal fungi affect both penetration and further life stage development of root-knot nematodes in tomato. *Mycorrhiza, 22,* 157–163.

Whipps, J. M. (2001). Microbial interactions and biocontrol in the rhizosphere. *Journal of Experiment Botany, 52,* 487 511. doi:10.1093/jexbot/52.suppl_1.487.

Zimmermann, H. G., & Moran, V. C. (1991). Biological control of prickly pear, *Opuntia ficus-indica* (Cactaceae) in South Africa. *Agriculture, Ecosystem and Environment, 37,* 29–35.

Chapter 12
Herbicide Resistant Weeds

Ian Heap

Contents

12.1	Introduction	282
12.2	Evolution of Resistance	283
12.3	Mechanisms of Resistance	284
12.4	The Occurrence of Herbicide Resistant Weeds	284
	12.4.1 ALS Inhibitor (B/2) Resistant Weeds	285
	12.4.2 Triazine (C1/5) Resistant Weeds	286
	12.4.3 ACCase Inhibitor (A/1) Resistant Weeds	288
	12.4.4 Synthetic Auxins (O/4)	289
	12.4.5 Glyphosate (G/9) Resistant Weeds	289
	12.4.6 Weed Resistance to Other Herbicides	290
12.5	Worst Herbicide Resistant Weeds	291
12.6	Herbicide Resistant Weeds in Major Crops	292
	12.6.1 Wheat	292
	12.6.2 Corn	294
	12.6.3 Soybean	294
	12.6.4 Rice	296
	12.6.5 Perennial Crops	296
	12.6.6 Non-Crop	296
12.7	Herbicide Resistant Crops	297
12.8	Integrated Weed Management is Herbicide Resistance Management	298
12.9	Summary	299
References		300

Abstract The International Survey of Herbicide-Resistant Weeds (www.weedscience.org) reports 388 unique cases (species x site of action) of herbicide-resistant weeds globally, with 210 species. Weeds have evolved resistance to 21 of the 25 known herbicide sites of action and to 152 different herbicides. The ALS inhibitors (126 resistant species) are most prone to resistance, followed by the triazines (69 species), and the ACCase inhibitors (42 species). Herbicide-resistant weeds first

---

I. Heap (✉)
Director of the International Survey of Herbicide-Resistant Weeds,
PO Box 1365, Corvallis OR, 97339 USA.
e-mail: IanHeap@weedscience.org

D. Pimentel, R. Peshin (eds.), *Integrated Pest Management*,
DOI 10.1007/978-94-007-7796-5_12,
© Springer Science+Business Media Dordrecht 2014

became problematic in the USA and Europe in the 1970s and early 1980s due to the repeated applications of atrazine and simazine in maize crops. Growers turned to the ALS and ACCase inhibitor herbicides in the 1980s and 1990s to control triazine-resistant weeds and then to glyphosate-resistant crops in the mid 1990s in part to control ALS inhibitor, ACCase inhibitor, and triazine-resistant weeds. The massive area treated with glyphosate alone in glyphosate-resistant crops has led to a rapid increase in the evolution of glyphosate-resistant weeds. Glyphosate-resistant weeds are found in 23 species and 18 countries and they now dominate herbicide-resistance research, but have not yet surpassed the economic damage caused by ALS inhibitor and ACCase inhibitor resistant weeds. *Lolium rigidum* remains the world's worst herbicide-resistant weed (12 countries, 11 sites of action, 9 cropping regimes, over 2 million hectares) followed by *Amaranthus palmeri, Conyza canadensis, Avena fatua, Amaranthus tuberculatus,* and *Echinochloa crus-galli*. In the years ahead multiple-resistance in weeds combined with the decline in the discovery of novel herbicide modes of action present the greatest threat to sustained weed control in agronomic crops. The discovery of new herbicide sites of action and new herbicide-resistant crop traits will play a major role in weed control in the future however growers must make the transition to integrated weed management that utilizes all economically available weed control techniques.

Keywords Herbicide-resistance · Resistant weeds · Resistance management · ALS inhibitors · ACCase inhibitors · Glyphosate · Survey · Integrated weed management, Herbicide tolerant crops · Herbicides

12.1 Introduction

Weeds impact crop production through direct competition for nutrients, moisture and light, and if left uncontrolled weeds can cause over 80% crop yield loss (Oerke, 2002). Prior to the introduction of modern herbicides man relied upon hand weeding, hoeing, tillage, crop rotations, cover crops, crop management (crop competition, seeding rates and times, row spacing, nutrition etc.), biological controls and burning as the primary methods of weed control. The first modern herbicides, the synthetic auxins (2,4-D, MCPA), were developed during world war II and first marketed in 1944 for broadleaf weed control in cereals. Their success spawned a new era in weed control and herbicide discovery. In the last 65 years agricultural chemical companies have brought more than 300 herbicide active ingredients to the market. Herbicides became the most reliable and least expensive weed control method in crop production and they are a major contributor to the dramatic increases in crop yields achieved over the last 65 years. Although highly successful, herbicides also face challenges to do with their safety to humans and the environment but their biggest challenge is that of weeds evolving resistance to them. Scientists foresaw the potential of herbicide-resistant weeds (Harper 1956) however it took until 1970 before the first well documented report of a herbicide-resistant weed. Until recently growers have been fortunate enough to have a steady supply of new herbicides to deal with the inevitable appearance of herbicide-resistant weeds. In the last few decades this steady supply of new herbicides has ceased, and growers are now faced with the harsh reality that herbicide-resistance can no longer be dealt with as it has in the past. Modern crop production is dependent on effective herbicides and now, more than ever, the sustained use of herbicides is threatened by herbicide-resistant weeds. This chapter aims to introduce herbicide-resistance, report its current status, and provide practical management strategies that will help stave off resistance until new herbicides and weed control practices arrive.

12.2 Evolution of Resistance

Weed resistance is the evolved capacity of a previously herbicide-susceptible weed population to survive a herbicide and complete its life cycle when the herbicide is used at its normal rate in an agricultural situation. Herbicide-resistance is a normal and predictable outcome of natural selection. Rare mutations that confer herbicide resistance exist in weed populations prior to any herbicide exposure and they increase in proportion over time after each herbicide application until they predominate at which time the population is called resistant. There are many factors that influence how long it takes for a weed population to evolve resistance to herbicide applications. The initial frequency of herbicide resistant mutations found in a weed population is dependent on the weed species and the herbicide mechanism of action. Some weed species, such as *Lolium rigidum* and *Amaranthus tuberculatus*, have a great propensity to evolve resistance partly due to their innate genetic variability. Herbicides also vary dramatically in their risk level for resistance. For some herbicides, like the ALS inhibitors and ACCase inhibitors, there are numerous mutations that can confer target site resistance, making these herbicides very prone to resistance. For other herbicides, like the synthetic auxins, or glyphosate, there are few target site mutations that confer resistance, making them relatively low risk herbicides for resistance. Resistance can also occur through the quantitative selection of multiple low level resistance genes (polygenic resistance) resulting in a progressive shift towards resistance in the population as a whole. These low level resistance genes may confer enhanced metabolism, decreased translocation, sequestration, and gene amplification and they are the cause of many of the cases of glyphosate resistance. Other key factors that influence the rate of evolution of resistance are the selection pressure (frequency and efficacy of herbicide use), the residual activity of the herbicide, the genetic basis of resistance (degree of dominance of the resistance trait and the breeding system of the weed), how prolific the weed is at producing seed, seed longevity in the soil, and the fitness of the resistance trait. Of these factors it is the selection pressure (in particular the frequency of herbicide use) that we can influence the most, and decreasing the selection pressure is the basis of herbicide resistance management strategies.

12.3 Mechanisms of Resistance

There are five primary mechanisms of herbicide resistance.

1. **Target site resistance** is the result of a modification of the herbicide binding site (usually an enzyme), which precludes a herbicide from effectively binding. If the herbicide cannot bind to the enzyme then it does not inhibit the enzyme and the plant survives. Target site resistance is the most common resistance mechanism. Most but not all cases of resistance to ALS inhibitor, ACCase inhibitor, dinitroanaline, and triazine herbicides are due to modifications of the site of action of the herbicide.
2. **Enhanced metabolism** occurs when the plant has the ability to degrade the herbicide before it can seriously affect the plant.
3. **Decreased absorption and/or translocation** can cause resistance because herbicide movement is restricted and the herbicide does not reach its site of action in sufficient concentration to cause death.
4. **Sequestration** of a herbicide into vacuoles or onto cell walls can keep the herbicide from the site of action resulting in resistance.
5. **Gene amplification/over-expression** is the most recently identified herbicide resistance mechanism, and causes resistance by increasing the production of the target enzyme, effectively diluting the herbicide in relation to the target site.

From a herbicide resistance management perspective it is important to note that weeds can exhibit cross-resistance and multiple resistance.

Cross-resistance occurs where a single resistance mechanism confers resistance to several herbicides. The most common type of cross-resistance is target site cross resistance, where an altered target site (enzyme) confers resistance to many or all of the herbicides that inhibit the same enzyme.

Multiple resistance occurs when two or more resistance mechanisms occur within the same plant, often due to sequential selection by different herbicide modes of action. A diagnosis of multiple resistance requires knowledge of the resistance mechanisms (Heap and LeBarron 2001).

12.4 The Occurrence of Herbicide Resistant Weeds

The data used in the tables and figures of this chapter come from the International Survey of Herbicide-Resistant Weeds website which is located at www.weedscience.com (Heap 2012). As of August 2012 the survey recorded 388 unique cases of resistant weeds. A unique case refers to the first instance of a weed species evolving resistance to one or more herbicides in a herbicide group (herbicides that act on the same site of action). There are 210 weed species (123 dicots and 87 monocots) that have evolved resistance to one or more herbicides. The rate of discovery of new types of herbicide resistant weeds is remarkably consistent at about 11 new cases per year (Fig. 12.1). Weeds have evolved resistance to 21 herbicide groups (Table 12.1) in

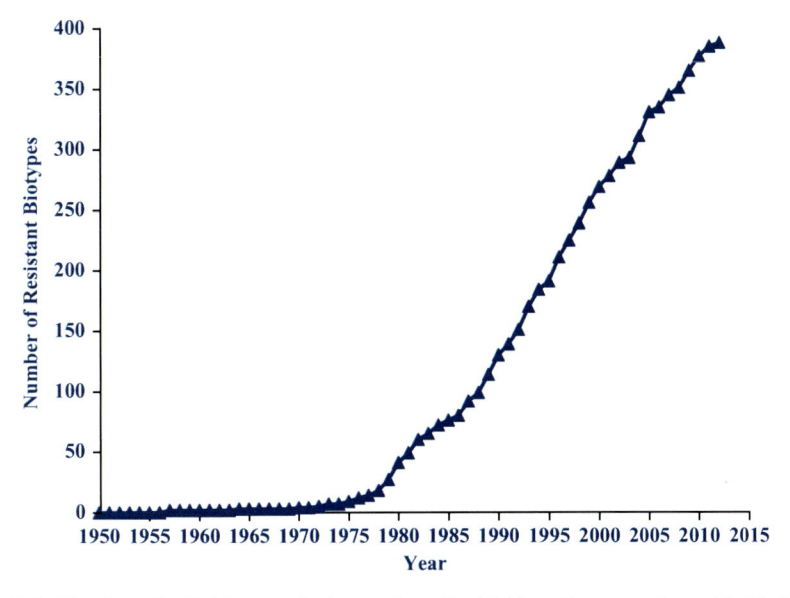

Fig. 12.1 The chronological increase in the number of herbicide-resistant weeds worldwide. Data accessed from the www.weedscience.com website on August 10, 2012. (Heap 2012)

61 countries (Table 12.2) and 152 herbicide active ingredients. Fig. 12.2 shows the clear difference in the propensity of weeds to evolve resistance to different herbicide groups.

12.4.1 ALS Inhibitor (B/2) Resistant Weeds

ALS inhibitor herbicides prevent the biosynthesis of the branched-chain amino acids (valine, leucine, and isoleucine) by inhibiting the acetolactate synthase enzyme in plants (Ray 1984). The ALS inhibitor herbicides are the highest risk herbicide group for the development of resistance (Fig. 12.2), with the first cases of resistance being identified only four years after their introduction. The first reported case of ALS inhibitor resistance was metabolic resistance to chlorsulfuron in *Lolium rigidum* in Australia (Heap and Knight 1982, 1986), followed by target site resistance in 1987 in *Lactuca serriola* (Mallory-Smith et al. 1990) in the USA. There are now 126 weed species that have evolved resistance to the ALS inhibitors (Table 12.1). There are a number of reasons why ALS inhibitors have selected more resistant weeds than any other herbicide group. One major factor is that there are more ALS inhibitor herbicides (over 55 actives in 5 chemical classes, twice as many as any other herbicide group) and they are used on a greater area annually than any other herbicide group. Another is that ALS inhibitors exert a strong selection pressure because they have very high activity on sensitive biotypes and they also have soil residual activity. However these factors alone do not account for the nearly 5 new weed species identified with ALS inhibitor resistance each year (Fig. 12.2). The

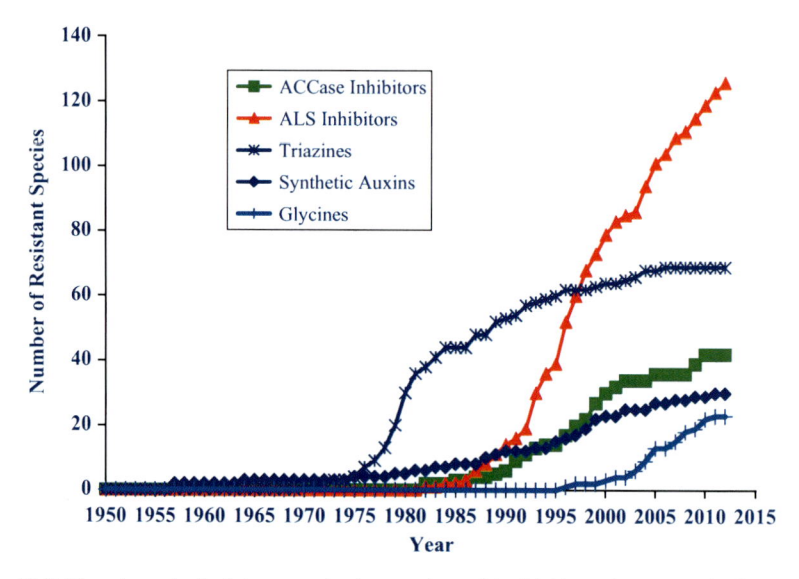

Fig. 12.2 The chronological increase in the number of herbicide-resistant weeds for several herbicide classes. Data accessed from the www.weedscience.com website on August 10, 2012. (Heap 2012)

Achilles heel of the ALS inhibitors is the ability of the target enzyme (acetolactate synthase) to undergo many mutations but still remain functional, and most cases of ALS inhibitor resistance are due to an alteration in this enzyme. At present there are eight amino acids (Ala 122, Pro 197, Ala 205, A sp 376, Arg 377, Trp 574, Ser 653, and Gly 654) on the ALS gene that resistance-conferring substitutions have been identified (www.weedscience.com ALS inhibitor mutation table Heap, 2012). There are many substitutions that can occur for each of these amino acids, in fact for Pro 197 there 9 different substitutions (Ala, Arg, Asn, Gln, His, Ile, Leu, Ser, and Thr) shown to cause resistance, and there are 24 substitutions in total that cause resistance. It is because there are so many variations in the ALS gene that confer resistance to the ALS enzyme that the ALS inhibitor herbicides are so prone to resistance. There are 5 different classes of ALS inhibitor herbicides and there are different patterns of cross-resistance to these classes depending on the particular mutation. This presents a problem to the grower, as without identifying the mutation that they are dealing with they do not know which classes of ALS inhibitor herbicides may still be effective on their particular resistant population.

12.4.2 Triazine (C1/5) Resistant Weeds

The triazines (PSII inhibitors) became heavily used in corn production in the USA and Europe in the 1960s and 1970s. Twenty six PSII inhibitor herbicides have been commercialized belonging to 6 chemical classes. Triazine herbicides inhibit

Table 12.1 The occurrence of herbicide-resistant weed species to herbicide groups

#	Herbicide group	HRAC/WSSA	Example herbicide	Dicots	Monocots	Total
1	ALS inhibitors	B/2	Chlorsulfuron	77	49	126
2	Photosystem II inhibitors	C1/5	Atrazine	47	22	69
3	ACCase inhibitors	A/1	Diclofop-methyl	0	42	42
4	Synthetic Auxins	O/4	2,4-D	23	7	30
5	Bipyridiliums	D/22	Paraquat	17	9	26
6	Glycines	G/9	Glyphosate	10	13	23
7	Ureas and amides	C2/7	Chlorotoluron	8	14	22
8	Dinitroanilines and others	K1/3	Trifluralin	2	9	11
9	Thiocarbamates and others	N/8	Triallate	0	8	8
10	PPO inhibitors	E/14	Oxyfluorfen	5	0	5
11	Triazoles, ureas, isoxazolidiones	F3/11	Amitrole	1	4	5
12	Nitriles and others	C3/6	Bromoxynil	3	1	4
13	Chloroacetamides and others	K3/15	Butachlor	0	4	4
14	Carotenoid biosynthesis inhibitors	F1/12	Flurtamone	2	1	3
15	Glutamine synthase inhibitors	H/10	Glufosinate-ammonium	0	2	2
16	Arylaminopropionic acids	Z/25	Flamprop-methyl	0	2	2
17	Unknown	Z/27	(chloro)—flurenol	0	2	2
18	4-HPPD inhibitors	F2/27	Isoxaflutole	1	0	1
19	Mitosis inhibitors	K2/23	Propham	0	1	1
20	Cellulose inhibitors	L/27	Dichlobenil	0	1	1
21	Organoarsenicals	Z/17	MSMA	1	0	1

HRAC Group—Herbicide grouping system developed by the Herbicide Resistance Action Committee
WSSA Group—Herbicide grouping system developed by the Weed Science Society of America
Data accessed from the www.weedscience.com website on August 10, 2012. (Heap 2012)

photosynthesis by competing with plastoquinone at its binding site which is located on the D1 protein in the photosystem two complex in chloroplasts (Gronwald 1994). The first well documented case of herbicide resistance was that of triazine-resistant *Senecio vulgaris* that appeared as a result of repeated use of simazine in a plant nursery in Washington State and was reported by Ryan (1970). The case itself was of little economic significance however it did alert weed researchers working in corn to the potential of triazine-resistant weeds and shortly thereafter in the mid 1970s there was an explosion of research documenting triazine-resistant weeds in corn production in both the United States and in Europe. There are some cases of triazine resistance due to enhanced metabolism (Gronwald 1997) however the majority of triazine resistance cases are due to a mutation ($Ser_2 64$ to Gly) in the psbAgene, which codes for the Dl protein and reduces the binding of triazine herbicides to the thylakoid membrane in chloroplasts. There are currently 69 documented cases of triazine resistance and the majority of them were identified prior to 1995 despite the continued widespread use of herbicides like atrazine, simazine, and metribuzin today. *Amaranthus*, *Chenopodium* and *Solanum sp.* are particularly prone to evolve triazine resistance and infest large areas in corn producing regions of the USA and Europe.

Table 12.2 The occurrence of herbicide-resistant weeds in countries

Rank	Country	# Resistant Weeds	Rank	Country	# Resistant Weeds
1	USA	140	32	Costa Rica	5
2	Australia	61	33	Mexico	5
3	Canada	58	34	Norway	5
4	France	33	35	Thailand	5
5	Spain	33	36	Bulgaria	4
6	China	30	37	India	3
7	Italy	29	38	Philippines	3
8	Brazil	27	39	Portugal	3
9	Israel	27	40	Austria	2
10	Germany	26	41	Paraguay	2
11	United Kingdom	24	42	Sri Lanka	2
12	Belgium	18	43	Sweden	2
13	Japan	18	44	Cyprus	1
14	Malaysia	17	45	Ecuador	1
15	Czech Republic	16	46	Egypt	1
16	Chile	15	47	El Salvador	1
17	Turkey	15	48	Ethiopia	1
18	Poland	14	49	Fiji	1
19	South Africa	14	50	Guatemala	1
20	Switzerland	14	51	Honduras	1
21	South Korea	12	52	Hungary	1
22	Iran	11	53	Indonesia	1
23	New Zealand	10	54	Ireland	1
24	Argentina	9	55	Kenya	1
25	Venezuela	9	56	Nicaragua	1
26	Denmark	8	57	Panama	1
27	Bolivia	7	58	Saudi Arabia	1
28	Greece	7	59	Slovenia	1
29	The Netherlands	7	60	Taiwan	1
30	Colombia	6	61	Tunisia	1
31	Yugoslavia	6			

Data accessed from the www.weedscience.com website on August 10, 2012. (Heap 2012)

12.4.3 ACCase Inhibitor (A/1) Resistant Weeds

ACCase inhibitor herbicides target the enzyme acetyl co-enzyme A carboxylase which catalyzes the first step in fatty acid biosynthesisin grasses (Buchanan et al. 2000). ACCase inhibitor herbicides came into widespread use in the 1980s in both broadleaf and cereal crops and there are now 20 active ingredients in 3 chemical classes. Although cases of enhanced metabolism and over expression of the ACCase enzyme have been identified, the overwhelming cause of resistance to this herbicide group is due to an altered, insensitive form of the ACCase enzyme (Brown et al., 2002). There are now 42 monocot weed species with resistance to the ACCase inhibitors (Table 12.1). ACCase inhibitors are prone to resistance for the same reason that the ALS inhibitors are prone to resistance, there are many mutations that prevent ACCase inhibitors from binding to the ACCase enzyme (De´lye et al., 2005,

Liu et al., 2007, Hochberg et al., 2009). Even though there are more triazine resistant species than ACCase inhibitor resistant species the economic impact of ACCase inhibitor resistant weeds is far greater than that of triazine-resistant weeds. ACCase inhibitor resistant weeds are widespread wherever cereal crops (wheat, barley, etc.) are grown and continue to increase in area and severity. ACCase inhibitor resistant *Avena, Lolium, Phalaris, Setaria* and *Alopecurus sp.* infest over 20 million hectares globally and contribute significantly to reductions in crop yields. These grass species are particularly problematic because they have not only evolved resistance to ACCase inhibitors but to most of the effective grass herbicides available to wheat producers, leaving growers with dwindling options to manage them.

12.4.4 Synthetic Auxins (O/4)

Synthetic auxins (22 commercialized actives in 5 chemical classes) were the first herbicides to be used on a massive scale and continue to be among the most widely used herbicides today. Synthetic auxins mimic the natural plant hormone indole-3-acetic acid (IAA) and affect several aspects of plant growth, including cell division, elongation, and differentiation, resulting in physiological and morphological abnormalities, including severe epinasty, hypertrophy, faciation of the crown and leaf petioles, and premature abscission of leaves (Sterling and Hall 1997). Despite being used for longer and over a greater area than any other herbicide group there are only 30 weed species that have evolved resistance to the synthetic auxins, and of these 30 only a handful are more than scientific curiosities. We have not seen widespread resistance to the synthetic auxins the way we have with triazine, ACCase inhibitor, ALS inhibitor or even glyphosate-resistant weeds. Four of the 30 cases are not classic synthetic auxin resistance, they are grasses that have become resistant to quinclorac (an unusual synthetic auxin that acts on grasses through a novel mechanism), and in those cases we do see widespread resistance. Of the other 26 cases, only dicamba resistant *Kochia scoparia* in the USA and 2,4-D resistant *Raphanus raphanistrum* in Australia infest more than 1,000 hectares. This is intriguing and points to the synthetic auxins as being one of the least prone to resistance of any of the widely used herbicide groups. The agricultural chemical industry was fortunate to initially discover a herbicide group that has a very low risk for resistance.

12.4.5 Glyphosate (G/9) Resistant Weeds

Glyphosate is the most widely used herbicide in the world, has been in commercial use since 1974 and is an extremely valuable resource (Baylis 2000, Woodburn 2000). Glyphosate inhibits the chloroplast enzyme 5-enolpyruvylshikimate-3-phosphate synthase (EPSPS) which disrupts the shikimate pathway resulting in the inhibition of aromatic amino acid production. Weeds have evolved resistance to glyphosate through decreased translocation/sequestration (Feng et al. 2004), target site mutations (Kaundun et al. 2008), and gene amplification (Gaines et al. 2010). The first case of glyphosate resistance was that of *Lolium rigidum* from an apple orchard in Australia in 1996, coincidentally the same year that the first Roundup Ready crop was commercialized. There are now 23 glyphosate resistant weeds found in 18 countries (Table 12.3) with half of them evolving resistance in Roundup Ready cropping systems and the other half from orchards and non-crop situations. This statistic belies the extent of the problem of glyphosate-resistant weeds in Roundup Ready crops vs the other cases of glyphosate-resistant weeds. On an area basis the survey reports that the 11 glyphosate-resistant weeds in Roundup Ready crops account for 98% of the area infested with glyphosate-resistant weeds globally. Whilst glyphosate-resistant *Conyza canadensis* is the most widespread glyphosate-resistant weed it is easily managed with synthetic auxins and other herbicides. It is the *Amaranthus* species (*Amaranthus palmeri* and *Amaranthus tuberculatus*) that present the greatest economic threat of any of the glyphosate-resistant weeds. Glyphosate-resistant *Amaranthus palmeri* was first identified in cotton in Georgia in 2005 and now infests 12 states primarily in cotton and soybean in the southern cropping belt of the USA. Similarly glyphosate-resistant *Amaranthus tuberculatus* was first identified in 2005 in Missouri and now infests 11 states, primarily in corn and soybean in the mid-west corn/soybean cropping belt in the USA. Other economically important glyphosate-resistant weeds in the USA include *Ambrosia sp.*, *Kochia scoparia*, and *Sorghum halepense*. *Kochia scoparia* threatens to increase rapidly because of its efficient method of seed dispersal (tumbleweed), a key reason that helped it rapidly spread ALS inhibitor resistance in western states of the USA in the 1990s. In South America the most serious cases of glyphosate resistance are *Sorghum halepense* (Argentina) and *Digitaria insularis* (Paraguay and Brazil) in Roundup Ready soybeans. Glyphosate-resistant *Conyza sp.* are also prevalent in Brazilian soybean crops (Table 12.3). It is important to note that even though many weeds have evolved resistance to glyphosate, and there will be many more to come, Roundup Ready crops and glyphosate will continue to be used extensively for at least another decade because glyphosate provides economic broad spectrum weed control. Growers will add supplemental herbicide groups to control glyphosate-resistant weeds just as they did with the continued use of atrazine after the appearance of triazine-resistant weeds. An important strategy in the management of glyphosate-resistant weeds is the use of other herbicides and the PPO inhibitors and HPPD inhibitors are of major utility when used in rotation/sequence with glyphosate in the mid-west USA. Unfortunately we have already seen the appearance of 5 weed species with resistance to PPO inhibitors, two of them in corn/soybean rotations in the USA (*Amaranthus tuberculatus*, and *Ambrosia artemissiifolia*), as well as HPPD inhibitor resistant *Amaranthus tuberculatus*.

12.4.6 Weed Resistance to Other Herbicides

Other groups with significant resistant weed problems are the bipyridiliums (26 species), phenylureas and phenylamides (22 species), and the dinitroanalines (11 species). Although these groups are still used they are not as prominent as they once were, and few new cases of resistance to them have been identified in the last 10 years.

Table 12.3 The occurrence of glyphosate-resistant weeds worldwide

#	Species	First Year	Country and Year (states are in order of first recorded case)
1	*Amaranthus palmeri*	2005	USA (2005—GA, NC, AR, NM, AL, MS, MO, TN, IL, LA, MI, VA)
2	*Amaranthus tuberculatus*	2005	USA (2005—MO, IL, KS, MN, IN, IA, MS, ND, SD, OK, TN)
3	*Ambrosia artemisiifolia*	2004	USA (2004—AR, MO, OH, IN, KS, ND, SD, MN)
4	*Ambrosia trifida*	2004	USA (2004—OH, AR, IN, KS, MN, TN, IA, MO, MS, NE, WI), Canada (2008—ON)
5	*Bromus diandrus*	2011	Australia (2011—SA)
6	*Chloris truncata*	2010	Australia (2010—NSW, QLD, SA)
7	*Conyza bonariensis*	2003	South Africa (2003), Spain (2004), Brazil (2005), Israel (2005), Columbia (2006), USA (2007—CA), Greece (2010), Portugal (2010)
8	*Conyza canadensis*	2000	USA (2000—DE, KY, TN, IN, MD, MO, NJ, OH, AR, MS, NC, PA, CA, IL, KS, VA, NE, MI, OK, SD, IA), Brazil (2005), China (2006), Spain (2006), Czech Republic (2007), Canada (2010—ON), Poland (2010), Italy (2011)
9	*Conyza sumatrensis*	2009	Spain (2009), Brazil (2010)
10	*Cynodon hirsutus*	2008	Argentina (2008)
11	*Digitaria insularis*	2005	Paraguay (2005), Brazil (2008)
12	*Echinochloa colona*	2007	Australia (2007—NSW, QLD, WA), USA (2008—CA), Argentina (2009)
13	*Eleusine indica*	1997	Malaysia (1997), Colombia (2006), China (2010), USA (2010—MS, TN)
14	*Kochia scoparia*	2007	USA (2007—KS, SD, NE), Canada (2012—AB)
15	*Leptochloa virgata*	2010	Mexico (2010)
16	*Lolium multiflorum*	2001	Chile (2001), Brazil (2003), USA (2004—OR, MI, AR), Spain (2006), Argentina (2007)
17	*Lolium perenne*	2008	Argentina (2008)
18	*Lolium rigidum*	1996	Australia (1996—VIC, NSW, SA, WA) USA (1998—CA), South Africa (2001), France (2005), Spain (2006), Israel (2007), Italy (2007)
19	*Parthenium hysterophorus*	2004	Colombia (2004)
20	*Plantago lanceolata*	2003	South Africa (2003)
21	*Poa annua*	2010	USA (2010—MO, TN)
22	*Sorghum halepense*	2005	Argentina (2005), USA (2007—AR, MS, LA)
23	*Urochloa panicoides*	2008	Australia (2008—NSW)

Data accessed from the www.weedscience.com website on August 10, 2012. (Heap 2012)

12.5 Worst Herbicide Resistant Weeds

The International Survey of Herbicide-Resistant Weeds database can be useful in identifying which weeds have the greatest propensity to evolve resistance. Table 12.4 presents a list of the 20 worst herbicide-resistant weeds based on the countries they infest, the number of sites of action that they have become resistant to, the number of sites and area of infestation and the number of cropping regimes that the weed has become resistant in. These same criterion were used in 1996 to identify the 20 worst weeds with the aim of predicting which would become resistant to glyphosate. This was successful, with the analysis predicting the very first glyphosate-resistant weed (*Lolium rigidum*), and also predicting 12 of the current list of 23 glyphosate-resistant weeds. The current list (Table 12.4) is not much changed from the 1996 list however the order of the worst resistant weeds has changed to some degree. Species in bold have evolved resistance to glyphosate. There are 8 species on this list that have not evolved resistance to glyphosate. Six of them (*Avena fatua, Echinochloa crus-galli, Setaria viridis, Alopecurus myosuroides, Phalaris minor*, and *Raphanus raphanistrum*) are not common weeds in Roundup Ready crops and thus do not receive a high selection pressure by glyphosate. This leaves *Chenopodium album* and *Amaranthus retroflexus*, both prime candidates for evolving glyphosate-resistance and should be managed accordingly. When weeds evolve resistance to a herbicide group it is often not a major problem for growers to use alternative herbicide groups to control them. The real issue is when weeds evolve multiple-resistance, leaving growers few or no herbicidal options for weed control. Multiple resistance in weeds is increasing rapidly and will be the major source of crop failure and economic problems caused by herbicide-resistant weeds.

Table 12.5 presents the number and percentage of herbicide resistant species by family and the most notable aspect of this table is that five weed families, Poaceae, Asteraceae, Brassicaceae, Amaranthaceae, and Chenopodiaceae account for about 70% of all cases of herbicide resistance even though they represent only 50% of the world's principal weeds. It is apparent that the grasses (Poaceae) and crucifers (Brassicaceae) are very prone to the development of herbicide resistance compared to other families and their prevalence as weeds in general(Table 12.5).

12.6 Herbicide Resistant Weeds in Major Crops

Herbicide-resistant weeds occur in all major cropping systems wherever herbicides are used. Table 12.6 presents the occurrence of herbicide-resistant weeds in various cropping systems.

12.6.1 Wheat

Sixty-four weed species have evolved herbicide resistance in wheat, 38 are broadleaf species and 26 are grass species (Table 12.6). Nineteen grasses have evolved resistance to ACCase inhibitors in wheat. The most troublesome weeds of wheat are grasses that have evolved multiple-resistance, in particular *Lolium rigidum* (11 sites of action), *Alopecurus myosuroides* (9 sites of action),and *Avena fatua* (5 sites of action). Throughout large areas of the wheat producing regions of the world these three species have evolved target site resistance to the ACCase inhibitor and ALS inhibitor herbicides. In addition metabolism based resistance is common in both *Lolium rigidum* and *Alopecurus myosuroides* (and to a lesser extent *Avena fatua*)

12 Herbicide Resistant Weeds

Table 12.4 The top 20 worst herbicide-resistant weeds globally—weighted by propensities in countries, MOA's, sites, hectares, and cropping systems

#	Species	Common Name	Countries	SOA	Sites	Hectares	Regimes
1	**Lolium rigidum**	**Rigid Ryegrass**	12	11	25,000	2,174,000	8
2	**Conyza canadensis**	**Horseweed**	14	5	11,000	2,383,000	9
3	Avena fatua	Wild Oat	13	5	48,000	4,902,000	5
4	**Amaranthus tuberculatus**	**Common Waterhemp**	2	6	69,000	4,741,000	7
5	Chenopodium album	Lambsquarters	18	4	28,000	593,000	9
6	Echinochloa crus-galli	Barnyardgrass	17	9	7,000	624,000	5
7	**Amaranthus palmeri**	**Palmer Amaranth**	1	4	201,000	4,093,000	7
8	Amaranthus retroflexus	Redroot Pigweed	14	4	7,000	156,000	11
9	**Eleusine indica**	**Goosegrass**	7	7	3,000	52,000	11
10	**Echinochloa colona**	**Junglerice**	13	6	2,000	64,000	8
11	**Lolium multiflorum**	**Italian Ryegrass**	9	6	4,000	113,000	11
12	**Kochia scoparia**	**Kochia**	3	4	23,000	1,574,000	7
13	Alopecurus myosuroides	Blackgrass	12	6	2,000	15,000	4
14	**Poa annua**	**Annual Bluegrass**	10	9	1,000	6,000	4
15	Setaria viridis	Green Foxtail	5	4	8,000	1,082,000	7
16	Phalaris minor	Little Seed Canary	6	3	61,000	654,000	3
17	**Conyza bonariensis**	**Hairy Fleabane**	11	4	1,000	6,000	9
18	**Ambrosia artemisiifolia**	**Common Ragweed**	2	5	1,000	52,000	5
19	**Sorghum halepense**	**Johnsongrass**	7	4	1,000	70,000	5
20	Raphanus raphanistrum	Wild Radish	2	4	6,000	45,000	4

Species in bold have evolved resistance to glyphosate. These 20 weeds were chosen by cycling through the International Survey of Herbicide Resistant Weeds database 5 times summing the ranks for each of the 210 weed species. The weeds were then sorted and ranked separately by the number of countries, SOA's, etc. for each of the categories. The cumulative rank for each species for each of the five categories was determined and the 20 with the highest ranks are shown. The rest may be seen on www.weedscience.com

Data accessed from the www.weedscience.com website on August 10, 2012. (Heap 2012)

which has provided them the ability to survive all of the major wheat herbicides available for their control. *Avena fatua* is the most widespread resistant weed globally, estimated to infest around 5 million hectares (Table 12.4). It should be noted that the area estimates provided by scientists are often out of date and inaccurate, resulting in an underestimate of the true area infested by herbicide-resistant weeds. Eighteen grass species and 37 broadleaf species have evolved resistance to the ALS inhibitors in wheat. As mentioned above, the grasses present the most serious economic and practical problems because there is enough diversity in herbicide mechanisms to control the ALS inhibitor resistant broadleaf resistant weeds.

12.6.2 Corn

Fifty-eight weed species have evolved resistance to herbicides in corn, 41 are broadleaf species and 17 are grass species (Table 12.6). The widespread adoption of atrazine for weed control in corn in the 1960s and 1970s resulted in widespread triazine-resistant weeds in corn between 1975 and 1985. Today there are 35 broadleaf weeds and 10 grasses that have evolved resistance to triazine herbicides in corn, primarily in the USA and Europe. In the 1990s ALS inhibitor resistant weeds proliferated in corn (22 cases in total) and from 2000 onwards we saw 12 species evolve glyphosate-resistance in corn. The ALS inhibitor and glyphosate-resistant weeds in corn mainly occurred in the USA and not in Europe, because the Europeans did not use ALS inhibitors extensively and they did not grow Roundup Ready corn. *Amaranthus tuberculatus* is the most serious herbicide-resistant weed of corn, and it has evolved multiple resistance to ALS inhibitors, PSII inhibitors, PPO inhibitors, 4-HPPD inhibitors, glyphosate, and the synthetic auxins. Corn growers are fortunate in that they have many herbicide sites of action available to them to control resistant weeds once they appear.

12.6.3 Soybean

Forty-five weed species have evolved resistance to herbicides in soybean, 25 are broadleaf species and 20 are grass species (Table 12.6). While 6 species had evolved triazine-resistance in soybean the majority of herbicide-resistant weeds in soybean are to ALS inhibitors, ACCase inhibitors, and more recently glyphosate. Fourteen grasses have evolved resistance to the ACCase inhibitors with *Sorghum halapense, Setaria sp., Digitaria sp.*, and *Echinochloa sp.* presenting the biggest problems. In the 1990s the soybean growers in the USA were reliant on ALS inhibitors, such as imazethapyr, for weed control and 27 weed species evolved resistance to ALS inhibitors in soybean. The rapid adoption of Roundup Ready Soybean, first introduced in 1996, resulted in a reduction in the identification of new ALS inhibitor resistant weeds in soybean. The reliance on glyphosate as the primary weed control in Roundup Ready Soybean resulted in fifteen weed species evolving resistance to

Table 12.5 The number and percentage of herbicide-resistant species by family, and the percentage of species considered principal weeds by Holm et al. (1991, 1997) for each of these families

Family	Number of resistant species in family	Resistant Species (% of total)	Weed species (% of world's principal weeds)*
Poaceae	71	34	25
Asteraceae	36	17	16
Brassicaceae	17	8	4
Amaranthaceae	10	5	3
Chenopodiaceae	8	4	2
Polygonaceae	7	3	5
Scrophulariaceae	7	3	1
Cyperaceae	6	3	5
Caryophyllaceae	5	2	1
Alismataceae	5	2	1
Solanaceae	4	2	2
Lythraceae	4	2	1
23 other families pooled	30	14	18
Total	210	99	84

Data accessed from the www.weedscience.com website on August 10th, 2012 (Heap 2012)

Table 12.6 The occurrence of herbicide-resistant weeds in various cropping situations

Category	Crop	# Resistant Biotypes
Field Crops	Wheat	64
	Corn	58
	Soybean	45
	Rice	39
	Pulses	17
	Canola	13
	Cotton	11
	Sugarbeet	8
	Sugarcane	3
	Other Crops	73
Vegetables	Vegetables (carrot, lettuce, potato, etc.)	21
Perennial Crops	Orchard (apple, pear, peach,…including vineyard)	38
	Pasture (clover, alfalfa, pasture seed, etc.)	26
	Forestry	6
	Other Perennial (tea, coffee, rubber, mint, etc.)	12
Non Crop	Non Crop—(roadside, railway, industrial site)	35

Data accessed from the www.weedscience.com website on August 10, 2012 (Heap 2012)

glyphosate in soybean, eight of them are broadleaves and 7 of them grasses. *Amaranthus sp., Conyza sp.*, and *Ambrosia sp.*, are the most troublesome glyphosate-resistant broadleaf weeds in soybean in the USA. *Digitaria insularis* and *Sorghum halepense* in South America present the worst cases of glyphosate-resistant grasses in soybean. Multiple resistance in *Amaranthus palmeri* in the south and *Amaranthus tuberculatus* in the mid-west are the greatest threat to soybean production in the USA.

12.6.4 Rice

Thirty-nine weed species have evolved resistance to herbicides in rice, 26 are broadleaf species and 13 are grass species (Table 12.6). The majority of herbicide-resistance cases (31 species) in rice are to the ALS inhibitors. Twenty broadleaf species and 11 grasses have evolved resistance to ALS inhibitors in rice, with the worst cases being *Echinochloa sp., Lindernia sp., Sagittaria sp., Scripus sp., Monochoria sp., Ammania sp.,* and *Limnophila sp.* Eight grasses have evolved resistance to ACCase inhibitors in rice, the worst being *Echinochloa sp., Leptochloa sp., Ischaemum rugosum* and *Eleusine indica.* The Echinochloa sp. (*Echinochloa colona, Echinochloa crus-galli, Echinochloa oryzoides,* and *Echinochloa phyllopogon*) are an intractable problem in rice because they have evolved multiple resistance to most of the available rice herbicides, including ACCase inhibitors (Group A—fenoxaprop+many), ALS inhibitors (Group B—bensulfuron+many), Ureas and Amides (Group C2–propanil), Isoxazolidinones (Group F3–clomazone), Chloracetamides (Group K3–butachlor), Thiocarbamates (Group N—thiobencarb & molinate), and Synthetic auxins (Group O—quinclorac). It is estimated that there are over 2 million hectares infested with target site cross-resistance to butachlor and thiobencarb in *Echinochloa crus-galli* in China (Huang and Gressel, 1997).

12.6.5 Perennial Crops

Thirty-eight weed species have evolved resistance in orchards, 26 in pastures, 6 in forestry and 12 in other perennial crops like tea, coffee, rubber and mint. Orchards are particularly prone to resistance because growers often attempt to keep the ground bare through several (3–10) applications of herbicides annually. Common herbicides used in orchards are the triazines (atrazine, simazine, and metribuzin), the bipyridiliums (paraquat and diquat), glyphosate, and glufosinate in orchards. Eighteen weed species have evolved resistance to the triazines, 5 to the bipyridiliyums, 11 to glyphosate, and 2 to glufosinate. It is interesting to note that glyphosate and glufosinate are considered very low risk herbicides for selecting resistance and the 11 glyphosate resistance cases in orchards constitute half of all known cases of glyphosate resistance. Glufosinate resistance in *Eleusine indica* and *Lolium multiflorum* in orchards are the only known cases of glufosinate resistance worldwide and has implications for the use of glufosinate in glufosinate resistant corn and soybean in the USA.

12.6.6 Non-Crop

Herbicides are often used to keep ground bare in non-crop situations, particularly on roadsides, railways, and industrial sites. Thirty five weed species have evolved resistance in these situations, 20 of them broadleaves and 15 of them are grasses. These situations often require repeated applications annually of the same herbicide and often the use of herbicides with a high level of residual, which has led to significant problems with herbicide-resistant weeds. Herbicide-resistant weeds selected on roadsides and railways are often not confirmed until they move into farmers' fields and present an economic problem. Triazine and ALS inhibitor resistance in *Kochia scoparia* and *Conyza canadensis* became widespread along roadsides and railways in the USA through many years use of these inexpensive herbicides and both species have extremely efficient dispersal systems allowing them to spread quickly into farmers' fields.

12.7 Herbicide Resistant Crops

Herbicide resistant crops are both the cause and solution to many herbicide resistant weed problems. Roundup-Ready crops (crops genetically engineered to survive high rates of the herbicide glyphosate) have dominated the herbicide resistant crop market since the introduction of Roundup-Ready soybeans in 1996. At that time soybean growers were facing serious resistance problems with ALS inhibitor and ACCase inhibitor resistant weeds and they saw Roundup Ready soybeans as the solution to those problems. Indeed Roundup Ready crops did rescue many growers from crop failure due to herbicide resistant weeds. Roundup Ready crops were adopted at a greater rate than nearly any other agricultural technology and their utility in managing weeds with resistance to other herbicide groups is one of the reasons for this rapid adoption rate. In addition to soybeans, Roundup Ready alfalfa, canola, cotton, corn, and sugarbeet have been commercially used and Roundup Ready rice, wheat, and bentgrass are under development (Dill 2005; Dill et al. 2008). The very success of Roundup Ready crops has resulted in the biggest threat to their sustainability, that of glyphosate-resistant weeds. Glyphosate is a low risk herbicide for the development of herbicide resistant weeds, however the massive adoption of Roundup Ready crops, and the over reliance on glyphosate alone for weed control by many growers has made the rapid increase in glyphosate resistant weeds the biggest herbicide resistance problem that we face today. Roundup Ready crops made farmers lives much easier at first, they no longer needed to know what weeds were growing in their fields (as glyphosate controlled most species), and they no longer needed to worry as much about the timing of herbicide applications as glyphosate controls weeds at all growth stages. In reality the growers should be very worried about controlling weeds at an early stage because of the negative impact on yield if weeds are left to compete with crops even when the weeds are relatively small. One negative aspect of the success of Roundup Ready crops is that we now have a whole generation of growers that know little about weed control. Another negative aspect is that glyphosate itself is so inexpensive that it made the discovery and development of new herbicide modes of action uneconomical because new products would not be able to compete with glyphosate in the major herbicide markets (corn, soybean, cotton etc.). Certainly there were other factors at play, including the increasing regulations and costs associated with bringing new products to the market and the availability of cheap generic products other than glyphosate. Because of these factors discovery programs have been decimated over the past 15 years. All this would be fine if it were not for the appearance of glyphosate-resistant weeds. Now that companies can see that glyphosate is not sustainable on its own their discovery programs are being reignited. But there is always a lag phase of 7 to 15 years to identify, test, and register products, so growers are faced with the dilemma of dealing with resistance using existing herbicides.

This is where herbicide resistant crops can, to some extent, come to the rescue again. Given that we will not have new herbicide modes of action to deal with herbicide-resistant weeds for some time then the next best thing is to be able to use the existing herbicides in new ways. Glufosinate, dicamba, 2,4-D, HPPD inhibitor and PPO inhibitor resistant crops offer great promise to use these modes of action in new ways and will certainly be one part of the puzzle to combat herbicide-resistant weeds. There are other herbicide resistant crop traits such as the ALS inhibitor (primarily imidazolinone and sulfonylurea) resistant crops and ACCase inhibitor resistant crops but they suffer the problem that there are already many weeds that have become resistant to these herbicide groups, and they are prone to select resistance very quickly.

12.8 Integrated Weed Management is Herbicide Resistance Management

Integrated weed management (IWM) includes strategies for weed control that consider the use of all economically available weed control techniques, including: preventative measures, monitoring, crop rotations, tillage, crop competition, herbicide site of action rotation, herbicide resistant crops, biological controls, crop competition, nutrition, burning etc. Herbicide resistance is a very predictable outcome of evolution. In fact any weed control practice will be subject to the forces of evolution and no matter what practice, if done consistently and long enough, weeds will evolve to survive the practice. The best way to foil the forces of evolution is to challenge it with diversity such that any one practice is not used consistently enough to select resistance and avoidance mechanisms.

Integrated weed management requires a holistic look at all aspects of crop production. It begins with preventing the spread of weeds by cleaning farm machinery between fields, tarping grain trucks, using certified seed, controlling weed seed nurseries along fence lines, farm roads, irrigation ditches, and stockyards, and ensuring that hay and livestock is weed free before bringing them onto the property. Growers need to inventory their weed problems in order to craft effective IWM programs. The aim of IWM is to destabilize and disrupt weed populations so they don't become serious problems. Available cultural practices for weed control include crop rotations, crop management (use of vigorous seed, competitive varieties, stale seed beds, reduced row spacing, early seeding, high seeding rates, shallow seeding, good "on row" seed packing, good crop nutrition and soil conditions, intercropping, and cover crops), tillage (harrowing, spring and fall tillage, inter-row tillage, strip tillage, rotary hoeing and conventional tillage), mowing, burning, allelopathy, and biological controls. While these cultural controls are valuable, herbicides are often the backbone of an IWM program because they are the most cost effective and efficacious method of weed control in the IWM toolbox. It is very important to rotate herbicide sites of action to avoid the selection of herbicide-resistant weeds and/or herbicide weed shifts. Herbicide mixtures and sequences are also an effective resistance management strategy. In fact there is a growing body of evidence that herbicide mixtures may be more effective than herbicide rotations at delaying resistance (Beckie and Reboud 2009). Ideally each component of the herbicide mixture should be active at different target sites, have a high level of efficacy, and both herbicides should have efficacy on key problem weeds. The use of pre-plant and pre-emergence herbicides will continue to increase as a way to rotate herbicide sites of action in the cropping system. Herbicide resistant crops will also play a larger role in IWM programs in future because they facilitate the goal of rotating herbicide sites of action. To avoid the selection of polygenic low level resistance it is also important to use the full recommended herbicide rate and proper application timing for the hardest to control weed species present in the field.

12.9 Summary

Herbicides have provided farmers with unprecedented success in controlling weeds over the last 65 years. Without herbicides the world would face a major reduction in crop yields resulting in high food costs and food shortages. Herbicide-resistant weeds have been a fact of life for growers for over 40 years and they have been successful in overcoming resistance problems primarily because the agricultural chemical industry was able to provide a steady supply of new herbicide sites of action to combat resistant weeds. This is no longer the case, no new herbicide sites of action have been delivered to the market in over 20 years (Duke 2011) and there does not appear to be any on the near horizon. Many growers are reliant on using glyphosate in Roundup Ready crops for weed control and this is now known to be unsustainable. Until new herbicide sites of action are brought to the market the best strategy to manage herbicide resistant weeds is to implement integrated weed management practices that will include the use of different herbicide sites of action in rotation, sequence, and mixtures. Herbicide-resistant crops will enable growers to achieve more sustainable herbicide site of action rotations and move away from relying upon glyphosate as their primary weed control solution. The biggest problem that we face in the future is multiple resistance in weeds resulting in no herbicidal options for weed control in some crops. *Amaranthus, Conyza, Echniochloa,* and *Lolium* species are the most worrisome because of their ability to rapidly evolve resistance to a wide range of herbicide sites of action in addition to them being primary weeds in many cropping systems. It is clear from history that any consistent practice to control weeds year after year will result in directed evolution towards their survival. The solution is to vary weed control practices and destabilize evolution.

Acknowledgements The author would like to acknowledge the Herbicide Resistance Action Committee for their support of the International Survey of Herbicide Resistant Weeds as well as the data contributions from weed scientists in over 60 countries.

References

Baylis, A. D. (2000). Why glyphosate is a global herbicide: Strengths, weaknesses and prospects. *Pest Management Science, 56,* 299–308.
Beckie, H. J., & Reboud, X. (2009). Selecting for weed resistance: Herbicide rotation and mixture. *Weed Technology, 23,* 363–370.
Brown, A. C., Moss, S. R., Wilson, Z. A., & Field, L. M. (2002). An isoleucine to leucine substitution in the ACCase of Alopecurus myosuroides (black-grass) is associated with resistance to the herbicide sethoxydim. *Pesticide Biochemistry and Physiology, 72,* 160–168.
Buchanan, B. B., Gruissem, W., & Jones, R. L. (2000). *Biochemistry and Molecular Biology of Plants.* American Society of Plant Physiology. Rockville, Maryland, USA: Courier Companies.
De´lye, C., Zhang, X. Q., Michel, S., Matejicek, A., & Powles, S. B. (2005). Molecular bases for sensitivity to acetyl-coenzyme-A-carboxylase inhibitors in black-grass. *Plant Physiology, 137,* 794–806.
Dill, G. M. (2005). Glyphosate-resistant crops: History, status and future. *Pest Management Science, 61*(3), 219–224.
Dill, G.M., CaJacob, C.A., & Padgette, S.R. (2008). Glyphosate-resistant crops: adoption, used and future considerations. *Pest Management Science, 64*(4), 326–331.
Duke, S. O. (2011). Comparing conventional and biotechnology-based pest management. *Journal of Agricultural Food Chemistry, 59,* 5793–5798.
Feng, P. C. C., Tran, M., Sammons, R. D., Heck, G. R., & Cajacop, C. A. (2004). Investigations into glyphosate-resistant horseweed (Conyza Canadensis): Retention, uptake, translocation, and metabolism. *Weed Science, 52,* 498–505.
Gaines, T. A., Preston, C., Leach, J. E., Chisholm, S. T., & Shaner, D. L. (2010). Gene amplification is a mechanism for glyphosate resistance evolution. *Proceedings of the National Academy of Sciences U S A, 107,* 1029–1034.
Gronwald, J. W. (1994). Resistance to photosystem II inhibiting herbicides. In S. B. Powles. & J. A. M. Holtum, (Eds.), *Herbicide Resistance in Pants: Biology and Bochemistry* (pp. 276–280). Tokyo: Lewis Publ.
Gronwald, J. W. (1997). Resistance to PSII inhibitor herbicides. In R. De Prado, J. Jorrin, & L. Garcia-Torres, (Eds.)., *Weed and Crop Resistance to Herbicides* (pp. 53–59). Dordrecht, Netherlands: Kluwer Academic Publishers.
Harper, J. C. (1956). The evolution of weeds in relation to herbicides. *Proceedings of the British Weed Control Conference, 3,* 179–188.
Heap I and R. Knight (1982). A population of ryegrass tolerant to the herbicide diclofop-methyl. *Journal of the Australian Institute of Agricultural Science,* 48, 156–157.
Heap I and R. Knight (1986). The occurrence of herbicide cross-resistance in a population of annual ryegrass, Lolium rigidum, resistant to diclofop-methyl. *Australian Journal of Agricultural Research, 37,* 149–156.
Heap, I. M., & LeBarron, H. (2001). Introduction and overview of resistance. In S. B. Powles & D. L. Shaner (Eds.)., *Herbicide Resistance and World Grains* (pp. 1–22). Boca Raton, Florida, USA: CRC Press.

Heap, I. (2012). *The International Survey of Herbicide Resistant Weeds*. http://www.weedscience.com. Accessed 10 Aug 2012.

Hochberg, O., Sibony, M., & Rubin, R. (2009). The response of ACCase-resistant Phalaris paradoxa populations involves two different target site mutations. *Weed Research, 49,* 37–46.

Holm, L. J., Plucknett, D. L., Pancho, J. V., & Herberger, J. (1991). *The World's Worst Weeds: Distribution and Biology*. Malabar, Florida, USA: Krieger.

Holm, L., Doll, J., Holm, E., Pancho, J., & Herberger, J. (1997). *The World's Worst Weeds: Natural Histories and Distribution*. New York: Wiley.

Huang, B.-Q., & Gressel, J. (1997). Barnyardgrass (Echinochloa crus-galli) resistance to both butachlor and thiobencarb in China. *Resistant Pest Management, 9,* 5.

Kaundun, S. S., Zelaya, I. A., Dale, R. P., Lycett, A. J., & Carter, P. (2008). Importance of the P106S target-site mutation in conferring resistance to glyphosate in a goosegrass (*Eleusine indica*) population from the Philippines. *Weed Science, 56,* 637–646.

Liu, W. J., Harrison, D. K., Chalupska, D, et al. (2007). Single-site mutations in the carboxyltransferase domain of plastid acetyl-CoA carboxylase confer resistance to grass-specific herbicides. *Proceedings of the National Academy of Sciences U S A, 104*(9), 3627–3632.

Oerke, E. C. (2002). Crop losses due to pests in major crops. In *CAB International Crop Protection Compendium 2002. Economic Impact*. Wallingford, United Kingdom: CAB International.

Mallory-Smith, C. A., Thill, D. C., & Dial, M. J. (1990). Identification of sulfonylurea herbicide-resistant prickly lettuce (*Lactuca serriola*). *Weed Technology, 4,* 163–168.

Ray, T. B. (1984). Site of action of chlorsulfuron. *Plant Physiology, 75,* 827–831.

Ryan, G. F. (1970). Resistance of common groundsel to simazine and atrazine. *Weed Science, 18,* 614–616.

Sterling, T. M., & Hall J. C. (1997). Mechanism of action of natural auxins and the auxinic herbicides. In R. M. Roe, J. D. Burton, & R. J. Kuhr. (Eds.)., *Toxicology, Biochemistry and Molecular Biology of Herbicide Activity* (pp. 111–141). Amsterdam: IOS Press.

Woodburn A. T. (2000). Glyphosate: Production, pricing and use world-wide. *Pest Management Science, 56,* 309–312.

Chapter 13
Strategies for Reduced Herbicide Use in Integrated Pest Management

Rakesh S. Chandran

Contents

13.1	Introduction	304
13.2	Herbicides: A Valuable yet Controversial Tool in Modern Crop Production	305
13.3	Historical Perspective	305
13.4	Herbicide Use Pattern in the United States	306
13.5	Role of GE Crops in Weed Management	308
13.6	Public Perception	309
13.7	Biology of Weeds and Relative Susceptibilities	310
13.8	Spatial Dynamics and Weed Management	312
	13.8.1 Plants Growing Out of Place	312
	13.8.2 Plants with Unknown Virtues	313
13.9	Sustainable Weed Management	314
13.10	Advances in Biological Weed Control	315
13.11	Technology in Weed Management	316
13.12	Herbicide Use Reduction in Agronomic Crops	318
	13.12.1 Banded Herbicide Application—Herbicide Use Reduction in Corn	319
	13.12.2 Horticultural Crops	321
	13.12.3 Engagement of Industry—A Potential Opportunity	323
13.13	Conclusion	324
References		325

Abstract The persistence of weeds in crop production systems leads to significant yield reductions, diminishing profitability, and ultimately translating to higher consumer prices. Chemical weed control has proven to be an economical and cost-effective method to manage weeds in agricultural settings. While herbicides are considered to be valuable tools in pest management, they account for about two-thirds of the total pesticide use in the United States. As the number of hectares planted under row crops is on the rise, management of weeds, especially herbicide-resistant weed biotypes, in cropping systems is increasingly important. Current weed control programs employed in crop production maintain the fields mostly weed-free during

R. S. Chandran (✉)
West Virginia University, Agricultural Sciences Building,
1076 Morgantown, West Virginia, USA
e-mail: RSChandran@mail.wvu.edu

the growing season. Managing the flora of such vast expanses of land under high selection-pressure is somewhat unprecedented given the long history of agriculture, and is worthy of scientific inquiry. In response to health and environmental concerns, scientists are exploring new methods to apply herbicides that could reduce the amount of herbicides used. This chapter explores several strategies to reduce herbicide inputs in crop production systems. Strategies such as banding herbicides, precision application, cultural methods, and novel mechanical and biological methods are discussed.

Keywords Integrated weed management · Weed control · Sustainable agriculture · Reduced pesticide use · Non-chemical weed control · Herbicide mitigation · Floral biodiversity · Cultural weed control · Mechanical weed control · Biological weed control · Herbicide application timing · Herbicide banding

List of Abbreviations

USDA-ERS	United States Department of Agriculture-Economic Research Service
NASS	National Agricultural Statistics Service
GE-Crops	Genetically Engineered Crops
IPM	Integrated Pest Management
PRE	Pre-emergence (herbicide)
POST	Post-emergence (herbicide)
VRT	Variable Rate Technology

13.1 Introduction

Services provided by vascular plants to the ecosystems are affected by reductions in floral diversity (Chapin et al. 2000). This phenomenon is to be taken into consideration under the assumption that the dynamic nature of ecosystems, as an ideal environment for life to thrive on the planet, tends to remain somewhat static in the human mind. In this context, it could be recalled that biodiversity is a process that evolves continually i.e., the existing levels of biodiversity are simply a snapshot of continual change that occurs through time and space. Natural and manmade causes may bring about such changes (Thuiller 2007). The relative rate at which such changes have occurred may have varied historically. The question that behooves our attention, however, is whether such changes are occurring at an accelerated rate and how this rate of change could be mitigated. Assuming that the static nature of this dynamicity is the ultimate goal, certain well-characterized changes to human activities may be necessary. If so, a practical option is to identify practices that have the potential to cause significant impacts referenced above and delineate well characterized mitigation efforts. Such efforts may include changes in weed management practices in agricultural systems as we attempt to increase efficiency in producing food, fiber, and of late, energy. This chapter attempts to examine possibilities to reduce our overall dependence on herbicides for weed management and understand the benefits as a result of doing so.

13.2 Herbicides: A Valuable yet Controversial Tool in Modern Crop Production

As plants 'growing out of place' and competing with crops for resources, weeds are managed to maintain and enhance crop productivity. On the other hand, as 'plants whose virtues are not well-understood', weeds may provide indirect albeit important services to ecosystems. One of the primary goals of weed scientists is to contribute towards a safe, secure, and abundant supply of food to meet the growing human demands. Based on a global review on crop losses to agricultural pests, weeds are considered to cause the highest potential yield losses with moderate estimates of 34% (Oerke 2006). Total crop-losses could occur in fields infested by weeds coupled with other forms of stress (Ross and Lembi 2008).

In some instances, certain adverse crop responses may be the result of an interaction of herbicides with plant physiology. Oka and Pimentel (1976) documented increased levels of European corn borer (*Ostrinia nubilalis*) and southern corn leaf blight (caused by *Cochliobolus heterostrophus*) in corn (*Zea mays*) treated with 2,4-D (2,4-dichlorophenoxyacetic acid). They attributed these differences to higher levels of proteins in corn treated with 2,4-D, compared to untreated corn. Altman and Campbell (1977) presented a review of herbicides that can interact with crop plants and noted that a few commonly used herbicides such as 2,4-D, mecoprop, metribuzin, simazine, and trifluralin may predispose plants to disease pathogens upon exposure. The herbicides affected physiological processes of the crop such as wax formation and growth regulation, and certain metabolic pathways.

Doubtlessly, herbicides are a boon for farmers not only to keep production costs down but also to accommodate other cultural practices such as conservation tillage, crop rotation, efficient harvest, and as an integrated approach to manage cover crops, insects, and diseases. Apart from weed management in food and fiber production, herbicides now play a dominant role in managing weeds in biofuel production, turf and ornamentals, vegetation management and restoration in non-crop areas, aquatic systems, and woodlots for management of invasive weeds. While modern herbicides may pose minimal risks to the environment and human health, their indirect impacts on floral biodiversity, carbon sequestration, habitat for other living organisms, soil and nutrient run-off from cultivated fields are worth closer examination. Several effective herbicides are losing efficacy due to buildup of resistant weed biotypes. Judicious use of herbicides will help maintain their continued availability as a valuable tool in food production.

13.3 Historical Perspective

Weed science is considered an old art, yet a young science (Timmons 1970). Details of primitive tools used to control weeds remain sketchy. Drawings from 6000 B.C. show a 'Y'-shaped portion of a tree with a bronze tip similar to hoe or mattock but its use is unclear (Gittins 1959). In his classic book Horse Hoeing Husbandry, Jethro

Tull (1762) described the benefits of using a horse-drawn hoe to cultivate row-crops for weeds. More efficient mechanical tools were developed to control weeds during the 19th century and early 20th century.

Sodium chloride was perhaps he first chemical used to control weeds. Accounts of common salts used by Romans to kill bushes were mentioned in early recorded history (Ashton and Monaco 1991). In agriculture, chemicals were initially used to control plant diseases and insect pests prior to their use to control weeds (Anonymous 1958). Common salt was also documented to control orange hawkweed (*Hieracitum aurantiacium* L.) in 1896 (Jones and Orton 1896). Other chemicals such as copper sulfate, iron sulfate, and sulfuric acid were documented for their weed control attributes shortly thereafter (Bolley 1901; Anonymous 1907; Groh 1922). Apart from these compounds, various persistent chemicals such as arsenicals, chlorates and borates were used for weed control in the early 20th century (Wunderlich 1961; Ross and Lembi 2008).

The advent of modern weed control began with the discovery of 2,4-D in 1941 followed by the discovery of other compounds such as silvex, 2,4,5-T, amitrole, diuron and monuron in the 1950s. Several effective herbicides such as atrazine, ETPC, alachlor, trifluralin, and paraquat were subsequently developed and proved successful in controlling weeds in a broad range of crops. More than 75 herbicides were synthesized in the following two decades, a three-fold increase to the number of herbicides known till then (Timmons 1970). The area of land treated with herbicides in the United States also witnessed an exponential growth to 48.6 million hectares during this period. Glyphosate, introduced in early 1970s, was considered to be an 'ideal' herbicide resulting in its worldwide adoption in the subsequent decades. The 1980s also witnessed a reduction in soil erosion in the U.S. as a result of conservation tillage practices owing to herbicide use. This period also witnessed the introduction of several new classes of selective herbicides such as acetyl CoA carboxylase inhibitors, protoporphyrinogen inhibitors, diphenylethers and acetolactate synthase inhibitors. As the demand for food and fiber increased along with simultaneous advances in science and technology, chemical weed control became a mainstay to manage weeds in various crop production systems.

13.4 Herbicide Use Pattern in the United States

An examination of herbicide use patterns in the United States from 1980 to 2007 reveals that about 48% of pesticide active ingredients used by agricultural producers were herbicides, which fluctuated ($\pm 4\%$) but remained steady otherwise (USDA-ERS 2012). The total amount of herbicide used decreased by 12% during this period from 504 to 442 million pounds. It should be noted however that drastic reductions were noted since the mid-1980s. This could be attributed to new classes of herbicides especially the sulfonyl ureas and the imidazolinones, effective at extremely low use rates.

Based on publicly available USDA data, Benbrook (2012), however, projected an increase of 527 million pounds of herbicide use in the U.S between 1996 and

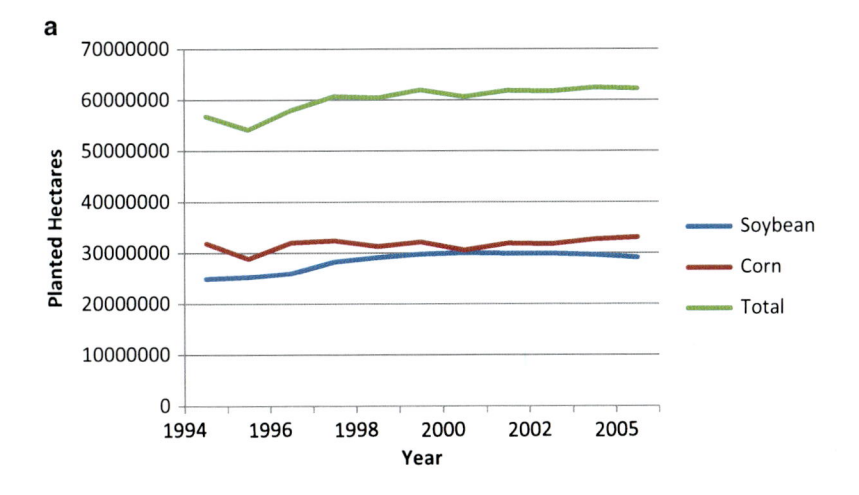

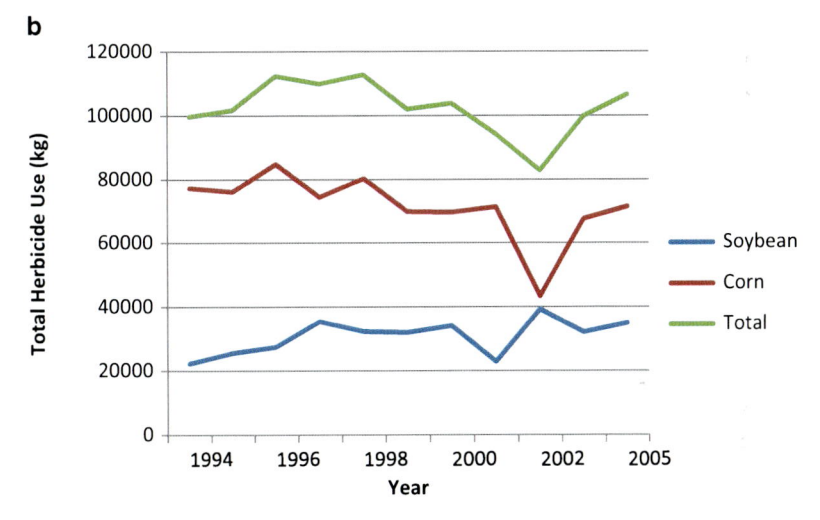

Fig. 13.1 a) Area planted to corn and soybean in the United States (*above*) **b)** compared to total herbicide use (*below*) prior to and after the introduction of genetically engineered crops

2011 as a result of weed management practices in herbicide-resistant crops. This was attributed primarily to the increased reliance on glyphosate in such crops. Benbrook also projected a two-fold rate of increase (2.7 % per year) in glyphosate use in soybeans resistant to glyphosate, per year from 2006 to 2011, compared to 1.3 % rate of increase in glyphosate use in conventional soybeans during the same period. The author also warned that such a trend could cause additional increases in herbicide use by approximately 50 % if herbicide resistant crops capable of tolerating growth-regulator herbicides are introduced into the market. In the United States, despite modest increases in area planted to corn and soybean, herbicide use in these crops began to rise since 2002 after a long-term decline (Fig. 13.1a, b).

Until the advent of glyphosate-tolerant soybean, less than 3 million kg of glyphosate was used in soybean production (Young 2006). By 2002, 30 million kg of glyphosate was used in soybean alone reducing the number of sites of action from seven to essentially one. The primary shift was from imidazolinone and dinitroaniline herbicides to glyphosate during this period. A similar trend was noted in cotton during the same period. Atrazine continued to dominate as the primary herbicide in corn as a cost-effective, broad-spectrum herbicide although glyphosate-resistant corn was introduced in 1998. However, by 2010, total glyphosate use exceeded that of atrazine by 2.9 million kg in corn (atrazine use in corn during 2010 was 23.3 million kg). Unlike soybean, at least three sites of action are still employed in corn production (USDA-NASS 2012). Due to increased adoption rates of glyphosate resistant crops and the availability of generic formulations of glyphosate in the market, the overall expenditure by U.S. agricultural producers of herbicides fell by 23 % between 2000 and 2007. Interestingly, Mortensen et al. (2012) pointed out that "agricultural weed management has become entrenched in a single tactic– herbicide resistant crops" as the ultimate result of such a trend.

13.5 Role of GE Crops in Weed Management

One of the most significant advances in agriculture towards the end of 20th century was the introduction of genetically engineered crops (GE crops). Genetically engineered crops have simplified weed management methods in most major field crops (Reddy and Koger 2006). Farmers in the United States have rapidly adopted GE crops that resist herbicides ever since their inception in the mid-1990s. The concomitant engagement of a narrow spectrum of herbicides in major crops resulted in an exponential increase in the use of otherwise benign pesticides such as glyphosate. The use of pesticides with benign attributes has increased to extremes that resulted in the engagement of a narrow spectrum of herbicides in major crops. A few applications of such herbicides can effectively control a broad spectrum of weeds causing no phytotoxic effect to the crop (Fig. 13.2). Farmers embraced this new tool not only based on simplicity but also based on cost-effectiveness.

Speculations were made by the scientific community about genetically engineered crops as a plausible tool in integrated pest management (IPM) and the resultant reduction in pesticide use. In the prevention, avoidance, monitoring, suppression (PAMS) strategy of IPM, use of GE crops was considered to fit under 'avoidance', where crops may be selected based on their genetic resistance to pests (North Central IPM Center 2010). Although such traits pertain more to insect pests and diseases, it may be applicable to weeds indirectly where GE crops that resist herbicides utilize such traits to attain selective weed control.

Today, GE crops capable of resisting glyphosate, glufosinate, bromoxynil, imidazolinone herbicides, and sethoxydim are used in major field crops such as corn, soybean, cotton and canola. Other crops such as alfalfa and sugarbeet have also been genetically engineered to resist glyphosate. However, biotypes of glyphosate-resistant weeds are reported to have increased exponentially since 2004 (Heap

Fig. 13.2 Progression of phytotoxicity symptoms in weeds following application of glyphosate (1.12 kg ai/ha) in alfalfa (*Medicago sativa*) genetically-modified to resist glyphosate

2012). Herbicide-resistant weed biotypes, especially those in row crops, continue to make headlines in weed management. Lately, biotypes of certain weeds—including Palmer amaranth (*Amaranthus palmeri*), water hemp (*Amaranthus rudis*), common and giant ragweed (*Ambrosia spp.*), horseweed/marestail (*Conyza candensis*), and johnsongrass (*Sorghum halepense*)—have been reported to be resistant to glyphosate in various parts of USA.

This technology continues to generate public interest as well as controversy. Scientific evidence to validate harmful health effects is yet to be documented (recently, a study demonstrated higher incidence of tumors in rats fed genetically engineered corn over a two year period, compared with those fed conventional corn during the same period, however, these findings have been refuted by the scientific community at the time of preparing this manuscript) (Séralini et al. 2012). Regardless, overdependence on this technology and related indirect effects on cropping systems and the ecosystem appear to be primary concerns among scientists. Conscientious use of this otherwise effective tool in the IPM toolbox will ensure its continued availability. Management practices to avoid the buildup of resistant biotypes of weeds will also ensure that such cost-effective herbicides remain available.

A minor problem encountered in row crops dedicated to the same crop or rotated to different crops capable of resisting the same herbicide is the periodic occurrence of volunteer plants from the previous crop interfering with the current crop. Management of such volunteers often requires broadcast application of otherwise unnecessary pre-emergence herbicides or spot treatment with limited options of post-emergence herbicides.

13.6 Public Perception

Herbicide use patterns and the buildup of herbicide-resistant weed biotypes since the advent of GE crops are alarming since the outcome has been contrary to expectations. Gasser and Fraley (1989), while explaining the benefits of genetic engineering tools to improve crops, had predicted that a shift in herbicide use towards more safe and environmentally benign chemicals, as opposed to an increase in overall use of herbicides, would be the driving force for the development of traits to resist herbicides. They also noted that the impact of GE crops would also be determined by factors including public perception. Goldburg (1992) recommended that public funds not be used to carry out research to develop herbicide-tolerant crops and that herbicide-tolerant crops should be regulated by governmental agencies especially if they pose a risk to human health and the environment. While there is a gap between "scientific truth" and "public perception", it is critical to base important policy and regulatory decisions on sound knowledge. Generation of such information has not kept pace with technological advances over the past two decades. Long-term studies to determine various indirect effects of such innovative strategies will help us gain a better understanding. Until we have sufficient knowledge, such decisions may have to be made conservatively.

13.7 Biology of Weeds and Relative Susceptibilities

A sound understanding of the biology of weeds, their life-cycles, and their relative periods of susceptibility is essential to delineate effective control options and to optimize herbicide use. Weeds compete with crops during the crops' active growth phase whether the crop is annual or perennial by nature. If the demand for resources coincides with the crops' active growth phase, weed competition could significantly affect crop yields. Application timing of herbicide relative to the weed life-cycle/growth stage means applying the herbicide at the proper time of the year or crop stage is critical to maximize efficiency. A few common misapplications include applying pre-mergence herbicides after weed emergence without a post-emergence herbicide, applying systemic herbicides to actively growing annual weeds intensifying selection pressure or contact herbicides to manage perennial weeds, or systemic herbicides being applied during the time of the year when preferential flow of sugars is acropetal resulting in poor translocation to the below-ground vegetative parts. Herbicides are applied occasionally when the weeds have surpassed their competitive stage.

Substantial research has been carried out to optimize herbicide use. A summary of relevant literature related to herbicide application timings as they affect weed control is presented in Table 13.1. Certain general conclusions can be made based on these research findings. Annual weeds were most susceptible to herbicides earlier on during the growing season when weeds were young and actively growing. Systemic herbicides were usually effective to control perennial weeds as they become mature. The competitive phase often coincided with the maximum period of growth of crops in most instances. Control of weeds during this window was found to be most effective. In soybean, however, late season weed control was also considered to be important (Van Acker et al. 1993).

Table 13.1 Summary of relevant literature to reduce herbicide inputs by following proper application timings for effective weed control

Situation	Herbicide	Strategy	Reference
Grass and broadleaf control in *Zea mays*	Nicosulfuron + bromoxynil	Weed control optimal up to 15-cm weed height. Control and yield affected after 20-cm height	Carey and Kells 1995
Orobanche control in *Trifolium pratense*	Imazamox	Optimal small broomrape control attained when herbicide was applied at 1000 growing degree days (GDD)	Eizenberg et al. 2006
Giant foxtail control in glyphosate-tolerant *Zea mays*	Glyphosate; atrazine and acetochlor	Applying glyphosate after weed was >15-cm affected yield; applying residual herbicides did not increase corn yield	Gower et al. 2002
Eclipta prostrata, Ipomoea lacunose control in *Arachis hypogaea*	2,4-DB, acifluorfen, bentazon, imazapic, and lactofen,	Early POST (5-cm tall eclipta, and 8-cm long morningglory) herbicide application provided optimal weed control and peanut yields	Grichar 1997
Control of *Amaranthus rudis* in *Glycine max*	Diphenylether herbicides	Application to 5-cm tall weed provided better control compared to that to 10-cm tall weed	Hager et al. 2003
Ligustrum sinense control in forests	Glyphosate and triclopyr	October application of glyphosate provided 100% control, followed by April application (93%). Summer applications of glyphosate and fall application timings of triclopyr provided lower control levels	Harrington and Miller 2005
Translocation of herbicides to *Agropyron repens*r hizomes	Glyphosate, sethoxydim, fluazifop, and haloxyfop	Translocation of systemic herbicides to rhizomes was similar during all growth stages	Harker and Dekker 1988
Xanthium strumarium- control in *Zea mays*	Mesotrione	Control highest when herbicide was applied to 3–8 cm tall weeds (3-lf stage of corn)	Johnson et al. 2002
Xanthium strumarium, Chenopodium album, Panicum dichotomiflorum, Setaria faberi, and *Abutillon theophrasti* control in *Zea mays*	Atrazine, metolachlor	Applications made closer to planting time improved weed control and corn yields compared to those made more than 15 d before planting.	Johnson et al. 1997
Microstegium vimineum control in forests	Fenoxaprop-P, imazapic, sethoxydim	Weed control was not affected by early-, mid-, or late-season herbicide application timings.	Judge et al. 2005
Weed control in IMI-tolerant *Oryza sativa*	Imazethapyr	Rice yields were higher from herbicide application timings (PRE and POST) up to 2- to 4-lf stage	Masson et al. 2001
Weed control in glyphosate-tolerant *Zea mays*	Glyphosate	V4 stage of corn considered ideal timing for glyphosate applied once for all weed densities	Myers et al. 2005

Table 13.1 (continued)

Situation	Herbicide	Strategy	Reference
Control of rhizome-Sorghum halepense	Nicosulfuron	Application to johnsongrass with >5 leaves controlled the weed better than when applied to johnsongrass with <5 leaves	Obrigawitch et al. 1990
Sorghum halepense and *Ipomoea lacunosa* control in *Glycine max*	Imazethapyr and Fluazifop	Both weeds better controlled by imazethapyr at 15-cm stage of johnsongrass. Fluazifop controlled johnsongrass up to 60-cm	Shaw et al. 1990
Control of the perennial weed *Brunnichia ovata*	Clopyralid, dicamba, glyphosate	Early October application timing found to provide highest weed control	Shaw and Mack 1991
Avena fatua control in spring *Hordeum vulgare*	Imazamethabenz	Barley yield was higher when herbicide was applied 1 wk after emergence compared to 2 and 3 wks	Stougaard et al. 1997
Control of annual grasses in *Zea mays*	Nicosulfuron	Application at 5–10 cm height of annual grasses provided similar or higher yields compared to application of PRE herbicides	Tapia et al. 1997
Killing *Vicia villosa* cover crop prior to planting no-till *Zea mays*	Burndown herbicides	Killing the cover crop before planting the crop optimized crop yield	Teasdale and Shirley 1998
Early POST vs. Late POST application of systemic non-selective herbicides in *Zea mays*	Glufosinate and glyphosate	Herbicide application 28 d after planting resulted in better weed control compared to that 35 d after planting.	Tharp and Kells 1999
Critical periods for weed control in *Glycine max*	Residual and POST herbicides	Weed control up to fourth node stage of soybean necessary to prevent yield loss; subsequent weed removal necessary from bloom to seed stage	Van Acker et al. 1993
Control of the perennial weeds—*Rubus* sp., *Lonicera japonica, Toxicodendron radicans*,and *Lespedeza cuneata*	Glyphosate	Optimal timings for control were: blackberry—mid-June to August; Japanese honeysuckle—August; poison-ivy—mid June to mid-August; sericea lespedeza—flowering time	Yonce and Skroch 1989

13.8 Spatial Dynamics and Weed Management

13.8.1 *Plants Growing Out of Place*

Weeds are typically considered as "plants growing out of place". This definition takes into consideration its role as a pest that interferes with human activities including agriculture. Conventionally, agricultural systems are intensively managed to maximize productivity of crops. In such systems the tolerance level of weeds is close to "zero" based on the above definition. Due to the high competitive and reproductive characteristics of weeds, farmers make all possible efforts to minimize their incidence and subsequent infestations in crop fields. In field crops, for instance, a mixture of three or four herbicides is typically applied to obtain a broad spectrum of weed control, and to manage the development of resistant weed biotypes (Hagood et al. 2010).

Radosevich (1987) examined the interactions between crops and weeds and determined factors such as plant density, species proportion, and spatial arrangement to play roles in competition. He noted that competition be considered based on plant proximity responses as determined by germination, growth, and reproductive characteristics of individual species rather than inherent differences in fitness. Soybean yield was affected by common cocklebur, Palmer amaranth growing only within 12.5 cm of the crop, and by tall morningglory growing within 25 cm of the crop (Monks and Oliver 1988). In their study, the proximity of johnsongrass and sicklepod did not affect soybean yield. Weed competition based on spatial arrangement of weeds with respect to crops have also been referred to as "area of influence" or "zone of exploitation" by researchers. These areas or zones may also be affected by weed canopy diameter (Wilkerson et al. 1989). Besides, tall-growing weeds such as common cocklebur (*Xantbium strumarium*), velvetleaf (*Abutilon theophrasti*), and jimsonweed (*Datura stramonium*) can successfully compete with shorter crops such as soybean for light with densities of 0.7 to 2.5 plants/m^2 causing yield reductions of 12 to 51% (Stoller and Woolley 1985).

A broader understanding of competitive zones of weeds will be of immense value to delineate site-specific weed management programs. Currently, we have a general understanding of the competitive nature of common weeds based on their ability to reduce yields, produce seeds, allelopathic attributes, etc. (Ross and Lembi 2008). Additional information on areas of influence of specific weeds will also be useful for targeted application of herbicides based on their prevalence and crop row spacing. Crop row spacing also plays a critical role in the ability of weeds to compete. Based on a mathematical model, crop plants grown in a square lattice, when all other factors are kept constant, provided optimal weed suppression (Fischer and Miles 1973).

13.8.2 Plants with Unknown Virtues

A weed is also known to many as "a plant whose virtues have not yet been discovered" (Blatchley 1912); or "considering all weeds as bad is nonsensical" (Cocannouer 1950); or "weeds have always been condemned without a fair trial" (King 1951). The relationship between weeds and crops growing side by side, and their mutual roles in the overall fabric of the ecosystem is complex and not well understood. In agricultural systems, crops are plants selected for survival whereas weeds are their cousins displaced gradually in the process. The role of weeds in improving soil quality and fertility, managing populations of herbivorous arthropod pests and their natural enemies, as self-sowing cover crops, as agents of biological tillage, as having edible value, as having an indirect role in plant breeding etc., have been described by Jordan and Vatovec (2004). The value of weeds as medicinal plants is yet another promising discipline worthy of renewed interest and due consideration.

Effective weed control methods developed in the recent decades, capable of managing the flora of large expanses of land are somewhat unprecedented given the long history of crop production. Harlan (1965) explained that weeds have served as reservoirs of germplasm and have periodically "injected portions of it" into crops to favor variability, herterozygosity and heterosis. According to the author, a biological significance exists between cultivated plants and their wild biotypes (weeds) and concluded that cultivated plants would never have succeeded without genetic support of their companion weeds. The implications of such phenomena are especially intriguing in the current era of genetically engineered crops where such "injected portions" belong to distantly related species and the two have essentially no survival tactics in common. Under this context, efforts to fill such voids in the literature will compliment current and future efforts to raise crops sustainably.

13.9 Sustainable Weed Management

Given the challenges faced by modern agricultural systems with shrinking levels of labor or human capital as a primary input in production, maintaining sustainability while remaining profitable can be a challenging task. This may apply to all activities related to agricultural production including weed management. This phenomenon can be explained by the bimodal nature of farms in the United States (Duffy 2006). The number of small farms (sales<$1,000/yr), increased by 37% during 1997 to 2002, and the number of the large farms (sales>$1 million/yr), increased by 8%. The numbers in all other farm size categories decreased during this period. In 2002, the large farms represented 3% of total US farms but accounted for 61% of produce sales. In such a situation, technology plays a critical role to maximize productivity and the expectation to shift from chemical to non-chemical methods for pest management could be largely unrealistic. On the other hand, sustainable practices may be more readily adopted in smaller farms where more intensive pest management practices can be carried out.

Wyse (1994) pointed out that weeds are a major deterrent to the development of sustainable agriculture systems since they dictate several crop production practices. He urged weed scientists to become leaders of collaborative integrated approaches to manage weeds in agricultural systems. Several strategies may be considered to manage weeds sustainably in agriculture. Developing cover/smother crops to suppress weeds, crop varieties with enhanced interference potential, biological weed control, and use of technology are a few areas of focus that would benefit from research.

The use of cover crops to manage weeds in agricultural systems continues to grow. Teasdale (1996) emphasized the viability of such crops in sustainable systems because of contributions to soil fertility and improved crop performance. Apart from this, crop residues from annual cover crops provide early-season weed suppression. The author also indicated that cover crops may also serve as living mulches that are effective to control weeds but may require chemical management to reduce competition with the crop.

Temporal and spatial diversification by adopting practices such as crop rotation and intercropping are strategies worthy of consideration to manage weeds in sustainable systems (Liebman and Dyck 1993). In a long-term study that lasted eight years, weed biomasses were recorded in four different rotations, which included two or three crops followed by fallow compared to a single continuous crop of proso millet (Anderson 2006). At the end of the study the weed biomass was 85 % lower in the wheat-millet-fallow rotational sequence compared to continuous proso millet. Carruthers et al. (1998) determined weed control levels comparable to conventional methods by intercropping corn with legumes compared to a monocrop of corn alone.

In an extensive study carried out in the Canadian prairies which spanned 56-site years, fewer perennial and biennial weeds were associated with minimum and zero-tillage compared to conventional tillage (Blackshaw et al. 2006). Several summer annuals were also less common under conservation tillage compared to conventional tillage. Winter annuals which germinated in fall and summer annuals dispersed by wind were higher in conservation tillage compared to conventional tillage. Melander et al. (2005) described the use of thermal and various mechanical devices to manage weeds in row crops in a number of investigations. Improved devices such flamers, harrows, brushes, hoes, torsion weeders, and finger weeders as well as certain novel devices such as robots were also reviewed. The authors indicated that such implements may be effective as an integrated approach to manage weeds that may include other approaches at the cropping systems level.

Sustainable approaches may be more readily applicable in non-crop situations such as turfgrasses, where weeds are primarily of aesthetic concerns. In turfgrasses, providing good growth conditions for the turf can reduce the opportunities for weed infestation (Chandran 2006). A fully functional turf with few weeds can be maintained sustainably. Occasional use of herbicides may be necessary to bring down the weed population to manageable levels prior to initiating or continuing a sustainable weed management program. Maintaining a dense turf with a competitive ability to reduce the emergence and establishment of weeds is perhaps the best strategy to minimize weed infestation in lawns. A good understanding of factors such as soil pH, species and cultivar selection, proper turf establishment, cultural requirements, etc., is essential to manage weeds proactively in turfgrasses. A summary of research findings related to weed management strategies based on reduced use of herbicides in various crops is presented in Table. 13.2.

13.10 Advances in Biological Weed Control

Biological control of weeds, which involves the use of other living organisms, is best regarded as a technique to be used in conjunction with other efforts in integrated weed management systems (Zimdahl 1999). While certain risks such as inconsistent results, possible escape to become a pest as a result of mutations, slow weed control etc., this method is considered to be more sustainable with a high ratio of benefit: cost. This is especially true in the case of managing certain invasive weeds that are widespread and chemical control methods are not feasible. While insects and fungi are the more commonly used biological control agents, fish,

Table 13.2 Summary of relevant literature on reduced herbicide application rates for weed management

Situation	Herbicide	Strategy	Reference
Post-emergence control of *Xanthium strumarium* and *Setaria faberi*, in Glycine max	Acifluorfen, bentazon, chlorimuron, and sethoxydim	Two sequential applications of tank-mixtures at 0.25X labeled rates of first three herbicides applied with sethoxydim (0.5X rate) provided similar weed control and yield as full rate of herbicides applied once	Defelice et al. 1989
Post-emergence control of *Xanthium strumarium*, *Ambrosia trifida*, *Helianthus annus*, *Amaranthus hybridus*, and *Abutilon theophrasti* in Glycine max	Acifluorfen, bentazon, and chlorimuron	Application of herbicides at 0.5X rate at 2 wk after planting controlled weeds similar to that of standard rate at 4 wk after planting; in some cases 0.25X rate provided similar results	Devlin et al. 1991
Pre-emergence weed control in Zea mays	Atrazine	Banding herbicide along with mechanical weeding as effective as broadcast application; reduced herbicide by 73% and quantified lower atrazine residues in soil	Heydel et al. 1999
Early season weed control in Glycine max	Acifluorfen, bentazon, chlorimuron, and imazaquin	Reduced rates of herbicides provided 90% weed control when applied 6–12 d after weed emergence	King and Oliver 1992
Pre-emergence weed control in Zea mays	Atrazine and metolachlor	Herbicide use was reduced by 50 to 75% with minimal loss of corn yield or weed control by integrating mechanical control and banded application of herbicides	Mudler and Doll 1993
Broadleaf weed control in Glycine max	Bentazon, chlorimuron, imazaquin, imazethapyr	Single and sequential application of herbicides at reduced rates did not affect yield compared to full rates	Steckel et al. 1990

aquatic mammals, and vertebrates have also been effectively used to control weeds. The United States, Australia, South Africa, Canada, and New Zealand use biological agents to control weeds the most in natural ecosystems (McFadyen 1998). An updated list of invasive weeds in North America and potential biological control agents is provided in Table 13.3.

13.11 Technology in Weed Management

Herbicide application technology has improved considerably in recent years. Variable-rate technology (VRT) although used widely for fertilizer applications, has not yet been adopted widely for herbicide application. Variability in weed spectrum,

Table 13.3 Recently reported biological control agents with potential to control certain invasive weeds in North America

Weed/s	Potential biological control agent	Reference
Lythrum salicaria (purple loosestrife)	*Galerucella calmariensis* and *G. pusilla*	Blossey et al. 2001
Cirsium arvense (Canada thistle)	*Ceutorhynchus litura*	Collier et al. 2007
Persicaria perfoliata (mile-a-minute)	*Rhinoncomimus latipes*	Colpetzer et al. 2004
Tamarix spp. (salt cedar)	*Diorhabda elongata* Brulle *deserticola* Chen	DeLoach et al. 2003
Ailanthus altissima (tree-of-heaven)	*Eucryptorrhynchus brandti*	Ding et al. 2006
Microstegium vimineum (Japanese stiltgrass)	*Bipolaris* sp.	Kleczewski and Flory 2010
Melaleuca quinquenervia (melaleuca)	*Puccinia psidii*	Rayachhetry et al. 2001
Fallopia japonica (Japanese knotweed)	*Aphalara itadori* Shinji	Shaw et al. 2009
Phragmites australis (common reed)	*Rhizedra lutosa, Phragmataecia castaneae,* *Chilo phragmitella, Schoenobius gigantella.* *Archanara, Arenostola* and *Platycephala planifrons*	Tewksbury et al. 2002
Fallopia japonica (Japanese knotweed)	*Gallerucida bifasciata*	Wang et al. 2008

seed bank, age, and spatial distribution of emerged weeds in the field are some of the barriers to be overcome for this otherwise promising technology. In a field study to test the effectiveness of VRT in soybeans involving three herbicides rates (100, 67 and 33 % use rates), the medium rate provided weed control similar to that of the full rate (Thorp and Tian 2004) while the 33 % rate failed to provide acceptable levels of weed control. To overcome the difficulty associated with weed distribution differences, Dammer and Wartenberg (2007) designed a sprayer capable of applying variable rates of herbicides by detecting weeds using a sensor. In 13 field trails carried out in cereals and peas an average of 25 % herbicide reduction was achieved without causing any crop yield reduction.

Slaughter et al. (2008) reviewed the status of using autonomous robots to control weeds and concluded that detection and identification of weeds under a wide range of conditions was the greatest challenge in agricultural situations. However, the authors indicated that there is potential for adopting this technology in the field. The authors presented concept diagrams of futuristic robots fitted with multiple cameras on mobile robotic arms to allow multiple views of each plant. Devices with onboard electronics and herbicide reservoirs would be used to discriminate weeds from crops and to manage them. Similar devices are being field-tested and developed for crop-thinning and mechanical weed control by Blue River Technology, California, USA (Fig. 13.3a). It is envisioned that autonomous robots capable of performing such tasks will play a significant role in weed management in the future (Fig. 13.3b).

Fig. 13.3 **a** Field-testing of a prototype equipment developed by Blue River Technology, CA, USA, capable of mechanically thinning crops or rouging weeds (*top*); **b** a futuristic vision of autonomous robots performing such tasks (*bottom*). *Photo credit: J. Heraud*

13.12 Herbicide Use Reduction in Agronomic Crops

Among various crops in the United States, agronomic crops have historically ranked first in total amount of active herbicide ingredients used. Roughly 75 % of total herbicide use in the U.S. was in corn and soybean in 1990 (Zoschke 1994). This trend continues today for the relative amounts of herbicide use in agronomic crops compared to other crops such as horticultural crops, turf and ornamentals, aquatic and other non-crop areas. Significant reductions in herbicide use could be accomplished by identifying areas within agronomic crops where herbicide use reductions could be implemented. The strategy discussed below may have significant implications in reducing overall herbicide use.

13.12.1 Banded Herbicide Application—Herbicide Use Reduction in Corn.

In the United States, about 37 million hectares (92 million acres) were dedicated to corn production in 2011, generating revenue of $2,052/ha ($831/acre) (USDA-NASS 2011). About 98% of US corn acreage in 2011 received herbicide application. The herbicide atrazine was applied to 61% of the hectares, averaging 1.15 kg atrazine per hectare (1.03 lb/A). While the ecological attributes of atrazine are under public scrutiny (Hayes et al. 2002), this herbicide is a cost-effective weed management tool for corn producers (Williams et al. 2010). Measures to mitigate its use while optimizing its effectiveness may ensure the continued availability of this broad-spectrum pre-emergence herbicide.

Corn is most vulnerable to weed competition during the 3- to 14-leaf stage (Hall et al. 1992), which typically coincides with the first six weeks of crop growth or until canopy closure. Corn grown for grain, silage, or ethanol may be able to tolerate different levels of weed competition. Current weed control programs in corn typically provide close to 100% weed control. The conventional weed management practice in corn is the application of a mixture of pre-emergence herbicides, which typically includes atrazine, along with a non-selective post-emergence herbicide, as a broadcast treatment. This practice keeps vast expanses of land under corn hectarage, more or less as a monoculture. Reduced biodiversity, reduced soil cover, habitat loss, decline of beneficial insects, increased nutrient and pesticide runoff, and reduced carbon sequestration are few of the drawbacks associated with this practice. Providing limited space for weeds to co-exist with the crop without affecting crop yields may also reduce selection pressure and the resultant development of herbicide-resistant weed biotypes.

Buildup of the weed seed bank and resultant yield losses due to weed competition are presumed risks that deter growers from adopting this practice. Burnside et al. (1986; p. 248) questioned "As farmers reduce the weed seed bank in soils, can they reduce their weed control expenditures without adversely affecting crop yields?" and indicated that "These and other questions will occupy considerable time of weed scientists in the future". In their 6-yr long experiment, it was determined that viable weed seed levels in the soil declined 95% during a 5-yr period during which weed seed production was eliminated by providing total weed control. However, the weed seed buildup recovered to >90% level when weeds were left unmanaged during the 6th year, at two out of five locations. In the remaining three locations, the weed seed buildup during the 6th year in untreated plots was similar to that in treated plots. They also determined that corn-yields were unchanged during the 6th year with minimum weed management.

Literature on the effect of banding herbicides on corn yield is limited and is restricted to older classes of herbicides. Uremis et al. (2004) determined that banding was as effective as broadcast application. In their study, different bandwidths gave similar levels of weed control and corn yield, and noted that banding decreased herbicide use by up to 78%. In a Missouri study, Donald et al. (2004) determined that banding herbicides reduced application rates by 53% when averaged over three years and that significant yield reductions were not seen compared to broadcast application of the same herbicides. Hansen et al. (2000) compared broadcast and banded application of a tank-mixture of PRE herbicides in tilled corn. They noted reduced levels of nutrient runoff as a result of ground cover provided by weeds in banded treatments compared to that from broadcast applications. No yield differences were recorded between broadcast and banded application of herbicides in this study also. In a study to compare atrazine leaching following broadcast or banded applications in corn, Heydel et al. (1999) quantified reduced levels of atrazine residues in the soil associated with banded applications without affecting corn yields.

Field experiments were conducted by the author at three locations in West Virginia to compare banded and broadcast applications of pre-emergence herbicides on corn yield and weed biodiversity levels, from 2009 to 2011. The objective of this research was to determine the effect of banding newer classes of pre-emergence herbicide mixtures containing atrazine on corn yield compared to conventional broadcast application of the same at grower level locations to simulate field conditions. The floral biodiversity at one location was also monitored. Corn rows, planted 75 cm apart, were treated with a pre-emergence mixture of atrazine, metolachlor, and mesotrione at 1.702, 1.702, and 0.220 kg ai/ha applied either broadcast or in bands of width 38 cm over 10- to 20-cm tall corn. Corn yield was estimated after determining its moisture content. All data were subjected to analysis of variance (ANOVA) and means were separated using LSD ($P=0.05$). Floral biodiversity levels were calculated using Shannon's Index.

Banded application resulted in 50% reduction of atrazine, metolachlor, and mesotrione, respectively, on a per-hectare basis, compared to broadcast application (Table 13.4). Yield data indicated no significant differences between plots that received banded and broadcast treatments (Table 13.5). Excellent (>95%) weed control was observed within band- or broadcast-treated areas until canopy closure. When the yield data from the four studies were combined, statistical differences could not be determined (Fig. 13.4a, b). Shannon's Index for Biodiversity analysis generated H values>1.5 which were considered to be biologically-diverse (Chandran et al. 2011). Banded application allowed for natural populations of weeds to establish between corn-rows. Broadcast application of herbicides kept the entire cornfields relatively weed-free.

A field-day was organized in 2010 to discuss this practice with growers (WVU Press Release 2010). One of the concerns expressed by growers was the buildup of the weed seed bank if weeds were left uncontrolled in banded fields. The growers requested data from long-term (5-yr) studies under different weed population levels and weather conditions to gain confidence. Future research to determine which years to warrant broadcast or banded application based on weed seed bank analysis will also be considered useful. Harvest weed seed control (HWSC) systems being developed in Australia, where machinery capable of harvesting and destroying weed seeds at the time of grain harvest, holds promise for the widespread implementation of herbicide banding in the future (Walsh et al. 2013).

Our results imply that it may be economically feasible to band-apply herbicides in cornfields that are relatively weed-free as a result of employing good weed control programs over several years. This is because the low weed seed bank may cause

Table 13.4 Use pattern of broadcast and banded applications of herbicides in corn at grower locations

Application [cm (inch)]	Spray Fluid L/ha (gal/acre)	Atrazine	Metolachlor	Mesotrione
			[kg/ha (lb/A)]	
Banded-38 (15.0)	56.77 (15)	0.85 (0.65)	0.85 (0.65)	0.11 (0.04)
Broadcast -76 (30)	113.55 (30)	1.702 (1.3)	1.702 (1.3)	0.22 (0.163)
Control	0	0	0	0

Table 13.5 Corn yield comparisons between banded and broadcast treatments at grower locations in Charles Town (Location 1), Moorefield (Location 2), and Point Pleasant, (Location 3), West Virginia

	Corn Yield								
	Year 2010		Year 2011						
Herbicide application[a]	Location 1		Location 1		Location 2		Location 3		Average
	kg/ha (*bushels/A*)								
Broadcast	6552	(*104*)	6048	(*96*)	10080	(*160*)	7623	(*121*)	7560 (*120*)
Banded	6363	(*101*)	5040	(*80*)	9576	(*152*)	6867	(*109*)	6993 (*111*)
Control	5103	(*81*)	630	(*10*)	5418	(*86*)	6363	(*101*)	4410 (*70*)
LSD ($P=0.05$)	1260	(*20*)	1008	(*16*)	7245	(*115*)	3213	(*51*)	2079 (*33*)

[a] A mixture of atrazine, glyphosate, metolachlor, and mesotrione was applied at 1.702, 1.702, and 0.220 kg ai/ha; banded treatments received half this quantity per hectare.

minimal weed pressure in such fields. However, if the weed seed bank is high, broadcast application may be necessary. In such instances, carrying out a bioassay by collecting representative soil samples from the field and recording viable seeds by transferring them to a greenhouse and testing for germination would be an appropriate decision making tool (Brainard and Bellinder 2004). Simpler methods such as scouting the fields for weeds during the growing season may also help make decisions for the following year. Perhaps, banded applications can be carried out periodically, based on weed pressure, or herbicides such as atrazine that carry higher risks may be applied separately in bands using modified spray equipment with separate tanks for broadcast and band applications. The implication of this strategy to reduce the buildup of herbicide-resistant weed biotypes by reducing selection pressure is worthy of further investigation. If deemed to be an effective strategy, it could be adopted as a practice to manage herbicide resistance for newer classes of herbicides and in regions where resistant populations are not present currently.

13.12.2 Horticultural Crops

As discussed earlier, the resurgence of small farms producing high-value horticultural crops provides opportunities for non-chemical weed control methods to be carried out. Ashworth and Harrison (1983) evaluated a variety of organic and synthetic mulch

Fig. 13.4 a Application of preemergence herbicides in bands over corn-rows reduced herbicide use by 50% while maintaining a biologically-diverse cornfield (*top*) without affecting yield, **b** compared to conventional broadcast application resulting in a weed-free cornfield (*bottom*)

treatments used around vegetable crops and woody ornamental species. They found that organic mulches required application to a depth of at least 5 cm and that the most effective weed control was provided by black polyethylene because it remained intact throughout the summer. Similarly, field-grown tomatoes grown under black polyethylene had significantly higher total yield than those tomatoes without the mulch (Abdul-Baki et al. 1992). Additionally, the use of black polyethylene mulch greatly increased fresh and dry weight yields of basil (*Ocimum basilicum*) and rosemary (*Rosmarinus officinalis*) (Ricotta and Masiunas 1991; Davis 1994). Straw mulch at 16 tons per hectare has the capacity to reduce weed biomass by 30 to 83% and increase the yield of pointed gourd compared to unmulched plots (Ghorai and Bera 1998).

Field experiments conducted by the author in West Virginia evaluated hand cultivation, plastic mulch, and straw mulch for weed control, growth attributes, and yield of sweet pepper (*Capsicum annum*) in 2000–2001. In 2000, under rain-fed conditions, plastic mulch resulted in maximum pepper yield with increases of ~150% compared to 20 cm straw mulch and 50% compared to hand cultivation (Table 13.6). In this study root dry weights correlated positively to pepper yields.

The use of composted poultry litter as a mulch in orchard systems was documented not only to reduce weed competition in apples but was also determined to

Table 13.6 Yield, shoot and root weights of rain-fed sweet pepper (*Capsicum annum* L.var. "Ace") as affected by physical weed control methods (2000)

Treatment	Pepper yield	Pepper number	Shoot dry wt.	Root length	Root dry wt
	kg/plot	(per plot)	(g/plot)	cm	g/plant
Hand Cultivation	14.68	321	714	11.3	3.25
Plastic Mulch	23.47	655	1161	17.5	3.17
Straw Mulch (5 cm)	5.02	173	296	11.5	1.66
Straw Mulch (10 cm)	3.47	152	246	9.5	1.38
Straw Mulch (20 cm)	9.41	285	554	13.0	2.74
Control	1.21	21	62	7.2	1.65
L.S.D (P=0.05)	3.83	104	156	1.5	1.32

be beneficial in an orchard ecosystem to manage tree fruit diseases and insect pests (Brown and Tworkoski 2006). Tworkoski and Glenn (2012) determined from a 4-yr study that certain cool-season grasses grown in tree-rows successfully deterred weed competition without affecting apple and peach yield. The authors concluded that growing an annually-mowed grass in tree rows may be a viable option to reduce herbicide use in orchards but fruit size may be reduced.

13.12.3 Engagement of Industry—A Potential Opportunity

Undoubtedly, the chemical industry plays a major role in crop protection (Gasser and Fraley 1989). If it were not for useful chemistries and other technology developed by the researchers in the industry, the supply of food and fiber would not have been able to keep up with the demands of a growing world population. These are valuable services seldom appreciated by an average individual. To maintain the ability of industry to remain innovative and service-oriented, profitability in the marketplace is critical. Conventionally, such profits are generated through sales of pesticides, hybrid seeds, and similar products of value to their clientele. It may be worthwhile for the industry to consider marketing other services to foster sustainable agriculture.

Mechanisms to engage the industry in sustainable agriculture may be fruitful in the long-term. It may require a process of "thinking outside the box" to generate and implement viable ideas. Including an 'IPM', 'Eco-friendly', or 'Green' facility under the infrastructural umbrella may be worthy of consideration by major chemical companies. Such facilities may provide a diverse array of services such as consultancy to help growers implement proven sustainable practices, insurance to minimize any associated risks, mass production of biological pesticides and other bio-control agents, development of novel application technologies, scouting and monitoring, development and marketing of cultural tactics to manage resistant biotypes of weeds, etc. Such products may counteract any losses in revenue as a result of reduced pesticide sales.

In the United States, several incentives are available to growers to conserve resources in agricultural settings. The industry could facilitate the adoption of such practices and be compensated by growers for the services provided. If such services are included under the same umbrella of larger corporates, operational costs could be reduced as activities of different entities are coordinated in a concerted manner. Moreover, such a system would dramatically improve the public perception and credibility of the industry among stakeholders and help build positive relationships with environmental groups towards a productive rapport.

13.13 Conclusion

Weed management will continue to play an important role to ensure the supply of conventional food and fiber to meet the demands of a growing global population in years to come. Currently we are at the crossroads of cutting-edge technology and growing concerns related to implications of the same on sustainability. At this juncture, it is important to realize that this phenomenon is the inevitable cost of fewer hands feeding more mouths worldwide. Based on the growth pattern of most economies, humans shift from a farm-based livelihood to one that is based on services. Production agriculture continues to remain the burden of a shrinking fraction of the human population. Therefore producers have limited choices but to depend on cost-effective technologies to remain viable. Unless corrections are in place such trends are bound to continue.

Conscientious efforts favoring locally-grown produce to those shipped from elsewhere are gaining popularity in urban communities. Weed management in small farms could be more sustainable compared to that in industrialized agriculture. Some of the strategies discussed in this chapter may be more readily applicable to small scale production. Currently most of the research related to weed management at universities in the United States is geared towards large-scale production agriculture. The current structure of most universities which foster a climate of revenue generation to remain competitive also tends to encourage such tendencies.

Agriculture has never been in balance with Mother Nature. Ever since man raised crops to feed and clothe himself, disturbances to the ecosystem have occurred progressively. Such imbalances may be correlated to changes in human population, economic growth, and land use patterns. While the demand for organic food has increased recently, the average consumer may not be able to afford them. If current trends in economic disparities of society continue to grow, industrialized agriculture may emerge as the only solution to feed the masses while foods posing fewer risks to human health and produced in an eco-friendly manner may become the convenient choice for others.

Acknowledgments The author wishes to thank Robert Clemmer, Carl Coburn, Wayne Currey, Daniel Foglesong, Michael Harman, Samuel Harper, Irvin King, Harold Kiger, Fred Roe, Kimberly Salinas, Rodney Wallbrown, Dave Workman, and Craig Yohn, for their assistance in conducting field experiments. Funding from the National Institute of Food and Agriculture is also acknowledged.

References

Abdul-Baki, A., Spence, C., & Hoover, R. (1992). Black polyethylene mulch doubled yield of fresh-market field tomatoes. *Hort Science, 27*(7), 787–789.

Altman, J., & Campbell, C. L. (1977). Effect of herbicides on plant diseases. *Annual Review of Phytopathology, 15,* 361–85.

Anderson, R. L. (2006). A rotation design that aids annual weed management in a semiarid region. In H. P. Singh, D. R. Bathish, & R. K. Kohli (Eds.)., *Handbook of Sustainable Weed Management* (pp. 159–177). New York: Haworth Press.

Anonymous. (1907). Vermont weeds. *Vermont Agricultural Experiment Station Annual Report, 20,* 417–422.

Anonymous. (1958). *Open Door to Plenty.* Washington. D.C.: National Agricultural Chemical Association (NACA).

Ashton, F. M., & Monaco, T. J. (1991). Weed management practices. In J. L. Garraway (Ed.)., *Weed Science Principles and Practices* (3rd ed., pp. 34–67). New York: Wiley.

Ashworth, S., & Harrison, H. (1983). Evaluation of mulches for use in the home garden. *HortScience, 18,* 180–182.

Benbrook, C. M. (2012). Impacts of genetically engineered crops on pesticide use in the U.S.—the first sixteen years. *Environmental Sciences Europe, 24,* 24.

Blackshaw, R. E., Thomas, A. G., Derksen, D. A., Moyer, J. R., Watson, Legere, A., & Turnbull, G. C. (2006). Examining tillage and crop rotation effects on weed populations in the Canadian prairies. In H. P. Singh, D. R. Bathish, & R. K. Kohli (Eds.)., *Handbook of Sustainable Weed Management* (pp. 179–207). New York: Haworth Press.

Blatchley, W. S. (1912). *The Indiana Weed Book.* Indianapolis, Indiana, USA: Nature.

Blossey, B., Casagrande, R., Tewksbury, L., Landis, D. A., Wiedenmann, R. N., & Ellis, D. R. (2001). Nontarget feeding of leaf-beetles introduced to control purple loosestrife (*Lythrum salicaria* L.). *Natural Areas Journal, 21,* 368–377.

Bolley, H. L. (1901). Studies upon weeds in 1900. *North Dakota Agricultural Experiment Station Annual Report, 11,* 48–56.

Brainard, D. C., & Bellinder, R. R. (2004). Weed suppression in a broccoli-winter rye intercropping system. *Weed Science, 52*(2), 281–290.

Brown, M. W., & Tworkoski, T. (2006). Pest management benefits of compost mulch in apple orchards. *Agriculture, Ecosystems & Environment, 103,* 465–472.

Burnside, O. C., Moomaw, R. S., Roeth, F. W., Wicks, G. A., & Wilson, R. G. (1986). Weed seed demise in soil in weed-free corn (*Zea mays*) production across Nebraska. *Weed Science, 34,* 248–251.

Carey, J. B., & Kells, J. J. (1995). Timing of total postemergence herbicide applications to maximize weed control and corn (*Zea mays*) yield. *Weed Technology, 9,* 356–361.

Carruthers, K., Cloutier, D. Fe, Q., & Smith, D. L. (1998). Intercropping corn with soybean, lupin and forages: Weed control by intercrops combined with interrow cultivation. *European Journal of Agronomy, 8,* 225–238.

Chandran, R. S. (2006). Integrated turfgrass weed management. In H. P. Singh, Bathish, D. R., & Kohli, R. K. (Eds.)., *Handbook of Sustainable Weed Management* (pp. 791–812). New York: Haworth Press.

Chandran, R. S., Yohn, C. W., & Coburn, C. W. (2011). Effect of herbicide banding on yield and biodiversity levels of field corn. *Proceedings of the Northeastern Weed Science Society, 65,* 93.

Chapin III, F. S., Zavaleta, E. S., Eviner, V. T., Naylor, R. L., Vitousek, P. M., Reynolds, H. L., Hooper, D. U, Lavorel, S., Sala, O. E., Hobbie, S. E., Mack, M. C., & Díaz, S. (2000). Consequences of changing biodiversity. *Nature, 405,* 234–242.

Cocannouer, J. A. (1950). *Weeds Guardians of the Soil.* Greenwich, Connecticut, USA: The Devin-Adair Company Old.

Collier, T. R., Enloe, S. F., Sciegienka, J. K., & Menalled, F. D. (2007). Combined impacts of *Ceutorhynchus litura* and herbicide treatments for Canada thistle suppression. *Biological Control, 43,* 231–236.

Colpetzer, K., Hough-Goldstein, J., Ding, J., & Fu, W. (2004). Host specificity of the Asian weevil, *Rhinoncomimus latipes* Korotyaev (Coleoptera: Curculionidae), a potential biological control agent of mile-a-minute weed. *Polygonum perfoliatum* L. *(Polygonales: Polygonaceae). Biological Control, 30,* 511–522.

Dammer. K.-H., & Wartenberg, G. (2007). Sensor-based weed detection and application of variable herbicide rates in real time. *Crop Protection, 26*(3), 270–277.

Davis, J. M. (1994). *North Carolina Basil Production Guide.* North Carolina Agricultural Extension Service, AG-477. http://www.ces.ncsu.edu/depts/hort/consumer/agpubs/basil.pdf. Accessed 24 April 2013.

Defelice, M. S., Brown, W. B., Aldrich, R. J., Sims, B. D., Judy, D. T., & Guethle, D. R. (1989). Weed control in soybeans (*Glycine max*) with reduced rates of postemergence herbicides. *Weed Science, 37,* 365–374.

DeLoach, C. J., Lewis, P. A., Herr, J. C., Carruthers, R. I., Tracy, J. L., & Johnson, J. (2003). Host specificity of the leaf beetle, *Diorhabda elongata deserticola* (Coleoptera: Chrysomelidae) from Asia, a biological control agent for saltcedars (*Tamarix*: Tamaricaceae) in the Western United States. *Biological Control, 27,* 117–147.

Devlin, D. L., Long, J. H., & Maddux, L. D. (1991). Using reduced rates of postemergence herbicides in soybeans (*Glycine max*). *Weed Technology, 5,* 834–840.

Ding, J., Wu, Y., Zheng, H., Fu, W., Reardon, R., & Liu, M. (2006). Assessing potential biological control of the invasive plant, tree of heaven, *Ailanthus altissima. Biocontrol Science and Technology, 16,* 547–566.

Donald, W. W., Archer, D., Johnson, W. G., & Nelson, K. (2004). Zone herbicide application controls annual weeds and reduces residual herbicide use in corn. *Weed Science, 52,* 821–833. [Erratum: 2005. 53:138.]

Duffy, M. (2006). The clock is ticking for rural America. *Working Paper #06015.* Department of Economics, Iowa State University, Ames, Iowa, USA.

Eizenberg. H., Colquhoun, J. B., & Mallory-Smith, C. A. (2006). Imazamox application timing for small broomrape (*Orobanche minor*) control in red clover. *Weed Science, 54,* 923–927.

Fischer R. A., & Miles, R. E. (1973). The role of spatial pattern in the competition between crop plants and weeds- a theoretical analysis. *Mathematical Biosciences, 18,* 335–350.

Gasser, C. S., & Fraley, R. T. (1989). Genetically engineering plants for crop improvement. *Science, 244,* 1293–1299.

Ghorai, A. K., & Bera, P. S. (1998). Weed control through organic mulching in pointed gourd (*Trichosanthes dioica* L.). *Indian Journal of Weed Science, 30(1–2),* 14–17.

Gittins, B. S. (1959). *Land of Plenty* (2nd ed.). Chicago: Farm Equipment Institute.

Goldburg, R. J. (1992). Environmental concerns with the development of herbicide-tolerant plants. *Weed Technology, 6,* 647–652.

Gower, S. A., Loux, M. M., Cardina, J., & Harrison, S. K. (2002). Effect of planting date, residual herbicide, and postemergence application timing on weed control and grain yield in glyphosate-tolerant corn (*Zea mays*). *Weed Technology, 16,* 488–494.

Grichar, W. J. (1997). Influence of herbicides and timing of application on broadleaf weed control in peanut (*Arachis hypogaea*). *Weed Technology, 11,* 708–713.

Groh, H. (1922). A survey of weed control and investigation in Canada. *Scientific Agriculture, 3,* 415–420.

Hager, A. G., Wax, L. M., Bollero, G. A., & Stoller, E. W. (2003). Influence of diphenylether herbicide application rate and timing on common waterhemp (*Amaranthus rudis*) control in soybean (*Glycine max*). *Weed Technology, 17,* 14–20.

Hagood, E. S., Wilson, H. P., Ritter, R. L., Majek, B. A., Curran, W. S., Chandran, R. S., & VanGessel, M. J. (2010). Weed control in field crops. Section 5–23. In L. Guinn (Ed.)., *Pest Management Guide for Field Crops—2010, Publication 456-016.* Virginia, USA: Virginia Cooperative Extension Service. http://pubs.ext.vt.edu/456/456-016/456-016.html. Accessed 27 Feb 2014.

Hall, M. R., Swanton, C. J., & Anderson, G. W. (1992). The critical period of weed control in grain corn. *Weed Science, 40,* 441–447.

Hansen, N. C., Gupta, S. C., & Moncrief, J. F. (2000). Herbicide banding and tillage effects on runoff, sediment, and phosphorus losses. *Journal of Environmental Quality, 29,* 1555–1560.

Harlan, J. R. (1965). The possible role of weed races in the evolution of cultivated plants. *Euphytica, 14,* 173–176.
Harker, K. N., & Dekker, J. (1988). Effects of phenology on translocation patterns of several herbicides in quackgrass, *Agropyron repens. Weed Science, 36,* 463–472.
Harrington, T. B., & Miller, J. H. (2005). Effects of application rate, timing, and formulation of glyphosate and triclopyr on control of chinese privet (*Ligustrum sinense*). *Weed Technology, 19,* 47–54.
Hayes, T. B., Collins, A., Lee, M., Mendoza, M., Noriega, N., Stuart, A. A., & Vonk, A. (2002). Hermaphroditic, demasculinized frogs after exposure to the herbicide atrazine at low ecologically relevant doses. *Proceedings of the National Academy of Sciences U S A, 99,* 5476–5480.
Heap, I. (2012). International survey of herbicide resistant weeds. http://www.weedscience.com. Accessed 12 Oct 2012.
Heydel, L., Benoit, M., & Schiavon, M. (1999). Reducing atrazine leaching by integrating reduced herbicide use with mechanical weeding in corn (*Zea mays*). *European Journal of Agronomy, 11,* 217–225.
Jones, L. R., & Orton, W. G. (1896). The orange hawkweed or "paint-brush." *Vermont Agricultural Experiment Station Bulletin, 56.*
Johnson, W. G., Defelice, M. S., & Holman, C. S. (1997). Application timing affects weed control with metolachlor plus atrazine in no-till corn (*Zea mays*). *Weed Technology, 11,* 207–211.
Johnson, B. C., Young, B. G., & Matthews, J. L. (2002). Effect of postemergence application rate and timing of mesotrione on corn (*Zea mays*) response and weed control. *Weed Technology, 16,* 414–420.
Jordan, N., & Vatovec, C. (2004). Agroecological benefits from weeds. In Inderjit (Ed.)., *Weed Biology and Management* (pp. 137–158). Dordrecht, Netherlands: Kluwer Academic Publishers.
Judge, C. A., Neal, J. C., & Derr, J. F. (2005). Response of Japanese stiltgrass (*Microstegium vimineun*) to application timing, rate, and frequency of postemergence herbicides. *Weed Technology, 19,* 912–917.
King, F. C. (1951). *The Weed Problem-A New Approach.* London: Faber and Faber Ltd.
King, C. A., & Oliver, L. R. (1992). Application rate and timing of Acifluorfen, Bentazon, Chlorimuron, and Imazaquin. *Weed Technology, 6,* 526–534.
Kleczewski, N. M., & Flory, S. L. (2010). Leaf blight disease on the invasive grass *Microstegium vimineum* caused by a *Bipolaris* sp. *Plant Disease, 94,* 807–811.
Liebman, M., & Dyck, E. (1993). Crop rotation and intercropping strategies for weed management. *Ecological Applications, 3,* 92–122.
McFadyen, R. E. C. (1998). Biological control of weeds. *Annual Review of Entomology, 43,* 369–393.
Masson, J. A., Webster, E. P., & Williams, B. J. (2001). Flood depth, application timing, and imazethapyr activity in imidazolinone-tolerant rice (*Oryza sativa*). *Weed Technology, 15,* 315–319.
Melander, B., Rasmussen, I. A., & Bàrberi, P. (2005). Integrating physical and cultural methods of weed control: examples from European research. *Weed Science, 53,* 369–381.
Monks, D. W., & Oliver, L. R. (1988). Interactions between soybean (*Glycine max*) cultivars and selected weeds. *Weed Science, 36,* 770–774.
Mortensen, D. A., Egan, J. F., Maxwell, B. D., Ryan, M. R., & Smith, R. G. (2012). Navigating a critical juncture for sustainable weed management. *Bioscience, 62,* 75–84.
Mudler, T. A., & Doll, J. D. (1993). Integrated reduced herbicide use with mechanical weeding in corn (*Zea mays*). *Weed Technology, 7,* 382–389.
Myers, M. W., Curran, W. S., VanGessel, M. J., Majek, B. A., Scott, B. A., Mortensen, D. A., Calvin, D. D., Karsten, H. D., & Roth, G. W. (2005). The effect of weed density and application timing on weed control and corn grain yield. *Weed Technology, 19,* 102–107.
North Central IPM Center. (2010). *Integrated Pest Management—The PAMS Approach.* http://nrcs.ipm.msu.edu/uploads/files/46/PAMSapproach2010–9–1new.pdf. Accessed Oct 2012.
Obrigawitch, T. T., Kenyon, W. H., & Kurtale, H. (1990). Effect of application timing on rhizome johnsongrass (*Sorghum halepense*) control with DPX-V9360. *Weed Science, 38,* 45–49.

Oerke, E. C. (2006). Crop losses to pests. *The Journal of Agricultural Science, 144,* 31–43.

Oka, I. N., & Pimentel, D. (1976). Herbicide (2,4-D) increases insect and pathogen pests on corn. *Science, 193,* 239–240.

Radosevich, S. R. (1987). Methods to study interactions among crops and weeds. *Weed Technology, 1,* 190–198.

Rayachhetry, M. B., Van, T. K., Center, T. D., & Elliott, M. L. (2001). Host range of *Puccinia psidii,* a potential biological control agent of *Melaleuca quinquenervia* in Florida. *Biological Control, 22,* 38–45.

Reddy, K. N., & Koger, C. H. (2006). Herbicide-resistant crops and weed management. In H. P. Singh, D. R. Bathish, & R. K. Kohli (Eds.)., *Handbook of Sustainable Weed Management* (pp. 549–580). New York: Haworth Press.

Ricotta, J. A., & Masiunas, J. B. (1991). The effects of black plastic mulch and weed control strategies on herb yield. *HortScience, 26*(5), 539–441.

Ross, M. A., & Lembi, C. A. (2008). Weeds and their importance. In *Applied Weed Science* (3rd ed.). Upper Saddle River, New Jersey, USA: Prentice Hall.

Séralini, G. E., Clair, E., Mesnage, R., Gress, S., Defarge, N., Malatesta, M., Hennequin, D., & Spiroux de Vendômois J., (2012). Long term toxicity of a Roundup herbicide and a Roundup-tolerant genetically modified maize. *Food and Chemical Toxicology,* xxx- xxx–xxx (in press)

Shaw, D. R., & Mack, R. E. (1991). Application timing of herbicides for the control of redvine (*Brunnichia ovata*). *Weed Technology, 5,* 125–129.

Shaw, D. R., Ratnayake, S., & Smith, C. A. (1990). Effects of herbicide application timing on johnsongrass (*Sorghum halepense*) and pitted morningglory (*Ipomoea lacunosa*) control. *Weed Technology, 4,* 900–903.

Shaw, R. H., Bryner, S., & Tanner, R. (2009). The life history and host range of the Japanese knotweed psyllid, Aphalara itadori Shinji: Potentially the first classical biological weed control agent for the European Union. *Biological Control, 49,* 105–113.

Slaughter, D. C., Giles, D. K., & Downey, D. (2008). Autonomous robotic weed control systems: A review. *Computers and Electronics in Agriculture, 61,* 63–78.

Steckel, L. E., Defelice, M. S., & Sims, B. D. (1990). Integrating reduced rates of postemergence herbicides and cultivation for broadleaf weed control in soybeans (*Glycine max*). *Weed Science, 38,* 541–545.

Stoller, E. W., & Woolley, J. T. (1985). Competition for light by broadleaf weeds in soybeans (*Glycine max*). *Weed Science, 33,* 199–202.

Stougaard, R. N., Maxwell, B. D., & Harris, J. D. (1997). Influence of application timing on the efficacy of reduced rate postemergence herbicides for wild oat (*Avena fatua*) control in spring barley (*Hordeum vulgare*). *Weed Technology, 11,* 283–289.

Tapia, L. S., Bauman, T. T., Harvey, R. G., Kells, J. J., Kapusta, G., Loux, M. M., Lueschen, W. E., Owen, M. D. K., Hageman, L. H., & Strachan, S. D. (1997). Postemergence herbicide application timing effects on annual grass control and corn (*Zea mays*) grain yield. *Weed Science, 45,* 138–143.

Teasdale, J. R. (1996). Contribution of cover crops to weed management in sustainable agricultural systems. *Journal of Production Agriculture, 9,* 475–479.

Teasdale, J. R., & Shirley, D. W. (1998). Influence of herbicide application timing on corn production in a hairy vetch cover crop. *Journal of Production Agriculture, 11,* 121–125.

Tewksbury, L., Casagrande, R., Blossey, B., Hafliger, P., & Schwarzlander, M. (2002). Potential for biological control of *Phragmites australis* in North America. *Biological Control, 23,* 191–212.

Tharp, B. E., & Kells, J. J. (1999). Influence of herbicide application rate, timing, and interrow cultivation on weed control and corn (*Zea mays*) yield in glufosinate-resistant and glyphosate-resistant corn. *Weed Technology, 13,* 807–813.

Thuiller, W. (2007). Climate change and the ecologist. *Nature, 448,* 550–552.

Thorp, K. R., & Tian, L. F. (2004). Performance study of variable-rate herbicide applications based on remote sensing imagery. *Biosystems Engineering, 88,* 35–47.

Timmons, F. L. (1970). A history of weed control in the United States and Canada. *Weed Science, 18,* 294–307.

Tull, J. (1762). Of hoeing. In *The Horse Hoeing Husbandry*. London: A. Millar.
Tworkoski, T. J., & Glenn, D. M. (2012). Weed suppression by grasses for orchard floor management. *Weed Technology, 26*, 559–565.
Uremis, I., Bayat, A., Uludag, A., Bozdogan, N., Aksoy, E., Soysal, A., & Gonen, O. (2004). Studies on different herbicide application methods in second-crop maize fields. *Crop Protection, 23*, 1137–1144.
USDA-NASS (National Agricultural Statistics Service). (2012). *Agricultural Chemical Use*. http://www.nass.usda.gov/Surveys/Guide_to_NASS_Surveys/Chemical_Use/FieldCrop-ChemicalUseFactSheet06.09.11.pdf. Accessed Oct 2012.
USDA-NASS (National Agricultural Statistics Service). (2011). *Agricultural Chemical Use*. http://www.nass.usda.gov/Surveys/Guide_to_NASS_Surveys/Chemical_Use/FieldCrop-ChemicalUseFactSheet06.09.11.pdf. Accessed 14 Feb 2012
USDA-ERS. (2012). Pesticide Use and Markets. June 2012. http://www.ers.usda.gov/topics/farm-practices-management/chemical-inputs/pesticide-use-markets.aspx. Accessed Oct 2012.
Van Acker, R. C., Swanton, C. J., & Weise, S. F. (1993). The critical period of weed control in soybean (*Glycine max*). *Weed Science, 41*, 194–200.
Wang, Y., Dinga, J., & Zhang, G. (2008). *Gallerucida bifasciata* (Coleoptera: Chrysomelidae), a potential biological control agent for Japanese knotweed (*Fallopia japonica*). *Biocontrol Science and Technology, 18*, 59–74.
Walsh, M., Newman, P., & Powles, S. (2013). Targeting weed seeds in-crop: A new weed control paradigm for global agriculture. *Weed Technology, 27*, 431–436.
Wilkerson, G. G., Jones, J. W., Coble, H. D., & Gunsolus, J. L. (1989). SOYWEED: A simulation model of soybean and common cocklebur growth and competition. *Agronomy Journal, 82*, 1003–1010.
Williams, M.M. II, Boerboom, C. M., & Rabaey, T. L. (2010). Significance of atrazine in sweet corn weed management systems. *Weed Technology, 24*, 139–142.
WVU Press Release. (2010). 'Reducing pesticide use': WVU field day set for Dri Lake farm June 24': http://wvutoday.wvu.edu/n/2010/06/21/reducing-pesticide-use-wvu-field-day-set-for-dri-lake-farm-june-24. Accessed 14 Feb 2013
Wunderlich, W. E. (1961). History of water hyacinth control. *U. S. Army Corps of Engineers*. New Orleans District: Mimeo Publication.
Wyse, D. L. (1994). New technologies and approaches for weed management in sustainable agriculture systems. *Weed Technology, 8*, 403–407.
Yonce, M. H., & Skroch, W. A. (1989). Control of selected perennial weeds with glyphosate. *Weed Science, 37*, 360–364.
Young, B. G. (2006). Changes in herbicide use patterns and production practices resulting from glyphosate-resistant crops. *Weed Technology, 20*, 301–307.
Zimdahl, R. L. (1999). *Fundamentals of Weed Science* (2nd ed.). San Diego, California, USA: Academic Press.
Zoschke, A. (1994). Toward reduced herbicide rates and adapted weed management. *Weed Technology, 8*, 376–386.

Chapter 14
Herbicide Resistant Crops and Weeds: Implications for Herbicide Use and Weed Management

George B. Frisvold and Jeanne M. Reeves

Contents

14.1	Introduction	332
14.2	Weed Management and Herbicide Use Before HR Crops	334
14.3	HR Crop Adoption and Effects	336
14.4	Reduced Diversity of Weed Control Tactics Leads to GR Weeds	339
14.5	Barriers to Resistance Management	347
14.6	Role and Limits of Stacked Trait HR Varieties	348
14.7	Conclusions	349
References		350

Abstract Since their introduction in the mid 1990s, growers adopted genetically modified (GM), herbicide resistant (HR) crop varieties quickly in the United States and they now account for most of the hectares planted to corn, soybeans, and cotton. Benefits to growers not captured in standard farm profit calculations appear to account for the popularity of HR varieties. HR crops have been credited with encouraging the adoption of conservation tillage and causing substitution to herbicides with lower toxicity and persistence in the environment. Evidence for the effect on conservation tillage is stronger than evidence for herbicide substitution. The latter has relied more on expert opinion surveys that are sometimes, but not always corroborated by careful farm-level studies. Adoption of HR crop varieties led to a dramatic reduction in the diversity of weed control tactics in U.S. agriculture and

G. B. Frisvold (✉)
Department of Agricultural & Resource Economics, University of Arizona,
319 Cesar Chavez Building,
Tucson, AZ 85721, USA
e-mail: frisvold@ag.arizona.edu

J. M. Reeves
Agricultural & Environmental Research Division, Cotton Incorporated,
6399 Weston Parkway Cary,
Cary, NC 27513, USA
e-mail: jreeves@cottoninc.com

D. Pimentel, R. Peshin (eds.), *Integrated Pest Management*,
DOI 10.1007/978-94-007-7796-5_14,
© Springer Science+Business Media Dordrecht 2014

the predictable evolution of HR weeds. Grower adoption of resistance management strategies has been limited and insufficient to delay resistance. Development of crop varieties resistant to multiple herbicides is being pursued as one strategy to respond to HR weeds. Debates remain over the potential of this approach relative to a more comprehensive integrated weed management strategy to successfully delay resistance.

Keywords Herbicides · Herbicide resistant weeds · Biotechnology · Genetically modified · Cotton · Maize · Soybeans

Abbreviations

ARMS	Agricultural Resources Management Survey
BMP	Best management practices (to delay weed resistance)
BXN	Bromoxynil (a herbicide)
CV	coefficient of variation
EIQ	Environmental Impact Quotient
EPA	U.S. Environmental Protection Agency
FIFRA	Federal Insecticide, Fungicide, and Rodenticide Act
GM	Genetically Modified
GR	Glyphosate resistant
HEL	Highly Erodible Land
HR	Herbicide resistant
LD_{50}	Amount of a material, given all at once that causes the death of 50% of a group of test animals (a measure of acute toxicity)
LL	Liberty Link® (gluphosinate resistant crop varieties)
MOA	Mechanism of action
MR	Multiple resistance
NASS	National Agricultural Statistics Service
NCFAP	National Center for Food and Agricultural Policy
USDA	U.S. Department of Agriculture

14.1 Introduction

Herbicide resistant crops are crops that have been genetically modified (GM) to be resistant to specific herbicides. That is, they are not damaged by those herbicide applications. HR crops have addressed a number of problems associated with earlier weed control practices. Because herbicides are designed to kill plants, they can cause injury to conventional crop varieties. This limits when and how herbicides can be applied, making them less effective for weed control. Because many herbicides are effective only against certain types of plants, growers face the complexity of choosing between multiple chemicals to apply for different weeds. Crops resistant to broad-spectrum herbicides overcome these problems by reducing crop injury and allowing applications of a single herbicide for most (or all) chemical weed control. Mechanical and hand tillage has become more costly over time as labor and fuel costs have risen relative to herbicide costs (Osteen and Fernandez-Cornejo 2013). There are additional economic and environmental problems with tillage, particularly on highly erodible land (HEL). Erosion can reduce the long-term productivity of soils and profitability of farming. Sediment from erosion can also reach water bodies contributing to numerous, costly environmental problems (Hansen and Ribaudo 2008). In the U.S., growers can lose eligibility for certain income support payments if they do not restrict certain tillage practices on HEL (Claassen 2006).

Crops genetically modified (GM) to be resistant to herbicides first became available in the United States in the mid 1990s. Transgenic bromoxynil-resistant (BXN) cotton was released in 1995. Bromoxynil controls many broadleaf weeds but not grasses (Carpenter and Gianessi 2001).

Round-up Ready® soybean varieties resistant to glyphosate became available in 1996. Glyphosate resistant (GR) cotton became available in 1997, followed by GR corn in 1998. Liberty Link® (LL) corn (genetically modified to be resistant to glufosinate) became available in 1997, followed by LL cotton in 2004 and LL soybeans in 2009. Both glyphosate and glufosinate are broad-spectrum, non-selective herbicides. BXN cotton was out-competed by GR cotton and was discontinued in 2005. Glyphosate resistant (GR) crops have accounted for the vast majority of HR crop hectares in the United States. Glyphosate controls more than 300 weed species (Green et al. 2008). Growers can control many broadleaf and grass weeds effectively using one herbicide instead of many different ones (Fernandez-Cornejo and McBride 2002). GR varieties of corn, soybeans, and cotton now account for most of the hectares planted to those crops. With rapid adoption of GR crops came a dramatic increase in glyphosate for chemical weed control, both in absolute terms and relative to other herbicides. This has led to the evolution of GR weeds in many parts of the United States, raising weed control costs significantly. Herbicide resistant weeds are not a new problem and GR weeds are not the result of GM crops *per se*. Rather, adoption of GM GR crops led to over-reliance on chemical means of weed control and within chemical control, over-reliance on a single mechanism of action (MOA). Both factors contributed to enormous selection pressure for GR weeds.

Problems with GR weeds have raised questions about the sustainability of GM HR crops. While many growers are adopting many Best Management Practices (BMPs) to manage weed resistance, adoption rates for a number of practices remain low. The current rate of BMP adoption has proved insufficient to delay herbicide resistance in many areas. Different entities, such as the U.S. Department of Agriculture (USDA), the Weed Science Society of America, the National Cotton Council, and the National Academies of Science, have been struggling to understand barriers to BMP adoption and to increase grower incentives to adopt them (Burgos et al. 2006; Price et al. 2011; NRC 2012; Norsworthy et al. 2012). Slowing resistance implies economic and environmental trade-offs. For growers, BMPs may entail reductions in short-run returns. No-till and reduced tillage practices have provided a number of environmental benefits that could be lost if growers revert to tillage in the face of weed resistance. Using herbicides with a different MOA than glyphosate may also delay resistance. This, however, may entail using more herbicides and using herbicides with greater persistence or toxicity than glyphosate. Thus, some of the practices to delay resistance to glyphosate may undercut some of the environmental benefits of glyphosate.

Another approach to manage resistance is to develop GM crops resistant to multiple herbicides. This way, growers could "rotate" applications between herbicides with different MOAs or combine herbicides in tank mixes. This, in principle, could slow the evolution of resistance to any one MOA. New crop varieties with multiple herbicide resistance (MR) traits are being released and scheduled for release over the next few years (Green et al. 2008; Green 2012). Whether MR varieties are a solution to resistance problems remains to be seen. Already, a number of weeds are resistant to multiple herbicide MOAs (undermining the MR trait strategy). In addition, no new MOAs have been developed since 1998 (Norsworthy et al. 2012). These new MR crop varieties rely on compounds that have been used for many years and have been generating selection pressure for resistance for some time. Some crop scientists have questioned whether MR varieties will ease or exacerbate weed resistance problems (Mortensen et al. 2012).

This chapter proceeds as follows. Section 14.2 discusses weed control and herbicide use in the United States before HR crops became available in 1996. We focus on corn, soybeans, and cotton, the main U.S. HR crops (HR canola is an important HR crop in Canada). Section 14.3 examines the trends in adoption of HR crops in the United States and some consequences of that adoption. Of particular interest are pecuniary and non-pecuniary returns to growers, implications for herbicide use and environmental impacts, and implications for conservation tillage adoption. Section 14.4 examines data to illustrate how introduction of glyphosate resistant (GR) crop varieties led to a dramatic reduction in the diversity of weed control tactics in U.S. agriculture. This reduction in diversity led to enormous selection pressure and predictable evolution of GR weeds. Section 14.5 discusses the barriers to adoption of best management practices (BMP) to delay weed resistance. Section 14.6 discusses the scope and limits of crop varieties with resistance to multiple herbicides to address GR resistant weed problems. Section 14.7 concludes by summarizing results, discussing future data needs, and identifying unresolved resistance management debates.

14.2 Weed Management and Herbicide Use Before HR Crops

In the 1960s, pre- or at-planting herbicides began to replace tillage and other cultural practices to control weeds in soybean production (Carpenter and Gianessi 1999). Growers would frequently follow herbicide applications with mechanical cultivation before soybean canopies closed over weeds (Pike et al. 1991). In the 1980s, post-emergence herbicides became available, allowing growers to control weeds without in-season tillage. Reducing the need for in-season cultivation time facilitated the operation of larger-scale farms. Post-emergence herbicides also facilitated use of narrow row spacing that increases yields per hectare, from more plants per hectare and better weed control.

From a weed management perspective, herbicides are not without their problems. Because they are designed to kill plants, herbicides can also injure crops (Padgette et al. 1996). Growers can limit injury by keeping application rates low. To be effective, however, low rates must be applied when weeds are small, which makes applications time-sensitive and can make delays in applications costly. Herbicide resistant weeds are another problem. Many weed populations have evolved resistance to ALS (acetolactate synthase) inhibitors. The persistence of many herbicides in the soil can create problems for crop rotations. For example, some herbicides used to control weeds in soybeans can hurt corn production when the two crops are grown in rotation.

Corn has upright leaves and is planted in wide rows, providing weeds with opportunities to grow without competition from corn early in the season. The wide row spacing also means that perennial weeds present special problems for corn growers. The first commonly used corn herbicide in the 1950s was 2,4-D. Atrazine was introduced in 1959. Because atrazine offered a broad spectrum of weed control, it largely replaced 2,4-D. Post-emergence herbicides are limited by their potential for injuring the corn crop. To avoid crop damage, herbicide applications on corn had to be made before the weeds reached a certain height. This made the effectiveness of herbicides time sensitive. Persistent herbicides applied in cornfields could also damage other crops grown in rotation with corn.

Before 1995, the available broadleaf herbicides that could be used over the top of a growing cotton crop could cause significant crop injury. To control weeds, cotton growers used cultivation and specialized application equipment to avoid herbicide contact with the cotton (Carpenter and Gianessi 2001). Here again, herbicide treatments had to made when weeds were still small, making application efficacy time sensitive. Both cultivation and directed spraying were relatively time-intensive.

U.S. herbicide use has risen dramatically since the mid 1960s. In 1966, 57% of corn, 52% of cotton, and 27% of U.S. soybean hectares were treated with herbicides (Osteen and Fernandez-Cornejo 2013). For all these crops, these percentages surpassed 91% by 1982 and have remained above that level since then. In 1964, herbicides accounted for 22% of total pesticide use in kilotons of active ingredient (a.i.) applied (Table 14.1). By 1995, this rose to 65% of all pesticide a.i. applied to U.S. crops. Corn has accounted for more than half of all herbicide kilotons of a.i. applied. From 1964 to 1995, kilotons of herbicide a.i applied to corn and cotton increased 7-fold. For soybeans, it increased 16-fold. These three crops accounted for 89% of all herbicides applied in the United States in 1995 and 2005 (Osteen and Fernandez-Cornejo 2013). Pesticide use estimates reported by Osteen and Fernandez-Cornejo (2013) in Table 14.1 are based on estimates from USDA's *Agricultural Chemical Usage* survey. While the survey samples major producing states for selected crops (usually accounting for about 90% of national production), its coverage is incomplete. Osteen and Fernandez-Cornejo (2013) derived national estimates by assuming non-sampled states had the same application rates as the average of sampled states.

Table 14.1 Herbicide applications in kilotons of active ingredient (a.i) applied for corn, cotton, and soybeans. (Source: Osteen and Fernandez-Cornejo 2013)

	1964	1995	2005
Total pesticides	97.5	235.7	222.8
Total herbicides	21.9	146.1	144.6
Corn	11.6	84.5	76.4
Cotton	2.1	14.7	13.1
Soybeans	1.9	30.9	38.9
Herbicide a.i./Total a.i. (%)	22	62	65
Crop herbicide a.i/total herbicides a.i.			
Corn (%)	53	58	53
Cotton (%)	10	10	9
Soybeans (%)	9	21	27
Three crops (%)	71	89	89

14.3 HR Crop Adoption and Effects

HR varieties of corn, cotton, and soybean first became commercially available in the mid-1990s. Adoption of HR soybeans has been particularly rapid, with HR varieties accounting for 90% of total soybean hectares by 2006 (Fig. 14.1). Adoption of HR cotton and corn proceeded more slowly, but HR cotton now accounts for 80% of U.S. cotton hectares, while HR corn accounts for >70% of U.S. corn hectares.

Adoption of HR crops has been rapid despite the evidence they increase farm profits has been mixed (Webster et al. 1999; Lin, et al. 2001; Marra et al. 2002; Bonny 2007). Researchers have suggested HR crops provide some benefits that are difficult to capture using standard farm profit estimates. These benefits include simplification of weed-management decisions, convenience, increased flexibility in timing, reduced crop damage, lower environmental risk, lower management time requirements (and higher off-farm income), and compatibility with conservation tillage (Carpenter and Gianessi 1999; Alston et al. 2002; Marra et al. 2004; Fernandez-Cornejo et al. 2005, 2007; Bonny 2007; Sydorovych and Marra 2007, 2008; Brookes and Barfoot 2008; Gianessi 2008; Piggott and Marra 2008; Gardner et al. 2009; Hurley et al. 2009a). These hard-to-measure benefits may thus account for rapid HR crop adoption.

Environmental benefits have also been claimed for HR crops. Adoption of HR crops appears to encourage conservation tillage (Carpenter et al. 2002; Fawcett and Towery 2002; Kalaitzandonakes and Suntornpithug 2003; Kim and Quinby 2003; Marra et al. 2004; Roberts et al. 2006; Fernandez-Cornejo and Caswell 2006; Frisvold et al. 2009a). This can reduce soil erosion and related water pollution (Brookes and Barfoot 2008; NRC 2010). The NRC (2010), however, notes that quantifying improvements in water quality attributable to HR crop adoption requires further research. Because conservation tillage reduces the number of machine passages over the field, it reduces fuel use. It thus contributes to soil carbon sequestration and reduced carbon emissions from fuel use (Brookes and Barfoot 2008; Horowitz et al. 2010).

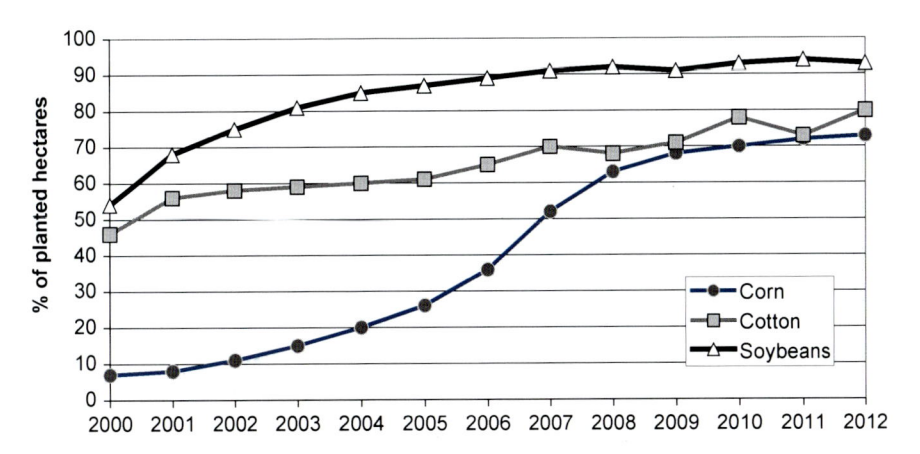

Fig. 14.1 Adoption of Herbicide Resistant Crops in the United States. Herbicide resistant (HR) corn, cotton, and soybeans first became commercially available in the mid 1990s. By 2000, about half of cotton and soybean hectares were planted to HR varieties, while fewer than 10% of corn hectares were planted to HR varieties. By 2012, >70% of US corn hectares were planted to HR varieties. For cotton, adoption surpassed 70% of total hectares. For soybeans, adoption surpassed 90% of total hectares. (Source: Adoption of Genetically Engineered Crops in the U.S. Data Set. USDA, Economic Research Service 2013b)

HR crops have also been credited with reducing environmental risks of herbicide applications. Attributing changes in environmental risks to HR crop adoption is difficult however. Table 14.1 shows that kilograms of herbicide active ingredient applied have declined for corn and cotton but increased for soybeans between the year 1995 (before HR crops) and 2005. Simple "before and after" comparisons are not an appropriate way to assess the impacts of HR crops on herbicide use. First, many things changed between 1995 and 2005 aside from HR crop adoption that could affect herbicide applications. These include changes in hectares planted, output and input prices, agricultural policies, and weather, for example. Estimates of the effect of HR crops on herbicide use must control at least for these factors. Second, growers are not randomly assigned to adopter and non-adopter groups in a controlled experiment. Growers choose whether or not to adopt HR crops. If adopters have fundamentally different characteristics than non-adopters comparing herbicide use across the two groups will suffer from sample selection bias (Fernandez-Cornejo et al. 2002, 2005, 2007). Third, kilograms of active ingredient applied are not a good measure of the environmental impact of herbicides. Herbicides vary in their toxicity to different species, persistence in the soil, half-life, leaching potential, run-off potential; that range of impact can have ecological effects as well as impacts on farm workers and consumers. Compared to many other herbicides it substitutes for, glyphosate has lower toxicity and persistence in the environment (Fernandez-Cornejo and Caswell 2006). This means a kilogram for kilogram substitution of glyphosate could reduce negative environmental risks of herbicides even if kilograms of a.i. remained unchanged. A number of studies have attempted to address this issue by weighting herbicide applications by factors such as mammalian toxicity, soil half-life, or the more comprehensive Environmental Impact Quotient (EIQ) (Kovach et al. 1992) that attempts to account for multiple types of risks.

Several studies have relied on estimates of changes in grower behavior published by the National Center for Food and Agricultural Policy (NCFAP). NCFAP carried out expert surveys of weed scientists across the United States. The scientists (primarily extension specialists) provided assessments of likely or recommended herbicide treatments that corn, soybean, and cotton growers would have made if they did not plant HR crops. Roughly, 50 experts were surveyed for each crop over different years. Various NCFAP studies then extrapolated the expert estimates to national changes in herbicide use (Carpenter and Gianessi 2001; Carpenter et al. 2002; Sankula and Blumenthal 2004). Other studies then took the extrapolated changes in herbicides used and weighted them using the Kovach et al.'s EIQ (Kleter and Kuiper 2003; Kleter et al. 2007; Brookes and Barfoot 2008; Green 2012). These studies tend to show a decline in the EIQ attributable to HR crop adoption. The expert survey approach is a rather clever and cost-effective way to conduct an *ex ante* assessment of the *potential* environmental impacts of HR crops. These estimates, however, are not statistically valid estimates of actual herbicide use, let alone estimates of changes in herbicide use attributable to HR crop adoption. To our knowledge, research has not been published that crosschecks these expert survey responses with detailed farm-level surveys to assess whether expert survey responses reasonably approximate actual on-farm behavior.

A few studies have estimated the impacts of HR crop adoption on the environmental impact of herbicide use, based on representative samples and using proper statistical methods to control for other factors affecting herbicide use and to address sample selection bias. These have made use of farm-level data from the Agricultural Resources Management Survey (ARMS) conducted by the USDA. Fernandez-Cornejo et al. (2002) found that farmers growing GR soybeans substituted glyphosate for other herbicides with measures of toxicity to humans that were three times higher and that persisted in the environment twice as long. Gardner and Nelson (2008) estimated impacts on GR varieties of corn, soybeans and cotton, weighting herbicide use by the LD_{50} dose for rats, a measure of acute mammalian toxicity. They found results depended critically on interactions with tillage practices. Under conventional tillage, use of GR varieties lowered the LD_{50} dose per hectare by 10% for soybeans, 17% for cotton and 98% for corn. Moving from conventional tillage and seeds to no-till and GR seeds would lead to a 94% decrease in LD_{50} doses for corn, but a 20% *increase* for soybeans, with no significant change for cotton. Alexandre et al. (2008) estimated that, adoption of HR corn contributed to a 5% reduction in impact-adjusted herbicide use, while HR soybean adoption contributed to a 13% reduction. Some of these farm-level results corroborate those of the NCFAP studies, but others do not.

Fernandez-Cornejo et al. (2012) used state-level data from 12 major soybean-producing states from 1996 to 2006 to examine the relationship between HR soybean adoption, conservation tillage adoption, and herbicide use, with herbicide applications weighted by toxicity and persistence. They found HR soybean adoption encouraged adoption of conservation tillage and decreased impact-adjusted herbicide use. This made use of data from the USDA *Agricultural Chemical Usage* survey, but used interpolation methods and data from other sources to fill in data gaps in the survey. Bonny (2011) also used the *Chemical Usage* survey to examine trends in the EIQ of herbicide applications to U.S. soybeans using the EIQ from Kovach et al. (1992). She found that the EIQ fell between the mid-1990s and 2001, the period of initial HR soybean adoption. The EIQ trended upward from 2001 to 2006, however, although it remained lower than in years before HR soybeans. The recent increase in the EIQ may be because GR weeds have necessitated the use of other herbicides.

14.4 Reduced Diversity of Weed Control Tactics Leads to GR Weeds

A fundamental means of delaying the evolution of weed resistance is to diversify control strategies (Duke and Powles 2009). This can be accomplished by using non-chemical control methods (such as tillage, row spacing, and crop rotations) along with chemical control (Beckie 2006; Beckie and Gill 2006). If herbicides are used, avoiding reliance on herbicides with the same mechanism of action (MOA) is also crucial (Beckie 2006; Beckie and Gill 2006; Green 2007; Green et al. 2008). The widespread adoption of GR crops, however, led to a pervasive reduction in the diversity of weed control tactics. Growers have relied less on non-chemical control methods and have relied heavily on a single mode of action for chemical control.

Data from the USDA NASS *Agricultural Chemical Usage* surveys illustrate the dramatic increase in reliance on glyphosate. Before the introduction of GR crop varieties, glyphosate use on corn, soybeans, and cotton was limited. Glyphosate accounted for 1% of all kilograms of herbicide active ingredient applied to corn in 1997, 11% of kilograms applied to soybeans in 1995 and 3% of kilograms applied to cotton in 1995 (Table 14.2). Soybean hectares treated with glyphosate rose from 20 to 95% from 1995 to 2006. Corn hectares treated with glyphosate rose from 4 to 66% from 1997 to 2010. For cotton, the share of hectares treated with glyphosate rose from 9 to 74% between 1995 and 2005, but then (possibly) fell slightly since then. The percent of hectares treated is a lower bound estimate because NASS records hectares treated with different types of glyphosate compounds. We have reported treatment shares for the most commonly applied compounds, but do know how much overlap there is in the use of different compounds. Thus, the share of hectares treated with glyphosate could be higher (by 20% points or more) than the lower bound estimates in Table 14.2. Nevertheless, Table 14.2 illustrates that growers began applying glyphosate much more extensively and relying increasingly on glyphosate relative to other compounds for chemical control.

Data from the USDA, ERS Agricultural Resources Management Survey (ARMS) tells a similar story. Table 14.3 lists the major herbicide families, and their MOAs, used in corn, soybean, and cotton production. It compares 1996 with years of the most recent ARMS data for each crop. For all three crops, there has been a reduction

Table 14.2 National trends in glyphosate use for U.S. corn, soybeans, and cotton. (Source: USDA, NASS Chemical Usage Survey—Field Crops)

Crop	Year	Hectares treated with glyphosate (%)	Glyphosate Kg of active ingredient (a.i.) applied as a % of total Kg of herbicide a.i. applied
Corn	1997	4	1
	1999	9	3
	2005	33	15
	2010	66	35
Soybeans	1995	20	11
	1999	62	54
	2006	95	89
Cotton	1995	9	3
	1999	36	20
	2005	74	57
	2010	68	62

in the diversity of MOAs employed. Glyphosate (a phosphinic acid herbicide and a 5-enolpyruvylshikimate-3-phosphate synthase inhibitor) has grown to dominate in soybeans and cotton. Over this time, use of several herbicide families has ceased or been dramatically reduced.

Three factors contributed to reliance on relatively few mechanisms of herbicide action. First, there was increasing resistance to MOAs that had been in general use for a long time, such as acetolactate synthesase (ALS—B2) and Photosystem II (Cs) herbicides. Second, glyphosate became attractive as a post-emergence herbicide because of its broad-spectrum efficacy and reliability. Low cost was also a factor after the patent on glyphosate expired in 2000 (allowing lower cost generics on the market). Third, herbicides with new MOAs have not been registered in the United States since 1998 (Norsworthy et al. 2012). Phosphinic acid herbicides accounted for 60% of hectare treatments for cotton in 2007 and 77% of hectare treatments for soybeans in 2006. By 2005, triazine and phosphinic acid treatments in corn accounted for two-thirds of hectare treatments. Use of several herbicide families, such as the amides, benzoic, and the sulfonylureas, sharply declined between 1996 and 2005. Overall, there is a narrowing of herbicide families and herbicide MOAs use for all three crops, which greatly increased the selection pressure for herbicide resistance.

Table 14.4 reports ARMS data for national trends in some weed management indicators. For all three crops, the data show: (a) the rapid adoption of GM HR seed varieties, especially for soybeans (97% of hectares) and cotton (90% of hectares); (b) a modest increase in the rate of field scouting for weeds; (c) increased reliance on post-emergence weed control; (d) decreased reliance on pre-emergence weed control; (e) reduced reliance on cultivation for weed control; and (f) increased reliance on burndown herbicides in soybeans and cotton.

Table 14.3 Herbicides, by herbicide family, applied to corn, soybean and cotton hectares. (Source: Agricultural Resource Management Survey (ARMS), USDA, Economic Research Service (2013a)

Herbicide family	Mechanism of action[1]	Percent of total herbicide hectare-treatments by survey year	
Corn		1996	2005
Phosphinic acid	$G^{(9)}$	2	19
Triazine	$C_1^{(5)}$	38	48
Amides	$K_3^{(15)}$	27	4
Benzoic/Phenoxy	$O^{(4)}$	15	5
Sulfonylurea	$B^{(2)}$	11	5
Pyridine	$F_1^{(12)}$	0	6
Other herbicides		8	9
Soybeans		1996	2006
Phosphinic acid	$G^{(9)}$	10	77
Dinitroaniline	$K_1^{(3)}$	20	3
Imidazolinone	$B^{(2)}$	21	2
Sulfonylurea	$B^{(2)}$	9	NA[2]
Diphenyl ether	$E^{(14)}$	8	1
Oxime	$A^{(1)}$	7	1
Aryloxyphenoxy propionic acid	$A^{(1)}$	7	NA
Phenoxy	$O^{(4)}$	5	5
Amides	$K_3^{(15)}$	4	2
Triazine	$C_1^{(5)}$	4	1
Benzothiadiazole	$C_3^{(6)}$	4	NA
Other herbicides		2	6
Cotton		1996	2007
Phosphinic acid	$G^{(9)}$	3	60
Dinitroaniline	$K_1^{(3)}$	26	14
Urea	$C_2^{(7)}$	20	6
Triazine	$C_1^{(5)}$	13	2
Organic arsenical	$Z^{(17)}$	12	1
Benzothiadiazole	$C_3^{(6)}$	3	1
Other herbicides		23	17

[1] The capitalized letter is the Herbicide Resistance Action Committee classification for the herbicide family mechanism if action and the superscript number is the Weed Science Society of America classification. (Senseman 2007)
B – Acetolactate synthase of acetohydroxy acid synthase inhibitors.
C_1 – photosystem II inhibitors.
F – Carotenoid biosynthesis inhibitors.
G – Enolpyruvyl shikimate-3-phosphate synthase inhibitors.
K – Mitosis inhibitors.
O – Synthetic auxins.
[2] NA – Estimate does not comply with the USDA disclosure limitation practices, is not available, or is not applicable.

Table 14.4 National trends in weed management for corn, soybeans and cotton. (Source: Agricultural Resource Management Survey (ARMS), USDA, Economic Research Service (2013)

Practice	Corn			Soybeans			Cotton		
	1996	2000	2005	1996	2000	2006	1996	2000	2007
	% of total national hectares planted on which practice is used								
Genetically modified herbicide resistant seed	NA[a]	11	31	7	59	97	NA	58	90
Field scouted for weeds	81	83	89	79	85	91	71	82	92
Burndown herbicide used	9	12	18	33	27	31	6	23	41
Pre-emergence weed control	78	71	61	67	46	28	90	79	73
Post-emergence weed control	59	63	66	78	87	95	62	76	89
Cultivated for weed control	33	38	15	29	17	NA	89	63	38

[a] *NA* Estimate does not comply with the USDA-ERS disclosure limitation practices, is not available, or is not applicable

Use of a burndown herbicide indicates producers are attempting to plant into a weed-free field. This practice is used in reduced tillage systems where the burndown herbicide replaces pre-plant tillage to control existing weeds. Burndown herbicides were used on 31 % of the soybean hectares planted in 2006, 41 % of the cotton hectares planted in 2007, and 18 % of the corn hectares planted in 2005 (Table 14.4). Use of burndown treatments increased dramatically in cotton, expanding from 6 to 41 %. This reflects increased adoption of both GR cotton and reduced tillage practices. Burndown herbicide use is not an effective part of resistance management if the same herbicide is used for both burndown and for later post-emergence applications. This is likely for GR crops because glyphosate has been the preferred choice for both applications. Growers typically make more than one application of glyphosate on GR crops and, in some areas, more than three (Norsworthy 2003; Norsworthy et al. 2007; Wilson et al. 2011).

Field scouting was carried out for weeds on 89 % of the corn hectares in 2005, 91 % of the soybean hectares planted in 2006, and 92 % of the cotton hectares planted in 2007 (Table 14.4). These seem to be high adoption rates of scouting. However, the ARMS does not report whether the scouting was carried out before, after, or both before and after herbicide applications. Field scouting has little utility in detecting emerging resistance problems unless it is carried out following herbicide use. Second, the high levels of scouting may simply reflect a move from preventive weed management to a curative approach, one that greatly depends on post-emergence herbicide treatments. For soybeans and cotton, post-emergence herbicide use is positively correlated with adoption of GR varieties (Table 14.4). That trend in adoption of GR cultivars was less apparent for corn (Table 14.4), where HR cultivars constituted only 31 % of total corn hectares in 2005, much lower than it was for soybean or cotton (97 and 90 %).

Cultivation for weed control adds diversity to an herbicide-based weed management system. Cultivation was practiced on 38% of U.S. cotton hectares in 2007 and on 15% of the corn hectares in 2005 (Table 14.4). The last reliable estimate of cultivation for weed control in soybean was 13% in 2002. Weed management through cultivation has steadily decreased with the adoption of GR varieties and is now less than half the levels reported in the late 1990s (Table 14.4).

The data from Table 14.4 indicate that a combination of all five weed management practices (scouting, burndown, pre-emergence and post emergence herbicide applications, and tillage) were *at most* used on just 15% of corn hectares in 2005, 17% of soybean hectares in 2000, and 38% of cotton hectares in 2007. Analysis of state-level ARMS data suggests the actual percentages are even lower (Norsworthy et al. 2012). Most of the U.S. hectares planted to those three crops are not under a diversified weed management program that combines scouting and non-chemical control with use of diverse herbicide MOAs.

Another vehicle for assessing the extent of resistance management practices are large-scale surveys of grower practices. A series of papers (Shaw et al. 2009; Givens et al. 2009a, 2009b) described results of a grower survey conducted in 2005/6 of 1050 producers from the states of Illinois, Indiana, Iowa, Mississippi, Nebraska, and North Carolina. The primary cropping systems practiced on the farms were continuous GR soybean, continuous GR cotton, a GR soybean/GR corn rotation, or GR soybean/non-GR crop rotation. The traditional value of crop rotation to manage resistance must be reassessed for HR crops. If an HR crop follows another HR crop, there may be little effect on the herbicide(s) used from year to year and consequently the same herbicide selects the weed population. The majority of hectares in Mississippi and North Carolina were in continuous monocropping. Few cotton farmers practiced rotation. Lack of rotation in cotton raises concern about the potential continuous herbicide resistance selection in the fields with GR cotton. Growers planting continuous GR soybeans had been doing so for an average of 4.8 years, while growers who planted continuous GR cotton averaged 5 years. A GR soybean–non-GR crop rotation had been practiced for an average of 6.4 years.

Growers responded that glyphosate had replaced non-glyphosate based weed management programs (Givens et al. 2009b). Glyphosate was at least the foundation, if not the only, herbicide used to manage weeds. While the majority of growers made two or fewer glyphosate applications in a crop, between 30–40% of the GR cotton growers made three glyphosate applications, depending on farm size. In GR soybeans, 66–74% of the producers made two of more glyphosate treatments. Soybean hectares were more likely to receive only glyphosate applications. For example, 85% of those growing continuous GR soybeans applied glyphosate alone, while more than 80% of soybean hectares in rotation with corn or a non-GR crop received only glyphosate applications. Cotton hectares were most likely to receive herbicides besides glyphosate, with corn hectares intermediate. Continuous GR cotton hectares and continuous GR soybean hectares were most likely to receive two or more glyphosate applications.

Foresman and Glasgow (2008) reported on a 2006 telephone survey that collected information from 200 growers in the Corn Belt (North) and 200 from the Cotton

Table 14.5 Frequency of weed resistance BMP adoption among 1205 cotton, corn and soybean growers. (Source: Frisvold et al. 2009b)

BMP (best management practice)	Always	Often	Sometimes	Rarely	Never
	% respondents practicing				
1. Scout before applying herbicides	57	26	11	3	2
2. Scout after applying herbicides	51	29	15	2	1
3. Start with clean field	60	14	13	5	8
4. Control weeds early	54	35	9	1	0
5. Control weeds escapes	45	34	15	4	2
6. Clean equipment	15	11	20	22	31
7. Use new seed	87	7	3	1	2
8. Use multiple herbicides with different MOAs	18	21	33	15	13
9. Supplemental tillage	11	10	26	21	32
10. Use label rate	74	19	4	1	0

Belt (South). Many growers in both the North and South also grew soybeans. More than 90% of growers in both regions used GR seed varieties. The share of total area planted to GR crops was greater among Southern producers (83%) than Northern ones (53%), where more growers planted non-GR corn. Rotating GR with non-GR crops was more prevalent in the North (55% of growers) than the South (20%). In the South, 56% of cropped area was planted consecutively to GR crops in 2005 and 2006. A high percentage of growers made 2–3 glyphosate applications per year (70% in the North 75%; in the South). In the South, only 9% of growers responded that they would rotate out of GR crops in the event of glyphosate resistance.

About half of growers planting corn or cotton applied a pre-emergence herbicide followed by glyphosate. About a third of soybean growers did so. About a fifth of corn, cotton, and southern soybean growers applied glyphosate in tank mixes with other herbicides. A very small percentage of growers used herbicides other than glyphosate. In contrast, significant shares of growers applied only glyphosate, with shares higher among soybean growers (Foresman and Glasgow 2008). These findings are consistent with those of Givens et al. (2009b) and Frisvold et al. (2009b) who found evidence that soybean growers were less likely to use multiple herbicides with different MOAs. A significant number of growers used glyphosate only and even larger shares of growers are applying glyphosate 2–3 times per year. Together, these suggest significant selection pressure for glyphosate resistance.

Frisvold et al. (2009b) reported on a telephone 2007 survey of 1,205 corn, cotton, and soybean producers (at least 400 respondents for each crop). The survey asked growers about use of ten Best Management Practices (BMPs) to delay weed resistance (Table 14.5). Growers chose among five responses when asked how frequently they adopted a BMP: (1) always, (2) often, (3) sometimes, (4) rarely, and (5) never. Six BMPs were always practiced by a majority of growers (Table 14.5). A large share of growers rarely or never practiced three BMPs, however. These included cleaning equipment before moving between fields (53%), using multiple herbicides with different MOAs (28%), and supplemental tillage (53%). Table 14.5

Table 14.6 Planned glyphosate resistant (GR) crop plantings and residual herbicide use from a survey of 1,205 cotton, corn, and soybean growers. (Source: Hurley et al. 2009b)

Variable	Corn	Soybean	Cotton
2008 GR hectares planned (%)	73	96	92
2008 GR hectares with residual planned (%)	66	28	66
2008 GR hectares following GR hectares planned (%)	63	47	68

combines responses for all three producers because adoption patterns were remarkably similar across producer groups. More than 70% of corn, cotton, or soybean growers practiced the same seven BMPs often or always. All used multiple herbicides with different MOAs, cleaned equipment, or practiced supplemental tillage much less frequently. Fewer than half practiced these three BMPs often or always. More corn producers used multiple herbicides with different modes of action often or always (49%) than either cotton (38%) or soybean (28%) growers.

Frisvold et al. (2009b) also conducted multivariate regression analysis to evaluate the factors that contribute to more or less frequent use of BMPs. They found that growers used herbicides with different MOAs more frequently if they: (a) had more years of education; (b) the more their expected crop yield exceeded the 10-year average yield in their county; (c) they were in a county with reported weed resistance to glyphosate; and (d) also raised livestock. They used multiple herbicides less frequently if they (a) farmed more years (b) were soybean growers; (c) planted a higher percentage of their targeted crop to GR varieties; and (d) farmed in a county with a higher yield coefficient of variation over the previous 10 years.

The coefficient of variation (CV) is the standard deviation of yield divided by its mean. It serves as a measure of marginal production areas—areas with historically low yields, high yield variability, or both. Highly variable production outcomes may hinder the observability and trialability of BMPs (Pannell and Zilberman 2001). With greater yield variability, it may be more difficult for growers to assess outcomes or benefits of BMP adoption. This suggests that counties with high crop yield CVs may be areas to look for low BMP adoption and focus extension programs for resistance management.

Analyzing a sub-sample of this same survey data, Hurley et al. (2009a) found that while more than 65% of corn and cotton growers used a residual herbicide with glyphosate, fewer than 30% of soybean growers did so. About 70% of GR corn and GR cotton growers were planting their GR crop following a GR crop planted the previous year. Nearly half of GR soybean growers were doing so (Hurley et al. 2009a). Again, using data from the same survey, Hurley et al. (2009b) reported cotton and soybean growers both planned to plant more than 90% of their crop with GR cultivars, while corn growers planned to plant more than 70% with GR cultivars (Table 14.6). Compared to corn and cotton growers, soybean growers planned to treat a smaller share of their GR hectares with a residual herbicide. Soybean growers also planned to plant a lower percentage of their GR hectares following a GR crop (possibly, because they planned a rotation with non-GR corn).

Harrington et al. (2009) conducted an on-line survey of U.S. agricultural professionals (growers, researchers, educators, consultants, and administrators) about

Table 14.7 Perceived changes in weed management practices resulting from adoption of HR crops from an Internet survey of 54 agricultural professionals. (Source: Harrington et al. 2009)

Weed management practice	Percent of respondents who believed growers were following the practice "less" or "much less" as a result of HR crop adoption (%)
Combination of weed control methods	>60[a]
Crop rotation for weed control	>40
Annual rotation of herbicides	>50
Use of multiple herbicides	>60
Tillage for weed control	>80

[a] Numbers are derived from a graph in Harrington et al.; exact values were not reported

how GM crops were perceived to have affected pest and weed management. Most respondents believed growers were using the following methods "less or much less": (a) using a combination of weed control methods, (b) using diverse MOAs, and (c) using tillage (Table 14.7). Between 40 and 50 % believed growers were using crop rotations less. Among 13 serious, negative consequences of HR crop adoption, respondents rated shifts in weed species composition and development of weed resistance as the first and second most serious. Respondents were asked to rate serious on a scale of 1 (not serious) to 5 (very serious). Shifts in weed composition were rated at 4.04 on average, compared to weed resistance, 3.98. Ratings for other problems ranged from 2.02 to 3.6. Public sector respondents rated weed resistance as more serious (3.96) than private sector respondents did (2.93), with the difference significant at the 1 % level.

Before 1998, there were no reported glyphosate-resistant (GR) weed species in the United States. By 2013, however, glyphosate resistance had been confirmed for 13 species in the United States (Heap 2013). GR weed species are spread across 32 U.S. states. GR weeds have proven problematic for cotton, soybeans, peanuts in rotation with cotton, maize, and in California, perennial crops (VanGessel 2001; Culpepper et al. 2006, 2008; Foresman and Glasgow 2008; Steckel et al. 2008; Davis et al. 2010; Hanson et al. 2009; Webster and Sosnoskie 2010). Resistance to glyphosate has evolved in Palmer amaranth (*A. palmeri*) in GR cotton fields throughout the southeast United States (Culpepper et al. 2006, 2008, 2009; Nichols et al. 2008; Norsworthy et al. 2008, 2012; Steckel et al. 2008). By 2008, GR Palmer amaranth infested more than 240,000 ha of land in Georgia, North Carolina, and South Carolina (Culpepper et al. 2009). An additional 87,000 ha of cotton were infested in Arkansas (Doherty et al. 2008).

Costs of GR weeds can be significant, ranging from $ 5–$ 130/ha (Mueller et al. 2005; Scott and VanGessel 2007; Webster and Sosnoskie 2010). In severe cases, growers may opt to abandon fields altogether (Culpepper et al. 2008). Regarding glyphosate-resistant Palmer amaranth, one extension publication warned, "there are no economical programs to manage this pest in cotton" (Culpepper and Kichler 2009, p. 1). A national survey of weed specialists estimated the average additional costs to control glyphosate resistant Palmer amaranth was $ 74/ha for cotton, $ 52/ha for soybeans, and $ 40/ha for corn (Carpenter and Gianessi 2010).

14.5 Barriers to Resistance Management

Despite costs of GR weeds, adoption of BMPs by growers to delay resistance has been incomplete and insufficient to prevent the onset of resistance. While resistance management practices for Bt crops are federally mandated, management of weed resistance for HR crops has been voluntary. Bt crops have pesticides incorporated into them and are thus regulated under the Federal Insecticide, Fungicide, and Rodenticide Act (FIFRA). HR crops do not include pesticide compounds themselves, however, so the EPA has no clear authority to regulate HR crop varieties directly (Horne 1992). In principle, the EPA could exert influence over weed resistance management in two areas. First, under FIFRA, the EPA has authority to regulate uses of herbicides that complement HR crops. Second, the EPA could require resistance management procedures to be implemented as a condition of granting Sect. 18 exemptions. The Emergency Exemption Program mandated by Sect. 18 of FIFRA gives the EPA authority to authorize emergency, non-registered uses of pesticides. States often make requests for Sect. 18 exemptions in response to pest or weed resistance that reduces the usefulness of registered compounds.

It would be difficult to implement a mandatory resistance management program for HR crops, however. First, it is not clear what would constitute "compliance" with resistance management. There are multiple crop, planting, herbicide, tillage, and machinery-cleaning choices one could make to delay resistance. How would one define and measure compliance and enforce it in a legal setting? In contrast to a regulatory approach, Monsanto has begun offering price rebates to growers who purchase residual herbicides to be used in conjunction with glyphosate. These subsidies apply to herbicides with MOAs that differ from glyphosate and even apply to some herbicides sold by competing companies (Frisvold and Reeves 2010).

As of the mid-2000s, many growers held attitudes and perceptions that would discourage BMP adoption. Johnson and Gibson (2006) found only 36% of growers expressed a high level of concern about weed resistance, while 19% expressed low or no concern. However, as resistance to glyphosate became more apparent, concern has increased. Hurley et al. (2009a) reported that resistance was a weed management concern mentioned by 59% of cotton growers, 54% of soybean growers, and 48% of corn growers. Harrington et al. (2009) reported that agricultural professionals rated weed shifts and resistance as the two most serious concerns. Public sector respondents, however, rated resistance as a more serious concern than did private consultants or growers.

A significant share of growers appeared unaware of certain major factors contributing to the evolution of weed resistance as recently as 2005. In results from the Benchmark study, one in eight medium and large growers and one in four small growers were unaware of weeds' potential to develop herbicide resistance. Fewer than half of growers rated rotating herbicides or using tank mixes (to diversify exposure to MOAs) as highly effective methods of delaying resistance (Johnson et al. 2009). Johnson and Gibson (2006) reported that only 58% of growers surveyed mentioned repeated use of the same MOA as a major factor contributing to weed resistance.

Many growers may attribute infestation and spread of resistant weeds to factors beyond their control such as natural forces (e.g. wind, birds, animals) or poor weed management by their neighbors (Llewellyn and Allen 2006; Wilson et al. 2008). If growers perceive that preventing weed resistance is beyond their individual control and requires collective grower action, they will have less incentive to take individual actions that incur additional costs to delay resistance.

Many growers may also believe that new chemistries or cultivars will soon become available to address resistance problems (Llewellyn et al. 2002; Foresman and Glasgow 2008). Foresman and Glasgow (2008) reported 92 % of respondents were "somewhat confident" to "very confident" that chemical manufacturers would develop new products to address glyphosate resistance within 3–5 years. Growers have less incentive to conserve the efficacy of an herbicide if they believe substitutes will be available in the future.

14.6 Role and Limits of Stacked Trait HR Varieties

One approach to address resistance to GR crops is through plant breeding by "stacking" resistance traits to multiple herbicides in individual crop varieties (Green et al. 2008). Resistance can, in theory, be delayed by rotating between herbicides with different MOAs and by using herbicide mixtures. This would reduce selection pressure on any one compound. If a particular weed was resistant to one herbicide, it may be killed by another herbicide that relies on a different MOA. Companies are developing new crop varieties that combine glyphosate resistance with resistance to herbicides with different MOAs (Green et al. 2008). One example will be varieties that stack glyphosate resistance with resistance to different ALS-inhibiting herbicides. Varieties resistant to two more herbicides will soon be commercially available (Green 2012). These stacked varieties will be combined with homogeneous blends (herbicide mixtures with different MOAs). Because these blends will be mixtures of currently registered herbicides, they may receive regulatory approval relatively quickly.

Combining herbicide mixtures with multiple resistant (MR) crop varieties can reduce reliance on a single MOA. This strategy also avoids the high cost and lengthy delays in developing novel herbicides. This strategy raises questions, however. First, how many different MOAs need to be combined in one HR crop variety to delay resistance substantially? How high is the potential for delay, given that some weeds are resistant to the herbicides to be combined. For example, some weeds are already resistant to glyphosate, others are resistant to ALS inhibitors, and some are resistant to both (e.g., Legleiter and Bradley 2008). The list of weeds resistant to multiple herbicides continues to grow (Mortensen 2012; Heap 2013).

Crop varieties that have MR traits have emerged as the immediate response to GR resistant weeds in the United States. However, this represents a "buy and apply" approach where growers passively select market products (seed varieties and chemicals). It remains to be seen whether rotations between a limited set of herbicides will be sufficient to delay resistance.

14.7 Conclusions

Since their introduction in the mid 1990s, adoption of genetically modified (GM), herbicide resistant (HR) crop varieties for corn, soybeans and cotton proceeded rapidly in the United States. GM HR varieties—particularly glyphosate resistant (GR) varieties—now account for the majority of hectares planted to these three crops. Many benefits to growers that are not captured in standard farm profit calculations appear to account for the popularity of these varieties among growers. These include convenience, simplification of weed-management decisions, greater flexibility in herbicide application timing, reduced crop damage, lower environmental risk, lower management time requirements, and compatibility with conservation tillage. Weed resistance to herbicides aside from glyphosate was also a factor.

GR crops have been credited with two types of environmental benefits: encouraging adoption of conservation tillage and substitution of herbicides with lower toxicity and persistence in the environment. The empirical support for the complementarities between GR crops and conservation tillage is stronger than for herbicide substitution. The most widely cited estimates of grower shifts in herbicide use have come from surveys of extension specialists, not from actual farm-level data. In some cases, careful farm-level analyses corroborate the expert survey results, but in other cases do not. While expert surveys may be a reasonable and cost effective way to measure *potential* environmental impacts of HR crop varieties, it is less clear they accurately reflect actual grower behavior and environmental impacts. A fruitful area of research might be a retrospective study, comparing expert survey predictions with actual farm-level survey data. Assessment of weed management and herbicide use in the United States is also hampered by the fact that the USDA is conducting surveys less frequently and covering a smaller number of states in their sampling frames. Even careful studies are relying increasingly on numbers that are extrapolated over space and time. This means that assessments of herbicide use and weed management in the United States are increasingly made based on expert opinion surveys and extrapolations and less on actual farm-level data.

Increased reliance on GR crops and glyphosate as the dominant means of weed control generated enormous selection pressure for GR weeds, however. From the mid-1990s to the mid-2000s, there was a pervasive reduction in the diversity of weed control tactics. The widespread, complementary adoption of GR cultivars and conservation tillage provided a number of economic and environmental benefits. Yet, increasing reliance on purely herbicide-based weed management has reduced the diversity of weed management tactics. From the mid-1990s to the mid-2000s, the share of corn, soybean, and cotton hectares cultivated for weed control fell by 50% or more. There was also a shift away from pre-emergence weed control to post-emergence herbicide use. Post-emergence control often relied on use of glyphosate as the only herbicide and using glyphosate multiple times in a single season. While rotating crops can delay weed resistance, many growers began rotating between GR crops (e.g. GR corn/GR soybean rotations and GR cotton/GR soybean rotations) and the same hectares received repeated applications of a single chemistry, glyphosate. The evolution of GR resistant weeds has become a large and growing weed management problem throughout the United States.

In contrast to Bt crops, where resistance management followed a regulatory approach, weed resistance management in the United States has been purely voluntary. Given the complexity of weed resistance management, a regulatory approach would have been difficult to implement. Nevertheless, it was a "road not taken" and grower associations still oppose a regulatory approach. As of the mid-2000s, many growers maintained attitudes and perceptions that would discourage adoption of resistance management practices. While most growers are adopting many best management practices (BMPs) to delay resistance, this has proven incomplete and insufficient. Development of crop varieties resistant to multiple herbicides has emerged as the immediate response to glyphosate resistant weeds. Some have criticized this strategy because it may lead to greater herbicide use and negative environmental impacts in the short run and divert attention and resources away more comprehensive integrated weed management research and extension (Mortensen et al. 2012).

References

Alexandre, V., Nehring, R., Fernandez-Cornejo, J., & Grube, A. (2008). Impact of GMO crop adoption on quality-adjusted pesticide use in corn and soybeans: A full picture. *Selected Paper Prepared for Presentation at the American Agricultural Economics Association Annual Meeting*, Orlando, FL, July 27–29, 2008. http://ageconsearch.umn.edu/bitstream/6429/2/AAEA_Vialou_RN_JF_AG_July14.pdf. Accessed 27 Feb 2014.

Alston, J. M., Hyde, J., Marra, M. C., & Mitchell, P. D. (2002). An ex-ante analysis of the benefits from the adoption of corn rootworm resistant transgenic corn technology. *AgBioForum, 5*(3), 71–84.

Beckie, H. J. (2006). Herbicide-resistant weeds: Management tactics and practices. *Weed Technology, 20*, 793–814.

Beckie, H. J., & Gill, G. S. (2006). Strategies for managing herbicide-resistant weeds. In H. P. Singh, D. R. Batish, & R. K. Kohli (Eds.)., *Handbook of Sustainable Weed Management* (pp. 581–625). New York: Food Products Press.

Bonny, S. (2007). Genetically modified glyphosate-tolerant soybean in the USA: Adoption factors, impacts and prospects. A review. *Agronomy for Sustainable Development, 28*, 21–32.

Bonny, S. (2011). Herbicide-tolerant transgenic soybean over 15 years of cultivation: Pesticide use, weed resistance, and some economic issues: The case of the USA. *Sustainability, 3*, 1302–1322.

Brookes, G., & Barfoot, P. (2008). Global impact of biotech crops: Socio-economic and environmental effects, 1996–2006. *AgBioForum, 11*(1), 21–38.

Burgos, N., Culpepper, S., Dotray, P., Kendig, J., Wilcut, J., & Nichols, R. (2006). *Managing Herbicide Resistance in Cotton Cropping Systems*. Memphis, Tennessee, USA: National Cotton Council of America.

Carpenter, J. E., & Gianessi, L. (1999). Herbicide tolerant soybeans: Why growers are adopting roundup ready varieties. *AgBioForum, 2*(2), 65–72.

Carpenter, J. E., & Gianessi, L. P. (2001). *Agricultural Biotechnology: Updated Benefit Estimates*. Washington, D.C.: National Center for Food and Agricultural Policy.

Carpenter, J. E., & Gianessi, L. P. (2010). Economic impact of glyphosate resistant weeds. In V. K. Nandula (Ed.)., *Glyphosate Resistance in Crops and Weeds* (pp. 297–312). New York: Wiley.

Carpenter, J. E., Gianessi, L., Sankula, S., & Silvers, C. S. (2002). *Plant Biotechnology: Current and Potential Impact for Improving Pest Management in U.S. Agriculture, an Analysis of 40 Case Studies, Herbicide Tolerant Cotton*. Washington, D.C.: National Center for Food and Agricultural Policy (NCFAP).

Claassen, R. (2006). Compliance provisions for soil and wetland conservation. In K. Wiebe & N. Gollehon (Eds.), *Agricultural Resources and Environmental Indicators (2006 ed.) (Economic Information Bulletin No. 16)*. Washington, D.C.: US Department of Agriculture (USDA), Economic Research Service (ERS).

Culpepper, A. S., & Kichler, J. (2009). *University of Georgia Programs for Controlling Glyphosate Resistant Palmer Amaranth in 2009 Cotton* (College of Agricultural and Environmental Sciences, Circ. No. 924). Athens, Georgia, USA: University of Georgia.

Culpepper, A. S., Grey, T. L., Vencill, W. K., Kichler, J. M., Webster, T. M., Brown, S. M., et al. (2006). Glyphosate-resistant Palmer amaranth (*Amaranthus palmeri*) confirmed in Georgia. *Weed Science, 54*, 620–626.

Culpepper, A. S., Whitaker, J. R., MacRae, A. W., & York, A. C. (2008). Distribution of glyphosate-resistant Palmer amaranth (Amaranthus palmeri) in Georgia and North Carolina during 2005 and 2006. *Journal of Cotton Science, 12*, 306–310.

Culpepper, A. S., York, A. C., & Marshall, M. W. (2009). Glyphosate-resistant Palmer amranth in the Southeast. In T. M. Webster (Ed.)., *Proceedings of the Southern Weed Science Society, 62*, 371.

Davis, V. M., Kruger, G. R., Young, B. G., & Johnson, W. G. (2010). Fall and spring preplant herbicide applications influence spring emergence of glyphosate-resistant horseweed (*Conyza canadensis*). *Weed Technology, 24*, 11–19.

Doherty, R. C., Smith, K. L., Bullington, J. A., & Meier, J. R. (2008). Glyphosate-resistant Palmer amaranth control in Roundup Ready® flex cotton. In D. M. Oosterhuis (Ed.)., *Summaries of Arkansas Cotton Research 2008* (pp. 148–151). Fayetteville, Arkansas, USA: Arkansas Agricultural Experiment Station.

Duke, S. O., & Powles, S. B. (2009). Glyphosate-resistant crops and weeds: Now and in the future. *AgBioForum, 12*(3 & 4), 346–357.

Fawcett, R., & Towery, D. (2002). *Conservation and Plant Biotechnology: How New Technologies can Improve the Environment by Reducing the Need to Plow*. West Lafayette, Indiana, USA: Conservation Technology Information Center.

Fernandez-Cornejo, J., & Caswell, M. (2006). *The First Decade of Genetically Engineered Crops in the United States* (*Economic Information Bulletin 11*). Washington, D.C.: US Department of Agriculture (USDA), Economic Research Service (ERS).

Fernandez-Cornejo, J., & McBride, W. D. (2002). *Adoption of Bio-engineered Crops (Agricultural Economic Report No. 810)*. Washington, D.C.: USDA ERS.

Fernandez-Cornejo, J., Klotz-Ingram, C., & Jans, S. (2002). Farm-level effects of adopting herbicide-tolerant soybeans in the U.S.A. *Journal of Agricultural and Applied Economics, 34*, 149–163.

Fernandez-Cornejo, J., Hendricks, C., & Mishra, A. (2005). Technology adoption and off-farm household income: The case of herbicide-tolerant soybeans. *Journal of Agricultural and Applied Economics, 37*, 549–563.

Fernandez-Cornejo, J., Mishra, A., Nehring, R., Hendricks, C., Southern, M., & Gregory, A. (2007). *Off-farm Income, Technology Adoption, and Farm Economic Performance (ERS Report No. 36)*. Washington, D.C.: USDA ERS.

Fernandez-Cornejo, J., Hallahan, C., Nehring, R., Wechsler, S., & Grube, A. (2012). Conservation tillage, herbicide use, and genetically engineered crops in the United States: The case of soybeans. *AgBioForum, 15*(3), 231–241.

Foresman, C., & Glasgow, L. (2008). US grower perceptions and experiences with glyphosate-resistant weeds. *Pest Management Science, 64*, 388–391.

Frisvold, G. B., & Reeves, J. M. (2010). Resistance management and sustainable use of agricultural biotechnology. *AgBioForum, 13*(4), 343–359.

Frisvold, G. B., Boor, A., & Reeves, J. M. (2009a). Simultaneous diffusion of herbicide resistant cotton and conservation tillage. *AgBioForum, 12*(3 & 4), 249–257.

Frisvold, G. B., Hurley, T. M., & Mitchell, P. D. (2009b). Adoption of best management practices to control weed resistance by corn, cotton, and soybean growers. *AgBioForum, 12*(3 & 4), 370–381.

Gardner, J. G., & Nelson, G. C. (2008). Herbicides, glyphosate resistance and acute mammalian toxicity: Simulating an environmental effect of glyphosate-resistant weeds in the USA. *Pest Management Science, 64*(4), 470–478.

Gardner, J. G., Nehring, R. F., & Nelson, C. H. (2009). Genetically modified crops and household labor savings in US crop production. *AgBioForum, 12*(3 & 4), 303–312.

Gianessi, L. P. (2008). Economic impacts of glyphosate-resistant crops. *Pest Management Science, 64*, 346–352.

Givens, W. A., Shaw, D. R., Johnson, W. G., Weller, S. C., Young, B. G., Wilson, R. G., Owen, M. D. K., & Jordan, D. (2009b). A grower survey of herbicide use patterns in glyphosate-resistant cropping systems. *Weed Technology, 23*, 156–161.

Givens, W. A., Shaw, D. R., Krueger, G. R., Johnson, W. G., Weller, S. C., Young, B. G., Wilson, R. G., Owen, M. D. K., & Jordan, D. (2009a). Survey of tillage trends following the adoption of glyphosate-resistant crops. *Weed Technology, 23*, 150–155.

Green, J. M. (2007). Review of glyphosate and ALS-inhibiting herbicide crop resistance and resistant weed management. *Weed Technology, 21*, 547–558.

Green, J. M. (2012). The benefits of herbicide resistant crops. *Pest Management Science, 68*, 1323–1331.

Green, J. M., Hazel, C. B., Forney, D. R., & Pugh, L. M. (2008). New multiple-herbicide crop resistance and formulation technology to augment the utility of glyphosate. *Pest Management Science, 64*(4), 332–339.

Hansen, L., & Ribaudo, M. (2008). Economic measures of soil conservation benefits: Regional values for policy assessment. *Technical Bulletin TB-1922*. Washington, D.C.: USDA, Economic Research Service.

Hanson, B. D., Shrestha, A., & Shaner, D. L. (2009). Distribution of glyphosate resistant horseweed (*Conyza canadensis*) and relationship to cropping systems in the central valley of California. *Weed Science, 57*, 48–53.

Harrington, J., Byrne, P. F., Peairs, F. B., Nissen, S. J., Westra, P., Ellsworth, P. C., et al. (2009). Perceived consequences of herbicide-tolerant and insect-resistant crops on integrated pest management strategies in the western United States: Results of an online survey. *AgBioForum, 12*(3 & 4), 412–421.

Heap, I. M. (2013). *International Survey of Herbicide Resistant Weeds*. http://www.weedscience.org/. Accessed 15 April 2013.

Horne, D. M. (1992). EPA's response to resistance management and herbicide-tolerant crop issues. *Weed Technology, 6*, 657–661.

Horowitz, J., Ebel, R., & Ueda, K. (2010). "No-till" farming is a growing practice. U.S. Department of Agriculture, Economic Research Service, *Economic Information Bulletin No. EIB070*. USDA, Economic Research Service

Hurley, T. M., Mitchell, P. D., & Frisvold, G. B. (2009a). Effects of weed resistance concerns and resistance management practices on the value of Roundup Ready® crops. *AgBioForum, 12*, 291–302.

Hurley, T. M., Mitchell, P. D., & Frisvold, G. B. (2009b). Weed management costs, weed best management practices, and the Roundup Ready® weed management program. *AgBioForum, 12*, 281–290.

Johnson, W. G., & Gibson, K. D. (2006). Glyphosate-resistant weeds and resistance management strategies: An Indiana grower perspective. *Weed Technology, 20*, 768–772.

Johnson, W. G., Owen, M. D., Kruger, G. R., Young, B. G., Shaw, D. R., Wilcut, J. W., et al. (2009). U.S. farmer awareness of glyphosate-resistant weeds and resistant management strategies. *Weed Technology, 23*, 308–312.

Kalaitzandonakes, N., & Suntornpithug, P. (2003). Adoption of cotton biotechnology in the United States: Implications for impact assessment. In N. Kalaitzandonakes (Ed.)., *The Economic and Environmental Impacts of Agbiotech: A Global Perspective* (pp. 103–118). New York: Kluwer Academic/Plenum Publishers.

Kim, C. S., & Quinby, W. (2003). ARMS data highlight trends in cropping practices. *Amber Waves, 1*, 12–13.

Kleter, G. A., & Kuiper, H. A. (2003). Environmental fate and impact considerations related to the use of transgenic crops. In G. Voss & G. Ramos (Eds.)., *Chemistry of Crop Protection: Progress and Prospects in Science and Regulation: 10th IUPAC International Congress on the Chemistry of Crop Protection* (pp. 304–321). Weinheim, Germany: Wiley-VCH Verlag

Kleter, G. A., Bhula, R., Bodnaruk, K., Carazo, E., Felsot, A. S., & Harris, C. A., et al. (2007). Altered pesticide use on transgenic crops and the associated general impact from an environmental perspective. *Pest Management Science, 63,* 1107–1115.

Kovach, J., Petzoldt, C., Degni, J., & Tette, J. (1992). A method to measure the environmental impact of pesticides. *NYS Agricultural Experiment Station, Bulletin 139*, Cornell University, Ithaca, New York, USA.

Legleiter, T. R., & Bradley, K. W. (2008). Glyphosate and multiple herbicide resistance in common waterhemp (Amaranthus rudis) populations from Missouri. *Weed Science, 56,* 582–587.

Lin, W., Price, G., & Fernandez-Cornejo, J. (2001). *Estimating Farm-level Effects of Adopting Herbicide-Tolerant Soybeans: Oil Cops Situation and Outlook Yearbook.* Washington, D.C.: U.S. Department of Agriculture, Economic Research Service.

Llewellyn, R. S., & Allen, D. M. (2006). Expected mobility of herbicide resistance via weed seeds and pollen in a Western Australian cropping region. *Crop Protection, 25,* 520–526.

Llewellyn, R. S., Lindner, R. K., Pannell, D. J., & Powles, S. B. (2002). Resistance and the herbicide resource: Perceptions of Western Australian grain growers. *Crop Protection, 21,* 1067–1075.

Marra, M. C., Pardey, P. G., & Alston, J. (2002). The payoffs to transgenic field crops: An assessment of the evidence. *AgBioForum, 5*(2), 43–50.

Marra, M. C., Piggott, N. E., & Carlson, G. A. (2004). The net benefits, including convenience, of roundup ready® soybeans: Results from a national survey. *NSF Center for IPM Technical Bulletin 2004-3.* Raleigh, North Carolina, USA: Center for Integrated Pest Management.

Mortensen, D. A., Egan, J. F., Maxwell, B. D., Ryan, M. R., & Smith, R. (2012). Navigating a critical juncture for sustainable weed management. *Bioscience, 62,* 75–84.

Mueller, T. C., Mitchell, P. D., Young, B. G., & Culpepper, A. S. (2005). Proactive versus reactive management of glyphosate-resistant or -tolerant weeds. *Weed Technology, 19,* 924–933.

National Research Council (NRC). (2010). *Impact of Genetically Engineered Crops on Farm Sustainability in the United States.* Washington, D.C.: The National Academies Press.

National Research Council (NRC). (2012). *National Summit on Strategies to Manage Herbicide Resistant Weeds: Proceedings of a Symposium.* Washington, D.C.: The National Academies Press.

Nichols, R. L., Culpepper, A. S., Main, C. L., Marshall, M., Mueller, T., Norsworthy, J., et al. (2008). Glyphosate-resistant populations of *Amaranthus palmeri* prove difficult to control in the Southern United States. *Paper Presented at the International Weed Science Conference, (Paper 556)* (pp. 227), Vancouver, Canada

Norsworthy, J. K. (2003). Use of soybean production surveys to determine weed management needs of South Carolina farmers. *Weed Technology, 17,* 195–201.

Norsworthy, J. K., Smith, K. L., Scott, R. C. E., & Gbur, E. (2007). Consultant perspectives on weed management needs in Arkansas cotton. *Weed Technology, 21,* 825–831.

Norsworthy, J. K., Griffith, G. M., Scott, R. C., Smith, K. L., & Oliver, L. R. (2008). Confirmation and control of glyphosate-resistant Palmer amaranth (*Amaranthus palmeri*) in Arkansas. *Weed Technology, 22,* 108–113.

Norsworthy, J. K., Ward, S. M., Shaw, D. R., Llewellyn, R. S., Nichols, R. L., Webster, T. M., Bradley, K. W., Frisvold, G., Powles, S. B., & Burgos, N. R. (2012). Reducing the risks of herbicide resistance: Best management practices and recommendations. *Weed Science, 60,* 31–62.

Osteen, C. D., & Fernandez-Cornejo, J. (2013). Economic and policy issues of U.S agricultural pesticide use trends. *Pest Management Science, 69*(9),1001–1025.

Padgette, S. R., Re, D. B., Barry, G. F., Eichholtz, D. E., Delannay, X., Fuchs, R. L., et al. (1996). New weed control opportunities: Development of soybeans with a Roundup Ready gene. In S. O. Duke (Ed.)., *Herbicide-resistant Crops* (pp. 53–84). Boca Raton, Florida, USA: CRC Press.

Pannell, D. J., & Zilberman, D. (2001). Economic and sociological factors affecting growers' decision making on herbicide resistance. In S. Powles & D. Shaner (Eds.)., *Herbicide Resistance and World Grains* (pp. 251–277). Boca Raton, Florida, USA: CRC Press.

Piggott, N. E., & Marra, M. C. (2008). Biotechnology adoption over time in the presence of non-pecuniary characteristics that directly affect utility: A derived demand approach. *AgBioForum, 11,* 58–70.

Pike, D. R., McGlamery, M. D., & Knake, E. L. (1991). A case study of herbicide use. *Weed Technology, 5*(3), 639–646.

Price, A. J., Balkcom, K. S., Culpepper, S. A., Kelton, J. A., Nichols, R. L., & Schomberg, H. (2011). Glyphosate-resistant Palmer amaranth: A threat to conservation tillage. *Journal of Soil and Water Conservation, 66,* 265–275.

Roberts, R. K., English, B. C., Gao, Q., & Larson, J. A. (2006). Simultaneous adoption of herbicide-resistant and conservation-tillage cotton technologies. *Journal of Agricultural and Applied Economics, 38,* 629–43.

Sankula, S., & Blumenthal, E. (2004). *Impacts on US Agriculture of Biotechnology-derived Crops Planted in 2003: An Update of Eleven Case Studies.* Washington, D.C.: National Center for Food and Agriculture Policy.

Scott, B. A., & VanGessel, M. J. (2007). Delaware soybean grower survey on glyphosate-resistant horseweed (*Conyza Canadensis*). *Weed Technology, 21,* 270–274.

Senseman, S. A. (2007). *Herbicide Handbook, 2007* (9th ed.). Lawrence, Kansas, USA: Weed Science Society of America.

Shaw, D. R., Givens, W. A., Farno, L. A., Gerard, P. D., Jordan, D., Johnson, W. G., Weller, S. C., Young, B. G., Wilson, R. G., & Owen, M. D. K. (2009). Using a grower survey to assess the benefits and challenges of glyphosate-resistant cropping systems for weed management in U.S. corn, cotton, and soybean. *Weed Technology, 23,* 134–149.

Steckel, L. E., Main, C. L., Ellis, A. T., & Mueller, T. C. (2008). Palmer amaranth *(Amaranthus palmeri)* in Tennessee has low-level glyphosate resistance. *Weed Technology, 22,* 119–123.

Sydorovych, O., & Marra, M. C. (2007). A genetically engineered crop's impact on pesticide use: A revealed-preference index approach. *Journal of Agricultural and Resource Economics, 32,* 476–491.

Sydorovych, O., & Marra, M. C. (2008). Valuing the changes in herbicide risks resulting from adoption of roundup ready soybeans by U.S. farmers: A revealed preference approach. *Journal of Agricultural and Applied Economics, 40,* 777–787.

U.S. Department of Agriculture, Economic Research Service. (2013a). *ARMS Farm Financial and Crop Production Practices.* http://www.ers.usda.gov/data-products/arms-farm-financial-and-crop-production-practices.aspx#.UYbf06Lvu0o. Accessed 15 April 2013.

U.S. Department of Agriculture, Economic Research Service. (2013b). *Adoption of Genetically Engineered Crops in the U.S. Data Set.* http://www.ers.usda.gov/data-products/adoption-of-genetically-engineered-crops-in-the-us.aspx#.UYbex6Lvu0o. Accessed 15 April 2013.

U.S. Department of Agriculture, National Agricultural Statistics Service. *Agricultural Chemical Usage-Field Crops Summary, Issues 1990–2005.* http://usda.mannlib.cornell.edu/MannUsda/viewDocumentInfo.do?documentID=1560. Accessed April 15, 2013.

VanGessel, M. J. (2001). Glyphosate-resistant horseweed in Delaware. *Weed Science, 49,* 703–705.

Webster, E. P., Bryant, K. J., & Earnest, L. D. (1999). Weed control and economics in non-transgenic and glyphosate-resistant soybean (Glycine max). *Weed Technology, 13,* 586–593.

Webster, T. M., & Sosnoskie, L. M. (2010). The loss of glyphosate efficacy: A changing weed spectrum in Georgia cotton. *Weed Science, 58,* 73–79.

Wilson, R. G., Young, B. G., Mathews, J. L., Weller, S. C., Johnson, W. G., Jordan, D. L., Owen, M. D. K., Dixon, P. M., & Shaw, D. R. (2011). Benchmark study on glyphosate-resistant cropping systems in the United States. Part 4: Weed populations and soils seedbanks. *Pest Management Science, 67,* 771–780.

Wilson, R. S., Tucker, M. A., Hooker, N. H., LeJeune, J. T., & Doohan, D. (2008). Perceptions and beliefs about weed management: Perspectives of ohio grain and produce farmers. *Weed Technology, 22,* 339–350.

Chapter 15
Integrating Research and Extension for Successful Integrated Pest Management

Cesar R. Rodriguez-Saona, Dean Polk and Lukasz L. Stelinski

Contents

15.1	Introduction	357
	15.1.1 Research in Integrated Pest Management	357
	15.1.2 Extension in Integrated Pest Management	358
15.2	Case Studies	360
	15.2.1 Peaches	360
	15.2.2 Blueberries	368
	15.2.3 Citrus	375
	15.2.4 Apples	381
15.3	Conclusions	384
References		385

Abstract A successful integrated pest management (IPM) program requires the integration of both research and extension. Current restrictions on pesticide use have demanded research on reduced-risk practices. For instance, in the US, the US Environmental Protection Agency Food Quality Protection Act (EPA FQPA) of 1996 imposes restrictions and tolerance reassessments on the use of broad-spectrum insecticides. Reduced-risk pest management practices include the use of softer

---

C. R. Rodriguez-Saona (✉)
Associate Extension Specialist, Department of Entomology, Rutgers University, P.E. Marucci Center for Blueberry & Cranberry Research & Extension, 125A Lake Oswego Rd., Chatsworth, NJ 08019, USA
e-mail: crodriguez@aesop.rutgers.edu

D. Polk
Agricultural and Resource Management Agents, Rutgers University, Fruit & Ornamental Research & Extension Center, 283 Route 539,
Cream Ridge, NJ 08514, USA
e-mail: polk@njaes.rutgers.edu

L. L. Stelinski
Entomology and Nematology Department, University of Florida, Citrus Research & Education Center, 700 Experiment Station Rd.,
Lake Alfred, FL 33840, USA
e-mail: stelinski@ufl.edu

D. Pimentel, R. Peshin (eds.), *Integrated Pest Management*,
DOI 10.1007/978-94-007-7796-5_15,
© Springer Science+Business Media Dordrecht 2014

pesticides, mating disruption technologies, development of degree-day models, geo-spatial technologies, cultural and ground cover management, and methods that conserve biological control agents. Constant threats from newly introduced pests are a major obstacle for IPM implementation because they disrupt existing practices. In addition, increased economic pressures that growers face, such as increased pesticide and labor costs and grower market competition, provide another dimension to this situation. More than ever the integration of multiple pest management tactics is needed for the development and implementation of sustainable IPM programs. The adoption of new technologies into existing IPM programs will depend on a comprehensive extension program that combines traditional forms of communication (e.g., outreach presentations, on-farm demonstrations, newsletters, factsheets, etc.) with new internet-based tools (e.g., WebPages, blogs, and webinars). Here we discuss various ways in which research and extension efforts can be coordinated to develop a successful pest management program. In particular, we provide examples based on our own experiences in peaches, blueberries, citrus, and apples.

Keywords Pest management · Reduced-risk practices · Insecticide use · Fruit crops · Food Quality Protection Act · Outreach

List of Abbreviations

IPM	Integrated Pest Management
US	United States
OPs	Organophosphates
DDT	Dichloro-diphenyl-tri-chloroethane
EPA	Environmental Protection Agency
FQPA	Food Quality Protection Act
USDA	United States Department of Agriculture
PMSP	Pest Management Strategic Plan
GIS	Geographic Information Systems
RAMP	Risk Avoidance and Mitigation Program
PMAP	Pest Management Alternatives Program
NIFA	National Institute of Food and Agriculture
SCRI	Specialty Crop Research Initiative
NASS	National Agricultural Statistics Service
ERS	Economic Research Service
CAR	Crops At Risk
IR-4	Interregional Research Project No. 4
SARE	Sustainable Agriculture Research and Education
HLB	Huanglongbing; citrus greening disease
CHMAs	Citrus Health Management Areas
ACP	Asian Citrus Psyllid
APHIS-PPQ	Animal and Plant Health Inspection Service Plant Protection and Quarantine

15.1 Introduction

Insect pest management is a dynamic science, constantly changing, and research and extension efforts need to reflect these changes. As practitioners of integrated pest management (IPM), we have learned many lessons since synthetic pesticides first became available. In the years 1940–1960, broad-spectrum insecticides (organochlorines, organophosphates [OPs], and carbamates), led by chlorinated hydrocarbons like DDT (dichloro-diphenyl-tri-chloroethane), were heavily used for insect pest control, a period referred to as the "insecticide era" (Pedigo 2002) or the "dark ages" of pest control (Peshin et al. 2009). Broad-spectrum insecticides were considered highly effective, cheap, and easy to apply, and thus intensively used in agriculture. However, over-reliance of these insecticides led to the onset of resistant pest populations. It also eliminated the natural enemies that regularly kept secondary pests below an economic threshold causing secondary pest outbreaks (Smith and van den Bosch 1967; van den Bosch et al. 1982). As a result, as pesticides became less effective against the target pests, growers tended to use them more frequently, which in turn promoted secondary pest outbreaks, creating a "pesticide treadmill" cycle (van den Bosch 1978). Another drawback of heavy reliance on broad-spectrum insecticides is their negative effects on the environment especially human and wildlife health, a problem that was first brought to the public's attention by Rachel Carson in her book *Silent Spring* in 1962 (Carson 1962). DDT was subsequently banned in the United States (US) in 1972.

15.1.1 Research in Integrated Pest Management

Since the concept of "integrated pest management" was introduced to the scientific community (Stern et al. 1959; Smith and van den Bosch 1967; CEQ 1972), several IPM-based practices became adopted by farmers worldwide including the use of pheromones for monitoring insect pests, use of degree-day models for better timing of insecticide applications, use of sex pheromones for mating disruption technologies, use of companion plantings for conservation biological control, among others. Still, chemical control (and the use of broad-spectrum insecticides) continued to be a common and in many instances a dominant practice in agriculture because of its lower cost and effective control against a complex of insect pests as compared with other more selective practices. However, use of broad-spectrum insecticides in agriculture is now becoming more limited or eliminated altogether worldwide. In the US, the Environmental Protection Agency (EPA) implemented the Food Quality Protection Act (FQPA) in 1996 which imposes restrictions and tolerance reassessments on broad-spectrum insecticide availability (US EPA 1996). Implementation of the FQPA has caused important changes in pest management programs. Since its implementation, research and extension efforts have led to grower adoption of "reduced-risk" insecticides and significant reductions in broad-spectrum insecticide inputs. Reduced-risk insecticides are those insecticides that pose fewer risks to humans, non-target organisms, and the environment than conventional insecticides, and thus are more compatible with IPM. Several new classes of insecticides (OP-replacement and reduced-risk) have since become available and been registered in various agricultural crops, including neonicotinoids, insect growth regulators (including chitin synthesis inhibitors and ecdysone agonists), anthranilic diamides, and spinosyns, among others. These compounds are highly effective but also very selective. Cost of these newer insecticides is also usually 2–3 times higher than conventional insecticides (Shearer et al. 2006), and their use may significantly increase pest management costs. Consequently, current insect pest management research and extension programs must be designed to address issues related to insecticide efficiency, environmental concerns, and production costs.

To complicate matters, growers are constantly facing the threat of new pests. Invasive pests are a major obstacle for IPM implementation because they disrupt existing practices (e.g., Hoddle 2006). Development of new practices, whether against an invasive pest or an alternative approach, may take time to become popular among growers. Invasive species post a particular challenge to IPM programs because the absence of specific biological control agents allows invasive pests to build large populations quickly. Growers need to respond fast to these situations. However, because these insects are not considered pests in their native geographic range, researchers and extension specialists often do not have immediate answers for how to best control them. The immediate response is often to use broad-spectrum insecticides that are, as we discussed before, not sustainable. It might then take several years before a more sustainable pest management program for controlling invasive pests is achieved.

15.1.2 Extension in Integrated Pest Management

A solid extension program is critical for growers to adopt new pest management practices. Initiating an extension program can be, however, a challenging task for new extension professionals because courses on extension IPM are rarely taught at universities. Hence, those working as extension educators, specialists, and support personnel, IPM agents and consultants, and other related positions face the challenge of learning new skills quickly and efficiently. The most important skill in extension is the ability to effectively communicate with growers, both verbally and in writing. It is important, however, to emphasize that giving an extension presentation is different from giving a scientific one. Growers are not generally interested in hearing many of the specific details of a study. For example, lengthy methods and details of experimental and statistical designs, although important in scientific presentations, should be mostly avoided when giving extension talks. Growers are more interested in hearing about the reasons for the study (i.e., questions like: how are these studies going to affect them? or why should they care?), the results, and general conclusions as to how these results will benefit them in the short- and long-terms. Repeating the main points several times during a presentation, and in multiple occasions, often helps in getting the message across to an audience. In

Fig. 15.1 A successful extension program should include the following: **a** regular meetings with growers, particularly during the growing season; **b** regular newsletter articles; **c** annual field demonstrations; **d** workshops for training and getting growers' input

addition to the importance of writing regular newsletters, having regular meetings with growers, particularly during the growing season to provide timely recommendations, is essential (Fig. 15.1a–b).

Reaching growers is now being facilitated by wide accessibility of the internet and social media (e.g., e-mails, twitter, Facebook pages, blogs, YouTube videos, etc.) that has allowed for immediate delivery of information. Webpages and webinars (internet-based presentations) are becoming popular in extension programs. Timely IPM-based information can now be accessed by growers directly from the field via iPhones, iPads, and Smartphones. Use of these high-technology tools in extension IPM will likely increase as growers and educators become more familiar with them. Still, these technologies should not replace regular face-to-face meetings with growers. In fact, annual field demonstrations are extremely important for the implementation of new technologies into pest management programs (Fig. 15.1c).

Because growers are the clientele, extension professionals should always listen and welcome feedback from growers. Workshops and surveys are ideal venues for training and getting growers' input (Fig. 15.1d).

We discuss below how we and our colleagues have integrated research and extension into successful pest management programs in four economically important fruit crops in the US: peaches, blueberries, citrus, and apples.

15.2 Case Studies

15.2.1 Peaches

One of the oldest cultivated fruits in the world, the peach, *Prunus persica* (L.), originated in China with historical records dating at least to 3,300 BC. From China the fruit moved into Persia, then to southern and western Europe. The Spanish explorers brought the fruit to the New World where it was spread among the Aztecs in the mid 1500s. A second introduction was likely made at about the same time in Florida (US). By the late 1700s to 1800 a commercial peach market had been established in the mid-Atlantic US, with Baltimore as the first commercial hub (Faust and Timon 2011). In most of the US, peach production is targeted for the fresh market, although a significant canning and processing market exists in California. The leading peach producing states in the US by usual rank include: California, South Carolina, Georgia, New Jersey, and Pennsylvania (USDA NASS 2012a). Given the differences in fruit culture, markets, and industry size, most of this discussion will pertain to peaches grown east of the Mississippi.

In peach-production areas of the eastern US, the crop is susceptible to just over two dozen arthropod pests and about one dozen pathogens (Hogmire 1995; Howitt 1993). Peach trees are rather "weak" trees that can succumb to a combination of plant parasitic nematodes, borers, winter injury, and a number of diseases. In fact, in most production areas, plantings seldom live over 12–14 years (Ritchie and Clayton 1981). The weak tree, combined with the fact that virtually all eastern production is for the fresh market has led to intensive pesticide programs to control multiple pest complexes. In the northeast and mid-Atlantic areas of the US key arthropod pests include the direct pests such as the oriental fruit moth, *Grapholita molesta* (Busck), plum curculio, *Conotrachelus nenuphar* (Herbst), tarnished plant bug, *Lygus lineolaris* (Palisot de Beauvois), several species of stink bugs, *Euschistus* spp., *Acrosternum hilare* (Say), and indirect pests such as green peach aphid, *Myzus persicae* (Sulzer), European red mite, *Tetranychus urticae* (Koch), several species of scale insects, and peachtree/greater peachtree borer, *Synanthedon exitiosa* (Say), and lesser peachtree borer, *Synanthedon pictipes* (Grote & Robinson) (Hogmire 1995). In the mid-Atlantic to northeastern US, the oriental fruit moth has four generations and has been the primary driver for repeated use of insecticide applications. The northern strain of plum curculio, with one generation per year usually requires 1–2 applications, and plant bugs have required pesticide control as needed, partially based on favorable ground cover and the presence of other alternate hosts. In Georgia and South Carolina, the southern strain of plum curculio has two or more generations per year (Horton and Ellis 1989; Akotsen-Mensah et al. 2011), thereby making it the primary pest that drives most repeated insecticide use, and a challenge to southeastern peach IPM programs.

15.2.1.1 Historical Perspective of Peach Integrated Pest Management in the Mid-Atlantic and New Jersey

Given the many pests present in peaches, growers traditionally used 11–12 full cover (every row middle) applications of combined insecticide and fungicide sprays (Halbrendt 2012; Ward 2012). Most arthropod treatments consisted of broad-spectrum OP and carbamate materials. In the 1980s many apple growers readily adopted new practices for apple IPM. The system was complex and pesticide costs were increasing, leading to narrower margins for growers. By contrast, the peach system was simpler with fewer key pests, and the reliance on virtually one insecticide class, the OPs, most of this being ethyl parathion, and then later methyl parathion, used in the formulation Penncap-M®. In fact, in 1984, when growers participating with the New Jersey extension IPM program for apples were asked about their interest in a peach IPM program, a common response was "No thanks, parathion is cheap"[1]. During this time period growers either paid consultants for scouting and recommendation services as they did in Pennsylvania, or participated in an extension-sponsored IPM program as in this example from New Jersey. Since growers paid the cost of scouting and other services in both instances, they could make cost/benefit decisions based on their perception of associated costs and risks. For peaches, the IPM program costs did not outweigh the production costs and risks, since parathion had cost only $ 9.88–12.35 ha^{-1} ($ 4–5 acre^{-1}) per application and controlled every orchard arthropod except mites (Polk et al. 1990; Hopfinger 1990).

Ethyl parathion is extremely toxic to both target and non-target organisms (US EPA 2000). Largely because of farm worker and safety concerns, EPA issued a cancellation order for most uses of ethyl parathion including all fruit labels in December 1992 (US EPA 1992). Although safer for handling, encapsulated methyl parathion (Penncap-M®) was well known to be highly toxic to honeybees (Atkins et al. 1978, 1981), and in 1995–1996 came under severe pressure for use restrictions in fruit crops. As a result, sales and use of Penncap-M® in fruit crops were subsequently curtailed, and all labels voluntarily withdrawn in 1999 (US EPA 1999), partially as a result of the FQPA of 1996. Regulatory changes, market factors and ongoing research were spawned by the FQPA that changed the face of fruit pest management and IPM practices. FQPA's main objective was to redefine the dietary risks and overall exposure, especially for infants and children, associated with these older pesticide materials. As a result, EPA reexamined the registrations and use patterns for most OPs as a group, which led to severe restrictions and use cancellations. For example, azinphos-methyl (Guthion®), which became one of the principal insecticides used in peaches after the cancellation of Penncap-M®, is no longer registered for use in peaches. These product restrictions opened up research funding opportunities through the US Department of Agriculture (USDA) for replacement strategies and materials. FQPA also spurred research in the agrichemical industry for the

---

[1] Based on authors' personal experiences.

development of new chemistries, several of which had already been introduced to the market. Tebufenozide (Confirm®) was being tested by university researchers in 1994. Spinosad (Spintor®) was discovered in 1982 and first labeled in 1996. Other new products followed as "OP replacement" products. While regulatory issues like FQPA ultimately helped to accelerate the development and marketing of alternative chemistries, it was the development of resistance to conventional OPs, illustrated by oriental fruit moth resistance in peaches (de Lame et al. 2001; Kanga et al. 2003), that first changed individual grower practices away from OP and carbamate materials to new reduced risk chemistries.

The existence and use of new chemistries resulted in several new considerations for peach growers, and were evident in the pest management strategic plans (PMSPs) developed in peach growing areas (Horton et al. 2000; Brunner et al. 2004; CTFA 2006; USDA NIFA 2013). First, several new products were insect growth regulators, specifically targeting lepidopteran pests including leaf rollers, fruit worms, and internal worms like oriental fruit moth. Secondly, most of the new chemistries are narrower spectrum than the old OP compounds, and therefore complicate a grower's decision process when managing multiple pests. Third, the grower's learning curve is increased. Since many new materials are narrow spectrum and life-stage specific, grower education about pest biology and pest management becomes more important. Finally, most of the new insecticides were more expensive than the OP and carbamate materials. All of these factors combined to make IPM programming in peaches, not only desirable, but necessary. Our IPM surveys showed that while the relatively cheap broad-spectrum parathion was gone, in its place came regulation, insecticide resistance management, narrow-spectrum replacement products, and increased costs.

15.2.1.2 Changes in Insecticide Use Patterns in Peaches

Table 15.1 shows changes in insecticide usage in peaches across US and in New Jersey (US) before and after FQPA of 1996. Since FQPA, there has been an increasing adoption of registered OP-replacement and reduced-risk insecticides. For example, the neonicotinoid imidacloprid, an OP-replacement insecticide, was used in 23 % of all insecticide-treated hectares from Georgia, Michigan, New Jersey, North Carolina, Oregon, and Washington, 7 years after FQPA. The neonicotinoid thiamethoxam and the reduced-risk insecticide flubendiamide were used in 12 and 3 % of treated hectares, respectively, across the US 15 years after FQPA.

In New Jersey peaches, imidacloprid was used in 23 % of all insecticide-treated hectares 7 years after FQPA (USDA NASS 2012a). The neonicotinoids thiamethoxam, imidacloprid, acetamiprid, and dinotefuran were used in 44, 21, 9, and 2 % of treated hectares, respectively, 15 years after FQPA, whereas the reduced-risk insecticide indoxacarb was used in 10 % of treated hectares 15 years after FQPA.

Table 15.1 US peach insecticide use nationally[1] (*left*) and in New Jersey (NJ) (*right*) for 1995, 2003, and 2011

Average Values of Top 12 Peach Insecticides in the US - By Percent of Hectares Treated					Average Values of Top 12 Peach Insecticides in NJ - By Percent of Hectares Treated				
1995 (Total=18)[2]					1995 (Total=7)				
A.I.	% Ha Trt	No. Appl.	Kg/Ha/Yr (lb/A/Yr)	IRAC Class	A.I.	% Ha Trt	No. Appl.	Kg/Ha/Yr (lb/A/Yr)	IRAC Class
Methyl Parathion	50	4.2	2.56 (2.29)	1B	Azinphos-Methyl	92	6.7	2.81 (2.51)	1B
Permethrin	29	2.0	0.34 (0.30)	3A	Methyl Parathion	60	5.3	1.48 (1.32)	1B
Esfenvalerate	21	1.5	0.07 (0.06)	3A	Methomyl	60	1.9	0.68 (0.61)	1A
Diazinon	21	1.2	3.14 (2.80)	1B	Chlorpyrifos	21	1.0	1.92 (1.71)	1B
Azinphos-Methyl	20	4.0	2.05 (1.83)	1B	Permethrin	12	2.0	0.25 (0.22)	3A
Chlorpyrifos	17	1.3	1.89 (1.69)	1B	Endosulfan	11	1.5	1.37 (1.22)	2A
Phosmet	11	2.1	2.62 (2.34)	1B	Phosmet	9	3.6	4.12 (3.68)	1B
Propargite	11	1.1	2.02 (1.81)	12C					
Methomyl	9	2.3	0.89 (0.80)	1A					
Carbaryl	8	1.5	3.25 (2.91)	1A					
Fenbutatin-Oxide	8	1.2	0.75 (0.67)	12B					
Endosulfan	6	2.8	3.09 (2.76)	2A					
2003 (Total=22+5 pheromones)					2003 (Total=8+2 pheromones)				
A.I.	% Ha Trt	No. Appl.	Kg/Ha/Yr (lb/A/Yr)	IRAC Class	A.I.	% Ha Trt	No. Appl.	Kg/Ha/Yr (lb/A/Yr)	IRAC Class
Esfenvalerate	39	0.10	2.13 (1.90)	3A	Azinphos-Methyl	57	6.3	2.80 (2.50)	1B
Phosmet	37	4.72	3.81 (3.40)	1B	Phosmet	48	3.1	3.26 (2.91)	1B
Imidacloprid	23	0.004	1.79 (1.60)	4A	Methomyl	37	4.8	4.01 (3.58)	1A
Chlorpyrifos	21	1.61	1.34 (1.20)	1B	Endosulfan	31	4.4	5.64 (5.04)	2A
Carbaryl	19	2.42	1.79 (1.60)	1A	Imidacloprid	23	1.9	0.02 (0.02)	4A
Azinphos-Methyl	14	2.18	4.37 (3.90)	1B	Carbaryl	17	3.2	2.89 (2.58)	1A
Diazinon	14	1.56	1.46 (1.30)	1B	Chlorpyrifos	13	4.2	5.11 (4.56)	1B
Fenbutatin-Oxide	11	0.45	1.12 (1.00)	12B	Esfenvalerate	13	4.2	0.12 (0.11)	3A
Endosulfan	8	1.64	3.69 (3.30)	2A					
Lambda-Cyhalothrin	8	0.01	2.91 (2.60)	3A					
Permethrin	7	0.41	2.57 (2.30)	3A					
Bifenazate	7	0.27	1.12 (1.00)	UN					
2011 (Total = 57+ 5 pheromones)					2011 (Total = 25+ 5 pheromones)				
A.I.	% Ha Trt	No. Appl.	Kg/Ha/Yr (lb/A/Yr)	IRAC Class	A.I.	% Ha Trt	No. Appl.	Kg/Ha/Yr (lb/A/Yr)	IRAC Class
Esfenvalerate	31	2.3	0.12 (0.11)	3A	Thiamethoxam	44	5.3	0.51 (0.45)	4A
Lambda-Cyhalothrin	19	2.6	0.09 (0.08)	3A	Permethrin	37	5.0	1.31 (1.17)	3A
Phosmet	13	2.9	5.22 (4.66)	1B	Esfenvalerate	23	2.4	0.13 (0.12)	3A
Thiamethoxam	12	2.9	0.28 (0.25)	4A	Imidacloprid	21	1.7	0.08 (0.07)	4A
Cyfluthrin	11	3.2	0.12 (0.11)	3A	Cyfluthrin	20	1.7	0.04 (0.04)	3A
Permethrin	7	3.6	0.84 (0.75)	3A	Phosmet	15	3.3	3.51 (3.13)	1B
Chlorpyrifos	7	1.3	1.48 (1.32)	1B	Endosulfan	14	1.1	1.34 (1.20)	2A
Beta-Cyfluthrin	6	2.2	0.04 (0.04)	3A	Lambda-Cyhalothrin	13	4.8	0.21 (0.19)	3A
Imidacloprid	6	1.7	0.10 (0.09)	4A	Indoxacarb	10	2.0	0.21 (0.19)	22A
Chlorantraniliprole	5	1.3	0.11 (0.10)	28	Acetamiprid	9	1.9	0.21 (0.19)	4A
Carbaryl	4	1.4	2.49 (2.23)	1A	Chlorpyrifos	3	1.0	1.96 (1.75)	1B
Flubendiamide	3	1.6	0.19 (0.17)	28	Dinotefuran	2	1.4	0.33 (0.30)	4A

USDANASS. (2012a). (http://quickstats.nass.usda.gov/).

[1] National survey data includes records from California, Georgia, Michigan, New Jersey, New York, North Carolina, Pennsylvania, South Carolina, Texas, Virginia, and Washington.

[2] Denotes number of insecticides in record. If less than 12 are listed, then insufficient data exists to list at least 12 materials.

A.I. = active ingredient; % Ha Trt = percent hectarage treated; No. Appl. = number of applications; Kg/Ha/Yr = kilograms per hectare per year; lb/A/Yr = pounds per acre per year; IRAC Class = Insecticide Resistance Action Committee Classification.

15.2.1.3 Developing an Integrated Pest Management Program—The Evolution of Multiple Tools and Practices

As peach growers requested IPM information, tools were developed that could be combined and used in a cohesive program (Atanassov et al. 2002; Halbrendt 2012).

The following is a list of IPM concepts and practices adopted by the mid-Atlantic and New Jersey peach growers over the last 30 years.

Degree-day Models

Degree-day phenology models were first used in peaches for oriental fruit moth. First developed in Michigan and California (Croft et al. 1980; Rice et al. 1984), they were later validated and adjusted for eastern conditions, and found to vary depending on the host plant e.g., whether the insect was in peaches or apples (Myers et al. 2007). These models helped precisely time insecticide applications for oriental fruit moth, which in itself saves multiple insecticide sprays. Insecticide timing can be further refined depending on what type of chemistry was being used because some chemical classes, like the diamides and the IGRs, act on the larval stages as opposed to the adult stages that might be targeted by a pyrethroid (Borchert et al. 2004). Degree-day models for both oriental fruit moth and tufted apple budmoth are now widely used, and recommendations can be found in state recommendation guides and online (Halbrendt 2012; Ward 2012). While a degree-day model for plum curculio was developed in New York for use in apples (Reissig et al. 1998), plum curculio insecticide timing still relies on direct monitoring, making management more challenging. However, advances in traps and attractants (Leskey and Zhang 2007) are being made, and a model for the southern plum curculio strain is being developed in Alabama (Akotsen-Mensah et al. 2011).

Mating Disruption

Mating disruption was first used in western states for oriental fruit moth, and then adopted on a small scale in Michigan, New Jersey, Pennsylvania, and other eastern states (Weakley et al. 1987; Cardé and Minks 1995; Halbrendt 2012; Ward 2012; Wise 2013). Mating disruption acts to confuse male moths, inhibiting their mate-finding ability and thereby directly reducing or eliminating mated females in the orchard. The effectiveness of oriental fruit moth mating disruption has been well documented with hand-placed dispensers, sprayable formulations, and 'wax' type emulsions (Rice and Kirsch 1990; Pree et al. 1994; Trimble et al. 2004; Stelinski et al. 2005; Stelinski et al. 2007). Peachtree borer and lesser peachtree borer can also be controlled with mating disruption, but all labeled materials are currently hand applied.

Ground Cover Management

A ranking of peach arthropod pests in New Jersey and the mid-Atlantic area would usually place oriental fruit moth first, since it is a direct pest with multiple generations. Plum curculio and roughly three species of cat facing insects or true bugs would be grouped second most important, based on grower pesticide use and damage surveys from growers (Polk et al. 1995, 2010). Ground covers and wild hosts can have a profound influence on populations of tarnished plant bugs and stink bugs found in the orchard (Killian and Meyer 1984; Shearer et al. 1998; Hardman et al. 2004). Growers who manage orchards with weed-based ground covers instead of turf-based management are much more likely to have higher cat facing insect populations, which result in more insecticide use and/or higher damage levels.

'Soft' Chemistries and the Conservation of Beneficials

Classical IPM in tree fruit crops has historically relied on insecticides applied for key pests that had minimal impact on mite predators and other beneficial arthropods. Before FQPA, these consisted of repeated use of specific OP materials that had less impact on beneficials than carbamates or most pyrethroids. This in turn improved the chances of having increased biological control of mites and decreased use of miticides. These programs were well established through the 1970s, 1980s and early 1990s. Depending on the number and type of insecticides used, parasitoids for other pests were often conserved, helping to prevent secondary pest outbreaks (Croft and Brown 1975; Hill et al. 1998). As pesticide chemistries changed, additional tests led to recommendations encouraging growers to continue to use selective products that have minimal impact on beneficials. For example, indoxacarb (Avaunt®), spinosad (Spintor®), spinetoram (Delegate®), the diamide chemistries (chlorantraniliprole–Altacor® and flubendiamide–Belt®), and most insect growth regulators have very little effect on certain beneficials and fit well into IPM programs. By contrast, pyrethroids can be highly toxic to predaceous mites and many parasitoids, and have been shown to increase mite populations (Coats at al. 1979; Croft and Whalon 1982). Repeated use of pyrethroids in New Jersey and other mid-Atlantic states has caused secondary pest outbreaks of scale populations in peach and apple orchards.[2]

Geographic Information Systems (GIS), Maps, Sprayer Calibration, Coverage, Alternate Row Middle Spraying

While it may seem elementary, knowing how much hectarage a tank of spray is covering, the distribution of pesticide into the foliage, and amount of chemical needed per tank are often not well defined. Based on the authors' experience, growers can fill a spray tank based on the perceived size of a planting, or the "tax map" hectarage of the land, rather than the actual measured planted tree area. This is critical in terms of effective pest control and pest management costs. Some new insecticides can cost over $ 98.80 ha^{-1} ($ 40 acre^{-1}) application. While GIS systems have been in use over large hectarages in the mid-west to assess crops and pest impact, use

---

[2] Based on authors' personal experiences.

on individual fruit farms has been limited. One practice used in the New Jersey IPM program was to supply a geo-referenced map to all peach growers using IPM practices, with each block identified to exact hectarage. Some growers recalibrated their sprayers from being as much as 15 % off. Sprayer calibration should be routine and done each year, but many growers do not go through this process for years at a time, leading to worn nozzles and excessive and inaccurate application of pesticides (Salyani 2003; Anonymous 2009). Simple water sensitive cards can be placed in trees to examine spray deposition. Another method is the practice of driving down every other orchard aisle— alternate row middle spraying—covering half the hectarage but twice as often as if driving every aisle. Conventional pesticides were traditionally applied at 1/3 to 1/2 the full field rate, but since they are applied more often, there is less time to weather off the plant, leading to better control of some insects, and often less pressure on beneficials (Asquith and Hull 1979; Hull et al. 1983; Hopfinger 1990).

Resistance Management

Educating growers about resistance management has become an important IPM component. Most growers remember the internal worm and budmoth OP resistance in the early 1990s. They are also keenly aware of the high cost of new chemistries. Therefore, today's IPM practices include (1) rotation of chemistries, (2) use of other non-pesticide strategies like mating disruption, and (3) use of action thresholds when available or a combination of all these practices listed to minimize pesticide use (Brattsten et al. 1986; IRAC 2009).

15.2.1.4 Putting It Together–Research and Extension Delivery

New Jersey has had an extension-based delivery program for tree fruit growers since 1981, but since pest management options were changing, there was a need to assemble IPM practices in a research and demonstration project (Atanassov et al. 1999; Polk et al. 1999). In 1998, a USDA Pest Management Alternatives Program (USDA PMAP) project focused efforts on ground cover management for true bugs, or the various hemipteran insects with piercing sucking mouthparts that deform peaches, causing what is commonly referred to as "catfacing" damage, mating disruption for oriental fruit moth, and IPM scouting and recommendations for all pests (Atanassov et al. 2002). Ground cover was managed by using turf type fescue grass in the aisles in demonstration "reduced-risk" blocks, and maintained with herbicides. This eliminated clover and other weeds otherwise used by these insects as alternate hosts. Any insecticide use for plum curculio or other insects was based on newer reduced-risk chemistries. Demonstration blocks were compared with grower standard blocks. The reduced-risk blocks had 2.3–4.9 times fewer heteropteran insects, about equal fruit quality, but required fewer insecticides as measured in terms of the number of applications made and the amount of active ingredient used. This had been a "peach only" program, and most peaches are produced by growers who also grow apples, as is the case with other eastern growers. Therefore, the need remained to examine whole orchard systems under similar reduced-risk practices. Between 2002 and 2005, a multi-state USDA Risk Avoidance and Mitigation Program (USDA RAMP) project examined reduced risk practices in both apples and peaches with collaborators in Michigan, New Jersey, New York, North Carolina, Pennsylvania, Virginia, and West Virginia (Agnello et al. 2009). This integrated project involved 50 commercial apple orchards and 20 peach orchards. Mating disruption was used for oriental fruit moth, lesser peachtree borer, and peachtree borer in some orchards. Since growers are concerned with the costs of alternative programs, an economic analysis was included. By the end of the 4 years, growers had adopted many of the practices being demonstrated, and the reduced risk practices were shown to work. In general, both the standard and reduced practices were equally effective. The reduced-risk block received 79 % less insecticide than the standards. Mating disruption increased costs by $ 292 ha^{-1}, but over the 4 years net income was lower in the reduced-risk treatments in only 1 year (Agnello et al. 2009). The success of this project showed that reduced risk practices could be woven into commercial IPM practices, but that growers had to be very aware of the costs associated with these practices.

15.2.1.5 An Asian Invasion

During the mid 1990s the brown marmorated stink bug, *Halyomorpha halys* (Stål), was introduced to the Allentown, Pennsylvania area. It has since spread throughout the mid-Atlantic area and is now present in 38 states (Leskey et al. 2012). In 2010, the damage caused by these populations developed into a severe economic problem. Funds from the USDA National Institute of Food and Agriculture (USDA NIFA) Critical Issues and USDA Specialty Crop Research Initiative (USDA SCRI) programs were obtained to address the ongoing problem. The brown marmorated stink bug is unlike any other orchard pest for several reasons. First, every life stage can be present in the orchard, and all stages except 1st instars feed on the host plant (Nielsen and Hamilton 2009a). Secondly, there is a wide host range that includes most agronomic and horticultural hosts (Nielsen and Hamilton 2009b). Third, it is a strong flier and mobile in all life stages. Fourth, it is not susceptible to many of the new reduced-risk insecticides recently developed and labeled for peaches. Finally, it does not have a known, strong complex of parasitoids and predators that can keep it in check. Therefore, since 2010 peach growers in the mid-Atlantic have returned to broad-spectrum pyrethroids, methomyl, and limited use of selected neonicotinoids. These have been used on a 6–7 day intensive schedule, thus eliminating many of the IPM practices developed over the last 30 years (Leskey et al. 2012). As research continues with this pest, new monitoring procedures and practices will be assembled to return the peach system to more of an IPM approach.

15.2.2 Blueberries

Highbush blueberry (*Vaccinium corymbosum* L.) is a crop native to North America that has been under commercial cultivation since the 1930s (Eck and Childers 1966). In the US, highbush blueberries are grown in more than 30 states on 29,150 ha (72,000 acres) mainly in the states of Michigan, Oregon, New Jersey, and Georgia (total US utilized production valued at more than $ 788 million) (USDA NASS 2012b). In 2011, New Jersey blueberries were grown on 3,113 ha (7,700 acres), producing 28.2 million kg (62 million lb) valued at $ 94.7 million (USDA NASS 2012b), making it the highest valued food crop grown in the state.

Blueberry production is increasing worldwide due to a growing per capita consumption of fruit. For example, in the 1980s, the per capita consumption of frozen blueberries was about 0.1 kg (0.22 lb), but grew to 0.15 kg (0.39 lb) year^{-1} by the early 2000s; while consumption of fresh market fruit grew from 0.09 kg (0.2 lb) in early 1990s to 0.15 kg (0.34 lb) during 2000–2002 (USDA ERS 2003). This greater consumption is associated with the increasing public awareness about the many nutritional and health benefits of blueberries: berries are low in calories and rich in vitamin C, potassium, and fiber (USDA ERS 2003). They also contain antioxidants that help neutralize free radicals, which have been linked to the formation of cancers, cardiovascular diseases, urinary tract infections, and improved vision (e.g., Howell et al. 1998; Youdim et al. 2000; Joseph et al. 2003; Ofek et al. 2003).

Blueberries suffer major yield losses due to insects. The pest complex in blueberries is extensive, with pests attacking all parts of the plant (fruit, buds, leaves, roots, stems, and flowers) (Marucci 1966). In New Jersey, blueberries are host to over 17 species of insect pests (Hamilton 2001). Key pests include: blueberry maggot (*Rhagoletis mendax* Curran), aphids (*Illinoia* spp. and *Ericaphis* spp.), oriental beetle (*Anomala orientalis* (Waterhouse)), cranberry fruitworm (*Acrobasis vaccinii* Riley), plum curculio (*C. nenuphar*), and cranberry weevil (*Anthonomus musculus* Say). The blueberry marketplace demands a zero tolerance for pest defects or presence of insects in the final product, necessitating a very aggressive and intensive insecticide use program.

15.2.2.1 Historical Perspective of Integrated Pest Management in New Jersey Blueberries

As the blueberry industry continues to grow, new pest management strategies need to be implemented that are efficient, cost-effective, and safe to humans and the environment. This is especially critical for New Jersey growers because blueberries are grown in one of the most environmentally-sensitive areas of the state (Moore 1995). Blueberry production in New Jersey is highly localized in the ecologically sensitive "New Jersey Pinelands", a national reserve, which is characterized by porous soils with high water tables and subject to vertical movement of a number of agricultural chemicals. This area is a source of a number of streams that drain into watersheds.

The counties surrounding these watersheds are home to over 2 million people (US Census 2004), many of whom use this water for recreation and consumption.

Blueberry growers rely heavily on insecticides to manage pest problems (Drummond 2000). For instance, pest management practices in blueberries require up to 12 pesticide sprays per year depending on pest pressure and variety. The vast majority of these sprays are broad-spectrum OP and carbamate insecticides. For example, insecticide-use data collected by Rutgers Cooperative Extension IPM programs in New Jersey for blueberries indicate that about 90% of insecticide applications are with OP and carbamate insecticides (Polk and Samoil 1993; Dill et al. 1998).

Potential environmental risks associated with the use of non-selective, broad-spectrum insecticides such as surface water pollution, negative effect on wildlife, and worker exposure is a major concern among regulators that implemented the FQPA (US EPA 1996). Several of the most effective pest management tools are currently under review, scheduled for cancellation, or severely restricted under the FQPA, as seen for the recent phase-out plan for azinphos-methyl and restrictions on diazinon and expected reduction in availability of malathion. This tolerance reassessment of broad-spectrum OP and carbamate insecticides is likely to impact the blueberry industry more than any other crop because of their minor crop status, zero tolerance for insect pests, high potential for insect infestation, and quarantine and contamination concerns. Therefore, it is critical that new selective insect management strategies become available to blueberry growers. The development and adoption of novel selective, reduced-risk practices and their implementation into IPM programs in blueberries are expected to improve timing of insecticide applications and reduce applications of broad-spectrum insecticides, thus reducing input of these insecticides into the southern New Jersey wetlands[3].

Another concern among blueberry growers is the potential for secondary pest outbreaks[4]. Several secondary pests have been maintained below economic threshold with applications of broad-spectrum insecticides. As broad-spectrum insecticides are being replaced with reduced-risk insecticides, there is potential for secondary pests to become major pests. Increase of secondary pests is already being observed on many farms in New Jersey. For example, since adoption of reduced-risk chemicals, there has been a steady increase in the populations of leafhoppers, scales, thrips, and cranberry tipworm in blueberry farms. Leafhoppers, in particular the sharp-nosed leafhopper (*Scaphytopius magdalensis* (Provancher)), are of special concern because they can transmit diseases caused by a phytoplasma such as blueberry stunt (Chen 1971). No effective monitoring and economic alternatives to broad-spectrum insecticides currently exist for many of these secondary pests.

---

[3] Based on authors' experiences.

[4] Based on authors' experiences.

Table 15.2 US blueberry insecticide use nationally[1] (*left*) and in New Jersey (NJ) (*right*) for 1995, 2003, and 2011

Average Values of Top 12 Blueberry Insecticides in the US - By Percent of Hectares Treated

A.I.	% Ha Trt	No. Appl.	Kg/Ha/Yr (lb/A/Yr)	IRAC Class
1995 (Total= 7)[2]				
Malathion	56	3.8	3.98 (3.56)	1B
Carbaryl	23	1.9	3.26 (2.91)	1A
Azinphos-Methyl	49	1.6	0.92 (0.82)	1B
Methomyl	44	1.6	1.05 (0.94)	1A
Phosmet	20	1.6	1.47 (1.31)	1B
Bt's	5	1.5	ND	11
Diazinon	7	1.4	1.51 (1.35)	1B
2003 (Total= 12)				
Malathion	37	2.5	3.75 (3.35)	1B
Phosmet	51	2.2	1.92 (1.72)	1B
Bt's	8	2.1	ND	11
Carbaryl	19	1.7	2.73 (2.44)	1A
Esfenvalerate	16	1.7	0.07 (0.06)	3A
Azinphos-Methyl	52	1.6	0.74 (0.66)	1B
Diazinon	17	1.6	1.19 (1.07)	1B
Methomyl	31	1.5	0.83 (0.74)	1A
Imidacloprid	11	1.3	0.07 (0.06)	4A
Spinosad	1	1.1	0.08 (0.07)	5
Endosulfan	2	1.0	0.55 (0.49)	2A
Tebufenozide	14	1.0	0.15 (0.13)	18
2011 (Total = 33)				
Phosmet	38	1.9	1.88 (1.68)	1B
Zeta-Cypermethrin	32	2.0	0.06 (0.05)	3A
Malathion	30	3.4	4.44 (3.96)	1B
Esfenvalerate	21	1.7	0.08 (0.07)	3A
Imidacloprid	16	1.3	0.12 (0.11)	4A
Methoxyfenozide	16	1.2	0.24 (0.21)	18
Diazinon	15	1.3	0.91 (0.81)	1B
Azinphos-Methyl	10	1.2	0.74 (0.66)	1B
Acetamiprid	9	1.3	0.12 (0.11)	4A
Methomyl	9	1.4	1.12 (1.00)	1A
Bt's	7	1.5	ND	11
Spinosad	6	1.5	0.19 (0.17)	5

Average Values of Top 12 Blueberry Insecticides in NJ - By Percent of Hectares Treated

A.I.	% Ha Trt	No. Appl.	Kg/Ha/Yr (lb/A/Yr)	IRAC Class
1995 (Total= 7)				
Methomyl	69	1.5	1.15 (1.03)	1A
Azinphos-Methyl	59	1.9	1.18 (1.05)	1B
Carbaryl	42	1.3	2.28 (2.04)	1A
Malathion	35	1.9	2.96 (2.64)	1B
Diazinon	22	1.5	1.68 (1.50)	1B
Phosmet	7	1.3	1.33 (1.19)	1B
Bt	4	1.0	ND	11
2003 (Total= 5)				
Phosmet	48	1.9	1.98 (1.77)	1B
Diazinon	31	1.5	1.32 (1.18)	1B
Imidacloprid	27	1.3	0.10 (0.09)	4A
Carbaryl	24	1.2	1.49 (1.33)	1A
Malathion	11	1.8	2.39 (2.14)	1B
2011 (Total = 23)				
Phosmet	51	1.9	1.90 (1.70)	1B
Methomyl	29	1.4	1.09 (0.97)	1A
Acetamiprid	22	1.3	0.12 (0.11)	4A
Imidacloprid	20	1.2	0.10 (0.09)	4A
Diazinon	19	1.1	0.71 (0.63)	1B
Esfenvalerate	14	1.9	0.07 (0.06)	3A
Carbaryl	4	1.5	3.00 (2.68)	1A
Malathion	4	1.9	2.22 (1.98)	1B
Pyriproxyfen	4	1.1	0.18 (0.16)	7C

USDANASS. (2012b). (http://quickstats.nass.usda.gov/).
[1] National survey data includes records from Georgia, Michigan, New Jersey, North Carolina, Oregon, and Washington.
[2] Denotes number of insecticides in record. If less than 12 are listed, then insufficient data exists to list at least 12 materials.
A.I. = active ingredient; % Ha Trt = percent hectarage treated; No. Appl. = number of applications; Kg/Ha/Yr = kilograms per hectare per year; lb/A/Yr = pounds per acre per year; IRAC Class = Insecticide Resistance Action Committee Classification; ND = not determined; Bt = *Bacillus thuringiensis*.

15.2.2.2 Changes in Insecticide Use Patterns in Blueberries

Table 15.2 shows changes in insecticide usage in blueberries across US and in New Jersey before and after FQPA. Across the US (Georgia, Michigan, New Jersey, North Carolina, Oregon, and Washington), the neonicotinoid imidacloprid and the reduced-risk insecticides spinosad and tebufenozide were used in 11, 1, and 14% of the insecticide-treated hectares, respectively, 7 years after FQPA. Imidacloprid, acetamiprid and the reduced-risk insecticide spinosad were used in 16, 9, and 6% of treated hectares, respectively, across the US 15 years after FQPA.

In New Jersey, imidacloprid was used in 27% of all insecticide-treated hectares 7 years after FQPA. After 15 years of FQPA, imidacloprid and acetamiprid, and the insect growth regulator pyriproxyfen were used in 20, 22, and 4% of treated hectares, respectively.

15.2.2.3 Developing a Reduced-Risk Integrated Pest Management Program for Blueberries

In response to the FQPA, a team from three of the leading blueberry-producing states (Michigan, New Jersey, and Maine) worked under a USDA RAMP project (2002–2006) to develop and implement reduced-risk IPM programs targeting insect pests of blueberries. This project resulted in grower adoption of several reduced-risk insecticides, including methoxyfenozide (Confirm 2F®) for cranberry fruitworm control, imidacloprid (Provado 2F®) for aphid control, and spinosad and imidacloprid for blueberry maggot control. Control of insect pests using reduced-risk programs was usually comparable to those using standard OP-based programs. Blueberries managed under the reduced-risk program also had between 45% and 58% lower amounts of insecticide active ingredient applied than those grown using grower standard programs, with even greater reductions in the total amount of insecticide residue detected on leaves and fruit at harvest (C. Rodriguez-Saona and D. Polk, unpublished data). In many cases, this control was achieved using more expensive insecticides. As a result, insecticide costs of the reduced-risk programs were often higher than the standard programs. Implementation of reduced-risk IPM programs is, however, expected to provide additional benefits to growers in the form of greater natural pest control, improved pollination, or other enhancements of ecosystem services.

15.2.2.4 Towards a Sustainable Ecologically-based Integrated Pest Management in Blueberries

Until recently, the majority of research and extension efforts in blueberries focused on replacing OP insecticides with newer chemistries. Yet, to maximize the long-term sustainability of blueberry production, it is critical to implement IPM programs that are based on a combination of integrated approaches. Current research and extension programs nationwide and at the Rutgers P.E. Marucci Center for Blueberry and Cranberry Research and Extension (Chatsworth, New Jersey) are moving towards the development and implementation of more sustainable, ecologically-based IPM programs.

Degree-day models

Ongoing research in Michigan and New Jersey under a USDA Crops-at-Risk (USDA CAR) project (2010–2012) focuses on the development and validation of a degree-day model for cranberry fruitworm, *A. vaccinii*. Monitoring traps are used in combination with growing degree days to better target the egg-laying period of this pest. Degree days can also be used to predict the development of other key blueberry pests. Blueberry growers are increasingly able to access weather information online and in the field through digital technologies[5]. For example, a degree-day model was developed and is currently available online for estimating thrips activity in blueberry farms in New Jersey (http://benedick.rutgers.edu/Blueberryweather/).

Mating Disruption

The oriental beetle, *A. orientalis*, is one of the most difficult insect pests of blueberries to control in New Jersey. Since its introduction sometime before 1920 (Vittum et al. 1999), this insect has become a problematic pest in blueberries and other crops in the northeastern US (Alm et al. 1999). The wide-ranging behavior of this invasive pest has led to its increasing population across New Jersey and the oriental beetle now threatens other blueberry-producing states. The insecticide imidacloprid (Admire®) is the only treatment available for oriental beetle control; however, having a single control method raises resistance management concerns. A promising ecologically-sound tool for controlling oriental beetles is the use of the sex pheromone to disrupt mating (Polavarapu et al. 2002; Sciarappa et al. 2005; Rodriguez-Saona et al. 2009). A formulation based on retrievable dispensers for oriental beetle mating disruption will soon become available to growers (expected in 2013). This research was supported by the EPA Region 2 and the USDA IR-4 Biopesticide and Organic Support Program.

'Soft' Chemistries and the Conservation of Beneficials

Reduced-risk insecticides are expected to help maintain natural enemy populations and minimize toxicity to native bees (Devine and Furlong 2007). In a recent study, Wise et al. (2010) examined the effect of various insecticides on non-target biocontrol agents in blueberries, showing that some reduced-risk insecticides are relatively safe to fruitworm, *A. vaccinii*, eggs parasitized by *Trichogramma minutum* Riley. Tuell and Isaacs (2010) found that increasing toxicity of a spray program applied after bloom is associated with declining abundance and diversity of bee communities in blueberries. This suggests that adoption of reduced-risk IPM programs will have additional benefits by potentially conserving natural enemies of pests and stabilizing crop pollination (Desneux et al. 2007; Gentz et al. 2010).

---

[5] Based on authors' experiences.

Geographic Information Systems (GIS)

In commercial blueberry fields, the vast majority of insect pest populations come from outside the fields. Fields that border a wooded area and other "edge" fields containing wild hosts, such as huckleberries or wild blueberries, are likely to have higher pest pressure from cranberry fruitworm, *A. vaccinii* (Mallampalli and Isaacs 2002), and blueberry maggot, *R. mendax* (Collins and Drummond 2004). Japanese beetles, *Popillia japonica* Newman, are most likely to move into crop fields from grassy regions outside, or from grassy areas where grubs overwinter (Szendrei et al. 2005). These edge fields and their perimeters are therefore at higher risk for fruit infestation by insects, requiring greater insecticide use. Use of site-specific pest mapping may help identify where treatments are needed, thereby optimizing and reducing insecticide use. For example, a project funded by the USDA Sustainable Agriculture Research and Education (USDA SARE) program in 2009–2011 revealed that GIS-based monitoring of blueberry maggot flies in New Jersey can result in some blueberry growers saving up to 4–5 insecticide applications and a 2- to 3-fold saving in insecticide use (Rodriguez-Saona and Polk, unpublished data). Research is underway (USDA CAR project) in Michigan and New Jersey to develop and implement whole-farm GIS-based IPM programs for several key fruit pests of blueberries.

15.2.2.5 Extension Delivery Methods–Experiences in New Jersey

New Jersey has active IPM programs that serve the needs of blueberry growers, consultants, scouts, and industry leaders within the state and in the surrounding regions. Both the research and extension programs work as a combined effort. Results of research conducted on campus, at experiment stations, and in farms are transferred to stakeholders through formal winter extension meetings, summer workshops, and through weekly newsletters. The Blueberry Bulletin (Pavlis 2012) provides a vehicle for rapid dissemination of information to growers across the region and beyond through direct mail, e-mail, and web presence. While current information is valuable during the growing season, it is also important to have a system to archive pest management guidelines and fact sheets. Pesticide information is provided annually to growers in the "Commercial Blueberry Pest Control Recommendations for New Jersey" (Oudemans et al. 2012).

The New Jersey Blueberry Industry Advisory Council funds the delivery program that involves scouting, data collection, individual grower consultations and reports, and pesticide recommendations for growers managing about 60% of the state's blueberry hectarage. Data from these 'primary participants' are summarized each week for the Blueberry Bulletin. Updates, presentations at grower meetings, along with broadcast e-mails, faxes and phone calls are the forms of information transfer to all growers. Our cooperative work between the IPM program, growers, and researchers fosters numerous on-farm research projects[6].

---

[6] Based on authors' experiences.

15.2.2.6 Another Asian Invasion

In 7 July 2011, the first adults of the spotted wing drosophila, *Drosophila suzukii* Matsumura, were found in blueberry farms in New Jersey. This insect pest is native to Asia and has rapidly spread from California, where it was first detected in 2008, to Oregon, Washington, British Columbia, Ontario, North Carolina, South Carolina, Michigan, and Florida, and other states of the US (Hauser 2011). The greatest potential economic impacts are expected in blueberry, peach, cherry, strawberry, raspberry, and blackberry crops, because soft-fleshed fruit are easier for the flies to lay eggs in and for larvae to develop. Currently, the blueberry IPM research and extension program in New Jersey is monitoring with apple cider-vinegar traps placed on over 40 farms (Rodriguez-Saona and Polk 2011). It is too early to predict what will happen to blueberry IPM programs after introduction of the spotted wing drosophila; however, this new invasive pest will undoubtedly change current pesticide use patterns and disrupt established IPM practices, as seen in 2012. For example, neonicotinoids (e.g. imidacloprid and acetamiprid) are effective against blueberry maggot (Liburd et al. 2003), and became adopted by several blueberry growers in the past decade to manage resistance to OPs. Neoniconitoids were recommended in IPM programs in rotation with other chemistries for blueberry maggot control. However, neonicotinoids are not very effective against the spotted wing drosophila and thus to combat these two pests, growers had no other choice but to use OP- (e.g. malathion) and pyrethroid-based programs. Also, unlike blueberry maggot that is distributed mainly in fields near wooded areas within blueberry farms (Collins and Drummond 2004), spotted wing drosophila is distributed throughout farms (C. Rodriguez-Saona and D. Polk, unpublished data). Many growers had started to adopt GIS-based management for blueberry maggot where only high-risk fields were treated, which in turn reduced pesticide use, as discussed above. This approach is less likely to work for the spotted wing drosophila. Research addressing landscape management of multiple fruit pests in blueberries is underway in Michigan and New Jersey (a project funded by USDA CAR).

15.2.2.7 Encouraging Integrated Pest Management Adoption—Experiences in New Jersey Fruit Crops

Adoption of IPM practices can be thought of on two levels: (1) the 'micro' or individual grower learning experience, and (2) the 'macro' or statewide, regional, or industry targeted experience. The two broad approaches are both distinct and dependent on each other. Economics is also the main driving force for adoption of IPM practices. While research and extension workers are very familiar with economic threshold levels and action levels, the entire approach for an IPM program needs to be based on the value and marketing of the commodity as a whole. Peach and blueberry systems are different and can be contrasted to each other.

The traditional extension 'macro' approaches of meetings, newsletters, and field research/demonstrations, and more recently web presence and blogs address whole grower groups, but individuals and individual farms differ. We have found that growers learn from each other, and often one or two progressive growers can be the spark that helps change practices for many more growers. Growers have risks by the nature of the business (Zeuli 1999), and individuals show various degrees of managing those risks. Demonstrating new practices is often one-on-one in the 'micro' sense, where repetition, demonstration and economics are combined. Individual data collection and scouting serves to both demonstrate and assure the grower that the new practice is working. Growers communicate among themselves, and if a new practice works, more people will use it, especially if there is an economic incentive, whether through reduced input costs or increased crop value.

System differences will affect both the type and extent of IPM practices used. In contrasting the wholesale peach and highbush blueberry industries, the peach system has more pests than in blueberries, has traditionally demanded more pesticide applications, and has a lower gross return. This makes economic incentives to lower production costs stronger in peaches than in blueberries. In both crops there is a '0' tolerance for direct fruit damage, but peaches are purchased as individual fruit where cosmetics are important, while blueberries are sold in pints, quarts, and other containers where cosmetic appearance is not as important, thus helping to contribute to the higher pest management costs associated with peaches[7]. Collection of pesticide use records and monitoring fruit quality and incidence of pest damage can be used to demonstrate to peach growers the results of specific IPM practices. The economic incentive of lower production costs can be demonstrated with peach growers. If grower "A" has 95% clean fruit and spent $ 618 ha^{-1} ($ 250 acre^{-1}) for pesticides, and grower "B" has 90% clean fruit and spent $ 1,235 ha^{-1} ($ 500 acre^{-1}) for pesticides, then grower "B" may be more willing to see what grower "A" did. However, this approach does not work as well in the blueberry system unless there is either a very large difference in production costs, or the grower is willing to be more progressive and views reduced spray applications and risk acceptance as a trade for easier management practices, e.g. not having to worry about more frequent sprays, resulting re-entry intervals and managing farm-worker crews for picking[8]. This is why the GIS approach to blueberry maggot management worked. It is also why the arrival of the spotted wing drosophila severely hinders IPM practices, since now acceptable risk levels previously associated with the newer blueberry IPM practices become too high for grower acceptance.

15.2.3 Citrus

15.2.3.1 Historical Perspective of Integrated Pest Management in Florida Citrus

Prior to 2008, citrus production in Florida may have been one of the most successful examples of implementing biological control in US agriculture (Rosen et al.

---

[7] Based on authors' experiences.

[8] Based on authors' experiences.

1994). Although citrus trees were introduced to and growing in Florida more than 400 years ago (Simanton 1996), interest in commercial citrus production began to expand in the 1920s, coinciding with significant real-estate expansion (Simanton 1996). Early commercial citrus production in Florida was marred by important pest and disease outbreaks, as well as cultural problems, because early land developers had limited knowledge of optimal citrus growing practices and often planted trees in areas poorly suited for commercial production (Simanton 1996). A need for citrus research facilities was recognized and both the USDA and the University of Florida Agricultural Experiment Station established such facilities where entomologists, plant pathologists, and horticulturalists began to generate information on best practices for citrus production. The University of Florida Citrus Research and Education Center in Lake Alfred, still a prominent citrus research facility today, was established in 1917. Significant inroads were made by 1950 to alleviate previously detrimental nutritional deficiencies in citrus production; however, insect pests and diseases have remained important areas of research well into the twenty-first century, because the Florida peninsula is highly susceptible to invasion by exotic pests and pathogens (Rogers et al. 2012). As early as 1950, the important role of biological control agents in regulating pest populations in Florida citrus was recognized (Fisher 1950).

The Florida citrus industry experienced a major expansion following the adoption of a new process for making frozen juice concentrate in 1946 and by 1950 there were approximately 28 million trees grown on over 177,000 ha (Simanton 1996). Showing tremendous foresight, in 1950, the director of the Florida Citrus Experiment Station, Dr. A.F. Camp, proposed initiation of a project to comprehensively survey the ecology of citrus groves in Florida statewide; the project would be called: "Ecological Survey of Citrus Pests and Disorders" (Simanton 1996). The project would include a detailed survey of living organisms in groves, both pest and beneficial, their interactions, and the impacts of pesticides on these organisms. The idea of reducing pesticide input, for both economic and ecological reasons, was therefore championed in Florida citrus production as early as 1950. There were 130 groves selected for this large survey effort throughout the state. The survey identified previously unknown pest species, including a spider mite, *Eutetranychus banski* (McGregor) (Muma et al. 1953). However, perhaps more important discoveries were made with respect to the existence and impact of biological control agents regulating pest populations. The accidental introduction of a parasitoid, *Aphytis lepidosaphes* Compere, of the purple scale, *Lepidosaphes beckii* (Newman), was discovered during the survey (Simanton 1960). Purple scale was the most important pest of citrus prior to 1959 and the realization that excessive use of sulfur caused outbreaks due to the impact of sulphur on natural enemies (Simanton 1960) was one of the early cases of implementing ideas that we would today consider as part of standard IPM practice. Advising growers to use sulfur more sparingly allowed populations of this parasitoid to rebound, which resulted in reduction of purple scale populations below economic thresholds. Moreover, this led to the implementation of classical biological control programs, including the importation of *Aphytis holoxanthus* Debach against Florida red scale, *Chrysomphalus ficus* (Ashmead), which essentially eliminated the status of this scale as a pest (Simanton 1974). The survey project lasted 16 years and likely developed one of the most comprehensive databases of seasonal cycles of organisms for a single crop in the US (Simanton 1996). The database was multifaceted and included descriptions of annual horticultural events and pesticide usage. From the standpoint of establishing IPM in citrus, it rigorously documented the importance of biological control in Florida citriculture, which spawned numerous successful classic biological control programs that persist to the present. This research-based program is perhaps the historical foundation of associated extension efforts that have significantly improved citrus production in Florida and made Florida a world leader of citriculture.

With roots of inception in the 1950s, formally defined IPM programs in Florida citrus production began in the 1970s (Knapp et al. 1996). This included the modern application of host-plant resistance, emphasis on classical and conservation biological control programs, development of meaningful horticultural/cultural control tactics, and emphasis on selective use of pesticides only when necessary (Knapp et al. 1996). This coincided with a significant increase in citrus production reaching over 330,000 ha of crop in 1971. From the 1970s and well into the 2000s, pesticide input in Florida citrus was limited to a few fungicide, herbicide, and horticultural oil sprays with significant de-emphasis of contact poison insecticides or miticides. Resistance rootstocks, cultural control practices for weeds, and significant reliance on the third trophic level (parasitoids, predators, and pathogens) for regulation of herbivores provided the bulk of pest management in a complex ecosystem that was continually faced with the prospect of new pest and disease introductions (Rogers et al. 2012). Growers have benefited from the knowledge generated by scientists researching various aspects of the Florida citrus ecosystem through traditional extension outlets characteristic of the land grant mission of the University of Florida. These have included extension publications, statewide extension training programs, on-farm educational visits by researchers and more recently the web-based Electronic Data Information Source (EDIS) (edis.ifas.ufl.edu) hosted by the University of Florida's Institute of Food and Agricultural Services.

15.2.3.2 Loss of Integrated Pest Management in Modern Florida Citrus Production and the Response in the Research/Extension Interface

While certain unique pests and pathogens such as citrus root weevil species (example: *Diaprepes abbreviatus* (L.)), citrus leafminer, *Pyllocnistis cytrella* Stainton, and citrus bacterial canker, *Xanthomonas axonopodis* (Hasse) have proven challenging to manage in Florida citrus and required unusual efforts of chemical input or tree removal as compared with most other pests and diseases in citrus, the arrival of the Asian citrus psyllid, *Diaphorina citri* Kuwayama, in Florida in the 1990s and the associated identification of a tree-killing disease, Huanglongbing (HLB), in Florida in 2005 (Grafton-Cardwell et al. 2013) essentially eliminated several decades of effort to develop effective IPM practices in Florida citrus. The disease is caused by a bacterial pathogen that this psyllid vector transmits. The disease transformed Florida citriculture due to the severity of this disease with respect to tree decline and possible tree death. The associated need for intensive vector management to maintain existence of productive trees has resulted in a near elimination of previously effective biological control (Grafton-Cardwell et al. 2013).

Citrus growers in Florida are finding it increasingly difficult to maintain low HLB incidence in their plantings and to remain profitable, even as they implement inoculum removal via infected tree removal, multiple insecticide application strategies believed to be effective in keeping disease rates low, and enhanced nutritional programs which aim to reduce the impact of disease symptoms (Grafton-Cardwell et al. 2013). The importance of effective vector control has increased as many growers are abandoning the strategy of inoculum removal and instead attempting to prolong the life and productivity of diseased trees with intense supplemental applications of micronutrients. These strategies are costly; moreover, with eroding success in controlling HLB incidence ostensibly due to high surrounding inoculum pressure, growers who were once optimistic that overcoming HLB was possible are now much less so. The loss of an insecticide due to development of resistance may be a future further blow to an industry that relies on insecticides for *D. citri* control and management of other significant pests (Grafton-Cardwell et al. 2013). Long-term HLB solutions such as resistant citrus cultivars and other methods of blocking HLB transmission by the vector may succeed in the future, but these methods will take time, perhaps decades, to develop (Grafton-Cardwell et al. 2013).

Diaphorina citri transmits three species of bacteria belonging to the genus *Candidatus* Liberibacter. *Ca.* Liberibacter asiaticus is the most likely causal agent of the citrus HLB disease in the US, although Koch's postulates have not yet been fulfilled (Tylor et al. 2009; Pelz-Stelinski et al. 2010). *D. citri* was first discovered in Florida in 1998 (Halbert 1998) and quickly became established throughout the state, making its eradication impossible. Currently, *D. citri* can be found in all citrus growing US states; i.e. Alabama, Arizona, California, Florida, Georgia, Louisiana, Mississippi, South Carolina, Texas, as well as Hawaii (Grafton-Cardwell et al. 2013). In 2005, plants with HLB infection were detected in Florida; within 4 years, all 32 citrus growing counties in the state had HLB-infected citrus (Grafton-Cardwell et al. 2013). HLB is considered the most destructive disease of citrus crops in the world. All known citrus cultivars are susceptible to HLB (Folimonova et al. 2009), and prevention of disease transmission has proven difficult worldwide (Grafton-Cardwell et al. 2013). Infected young trees die before they reach the fruit bearing stage (Grafton-Cardwell et al. 2013). Infected mature trees decline and die within 5–10 years of infection (Halbert and Manjunath 2004). Before their demise, infected mature trees produce unmarketable fruit; the fruit are small, misshapen, with uneven color development or remain green (Halbert and Manjunath 2004). Juice from infected fruit tastes bitter and unbalanced due to low soluble solids and high acid contents (Halbert and Manjunath 2004).

Given the efficient pathogen acquisition and transmission capabilities of *D. citri* (Pelz-Stelinski et al. 2010) aided by short (Boina et al. 2009a) and long-range dispersal capabilities, HLB spreads rapidly within and between groves. Continuous 30 min feeding by healthy *D. citri* on a HLB-infected host plant is necessary for acquiring the bacterium (Grafton-Cardwell et al. 2013). Once the *D. citri* acquires the bacterium, a latent period ranging from 7–25 days may exist before the *D. citri* is able to transmit the bacteria into another host plant (Pelz-Stelinski et al. 2010). On the other hand, an infected *D. citri* can transmit the bacteria into another host plant by continuous feeding for only 5–7 h (Xu et al. 1988). Time periods involved in acquisition, latency and transmission provide a window of opportunity for controlling infected *D. citri* and preventing disease spread. Gravid *D. citri* females lay eggs only on new flush, and early instar nymphs feed on new flush and go through five instars before becoming adults. This results in very high buildups of *D. citri* populations during flushing periods that are capable of acquiring and transmitting HLB bacteria. Furthermore, within perennially growing citrus in sub-tropical climates, the multigenerational and season-long occurrence of arthropod pests, such as *D. citri*, is common. The problem of insecticide resistance development in citrus is a recurrent worldwide problem and has been well documented. Use of OPs for more than 40 years in California to control the citrus thrips, *Scirtothrips citri* (Moulton), and the California red scale, *Aonidiella aurantii* (Maskell), resulted in insecticide failures in the 1980s and 1990s, respectively (Morse and Brawner 1986; Immaraju et al. 1989; Grafton- Cardwell and Vehrs 1995). Resistance to OPs in California has been found for the yellow scale (*Aonidiella citrine* (Coquillett)) and the citricola scale (*Coccus psuedomagnoliarum* Bartlet) (Grafton-Cardwell and Vehrs 1995). In Florida, the citrus rust mite (*Phyllocoptruta oleivora* (Ashmead)) showed resistance to dicofol, which could be managed by restricting use to a maximum of one application per year (Omoto et al. 1995).

Soil and foliar applied insecticides play a vital role in suppressing *D. citri* populations (Srinivasan et al. 2008; Boina et al. 2009b) that are infected or capable of acquiring and transmitting HLB bacteria. Recently, the widely used systemic insecticide, aldicarb, has been restricted for use in Florida citrus by the EPA (Rogers et al. 2011). Therefore, there is only one mode of action (neonicotinoids) available as a soil-applied, systemic treatment for control of *D. citri*; a situation that will encourage rapid development of resistance (Tiwari et al. 2011). Neonicotinoids are the most widely used insecticides for protecting newly planted groves or young tree resets (Rogers et al. 2011). After trees come into production, foliar applications are implemented (Rogers et al. 2011). Conventional spray application methods of high volume airblast sprays of more than 933 liters ha^{-1} (100 gallons acre^{-1}) are often insufficient to manage the need for 8–12 applications that target *D. citri* alone (Grafton-Cardwell et al. 2013). Control of *D. citri* to suppress HLB incidence costs growers an additional \$ 2,470 or more hectare^{-1} (\$ 1,000 acre^{-1}) each year (Muraro 2009). Given the extra expense, an optimized spray delivery method for these additional *D. citri* applications is a must for the US citrus industry (Grafton-Cardwell et al. 2013). In response, low volume, 'misting' based on cold fogging technology was widely investigated in Florida citrus and has now become widely accepted by the Florida citrus industry (Grafton-Cardwell et al. 2013).

Given the mobility of *D. citri*, HLB has rapidly spread in Florida despite efforts to control the disease (Boina et al. 2009a; Tiwari et al. 2010). *D. citri* adults disperse long distances during periods of peak activity and move the causal pathogen of HLB widely; particularly from unmanaged, abandoned areas into well managed groves (Tiwari et al. 2010). Grower neighbors often made independent decisions on production and pest management, and these tendencies are being overcome through outreach and demonstrations of the benefits of cooperative actions. A spray program effectively applied to a single block of citrus and not to neighboring blocks will be ineffectual, as untreated blocks serve as sources of infected *D. citri* and disease (Tiwari et al. 2010). Thus, it has become clear that area-wide, cooperative management of the vector has been necessary. Recently, the US National Academy of Sciences published a strategic plan for management of HLB and identified the development and implementation of area-wide cooperatives, so-called "Citrus Health Management Areas (CHMAs)", as the most important organizational priority for HLB management (National Research Council 2010). Significant progress has been made to develop these CHMAs and many, discussed below, already exist (www.flchma.org). Similarly, meetings have been held in other states, such as California, in 2010 to organize CHMAs in those citrus producing areas. Existing CHMAs will continue to improve in Florida and throughout the US in effectiveness and new ones will become established as new *D. citri* control and resistance management strategies are developed, as described below.

In Florida, baseline susceptibility data have been established to the majority of currently registered and used insecticides for *D. citri* and potential development of resistance has been monitored for four consecutive years. Resistance is already a reality and it is clear that it continues to worsen over time. In 2009, monitoring field populations of *D. citri* in Florida for insecticide susceptibility indicated decreases in susceptibility to several important insecticides, including up to a 34-fold decrease in susceptibility to the critical tool for young tree protection—imidacloprid (Tiwari et al. 2011). In 2010, the *D. citri* problem has worsened given decreased susceptibility to the majority of insecticides used for psyllid control throughout the state of Florida (Tiwari et al. 2011). In addition to field surveys of resistance over the past several years, laboratory investigations have been conducted to determine the mechanisms of resistance in *D. citri* (Tiwari et al. 2012). Based on an understanding of the fundamental mechanisms contributing to the development of resistance in *D. citri*, optimized rotation schedules of insecticides are being developed that will be eventually extended to grower practice (Tiwari et al. 2012).

15.2.3.3 Modern Outreach to Florida Citrus Growers in the Huanglongbing Era

In quick response to the recommendations made by the National Academy of Sciences, CHMAs were established in the fall of 2010 in Florida (Rogers et al. 2012). From the program's inception in 2010, the number of active CHMAs in Florida grew from approximately a dozen to 38 by the summer of 2012, encompassing a total of slightly over 200,000 ha (Rogers et al. 2012). The basic purpose of CHMAs is to encourage neighboring citrus growers to work collaboratively and on an area-wide basis to manage the *D. citri* vector of HLB through coordinated sprays of insecticides. Given the movement capability of *D. citri* and propensity for resistance development, area-wide control is considered an effective method for combating both the disease and its vector. In the summer of 2011, the "CHMA ACP (Asian citrus psyllid) Monitoring" program was officially initiated by scouts from the USDA's APHIS-PPQ Citrus Health Response Program and the Florida Department of Agriculture and Consumer Services' Division of Plant Industry (Rogers et al. 2012). This program employs staff to monitor approximately 6,000 blocks of citrus within all of the CHMAs throughout the state. All blocks are monitored at three-week intervals. The purpose of the monitoring is both to validate that the coordinated area-wide vector management program is working and perhaps more importantly to provide growers with real-time updates regarding psyllid populations to aid their pest management decision-making regarding the need for pesticide application (Rogers et al. 2012). From 2011 to 2012, 30–70% reductions in psyllid populations were recorded in citrus groves where growers were participating in CHMAs as compared with non-CHMA areas (Rogers et al. 2012). While the CHMA program is not a traditional IPM-based outreach program, it is analogous to such IPM programs and designed for management of a plant pathogen disease vector and a necessity for intense vector suppression in the HLB era of citriculture in Florida.

Citrus growers continue to receive the latest pest management information from the University of Florida extension service through the more traditional communications such as formally organized meetings, pamphlets, and handbooks[9]. However, web-based approaches are heavily emphasized and new tools have been developed in addition to the above-mentioned, online EDIS database. The CHMA program and associated literature on the latest citrus research is available on the CHMA website mentioned above. The website provides detailed information for how to join a local CHMA. The CHMA website has a directory of all currently functioning CHMAs, including maps and up-to-date scouting reports of psyllid populations, as well as, the most recent information regarding statewide CHMA meetings and planned coordinated sprays for the psyllid. It is also the single most comprehensive, web-based collection on modern extension literature regarding citrus production in Florida from diverse fields, including entomology/nematology, plant pathology, horticulture, economics, and food science and nutrition.

15.2.4 Apples

15.2.4.1 The Research/Extension Interface in Eastern US Apple Production and its Impacts on Integrated Pest Management

The saying "it's like comparing apples and oranges" is to a large extent appropriate when comparing and contrasting current research-based extension of IPM in these two very different tree fruit systems[10]. However, there are several parallels. In Florida citrus, the majority (80–85%) of fruit is processed for juice production

---

[9] Based on authors' experiences.
[10] Based on authors' experiences.

and thus the economic injury level for fruit by direct pest damage is high. Most of the loss occurs through insect transmitted diseases like HLB (which can kill entire trees) or mechanically transmitted diseases like bacterial canker, which can lead to fruit drop in certain juicing varieties and reduce marketability of the smaller fresh fruit market (grapefruit in particular). Although some of the northeast US apple production is processed for juice, cider, and baby food, a much larger proportion of it (as compared with Florida citrus) has lower tolerance for direct damage because of how these products are processed and developed (no tolerance for internally feeding pests), and there is also a sizable fresh fruit market for apples. Therefore, direct damage by pests plays a much larger role in affecting production of apples in the northeastern US than in production of citrus in Florida. Although pest management in Florida is currently dominated by two insect pests and associated diseases, apple production in the northeastern US is affected by a large complex of insect pests and diseases that cause direct damage to fruit, thus reducing marketability of the crop. The northeast US apple insect pest complex includes many Lepidoptera, Coleoptera, Diptera, and Hemiptera species, which results in significant pesticide inputs (Howitt 1993). Similarly to that observed in citrus, biological control can be a potent contributor to regulation of pest populations, particularly for pests such as European red mite, *Panonychus ulmi* Koch. Also, disruption of predatory mite populations due to pesticides targeting key insect pests can flare populations of the phytophagous mite pests.

A critical change began to occur and continues to this day in modern apple production in the northeast US following the passage of the FQPA. This act required the US EPA to develop and implement more stringent tolerances for pesticide exposure in a more holistic manner than anytime previously and with particular reference to reducing possible exposure of infants and children to pesticide residues in agricultural crops. The OP group of insecticides was the first to be scrutinized under these new guidelines because of a historic concern with this group of chemicals in terms of worker safety hazard and health concerns associated with residues on produce (Agnello et al. 2009). Certain heavily used insecticides, such as azinphosmethyl, were slated for eventual bans within 6–10 year periods, in some cases. This spawned significant research on OP replacement technologies and programs during the late 1990s and throughout the 2000s. The broad-spectrum activity of OPs, combined with their lesser effect on predaceous mites, as compared with pyrethroid and carbamate insecticides (Croft and Bode 1983), resulted in heavy use by apple growers that were faced with a large complex of insect and mite pests. In addition, a trend to develop so-called "reduced-risk" pesticides was under way in the pesticide industry with the specific goal of bringing so-called OP replacements to the market. Investigations were proving that these new chemistries were effective against a broad range of pests in northeastern US tree fruit (Reissig 2003; Villanueva and Walgenbach 2007); however, there was a need to develop new reduced-risk and IPM-compatible programs and to directly compare their efficacy and economics with the previous growers' standards that were heavily based on OP insecticides. Another important aspect of these new IPM programs was the integration of pheromone-based mating disruption for the complex of tortricid moth pests affecting apple production (Stelinski et al. 2009). Therefore, a four-year project was initiated in 2002 that involved collaboration between investigators in Michigan, New Jersey, New York, North Carolina, Pennsylvania, Virginia, and West Virginia to compare these new reduced risk programs with previous growers' standards that were heavily reliant on OPs (Agnello et al. 2009). The multi-state and multi-year project was funded primarily by USDA's RAMP program with material donations from pesticide and pest management industry cooperators (Agnello et al. 2009).

15.2.4.2 Four Years of "RAMP" Effort to Replace Organophosphate Insecticides in Northeast US Apples

The details of the four-year regional project that compared reduced-risk programs with previously standard OP-based management programs in northeast apple production is detailed by Agnello et al. (2009). Below is a summary of the project's major accomplishments and how research and extension efforts were integrated to change grower practices by adopting a more IPM-based approach. This project was truly regional in that similar programs were compared in all seven states involved, spanning the vast majority of apple production in the northeastern US. Trials were also conducted in peach orchards with similar overall results as discussed previously. Although there was some state-to-state variation in programs, necessary to accommodate certain variation in pest complexes between states, as well as, variation in local crop destination, the main objective was the same throughout. The goal was to compare a program defined as "reduced-risk" and without OP insecticides with the so-called "grower standard", the traditional program heavily reliant on OP insecticides (Agnello et al. 2009). A total of 65 locations were established across the region and monitored data included populations of pest and beneficial insects, fruit damage, and an economic comparison between the new and traditional programs. The reduced-risk and standard programs not only differed in insecticide modes of action, but also the reduced-risk treatment, in some instances, integrated large-scale use of pheromone-based mating disruption for lepidopteran pests (Agnello et al. 2009).

The general outcome of this large, region-wide project was that no significant difference in pest damage was determined between the reduced-risk and standard, OP-based programs, which included a complex of 10–12 insect species, depending on the state (Agnello et al. 2009). However, there was a marked reduction in pesticide input in the reduced-risk programs, compared with the standard comparison programs. After four years, the apple blocks under the reduced-risk programs received 86% less insecticide active ingredient input as compared with standard programs (Agnello et al. 2009), which could result in a profound reduction of environmental contamination when generally accepted by most growers. Another benefit of the alternative insecticide programs was overcoming the pre-existence of OP resistance in some pests in this region (Waldstein and Reissig 2000; Mota-Sanchez et al. 2008). This success in overall pesticide usage did not come without some additional cost directly related to greater expense of the newer pesticide chemistries, as well as, the additional investment in pheromone products. The resultant profitability gap at the project's end was approximately $ 100/ha^{-1}; however, the authors of the study suggested this may be further reduced in the future as generic versions of certain products become available and insecticide resistance to older-generation chemistries continues to increase (Agnello et al. 2009).

This four-year RAMP project is a successful model for integrating research and extension, which led to grower adoption of modern IPM programs fairly rapidly and, in this case, provided new tools before the expiration of OP insecticide availability in the northeastern US (Agnello et al. 2009). The research was conducted "on-farm" and therefore growers throughout the entire region received first-hand experience with these new programs throughout the duration of the project. Because of this, grower collaborators began to incorporate these new tools into their standard growing practices over the course of the project (Agnello et al. 2009). This was largely due to the concurrent extension effort of the investigators, who were able to provide individual growers with essentially real-time feedback on the efficacy and economics of the modern reduced-risk programs compared with the traditional OP-based standards (Agnello et al. 2009). This is an example of a modern, large-scale project that seamlessly integrated research and extension to change the behavior of growers in the northeast US apple-growing region in the face of the loss of a major tool (OP insecticides) that was relied on for years to successfully grow and market their crop.

15.3 Conclusions

Research and extension IPM programs need to constantly change and adapt to meet the growers' needs. History has shown that grower over-reliance on a single tool for insect pest management, i.e., broad-spectrum insecticides, is not sustainable and will likely lead to resistant pest populations—a lesson that entomologists learned many times over the past several decades (Dover and Croft 1986). However, developing alternative tools is not an easy task, often requiring many years of development, evaluation, training, registration, commercialization, and ultimately widespread grower adoption. Cost of new technologies, as they compare to conventional insecticides, is often a key factor in determining whether they will be adopted by growers or not (Cowan and Gunby 1996). Moreover, in many instances, growers are not comfortable with change, which might further limit the adoption of new technologies. Existing IPM programs also face the constant threat of incoming new pests (Pimentel et al. 2000) (e.g., brown marmorated stink bug, Asian citrus psyllid, and spotted wing drosophila)[11], an increasing trend seen in recent years as global markets for fruit products continue to expand. Researchers and extension professionals need to respond rapidly to these challenges. Educating growers on new IPM technologies as well as new invasive pests, from their basic biology to monitor-

---

[11] Based on authors' experiences.

ing and management practices, is an important component of any effective modern extension program. Newsletters, factsheets, on-farm demonstrations, and regular grower meetings are ways to transfer information to growers. Information transfer is now facilitated via internet and social media outlets. Funding for research and extension is also critical. In this book chapter we have discussed several sources of funding (e.g. USDA, EPA) that helped fruit growers' transition towards reduced-risk and ecologically-based IPM programs in the US.

Acknowledgments We thank the editors, Drs. Rajinder Peshin and David Pimentel, for the invitation to write this chapter. Drs. George Hamilton, Joyce Parker, and Anne Nielsen kindly provided comments on an earlier draft of the manuscript. We are very thankful to the following agencies and grower associations for supporting our research and extension programs: USDA, EPA, New Jersey Blueberry/Cranberry Research Council, and the Citrus Research and Development Foundation.

References

Agnello, A. M., Atanassov, A., Bergh, J. C., Biddinger, L. J., Gut, L. J., Haas, M. J., Harper, J. K., Hogmire, H. W., Hull, L. A., Kime, L. F., Krawczyk, G., McGhee, P. S., Nyrop, J. P., Reissig, W. H., Shearer, P. W., Straub, R. W., Villanueva, R. T., & Walgenbach, J. F. (2009). Reduced-risk pest management programs for eastern U.S. apple and peach orchards: A 4-year regional project. *American Entomologist, 55*(3), 184–197.

Akotsen-Mensah, C., Boozer, R. T., Appel, A. G., & Fadamiro, H. Y. (2011). Seasonal occurrence and development of degree-day models for predicting activity of *Conotrachelus nenuphar* (Coleoptera: Curculionidae) in Alabama peaches. *Annals of the Entomological Society of America, 104*(2), 192–201.

Alm, S. R., Villani, M. G., & Roelofs, W. L. (1999). Oriental beetles (Coleoptera: Scarabaeidae): Current distribution in the United States and optimization of monitoring traps. *Environmental Entomology, 92*, 931–935.

Anonymous. (2009). *Optimizing Your Spray System*. Wheaton: Spraying Systems Co., Wheaton, Illinois, USA. http://www.spray.com/Literature_PDFs/TM410B_Optimizing_Your_Spray_System.pdf. Accessed 27 Feb 2014.

Asquith, D., & Hull, L. A. (1979). Integrated pest management systems in Pennsylvania apple orchards. In D. J. Boethel & R. D. Eikenbarry (Eds.)., *Pest Management Programs for Deciduous Tree Fruits and Nuts* (pp. 203–222). New York: Plenum Press.

Atanassov, A., Shearer, P., Hamilton, G., & Polk, D. (1999). Reduced risk peach pest management program. *Proceedings 75th Cumberland-Shenandoah Fruit Workers Conference* (pp. 83–88). Winchester, Virginia, USA.

Atanassov, A., Shearer, P. W., Hamilton, G., & Polk, D. (2002). Development and implementation of a reduced risk peach arthropod management program in New Jersey. *Journal of Economic Entomology, 95*(4), 803–812.

Atkins, E. L., Kellum, D., & Atkins, K. W. (1978). Encapsulated methyl parathion formulation is highly hazardous to honey bees. *American Bee Journal, 118*, 483–485.

Atkins, E. L., Kellum, D., & Atkins, K. W. (1981). *Reducing Pesticide Hazards to Honey Bees: Mortality Prediction Techniques and Integrated Management Strategies*. Division of Agricultural Sciences University of California, Leaflet 2883. Riverside, California, USA.

Boina, D. R., Meyer, W. L., Onagbola, E. O., & Stelinski, L. L. (2009a). Quantifying dispersal of *Diaphorina citri* (Hemiptera: Psyllidae) by immunomarking and potential impact of unmanaged groves on commercial citrus management. *Environmental Entomology, 38*, 1250–1258.

Boina, D., Onagbola, E. O., Salyani, M., & Stelinski, L. L. (2009b). Antifeedant and sublethal effects of imidacloprid on Asian citrus psyllid, *Diaphorina citri*. *Pest Management Science, 65*, 870–877.

Borchert, D. M., Stinner, R. E., Walgenbach, J. F., & Kennedy, G. G. (2004). Oriental fruit moth (Lepidoptera:Tortricidae) phenology and management with methoxyfenozide in North Carolina apples. *Journal of Economic Entomology, 97*(4), 1353–1374.

Brattsten, L. B., Holyoke Jr., C. W., Leeper, J. R., & Raffa, K. F. (1986). Insecticide resistance: Challenge to pest management and basic research. *Science, 231,* 1255–1260.

Brunner, J, Dunley, J, Beers, E., & Doerr, M. (2004). *New Insecticides and Miticides for Apple and Pear IPM.* Washington State University Tree Fruit Research and Extension Center. http://entomology.tfrec.wsu.edu/New_Insecticides/New_Insecticides_IPM.pdf. Accessed 8 Jan 2013.

Cardé, R. T., & Minks, A. K. (1995). Control of moth pests by mating disruption: Successes and constraints. *Annual Review of Entomology, 40,* 559–585.

Carson, R. (1962). *Silent spring.* New York: Fawcett Crest.

CEQ [Council on Environmental Quality]. (1972). *Integrated Pest Management.* Washington, D.C.: U. S. Govt. Printing Office.

Chen, T. A. (1971). Mycoplasmalike organisms in sieve tube elements of plants infected with blueberry stunt and cranberry false blossom. *Phytopathology, 61,* 233–236.

Coats, S. A., Coats, J. R., & Ellis, C. R. (1979). Selective toxicity of three synthetic pyrethroids to eight coccinellids a eulophid parasitoid, and two pest chrysomelids. *Environmental Entomology, 8*(4), 720–722.

Collins, J. A., & Drummond, F. A. (2004). Field-edge based management tactics for blueberry maggot in lowbush blueberry. *Small Fruits Review, 3*(3/4), 283–293.

Cowan, R., & Gunby, P. (1996). Sprayed to death: Path dependence, lock-in and pest control strategies. *The Economic Journal, 106*(436), 521–542.

Croft, B. A., & Whalon, M. E. (1982). Selective toxicity of pyrethroid insecticides to arthropod natural enemies and pests of agricultural crops. *Biocontrol, 27*(1), 3–21.

Croft, B. A., & Brown, A. W. (1975). Responses of arthropod natural enemies to insecticides. *Annual Review of Entomology, 20,* 285–335.

Croft, B. A., & Bode, W. M. (1983). Tactic for deciduous fruit IPM. In B. A. Croft & S. C. Hoyt (Eds.)., *Integrated Management of Insect Pests of Pome and Stone Fruits* (pp. 270–291). New York: Wiley.

Croft, B. A., Michels, M. F., & Rice, R. E. (1980). Validation of a PETE timing model for the oriental fruit moth in Michigan and central California (Lepidoptera: Olethreutidae). *Great Lakes Entomologist, 13,* 211–217.

CTFA [The California Tree Fruit Agreement]. (2006). *A Pest Management Strategic Plan for Peach Production in California.* The California Tree Fruit Agreement, The California Canning Peach Association, and The California Minor Crops Council. http://www.ipmcenters.org/pmsp/pdf/CAPEACHPMSP.pdf. Accessed 10 Jan 2013.

de Lame, F. M., Hong, J. J., Shearer, P. W., & Brattsten, L. B. (2001). Sex-related differences in the tolerance of oriental fruit moth (*Grapholita molesta*) to organophosphate insecticides. *Pest Management Science, 57*(9), 827–832.

Desneux, N., Decourtye, A., & Delpuech, J. M. (2007). The sublethal effects of pesticides on beneficial arthropods. *Annual Review of Entomology, 52,* 81–106.

Devine, G. J., & Furlong, M. J. (2007). Insecticide use: Contexts and ecological consequences. *Agriculture and Human Values, 24,* 281–306.

Dill, J. F., Drummond, F. A., & Stubbs, C. S. (1998). Pesticide use on blueberry: A survey. Penn State Contract No.USDA-TPSU-UM-0051-1300. Orono, Maine, USA: University of Maine.

Dover, M. J., & Croft, B. A. (1986). Pesticide resistance and public policy. *Bioscience, 36*(2), 78–85.

Drummond, F. A. (2000). History of insect pest management for lowbush blueberries in Maine. *Trends in Entomology, 3,* 23–32.

Eck, P., & Childers, N. F. (1966). The blueberry industry. In P. Eck & N. F. Childers (Eds.)., *Blueberry Culture* (pp. 3–13). New Brunswick, New Jersey, USA: Rutgers University Press.

Faust, M., & Timon, B. (2011). Origin and dissemination of peach. In J. Janick (Ed.), *Origin and Dissemination of Prunus Crops Peach, Cherry, Apricot, Plum, Almond.* (pp. 11–54). Scripta Horticulturae 11. American Pomological Soc., ISHS, Leuven, Belgium.

Fisher, F. E. (1950). Entomogenous fungi attacking scale insects and rust mites on citrus in Florida. *Journal of Economic Entomology, 43,* 305–309.

Folimonova, S. Y., Robertson, C. J., Garnsey, S. M., Gowda, S., & Dawson, W. O. (2009). Examination of the responses of different genotypes of citrus to Huanglongbing (Citrus Greening) under different conditions. *Phytopathology, 99,* 1346–1354.

Gentz, M. C., Murdoch, G., & King, G. F. (2010). Tandem use of selective insecticides and natural enemies for effective, reduced-risk pest management. *Biological Control, 52,* 208–215.

Grafton-Cardwell, B. E., & Vehrs, S. L. C. (1995). Monitoring for organophosphate- and carbamate- resistance in San Joaquin Valley citrus. *Journal of Economic Entomology, 88,* 495–504.

Grafton-Cardwell, E. E., Stelinski, L. L., & Stansly, P. A. (2013). Biology and management of Asian citrus psyllid, vector of the huanglongbing pathogens. *Annual Review of Entomology, 58,* 413–432.

Halbert, S. E. (1998). Entomology section. *Tri-ology, 37,* 6–7.

Halbert, S. E., & Manjunath, K. L. (2004). Asian citrus psyllids (Sternorrhyncha: Psyllidae) and greening disease of citrus: A literature review and assessment of risk in Florida. *Florida Entomologist, 87,* 330–353.

Halbrendt, J. M. (2012). *Pennsylvania 2012–2013 Tree Fruit Production Guide.* Pennsylvania: Pennsylvania State University AGRS-045, University Park, Pennsylvania, USA.

Hamilton, G. H. (2001). *Crop Profile for Highbush Blueberry in New Jersey.* http://www.pestmanagement.rutgers.edu/njinpas/CropProfiles/2000blueberryprofileweedsextracted.pdf. Accessed 25 Sept 2012.

Hardman, J. M., Jensen, K. I., Moreau, D. L., & Bent, E. D. (2004). Effects of ground cover treatments and insecticide use on population density and damage caused by *Lygus lineolaris* (Heteroptera: Miridae) in apple orchards. *Journal of Economic Entomology, 97*(3), 993–1002.

Hauser, M. (2011). A historic account of the invasion of *Drosophila suzukii* (Matsumura) (Diptera: Drosophilidae) in the continental United States, with remarks on their identification. *Pest Management Science, 67*(11), 1352–1357.

Hill, T. A., & Foster, R. E. (1998). Influence of selective insecticides on population dynamics of European red mite (Acari: Tetranychidae), apple rust mite (Acari: Eriophyidae), and their predator *Amblyseius fallacis* (Acari: Phytoseiidae) in apple. *Journal of Economic Entomology, 91*(1), 191–199.

Hoddle, M. S. (2006). Challenges to IPM advancement: Pesticides, biocontrol, genetic engineering, and invasive pests. *New Zealand Entomologist, 29*(1), 77–88.

Hogmire, H. W. (1995). *Mid-Atlantic Orchard Monitoring Guide.* Northeast Regional Agricultural Engineering Service NRAES-75. Ithaca, New York, USA.

Hopfinger, A. (1990). *New Jersey Tree Fruit Production Guide, 1990.* Rutgers NJAES Cooperative Extension E002F. New Brunswick, New Jersey, USA.

Horton, D. L., & Ellis, H. C. (1989). Plum curculio. In S. C. Myers (Ed.)., *Peach Production Handbook* (pp. 169–170). Athens, Georgia, USA: Cooperative Extension Service, University of Georgia.

Horton, D., Bellinger, B., & Elworth, E. (2000). *Eastern Peach Pest Management Strategies for Adapting to Changing Management Options.* http://www.ipmcenters.org/pmsp/pdf/easternpeach.pdf. Accessed 10 Jan 2013.

Howell, A. B., Vorsa, N., Der Marderosian, A., & Foo, L. Y. (1998). Inhibition of the adherence of P-fimbriated *Escherichia coli* to uroepithelial-cell surfaces by proanthocyanidin extracts from cranberries. *New England Journal of Medicine, 339,* 1085–1086.

Howitt, A. H. (1993). *Common Tree Fruit Pests.* East Lansing, Michigan, USA: Michigan State University Extension NCR 63.

Hull, L. A., Hickey, K. D., & Kanour, W. W. (1983). Pesticide usage patterns and associated pest damage in commercial apple orchards of Pennsylvania. *Journal of Economic Entomology, 76,* 577–583.

Immaraju, J. A., Morse, J. G., & Kersten, D. J. (1989). Citrus thrips (Tysanoptera: Thripidae) pesticide resistance in the Coachella and San Joaquin valleys of California. *Journal of Economic Entomology, 82,* 374–380.

IRAC. (2009). *General Principles of Insecticide Resistance Management*. http://www.irac-online.org/content/uploads/2009/09/Principles-of-IRM.pdf. Accessed 13 Jan 2013.

Joseph, J. A., Denisova, N. A., Arendash, G., Gordon, M., Diamond, D., Shukitt-Hale, B., & Morgan, D. (2003). Blueberry supplementation enhances signaling and prevents behavioral deficits in an Alzheimer disease model. *Nutritional Neuroscience, 6,* 153–162.

Kanga, L. H., Pree, D. J., van Lier, J. L., & Walker, G. M. (2003). Management of insecticide resistance in oriental fruit moth (*Grapholita molesta*; Lepidoptera: Tortricidae) populations from Ontario. *Pest Management Science, 59*(8), 921–927.

Killian, J. C., & Meyer, J. R. (1984). Effect of orchard weed management on catfacing damage to peaches in North Carolina. *Journal of Economic Entomology, 77*(6), 1596–1600.

Knapp, J. L., Noling, J. W., Timmer, L. W., & Tucker, D. P. H. (1996). Florida citrus IPM. In D. Rosen, F. F. Bennett, & J. L. Capinera (Eds.)., *Pest Management in the Subtropics: Integrated Pest Management—A Florida Perspective* (pp. 317–347). Andover, Hants, United Kingdom: Intercept.

Leskey, T. C., & Zhang, A. (2007). Impact of temperature on plum curculio (Coleoptera: Curculionidae) responses to odor-baited traps. *Journal of Economic Entomology, 100*(2), 343–349.

Leskey, T. C., Hamilton, G. C., Nielsen, A. L., Polk, D. F., Rodriguez-Saona, C., Bergh, J. C., Herbert, D. A., Kuhar, T. P., Pfeiffer, D., Dively, G., Hooks, C. R. R., Raupp, M. J., Shrewsbury, P. M., Krawczyk, G., Shearer, P. W., Whalen, J., Koplinka-Loehr, C., Myers, E., Inkley, D., Hoelmer, K. A., Lee, D.-Y., & Wright, S. E. (2012). Pest status of the brown marmorated stink bug, *Halyomorpha halys* (Stål), in the USA. *Outlooks on Pest Management, 23,* 218–226.

Liburd, O. E., Finn, E. M., Pettit, K. L., & Wise, J. C. (2003). Response of blueberry maggot fly (Diptera: Tephritidae) to imidacloprid-treated spheres and selected insecticides. *The Canadian Entomologist, 135*(3), 427–438.

Mallampalli, N., & Isaacs, R. (2002). Distribution of egg and larval populations of cranberry fruitworm (Lepidoptera: Pyralidae) and cherry fruitworm (Lepidoptera: Tortricidae) in highbush blueberries. *Environmental Entomology, 31,* 852–858.

Marucci, P. E. (1966). Insects and their control. In P. Eck & N. F. Childers (Eds.)., *Blueberry Culture*. New Brunswick, New Jersey, USA: Rutgers University Press.

Moore, T. D. (1995). The pinelands national reserve: An experiment in land management. In J. Endter-Wada & R. J. Lilieholm (Eds.)., *Conflicts in Natural Resources Management: Integrating Social and Ecological Concerns* (pp. 57–61). Logan, Utah, USA: College of Natural Resources, Utah State University.

Morse, J. G., & Brawner, O. L. (1986). Toxicity of pesticides to *Scirtothrips citri* (Thysanptera: Thripidae) and implications to resistance management. *Journal of Economic Entomology, 79,* 565–570.

Mota-Sanchez, D., Wise, J. C., Van der Poppen, R., Gut, L. J., & Hollingworth, R. M. (2008). Resistance of codling moth, *Cydia pomonella*, (L.) (Lepidoptera: Tortricidae), larvae in Michigan to insecticides with different modes of action and the impact of field residual activity. *Pest Management Science, 64,* 881–890.

Muma, M., Holtsberg, H., & Pratt, R. (1953). *Eutetranychus banski* (McG) recently found on citrus in Florida (Acarina: Tetranychidae). *Florida Entomologist, 36,* 141–144.

Muraro, R. P. (2009). Summary of 2008–2009 citrus budget for the Southwest Florida production region. http://www.crec.ifas.ufl.edu/extension/economics/. Accessed 25 Sept 2012.

Myers, C. T., Hull, L. A., & Krawczyk, G. (2007). Effects of orchard host plants (apple and peach) on development of oriental fruit moth (Lepidoptera: Tortricidae). *Journal of Economic Entomology, 100*(2), 421–430.

National Research Council. (2010). *Strategic planning for the Florida Citrus Industry: Addressing Citrus Greening Disease (Huanglongbing)*. National Academies Press. http://www.nap.edu/catalog.php?record_id=12880. Accessed 25 Sept 2012.

Nielsen, A. L., & Hamilton, G. C. (2009a). Seasonal occurrence and impact of *Halyomorpha halys* in tree fruit. *Journal of Economic Entomology, 102*(3), 1133–1140.

Nielsen, A. L., & Hamilton, G. C. (2009b). Life history of the invasive species *Halyomorpha halys* (Hemiptera: Pentatomidae) in northeastern United States. *Annals of the Entomological Society of America, 102*(4), 608–616.

Ofek, I., Hasty, D. L., & Sharon, N. (2003). Anti-adhesion therapy for bacterial diseases: Prospects and problems. *FEMS Immunology and Medical Microbiology, 38,* 181–191.

Omoto, C., Dennehy, T. J., McCoy, C. W., Crane, S. E., & Long, J. W. (1995). Management of citrus rust mite (Acari: Eriophyidae) resistance to dicofol in Florida citrus. *Journal of Economic Entomology, 88,* 1120–1128.

Oudemans, P., Ward, D., Majek, B., Polk, D., & Rodriguez-Saona, C. (2012). *2012 Commercial Blueberry Pest Control Recommendations for New Jersey.* Rutgers Cooperative Extension, NJAES E265. http://njaes.rutgers.edu/pubs/publication.asp?pid=E265. Accessed 25 Sept 2012.

Pavlis, G. C. (2012). *The Blueberry Bulletin* (multiple issues). Rutgers cooperative extension of Atlantic county. Mays Landing, New Jersey, USA. http://www.njaes.rutgers.edu/pubs/blueberrybulletin. Accessed 25 Sept 2012.

Pedigo, L. R. (2002). *Entomology and Pest Management* (4th ed.). New Jersey: Prentice Hall.

Pelz-Stelinski, K. S., Brlansky, R. H., Ebert, T. A., & Rogers, M. E. (2010). Transmission parameters for *Candidatus Liberibacter asiaticus* by Asian citrus psyllid (Hemiptera: Psyllidae). *Journal of Economic Entomology, 103,* 1531–1541.

Peshin, R., Bandral, R. S., Zhang, W., Wilson, L., & Dhawan, A. K. (2009). Integrated pest management: A global overview of history, programs and adoption. In R. Peshin & A. K. Dhawan (Eds.), *Integrated Pest Management: Innovation-Development Process, Vol. 1.* (pp. 1–49). United Kingdom: Springer.

Pimentel, D., Lach, L., Zuniga, R., & Morrison, D. (2000). Environmental and economic costs of nonindigenous species in the United States. *Bioscience, 50*(1), 53–65.

Polavarapu, S., Wicki, M., Vogel, K., Lonergan, G., & Nielsen, K. (2002). Disruption of sexual communication of Oriental beetles (Coleoptera: Scarabaeidae) with a microencapsulated formulation of sex pheromone components in blueberries and ornamental nurseries. *Environmental Entomology, 31,* 1268–1275.

Polk, D. F., & Samoil, K. S. (1993). Blueberry pesticide use and fruit quality 1992. *Proceedings of the Blueberry Openhouse* (pp. 8–10). Hammonton, New Jersey, USA.

Polk, D. F., Tietjen, B., & Beatty, K. (1990). The cost of pest control in New Jersey peaches. *Horticultural News, 70*(4), 12–13.

Polk, D. F., Schmitt, D., Rizio, E. F., & Petersen, K. (1995). Key pest problems in New Jersey tree fruit—1995. *Proceedings, 71st Annual Cumberland-Shenandoah Fruit Workers Conference* (pp. 221–225). Winchester, Virginia, USA.

Polk, D., Shearer, P. Majek, B, Belding, B., Lalancette, N., & Halbrendt, J. (1999). Minimizing catfacing insect pressure in peaches through ground cover management. *Horticultural News, 79*(3), 15–17.

Polk, D. F., Schmitt, D., & Atanassov, A. (2009). Fruit quality and spray programs in NJ orchards. *Proceedings, 85th Annual Cumberland-Shenandoah Fruit Workers Conference* (pp. 63–67). Winchester, Virginia, USA.

Pree, D. J., Trimble, R. M., Whitty, K. J., & Vickers, P. M. (1994). Control of oriental fruit moth by mating disruption using sex pheromone in the Niagara Peninsula, Ontario. *Canadian Entomologist, 126*(6), 1287–1299.

Reissig, H. (2003). Internal Lepidoptera problems in apple orchards: From the world to New York. *Internal Dwarf Fruit Tree Association, Compact Fruit Tree, 36,* 26–27.

Reissig, W. H., Nyrop, J. P., & Straub, R. (1998). Oviposition model for timing insecticide sprays against plum curculio (Coleoptera: Curculionidae) in New York state. *Environmental Entomology, 27*(5), 1053–1061.

Rice, R. E., & Kirsch, P. (1990). Mating disruption of oriental fruit moth in the United States. In R. L. Ridgeway, R. M. Silverstein, & M. N. Inscoe (Eds.), *Behavior-modifying Chemicals for Insect Management* (pp.193–211). New York: Marcel Dekker.

Rice, R. E., Weakley, C. V., & Jones, R. A. (1984). Using degree-days to determine optimum spray timing for the oriental fruit moth (Lepidoptera: Tortricidae). *Journal of Economic Entomology, 77,* 698–700.

Ritchie, D. F., & Clayton, C. N. (1981). Peachtree short life: A complex of interacting factors. *Plant Disease, 65*(6), 462–469.

Rodriguez-Saona, C., & Polk, D. (2011). Spotted wing drosophila—A potential pest of New Jersey blueberries and other soft fruit. In G. C. Pavlis (Ed.) *The Blueberry Bulletin 27*(16), 6–7. Rutgers Cooperative Extension of Atlantic County, Mays Landing, New Jersey, USA.

Rodriguez-Saona, C., Polk, D. F., & Barry, J. D. (2009). Optimization of pheromone rates for effective mating disruption of oriental beetle (Coleoptera: Scarabaeidae) in commercial blueberries. *Journal of Economic Entomology, 102*, 659–669.

Rogers, M. E., Stansly, P. A., & Stelinski, L. L. (2011). Florida citrus pest management guide: Asian citrus psyllid and citrus leafminer, ENY-734. In M. E. Rogers, L. W. Timmers, & T. M. Spann (Eds.)., *2011 Florida Citrus Pest Management Guide*. Gainesville, Florida, USA: University of Florida, Institute of Food and Agriculture Science Extension Publication No. SP-43.

Rogers, M. E., Carlton, G., & Riley, T. D. (2012). Results from the "CHMA ACP Monitoring" program: Statewide psyllid populations are considerably lower in 2012 compared to one year ago. *Citrus Industry, 93,* 12–16.

Rosen, D., Bennett, F. D., & Capinera, J. L. (1994). *Pest Management in the Subtropics: Biological Control-A Florida Perspective*. Andover, Hants, United Kingdom: Intercept.

Salyani, M. (2003). *Calibration of Airblast Sprayers*. University of Florida IFAS Extension Circular1435. Gainesville, Florida, USA. http://edis.ifas.ufl.edu/ae238. Accessed 27 May 2013.

Sciarappa, W. J., Polavarapu, S., Holdcraft, R. J., & Barry, J. D. (2005). Disruption of sexual communication of oriental beetles (Coleoptera: Scarabaeidae) in highbush blueberries with retrievable pheromone sources. *Environmental Entomology, 34,* 54–58.

Shearer, P. W., Majek, B., Polk, D., Belding, B., & Lalancette, N. (1998). Orchard ground cover management affects peach insect damage. In *Proceedings 74th Cumberland-Shenandoah Fruit Workers Conference* (pp. 278–279). Winchester, Virginia, USA.

Shearer, P. W., Atanassov, A., & Rucker, A. (2006). Eliminating organophosphate and carbamate insecticides from New Jersey, USA, peach culture. *Acta Horticulturae, 713,* 391–395.

Simanton, W. A. (1960). The reduced status of purple scale as a citrus pest. *Proceedings of the Florida State Horticultural Society, 73,* 64–69.

Simanton, W. A. (1974). Occurrence of insect an mite pests of citrus, their predators and parasitism in relation to spraying operations. *Proceedings of the Tall Timbers Conference on Ecological Animal Control by Habitat Management, 6,* 135–163.

Simanton, W. A. (1996). Foundations of IPM: Ecological survey of citrus groves. In D. Rosen, F. F. Bennett, & J. L. Capinera (Eds.)., *Pest Management in the Subtropics: Integrated Pest Management—A Florida Perspective* (pp. 11–20). Andover, Hants, United Kingdom: Intercept.

Smith, R. F., & van den Bosch, R. (1967). Integrated control. In W. W. Kilgore, & R. L. Doutt (Eds.), *Pest Control: Biological, Physical, and Selected Chemical Methods* (pp. 295–340). New York: Academic Press.

Srinivasan, R., Hoy, M. A., Singh, R., & Rogers, M. E. (2008). Laboratory and field evaluations of silwet L-77 and kinetic alone and in combination with imidacloprid and abamectin for the management of the Asian citrus psyllid, *Diaphorina citri* (Hemiptera: Psyllidae). *Florida Entomologist, 91,* 87–100.

Stelinski, L. L., Gut, L. J., Mallinger, R. E., Epstein, D., Reed, T. P., & Miller, J. R. (2005). Small plot trials documenting effective mating disruption of oriental fruit moth by using high densities of wax-drop pheromone dispensers. *Journal of Economic Entomology, 98*(4), 1267–1274.

Stelinski, L. L., Gut, L. J., Haas, M., McGhee, P., & Epstein, D. (2007). Evaluation of aerosol devices for simultaneous disruption of sex pheromone communication in *Cydia pomonella* and *Grapholita molesta* (Lepidoptera: Tortricidae). *Journal of Pest Science, 80,* 225–233.

Stelinski, L. L., Il'ichev, A. L., & Gut, L. J. (2009). Efficacy and release rate of reservoir pheromone dispensers for simultaneous mating disruption of codling moth and oriental fruit moth (Lepidoptera: Tortricidae). *Journal of Economic Entomology, 102,* 315–323.

Stern, V. M., Smith, R. F., van den Bosch, R., & Hagen, K. S. (1959). The integrated control concept. *Hilgardia, 29,* 81–101.

Szendrei, Z., Mallampalli, N., & Isaacs, R. (2005). Effect of tillage on abundance of Japanese beetle, *Popillia japonica* Newman (Col., Scarabaeidae), larvae and adults in highbush blueberry fields. *Journal of Applied Entomology, 129,* 258–264.

Tiwari, S., Lewis-Rosenblum, H., Pelz-Stelinski, K., & Stelinski, L. L. (2010). Incidence of *Candidatus* Liberibacter asiaticus infection in abandoned citrus occurring in proximity to commercially managed groves. *Journal of Economic Entomology, 103,* 1972–1978.

Tiwari, S., Mann, R. S., Rogers, M. E., & Stelinski, L. L. (2011). Insecticide resistance in field populations of Asian citrus psyllid in Florida. *Pest Management Science, 67,* 1258–1268.

Tiwari, S., Stelinski, L. L., & Rogers, M. E. (2012). Biochemical basis of organophosphate and carbamate resistance in Asian citrus psyllid. *Journal of Economic Entomology, 105,* 540–548.

Trimble, R. M., Pree, D. J., Barszez, E. S., & Carter, N. J. (2004). Comparison of a sprayable pheromone formulation and two hand applied pheromone dispensers for use in the integrated control of oriental fruit moth (Lepidoptera: Tortricidae). *Journal of Economic Entomology, 97*(2), 482–489.

Tuell, J. K., & Isaacs, R. (2010). Community and species-specific responses of wild bees to insect pest control programs applied to a pollinator-dependent crop. *Journal of Economic Entomology, 103,* 668–675.

Tylor, H. L., Roesch, L. F. W., Gowda, S., Dawson, W. O., & Triplett, E. W. (2009). Confirmation of the sequence of '*Candidatus* Liberibacter asiaticus' and assessment of microbial diversity in Huanglongbing-infected citrus phloem using a metagenomic approach. *Molecular Plant-Microbe Interactions, 22,* 1624–1634.

US Census. (2004). *State & County Quick Facts.* http://quickfacts.census.gov/qfd/states/34/34029.html. Accessed 25 Sept 2012.

US EPA [United States Environmental Protection Agency, food quality protection act of 1996]. (1996). U.S. Public Law No. 104-170. *U.S. Congressional Record, 142,* 1489–1538.

US EPA [United States Environmental Protection Agency]. (2000). *R.E.D. Facts Ethyl Parathion.* EPA-738-F00-009. http://www.epa.gov/oppsrrd1/REDs/factsheets/0155fct.pdf. Accessed 8 Jan 2013.

US EPA [United States Environmental Protection Agency]. (1992). *Ethyl Parathion, Correction to the Amended Cancellation Order.* Washington D.C.: Office of Pesticide Programs (OPP), USEPA.

US EPA [United States Environmental Protection Agency]. (1999). *Methyl Parathion (Penncap-M)—Cancellation Request 10/99.* Federal Register: October 27, 1999. http://pmep.cce.cornell.edu/profiles/insect-mite/fenitrothion-methylpara/methyl-parathion/methpara_can_1099.html.

USDA ERS [United States department of agriculture, economic research service]. (2003). *Fruit and Tree Nuts Outlook, FTS-305.* http://usda01.library.cornell.edu/usda/ers/FTS//2000s/2003/FTS-07-30-2003.pdf. Accessed 15 Jan 2013.

USDA NASS [United States department of agriculture, National agricultural statistics service]. (2012a). *Quick Stats.* http://quickstats.nass.usda.gov/#C279CCC7-D259-381D-AE23-D353D5D0B80B. Accessed 25 Sept 2012.

USDA NASS [United States department of agriculture, National agricultural statistics service]. (2012b). *Blueberry Statistics.* NJ Agricultural Statistics Service. http://www.nass.usda.gov/Statistics_by_State/New_Jersey/index.asp. Accessed 25 Sept 2012.

USDA NIFA [United States department of agriculture, National Institute of food and agriculture]. (2013). *National Information for the Regional IPM Centers.* http://www.ipmcenters.org/pmsp/. Accessed 9 Jan 2013.

van den Bosch, R. (1978). *The Pesticide Conspiracy.* New York: Doubleday.

van den Bosch, R., Messenger, P. S., & Gutierrez, A. P. (1982). *An Introduction to Biological Control.* New York: Plenum Press.

Villanueva, R. T., & Walgenbach, J. F. (2007). Phenology, management, and effects of Surround on behavior of the apple maggot (Diptera: Tephritidae) in North Carolina. *Crop Protection, 26,* 1404–1411.

Vittum, P. J., Villani, M. G., & Tashiro, H. (1999). *Turfgrass Insects of the United States and Canada.* Ithaca, New York, USA: Cornell Univ. Press.

Waldstein, D. E., & Reissig, W. H. (2000). Synergism of tebufenozide in resistant and susceptible strains of obliquebanded leafroller (Lepidoptera: Tortricidae) and resistance to new insecticides. *Journal of Economic Entomology, 93,* 1768–1772.

Ward, D. (2012). *New Jersey Tree Fruit Production Guide, 2012*. Rutgers NJAES Cooperative Extension E002, New Brunswick, New Jersey, USA.

Weakley, C. V., Kirsch, P., & Rice, R. E. (1987). Control of oriental fruit moth by mating disruption. *California Agriculture, 41*(5), 7–8.

Wise, J. (2013). *Michigan Fruit Management Guide 2013*. Michigan State University Extension E0154. East Lansing, Michigan, USA.

Wise, J. C., Jenkins, P., Van der Poppen, R., & Isaacs, R. (2010). Activity of broad spectrum and reduced-risk insecticides on various life stages of cranberry fruitworm (Lepidoptera: Pyralidae) in highbush blueberry. *Journal of Economic Entomology, 103,* 1720–1728.

Xu, C. F., Xia, Y. H., Li, K. B., & Ke, C. (1988). Further study of the transmission of citrus huanglongbing by a psyllid, *Diaphorina citri* Kuwayama. In L. W. Timmer, S. M. Garnsey, & L. Navarro (Eds.)., *Proceedings of the 10th Conference of the International Organization of Citrus Virologists* (pp. 243–248). Riverside: International Organization of Citrus Virologists

Youdim, K. A., Shukitt-Hale, B., MacKinnon, S., Kalt, W., & Joseph, J. A. (2000). Polyphenolics enhance red blood cell resistance to oxidative stress: In vitro and in vivo. *Biochimica et Biophysica Acta, 1523,* 117–122.

Zeuli, K. A. (1999). New risk-management strategies for agricultural cooperatives. *American Journal of Agricultural Economics, 81*(5), 1234–1239.

Chapter 16
Promotion of Integrated Pest Management by the Plant Science Industry: Activities and Outcomes

Keith A. Jones

Contents

16.1	Introduction..	394
16.2	Promotion of IPM by CropLife International...	397
16.3	Impact Assessment...	401
	16.3.1 IPM Uptake..	401
	16.3.2 IPM Impacts...	402
16.4	Examples of IPM Interventions..	402
	16.4.1 Kenya...	402
	16.4.2 India...	402
	16.4.3 Philippines...	404
	16.4.4 Thailand...	405
	16.4.5 Bangladesh...	406
	16.4.6 Nicaragua...	406
	16.4.7 Guatemala..	407
16.5	Training Materials..	407
16.6	Concluding Remarks..	407
References..		408

Abstract The Plant Science Industry, as represented by CropLife International, promotes a lifecycle approach to the stewardship of crop protection and plant biotechnology products. Stewardship is the responsibility of all stakeholders, and within stewardship a major element is the promotion of Integrated Pest Management (IPM). The Plant Science Industry follows the description of IPM given in The International Code of Conduct on Pesticide Management, which recognizes that a range of tools can be used in IPM, including chemical pesticides and resistant plant varieties. The Plant Science Industry has been actively promoting IPM since the 1980s through education and training programs implemented in partnership with a wide group of stakeholders. Since 2005, when more accurate measurement of training impacts was implemented, almost two million people have been directly

K. A. Jones (✉)
CropLife International, 326 Avenue Louise,
Box 35, 1050 Brussels, Belgium
e-mail: keith.jones@croplife.org

trained and ten times this figure reached. CropLife International is putting in place more sophisticated monitoring and evaluation tools that will clearly distinguish between IPM uptake and IPM impact, which will be used across its programs. A range of guidelines and training materials are available from CropLife International to assist IPM implementation. IPM is promoted by CropLife associations and member companies; this chapter gives several examples of programs implemented by the association.

Keywords IPM impacts · Responsible use · Pesticide use context · Stewardship · Training

16.1 Introduction

The term Integrated Pest Management (IPM) has been used since the 1970s promoting a strategy that uses a range of tools to manage pest populations (Geier 1970). However, the understanding of the need to not rely on a single means of pest control (Integrated Pest Control) has been around much longer—for example, in the 1960s the United Nations Food and Agricultural Organization (FAO) held a symposium on Integrated Pest Control (Geier and Clark 1961; Reynolds 1965). The concept has been continually developing since then, as a better understanding of agroecology, including the impact of general agronomic practices on pest numbers, their impact on crop plants and on the predators and parasites that can regulate pest populations, has been gained. It is now recognized that the starting point of IPM is growing a healthy crop, so the plants can withstand a level of pest attack. This includes choice of pest resistant varieties derived through traditional breeding methods or genetic manipulation. It is also now recognized that with some crops, e.g. cotton, some damage can result in increased yield rather than a loss (Russell et al. 1993). This is then supplemented by maintaining naturally occurring mechanisms that can help to suppress, or keep in balance, pest numbers. If pest numbers do reach a level that requires intervention, the most effective tools, applied in a way that minimize impacts on human health and the environment should be used. Essentially, this means that non-chemical control measures should first be considered, and if not suitable either from an availability, cost or effectiveness point of view, chemical control should be used. If the chemical route is chosen then the chemicals should be handled and applied safely and in accordance with the pesticide label instructions. In fact, the responsible use of pesticides is an essential tool in IPM, and in reality the most commonly used. This description of IPM is in-line with the definition in The International Code of Conduct on Pesticide Management (FAO 2012):

> Integrated Pest Management (IPM) means the careful consideration of all available pest control techniques and subsequent integration of appropriate measures that discourage the development of pest populations and keep pesticides and other interventions to levels that are economically justified and reduce or minimize risks to human health and the environment. IPM emphasizes the growth of a healthy crop with the least possible disruption to agroecosytems and encourages natural pest control mechanisms.

This definition, as part of the CoC, has been adopted by the FAO Conference, which represents the 191 member countries (plus the European Union) of FAO, as well as non-governmental organizations (NGOs), such as the Pesticide Action Network, the research-based crop protection and plant biotechnology industry, represented by CropLife International[1] and the generic pesticide industry, represented by Agro-Care. Unfortunately, there are also many other definitions (Bajwa and Kogan 2002). Most of these define IPM as being a combination of control methods or tactics with minimal environment impact, and most include chemical pesticides as one of the options, albeit the choice of last resort. However, some groups have increasingly promoted IPM as an approach that is separate from the use of chemical pesticides, which is likely to lead to confusion among farmers and can threaten yields, as effective alternatives to the use of chemical pesticides are often not readily available. Promoting IPM as an approach that does not include the use of chemical pesticides also gives the impression that there is less of a need to train farmers in the safe handling and responsible use of chemical pesticides, resulting in this training being neglected and increasing the potential for misuse of pesticides. There is also a worrying trend of describing the goal of IPM as pesticide reduction, which is clearly not the case: as the name indicates, the goal of IPM is *to manage pests*; adoption of IPM may often result in reduced pesticide use, due to reduction in the overuse and misuse of pesticides, but reduced pesticide use should not be the goal. In some cases the adoption of IPM may result in increased pesticide use because a previously unknown problem (e.g. presence of damaging levels of nematodes) is identified as needing treatment. Placing pesticide reduction as the goal for IPM will likely result in a breakdown in pest management as farmers' access to an effective range of pest management tools will be limited and could result in the abandonment of IPM and reversion to practices such as calendar or unnecessary preventative application of pesticides. A further threat to pest management from limiting available tools is the increased likelihood of development of resistance to key pesticides; rotation of a range of pesticide modes of action or appropriate use of mixtures helps slow resistance development in many key pests. Managing pest resistance to pesticides and to IPM tools, such as biotechnology-derived pest resistant varieties, is an integral part of IPM[2].

---

[1] CropLife International is the global federation representing the research based Plant Science Industry that manufacture and sell plant biotechnology and crop protection products. The members of CropLife International are BASF, Bayer CropScience, Dow AgroSciences, DuPont, FMC, Monsanto, Sumitomo Chemical and Syngenta, as well as, regional biotechnology and crop protection associations. Through the regional members there are additional company members and CropLife national associations in 91 countries across the world.

[2] Further information on resistance management, including recommended approaches, can be found on the websites of the Fungicide, Herbicide, Insecticide and Rodenticide Resistance Action Committees (RACs). These are industry expert groups within the CropLife International structure that monitor resistance and develop recommendations and educational material to address the issue. Their websites can be accessed via www.croplife.org

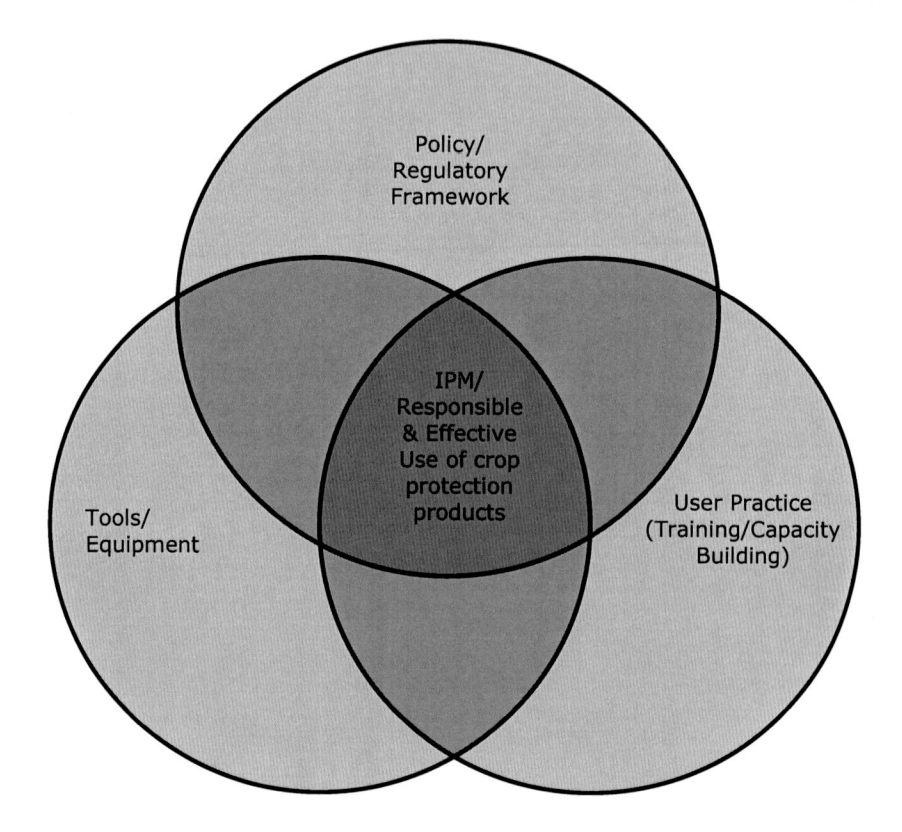

Fig. 16.1 Pesticide Use Context. (Adapted from Dobson and Jones 2007)

Although the general principles of IPM, as stated in the definition above, are universal, IPM is a knowledge intensive system and the actual practices employed will vary according to location and prevailing climatic and agronomic conditions. Furthermore, it depends also on local regulation, availability of tools (including chemical and non-chemical pesticides) and farmer education and training. This can be referred to as the Pesticide (or Pest Management) Use Context (see Fig. 16.1), which also determines the risk any intervention poses to human health and the environment. The regulatory/policy domain includes areas such as policies to promote IPM and appropriate regulation that helps ensure pest control products are available. The tools/equipment domain includes ensuring functional agricultural input distribution and sales channels that bring products to farmers and help to ensure that appropriate equipment and spares, including spray equipment and, where needed, personal protective equipment are available. User Practice (training/capacity building) includes the provision of information, demonstration and training to farmers so that they acquire the skills, knowledge and confidence to implement IPM, including using pesticides and other interventions, safely and effectively.

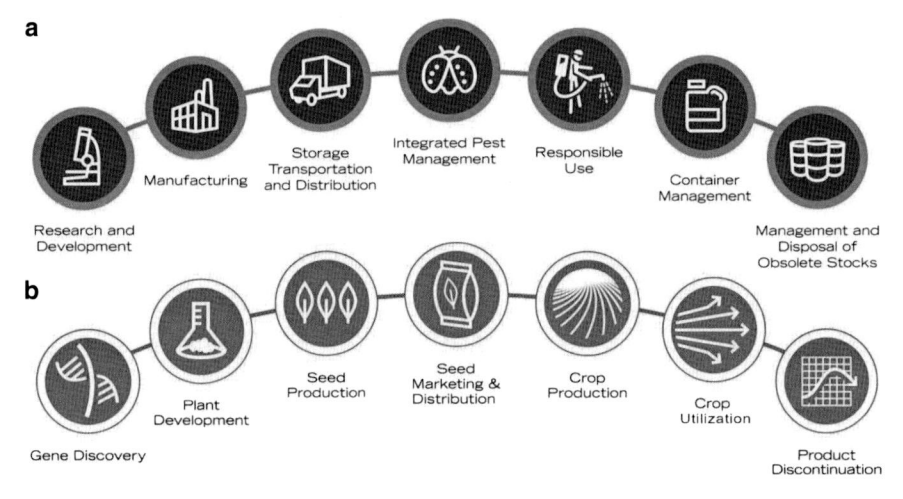

Fig. 16.2 Stewardship Lifecycle for (**a**) Crop protection products and (**b**) Plant biotechnology products (for further information see http://www.croplife.org)

16.2 Promotion of IPM by CropLife International

The member companies of CropLife International signaled their support for IPM as the underlying strategy for managing pests by signing an IPM declaration in 1995, although individual companies had developed alternative IPM tools and supported IPM programs well before this, e.g. Critchley et al. 1989. The support and promotion of IPM forms part of the industry's commitment to stewardship: the responsible and ethical management of a crop protection or plant biotechnology product[3] throughout its lifecycle, from initial research, through manufacture, distribution, use in the field and disposal of any waste or unused product (see Fig. 16.2).

Farmer orientated training programs, supported by CropLife International, started in 1991 with the 'Safe Use Initiatives' in Guatemala, Kenya and Thailand (CropLife International 2006), however, from 1995 these programs expanded and increasingly included IPM. IPM is promoted as a holistic approach, often as part of an Integrated Crop Management[4] approach that supports sustainable agriculture (see Fig. 16.3). Now IPM is the core of CropLife International's farmer training

---

[3] Crop protection products include chemical and non-chemical pesticides, as well as insect pheromones, repellents etc.

[4] Integrated Crop Management (ICM) is a whole farm strategy, which involves managing crops profitably, with respect for the environment, in ways which suit local soil, climatic and economic conditions. It safeguards the farm's natural assets in the long term. It includes practices that avoid waste, enhance energy efficiency and minimize pollution. ICM is not a rigidly defined form of crop production but is a dynamic system, which adapts and makes appropriate use of the latest research, technology, advice and experience.

```
                        ┌─────────────────────────────┐
                        │  Sustainable Development    │
                        └─────────────────────────────┘
                    ┌─────────────────────────────────────┐
                    │      Sustainable Agriculture        │
                    └─────────────────────────────────────┘
                ┌─────────────────────────────────────────────┐
                │    Integrated Crop Management (ICM)         │
                └─────────────────────────────────────────────┘
            ┌─────────────────────────────────────────────────────┐
            │    Integrated Pest Management (IPM)                 │
            └─────────────────────────────────────────────────────┘
```

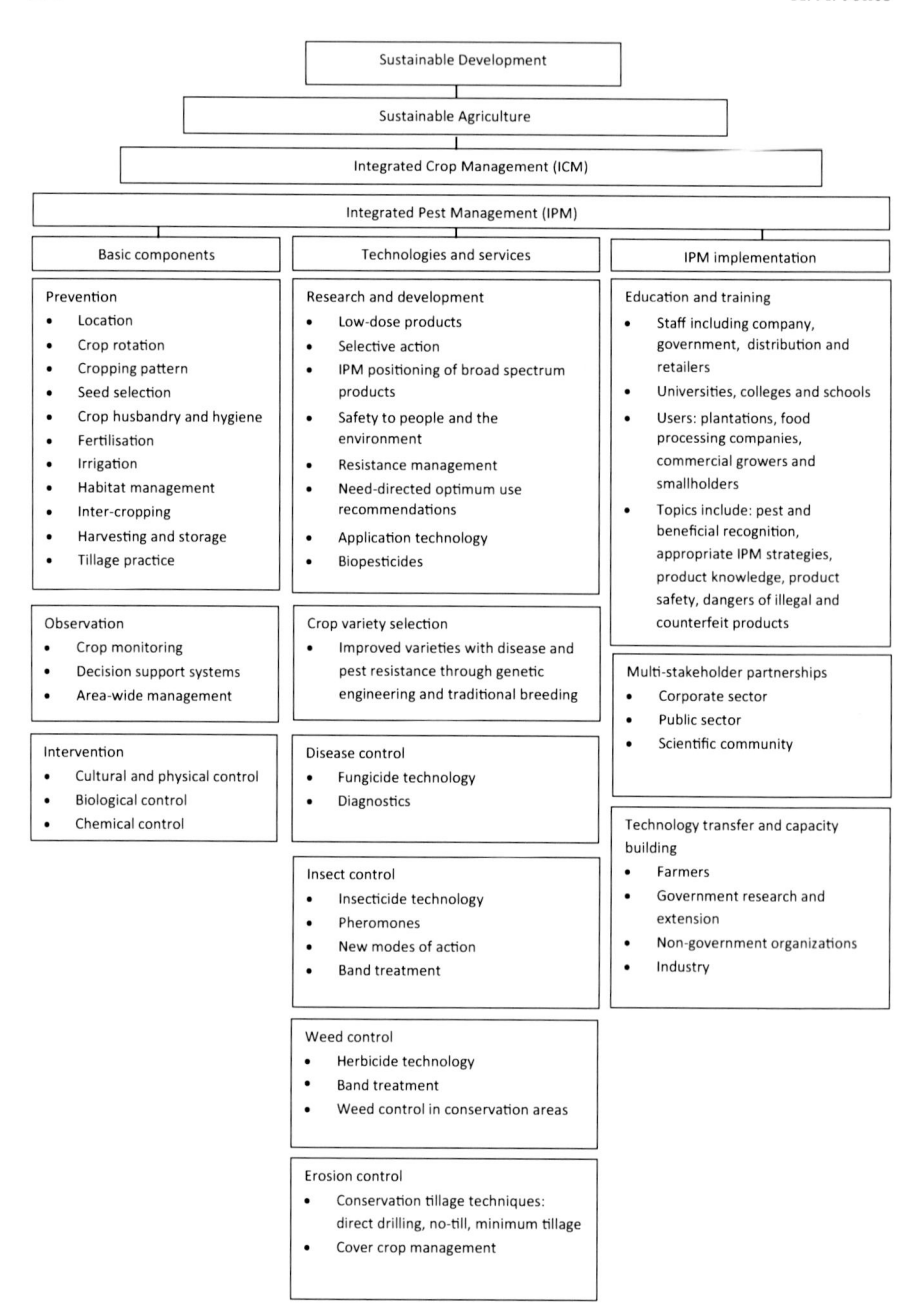

Basic components	Technologies and services	IPM implementation
Prevention • Location • Crop rotation • Cropping pattern • Seed selection • Crop husbandry and hygiene • Fertilisation • Irrigation • Habitat management • Inter-cropping • Harvesting and storage • Tillage practice	**Research and development** • Low-dose products • Selective action • IPM positioning of broad spectrum products • Safety to people and the environment • Resistance management • Need-directed optimum use recommendations • Application technology • Biopesticides	**Education and training** • Staff including company, government, distribution and retailers • Universities, colleges and schools • Users: plantations, food processing companies, commercial growers and smallholders • Topics include: pest and beneficial recognition, appropriate IPM strategies, product knowledge, product safety, dangers of illegal and counterfeit products
Observation • Crop monitoring • Decision support systems • Area-wide management	**Crop variety selection** • Improved varieties with disease and pest resistance through genetic engineering and traditional breeding	**Multi-stakeholder partnerships** • Corporate sector • Public sector • Scientific community
Intervention • Cultural and physical control • Biological control • Chemical control	**Disease control** • Fungicide technology • Diagnostics	**Technology transfer and capacity building** • Farmers • Government research and extension • Non-government organizations • Industry
	Insect control • Insecticide technology • Pheromones • New modes of action • Band treatment	
	Weed control • Herbicide technology • Band treatment • Weed control in conservation areas	
	Erosion control • Conservation tillage techniques: direct drilling, no-till, minimum tillage • Cover crop management	

Fig. 16.3 Elements of IPM. (Source: CropLife International 2004)

Fig. 16.4 Location of training programs supported by the CropLife association network (1990–2011). (Source: http://www.croplife.org/presentations. Stewardship Dialogue 2011, K. Jones)

programs and all trainings on the responsible use of crop protection products are given within the context of IPM[5].

Since 1991 CropLife International has supported programs that have directly trained more than three million people. Each year programs are implemented in around 60 countries in Africa, Asia, Europe, Latin America and North America[6] and programs have been implemented in 70 countries since 1991 (see Fig. 16.4). These programs are undertaken in partnership with various groups—up to 350 different ones each year, including national and local governments (including extension services), international donors, universities, schools, national and international research organizations and NGOs. The number of people trained from 2005 to 2011 is shown in Fig. 16.5, which shows that almost two million were trained during this period—between 110,000 and 670,000 being trained each year. Several different target groups were trained (Fig. 16.6), including farmers, pesticide retailers, government extension officers and NGO staff. Significantly, some 20,000 trainers and educators have been trained, who will continue to train and reach more people (see Fig. 16.6). Training is backed up through continued messaging and provision of information through various media, including television, radio, newspapers, magazines, posters and wall paintings.

---

[5] CropLife International has promoted stewardship of crop protection products, including IPM for more than 20 years. This paper concentrates on stewardship of crop protection products rather than stewardship of plant biotechnology products.

[6] In North America training is undertaken by other bodies, such as independent, certified consultants or the Land Grant Universities, as well as the member companies of CropLife America and CropLife Canada, rather than by the industry association.

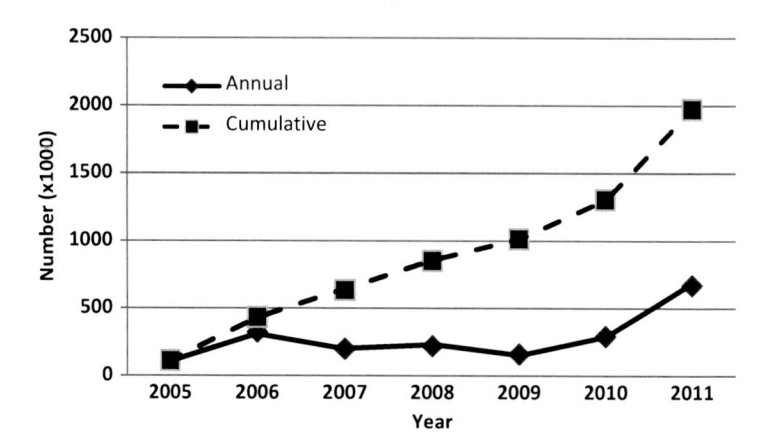

Fig. 16.5 Number of individuals directly trained by the CropLife association network and partners, 2005–2011. (Source: http://www.croplife.org/presentations. Stewardship Dialogue 2011, K. Jones)

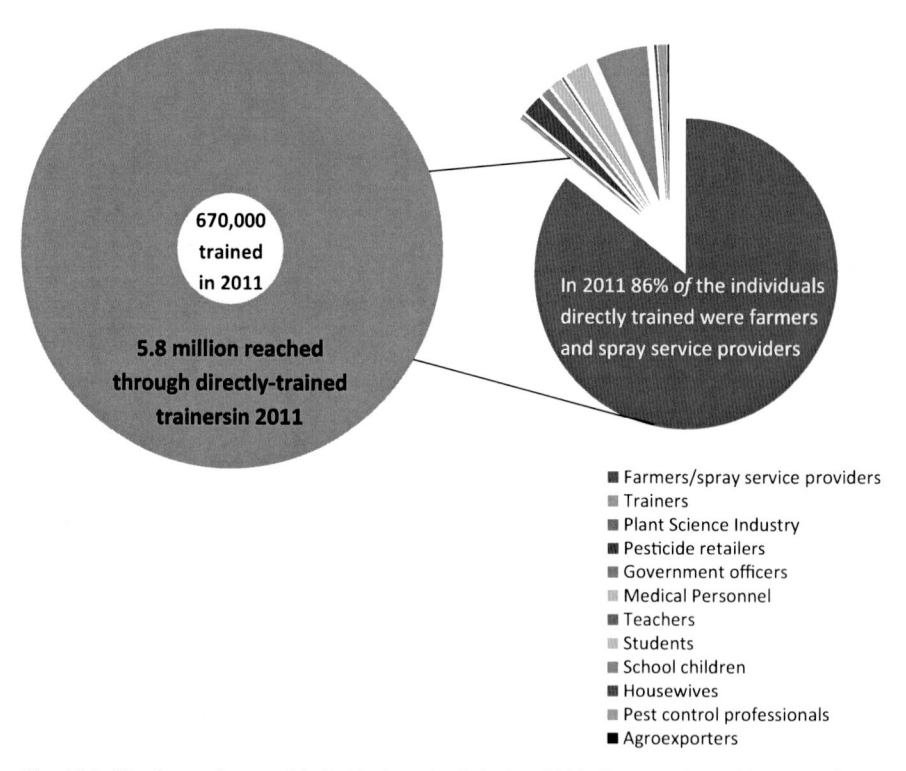

Fig. 16.6 Number and type of individuals trained during 2011. (Source: CropLife International 2011)

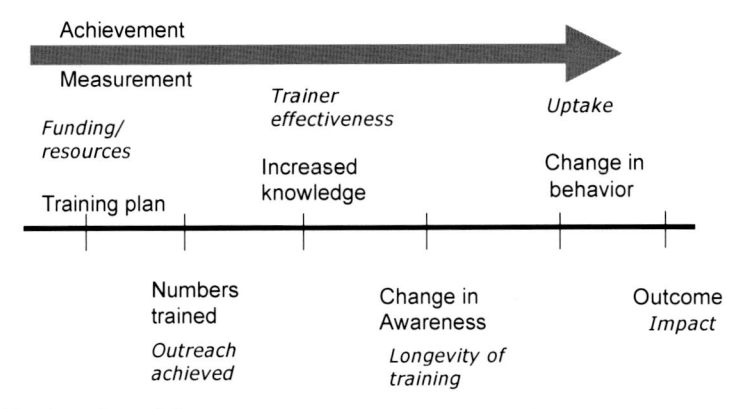

Fig. 16.7 Measuring training outcomes

16.3 Impact Assessment

The figures given above measure the number of individuals trained: an *activity* indicator. However, the figures do not indicate whether the training resulted in a change in farming practices. In common with many other groups, CropLife international has been trying to improve its measurement of the outcomes of its programs, through measurements of changes in behavior i.e. adoption of IPM by farmers and subsequent impacts, such as pesticide residue levels on harvested crops.

Thus, assessments can include *uptake* indicators i.e., IPM adoption, and *impact* indicators i.e., the impact of adoption, which are assessed on the basis of a subsample. This combined with an activity indicator, such as number trained or size of training intervention provides an estimate of overall outcomes. All types of indicators can be useful, the choice will depend on relevance, ease of assessment and cost (see Fig. 16.7).

16.3.1 IPM Uptake

IPM is site specific and can include a range of different actions. Thus, a single measure of adoption is difficult or impossible to make. One way is to have a range of uptake indicators for various actions that are considered part of the IPM toolbox. This may include monitoring pest and beneficial organism populations, agronomic practices such as crop rotation or intercropping, use of resistant plant varieties, maintenance of uncultivated areas to promote build-up of beneficial insects, use of biocontrol measures and responsible use of pesticides (product choice, personal protective equipment, targeted application, etc.). Adoption of a minimum number of these actions—some of which, for example, monitoring the crop for pest and beneficial organism numbers, would be mandatory—indicates that the farmer is implementing IPM. Although IPM actions will vary, overall measurement of IPM uptake is independent of location and of season, i.e., IPM adopters will utilize tools according to IPM principles no matter what the prevailing conditions are.

16.3.2 IPM Impacts

Impacts include measures such as yields, incomes, pesticide application frequency and pesticide residue levels. Comparisons, however, need to be carefully made as these impacts vary from year to year and location to location due to factors such as weather, soil quality, pest pressure and availability of labor and/or inputs.

16.4 Examples of IPM Interventions

The following section provides a few examples of the IPM training programs implemented each year. Programs are supported by CropLife International, through its national and regional associations and in many cases by partner organizations, including USAID (Global Development Alliance). The examples chosen cover a range of program objectives and target groups. Measurements are based on activity, uptake or impact indicators.

16.4.1 Kenya

Several different programs have been undertaken in Kenya. One, in partnership with the Ministry of Agriculture, consisted of six training courses (lasting 3–5 days) in IPM, including responsible pesticide use, in which 183 extension officers participated (98 males and 85 females). These officers and previously trained master trainers went on to train an estimated 11,675 farmers. In partnership with the International Fertilizer Development Center (IFDC), the Agricultural Market Development Trust (AGMARK) and the Ministry of Agriculture, 455 pesticide retailers (321 males and 134 females) were also trained in IPM principles incorporating responsible pesticide use.

16.4.2 India

A number of different initiatives have been implemented in India, all aimed at improving farmers' livelihood.

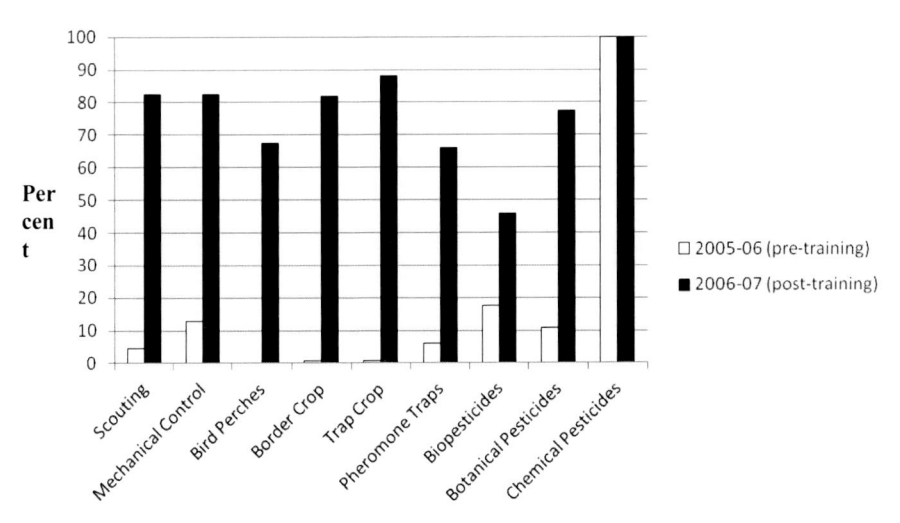

Fig. 16.8 Farmers' adoption of IPM components, Guntar, Andhra Pradesh. (Source: Reddy et al. 2011a)

16.4.2.1 IPM in Chilli, Guntar, Andhra Pradesh

Farmers faced several challenges, including crop losses from fruit borers, sucking pests (mites and thrips) and diseases. Also the crops were being rejected for export due to high pesticide residues. A program was implemented in collaboration with the Indian Spice Board aimed at improving pest control practices and reducing pesticide residue levels enabling farmers to reach the standards set for the export of their crop. Other partners involved in the program included the Regional Agricultural Research Station (Guntar), State Agricultural University, State Horticultural Department and the State Department of Health. These groups provided information on non-chemical pest control techniques, a subsidy for improved spray machinery and guidelines on chilli cultivation.

The program ran for 4 years from 2006 and trained more than 5,000 chilli farmers. Additionally, over 2,000 schoolchildren were also provided training in IPM principles in the schools within the villages where the chilli farmers were located. Finally, 450 women were trained through special women group training programs on the safe handling and storage of pesticides and health management. The uptake indicators from 1 year are shown in Fig. 16.8. Impact indicators included reduced cost of production, increased yields and, as a result of reduced pesticide residues and higher chilli quality, nearly all farmers were able to sell their crop for export. Overall, there was a 52% increase in return on investment (net return of 83,576.50 Indian rupees compared to 54,949) for IPM farmers compared to non-IPM farmers (Reddy et al. 2011b). Finally, although the data shows 100% of the farmers continue to use chemical pesticides, the average number of applications per season has been shown to drop from 30 to 15 following adoption of IPM.

The learning and responsibility for continuing activities has been handed to the Spice Board and Exporters Association, who are now working with a local NGO to continue and expand the training.

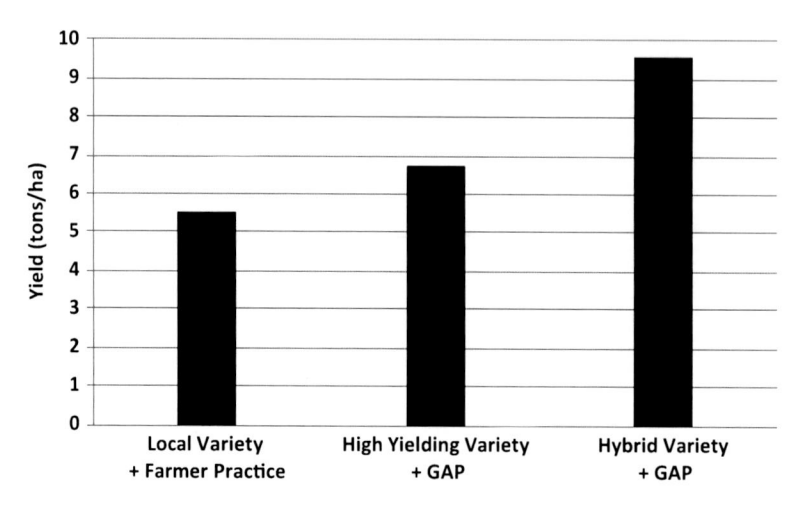

Fig. 16.9 Impact of adoption of good agricultural practices (GAP) and hybrid rice varieties on rice yields, West Bengal, India. (Source: http://www.croplifeindia.org/31-st-Annual-General-Meeting.pdf)

16.4.2.2 Green Revolution in Eastern States

This program is aimed at training farmers in good agricultural practices (GAP), which includes IPM, as well as demonstrating the benefits of using good-quality hybrid seed. The program initially covered almost a hundred villages in West Bengal, where demonstration plots were established. Approximately 1,000 farmers directly participated in the demonstrations and GAP training sessions. Many more were able to see the impacts of the interventions at the demonstration plots and attended farmer meetings where the project activities and its impacts were described. Partners involved included the Ministry of Agriculture, State Department of Agriculture for West Bengal, State Agricultural University—West Bengal, Rice Mills' Association and local government (Panchayats). Yield results from the demonstration plots show a highly significant increase in productivity from GAP and hybrid rice adoption (Fig. 16.9). The program has now been handed over to the government of West Bengal which intends to expand the program to further villages.

16.4.3 Philippines

An IPM program focusing on insect resistance management in rice and vegetables was initiated in 2011. This three-year project aims to deal with the following issues:-

- Development of insecticide resistance of high-risk pests such as diamond-backed moth, whitefly and rice plant hoppers.

- Lack of understanding among stakeholders, including farmers, of IPM and integrated resistance management (IRM): insecticide mode of action rotation, based on Ministry of Agriculture recommendations.
- The need to incorporate insecticide mode of action on pesticide container labels.

The program was initiated by training of trainer sessions in 2011, in which 722 individuals from CropLife Philippines member companies, government and the academic sector participated. This was followed up by more intensive training for 34 individuals in early 2012. The trainers then targeted 5,600 farmers for training during the remainder of 2012 (3,600 on rice and 2,000 on vegetables), using training manuals, pocket information books on mode of action and other materials developed in 2011.

16.4.4 Thailand

Training programs have been carried out in several districts across the country. Interventions were targeted at farmers, retailers and school children (for IPM). The following interventions were undertaken between 2009 and 2011:

- Training 16,731 farmers (including 1,385 farmer leaders) in IPM and GAP.
- Training of 1,580 farmers through the community pest management center and as part of the National Mealybug Control Project and 940 farmers as part of the National Brown Plant Hopper Control Project.
- Follow-up re-enforcement sessions for 3,654 farmers trained in previous years.
- Training of 100 contract farmers in IPM strategies, including responsible use of pesticides, at the River Kwai International Food Industry Company.
- Training of 183 medical personnel on diagnosis and management of pesticide poisoning.
- Training of 1,720 retailers in the safe handling and responsible use of pesticides.
- Training of children at three schools in IPM practices (estimated 1,500 children), which included a practical element where seasonal vegetables were grown by students under an IPM regime.
- Eight exhibitions on IPM activities, including responsible use of pesticides, across the country reaching at least 5,000 individuals.
- The Thai Crop Protection Association (TCPA) member company staff, which had previously received training and materials from TCPA, trained 123,550 farmers.
- Radio programs on IPM were broadcast in Chanthaburi province (total 730 'slots').

Partners included the Department of Agriculture, Department of Agricultural Extension, Royal Project Foundation, Western GAP Cluster and local schools and universities.

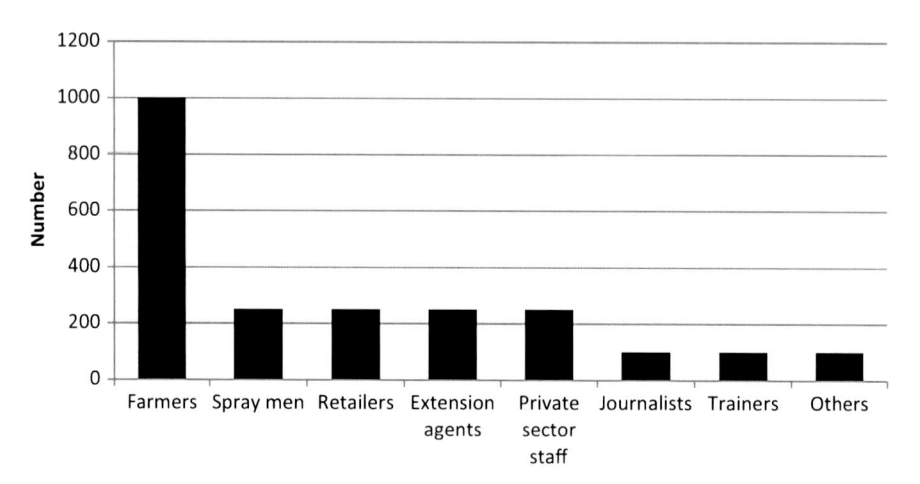

Fig. 16.10 Breakdown of trainees, Bangladesh 2010

16.4.5 Bangladesh

A country-wide program on IPM, including responsible use of pesticides was carried out in collaboration with the country's largest NGO, BRAC and the Bangladesh Department of Agricultural Extension. The program trained 2,300 people directly, including trainers and farmers. The breakdown of training groups is shown in Fig. 16.10.

Re-enforcement of training was achieved through training messages that were displayed as posters (for 4 months) and documentary film shows (90 showings) at BRAC centers, film shows at local bazaars (200 showings) and 12 TV and newspaper advertisements. Additionally, personal protective equipment (e.g. gloves) was distributed to trainees. A community-based NGO, Katalyst, conducted surveys in the areas where training had been undertaken and concluded that 'the training had been very effective' in informing farmers of proper pest management and the safe handling of pesticides.

Member companies of the Bangladesh Crop Protection Association also carried out their own training, which resulted in a further 183,000 farmers and 26,150 retailers being trained in 2010 alone.

16.4.6 Nicaragua

The multi-year program, which started in 2008, was carried out in the provinces of Jalapa, Condega, Esteli, El Viejo and Chinandega. The program's principal aim was to ensure that pesticide residues were minimized and to minimize pesticide contamination among workers of tobacco, vegetables, basic grains, peanut, soybean and sesame crops. The program was carried out in collaboration with the Ministry Of Health and the Nicaraguan Institute of Agricultural Technology (INTA). In 2010 alone, a total of 12,747 people were trained.

Re-enforcement of training was achieved through radio spots and distribution of brochures, posters etc. (10,628 units were distributed).

As a result of training, there was a 76 % reduction in reported intoxication cases amongst agricultural workers in 2010 compared to 2007. The rejection of vegetables from the project area at ports of entry into the USA in 2010 was zero.

16.4.7 Guatemala

The CropLife national association in Guatemala, Agrequima, implemented a training program in the Solalá region of Guatemala on GAP. The program targeted a local farmer association, K'aqchikel Indigenous Association for Integral Development (Asinkad), whose 45 members cultivate 52.6 ha of land, mainly to sugar snap peas. The training focused on improved pest management, including the responsible use of pesticides and fertilizers and the use of biodeps (a biological bed that breaks down surplus pesticide where mixing and sprayer filling takes place). The training provided the skills required to obtain GAP accreditation required for export.

16.5 Training Materials

In 2003, CropLife Asia developed a web-based e-learning course, which can be freely downloaded (www.aglearn.net). The course includes modules covering an introduction to IPM, IPM in cotton, IPM in rice, and IPM in vegetables. CropLife Latin America and CropLife Africa Middle East have also supported post-graduate (or equivalent) training in IPM at universities and colleges. In 2008, the promotion of IPM was further consolidated by the publication of an IPM training manual, which has since been rolled out across the CropLife International network. The manual, which includes material for course facilitators and for trainees, is freely available on the CropLife International website (www.croplife.org). Guidelines on the responsible use of pesticides, as well as training posters are also freely available on the website.

16.6 Concluding Remarks

The Plant Science Industry, as represented by CropLife International, recognizes, supports and promotes IPM. It recognizes that our general understanding of IPM strategies and tools has developed, and will continue to develop, over time and that traditional and new technologies have a role to play. An understanding of agro-ecology is an important element and is essential not only to growing a healthy crop, but also to understand the impacts of any interventions that the farmer may make. The important role of chemical pesticides in IPM, now and in the near future, needs to be recognized, as well as the need to train farmers in their proper and safe handling and responsible use, as well as when and when not to use them—'as little as possible, as much as necessary.'

References

Bajwa, W. I., & Kogan, M. (2002). *Compendium of IPM Definitions (CID)-What is IPM and How is it Defined in the Worldwide Literature*. Publication No. 998. Corvallis, Oregon, USA: Integrated Plant Protection Centre (IPPC), Oregon State University.

Critchley, B. R., Campion, D. G., & McVeigh, L. J. (1989). Pheromone control in the integrated pest management of cotton. In M. B. Green, & D. J. de B. (Eds.)., *Pest Management in Cotton* (pp. 83–92). Chichester, United Kingdom: Ellis Horwood.

CropLife International. (2004). *IPM—The Way Forward for the Plant Science Industry*. http://www.croplife.org/view_document.aspx?docId=420. p. 26. Accessed 23 March 2013.

CropLife International. (2006). *IPM Responsible Use Case Studies*. http://www.croplife.org/view_document.aspx?docId=422. p. 46. Accessed 23 March 2013.

CropLife International. (2011). *A Stocktaking Report: Crop Protection Stewardship Activities of the Plant Science Industry 2005–2011*. http://www.croplife.org/view_document.aspx?docId=3317. p. 59. Accessed 23 March 2013.

Dobson, H. M., & Jones, K. A. (2007). *Pesticides and Poverty—Analysing Pesticide Use Context (PUC) to Unleash the Benefits without the Backlash*. http://www.croplife.org/view_document.aspx?docId=4091. Accessed 7 May 2013.

FAO. (2012). *The International Code of Conduct on Pesticide Management* (Revised ed.). Rome: Food & Agriculture Organization of the United Nations.

Geier, P. W. (1970). Organizing large scale projects in pest management. In *Meeting on Cotton Pests*. Rome: Panel of Experts on Pest Control, FAO. Sept. 1970.

Geier, P. W., & Clark, L. R. (1961). An ecological approach to pest control. In *Proceedings of the Eighth Technical Meeting* (pp. 10–18). Warsaw: International Union of Conservation of Nature and Natural Resources, 1960.

Reddy, K. G., Reddy, A. S., Babu, J. S., & Reddy, M. C. S. (2011a). Adoption of integrated pest management (IPM) in chilli (*Capsicum annuum* L.): A case study in Guntur district of Andhra Pradesh. *International Journal of Applied Biology and Pharmaceutical Technology, 2*(2), 117–122. http://www.ijabpt.com/pdf/92017-Gurava%20reddy-IPM%20Chillies[1].pdf. Accessed 7 May 2013.

Reddy, M. C., Reddy, K. G., Tirupamma, K., & Reddy, K. V. S. (2011b). Economics of integrated pest management (IPM) in chilli in Guntur district of Andhra Pradesh. *International Journal of Plant, Animal and Environmental Sciences, 1*(1), 140–143. http://www.ijpaes.com/admin/php/uploads/104_pdf.pdf. Accessed 13 May 2013.

Smith, R. F., & Reynolds, H. T. (1965). Principles, definition and scope of integrated pest control. In *Proceedings, FAO Symposium on Integrated Pest Control* (Vol. 1, pp. 11–17). Rome: Food and Agriculture Organization of the United Nations.

Russell, D. A., Radwan, S. M., Irving, N. S., Jones, K. A., & Downham, M. C. A. (1993). Experimental assessment of the impact of defoliation by *Spodoptera littoralis* on the growth and yield of Giza '75 cotton. *Crop Protection, 12*(4), 303–309.

Chapter 17
From the Farmers' Perspective: Pesticide Use and Pest Control

Seyyed Mahmoud Hashemi, Rajinder Peshin and Giuseppe Feola

Contents

17.1	Introduction	410
17.2	Toward an Integrative Perspective	411
17.3	Pesticide Use and IPM in Iran	412
	17.3.1 Pesticide Use in Agriculture from the Iranian Farmers' Perspective	414
17.4	Pesticide Problems and IPM in India	418
	17.4.1 Pesticide Use and Pest Problems in Punjab, India	419
	17.4.2 Integrated Pest Management in Cotton	420
17.5	Pesticide Use in the Colombian Andes	423
	17.5.1 Background and Research Problem	423
	17.5.2 Goals	424
	17.5.3 Methods	424
	17.5.4 Theoretical Background	424
	17.5.5 Results	426
17.6	Conclusions	427
References		429

S. M. Hashemi (✉)
Department of Agricultural Extension and Education, College of Agriculture, University of Tehran, Karaj, Iran
e-mail: seyyedmahmoodhashemi@gmail.com; s.m.hashemi@ut.ac.ir

R. Peshin
Division of Agricultural Extension Education, Sher-e-Kashmir University of Agricultural Sciences and Technology of Jammu, Chatha, Jammu 180 009, India
e-mail: rpeshin@rediffmail.com

G. Feola
Department of Geography and Environmental Science, University of Reading, Reading, UK
e-mail: g.feola@reading.ac.uk

Abstract Many studies have shown that farmers in developing countries often overuse pesticides and do not adopt safety practices. Policies and interventions to promote a safer use of pesticides are often based on a limited understanding of the farmers' own perspective of pesticide use. This often results in ineffective policies and the persistence of significant pesticide-related health and environmental problems, especially in developing countries. This chapter explores potentials and limitations of different approaches to study pesticide use in agriculture from the farmers' perspective. In contrast to the reductionist and mono-disciplinary approaches often adopted, this chapter calls for integrative methodological approaches to provide a realistic and thorough understanding of the farmers' perspective on pesticide use and illustrates the added value of such an approach with three case studies of pesticide use in Iran, India, and Colombia.

Keywords Integrative approach · Integrated pest management · Pest control · Pesticide use · Safe use of pesticides

List of acronyms and abbreviations

CICR	Central Institute for Cotton Research
FFS	Farmer Field School
IAC	Integrative Agent-centred
IPM	Integrated Pest Management
IRM	Insecticide Resistance Management
IRMIPM	Insecticide Resistance Management–based IPM
NGO	Non-governmental Organization
PPE	Personal Protective Equipment
SES	Social–Ecological Systems

17.1 Introduction

Pest control as a matter of concern is as old as agriculture itself. Given the present growing demand for food, however, food loss to pests is more critical today than ever (Pimentel 2009). The potential losses as a result of pest infestations may vary, depending on crop and pest, from less than 50% to more than 80% (Oerke and Dehne 2004). For decades, chemical pesticides have been used as one of the many pest control tools in agricultural production to ensure high-quality and quantity of safe and inexpensive food to meet the consumer demand (Ecobichon 2001; Damalas 2009).

Although current literature lacks accurate data on the impact of pesticides on public health and the environment (Pimentel 2009), their negative impacts are widely acknowledged. Acute poisonings by agricultural pesticides are currently considered to be an important cause of human morbidity and mortality worldwide, with some 26 million human pesticide poisonings and with about 220,000 deaths per annum in the world (Pimentel 2009; Kesavachandran et al. 2009). In addition, ecosystems are also being affected by pesticides (Dhawan and Peshin 2009). The negative impacts of pesticides are particularly severe in developing countries. Although only 20% of the world's agrochemicals are used in the developing countries, such countries suffer 99% of deaths from pesticide poisonings (Jeyaratnam and Chia 1994).

Many programs and initiatives for the safe use of pesticides have been initiated worldwide, but often fail to achieve their goals (e.g., Orr 2003; Wyckhuys and O'Neill 2007). This failure can be at least partially ascribed to the fact that policy-makers have only a limited understanding of how farmers conceptualize their farming systems and, consequently, of why farmers adopt certain pesticide use practices. Such a limited understanding on the part of policy-makers does not translate into effective pesticide use policies (Wyckhuys and O'Neill 2007).

Furthermore, policy-makers mostly rely on reductionist approaches to pesticide use in agriculture, understanding a phenomenon by identifying and addressing individual components of the phenomenon separately and each discipline coming to an understanding from its own perspective. This chapter, in contrast, contends that a more integrated methodological approach is necessary, that is, one inspired by a holistic paradigm for properly understanding and addressing pesticide use in agriculture as a real-world subject of research which is embedded in the societal context in which pesticide use occurs. This chapter originates from the premise that there may be significant differences between farmers' perspectives and scientific and policy communities' perspectives on such issues, not least because of each communities' different mental models. In this chapter, reductionist approaches to study pesticide use practices are briefly reviewed, and their limitations in providing a realistic and thorough understanding of pesticide use briefly discussed. In contrast with these approaches, holistic approaches are described which provide a more realistic and farmer-centered understanding of pesticide use. These approaches are illustrated with three case studies from Iran, India, and Colombia, respectively.

17.2 Toward an Integrative Perspective

In contrast to conventional practice which assumes that farmers are passive adopters (Bruin and Meerman 2001), farmers' adoption of technologies reflects a dynamic decision-making process (Feola and Binder 2010a). However, policy-makers and agricultural experts do not necessarily understand a farmer's decision-making process. Kalaugher et al. (2012) highlight the existence of divergent perceptions of a farming system and different approaches to solving a particular problem between researchers and farmers. For instance, with regard to risk perception of pesticide use, Schöll and Binder (2009a) showed that the mental models of farmers and experts differed significantly from each other. Such a lack of understanding of farmers' decision-making is one of the main causes of policy failure (Feola and Binder 2010a).

The social sciences can contribute to the study of the decisions of the actors involved and the related institutional context. However, reductionist and mono-disciplinary approaches have dominated this field. This can seriously limit the contribution of the social sciences because the diverse range of factors that determine a farmer's pesticide use behavior can hardly be captured without considering multiple social science disciplines simultaneously. As Costanza and Kubiszewski (2012; p. 1) puts it: "Real-world problems do not come in disciplinary-shaped boxes (Jeffrey 2003), and neither do the solutions associated with these problems".

As argued by Atreya et al. (2012), the global knowledge on pesticide issues has been shifting from "mono-disciplinary" to "interdisciplinary" sciences as the pesticide-induced impacts are complex and interconnected in nature. But, minimal efforts are being made at the local level to move from mono-disciplinary sciences to new perspectives that are interdisciplinary in nature. Similarly, van Huis (2009) states that, in connection with challenges facing integrated pest management (IPM) in Sub-Saharan Africa, "A disciplinary entry point when dealing with subsistence farmers without a proper identification of their needs and opportunities is a wrong approach" (p. 408).

The potential of methods of study based on interdisciplinary approaches has remained largely untapped by scholarly research on pesticide use in agriculture, although calls for methods based on interdisciplinary approaches to address linked social and agro-environmental issues are not new (Evans 1951; Wohl 1955 as cited in Miller et al. 2008). For example, pesticide use studies tend to address "hard" (natural sciences) and "soft" (social sciences) aspects separately, which is mirrored by the lack of interdisciplinary journals dealing with pest management issues. Most journals dealing with pest management issues, in general, tend to cover articles that look at the subject from a natural sciences perspective as their first and most important priority and those that cover a social science perspective tend to follow conventional disciplinary boundaries.

In addition, farmers decisions on pesticide use are not made in a vacuum, but in a broader context of risks (e.g., health, economic) and livelihoods, in which trade-offs might exist between crop protection and other objectives. Understanding pesticide use, therefore, requires considering the context in which decisions are made, including contextual factors that might act as barriers or facilitating factors, and multiple and potentially competing farming or livelihood objectives (Schöll and Binder 2009a; Feola and Binder 2010a).

In sum, to fill the gap of understanding farmers' pesticide use practices, reductionist and mono-disciplinary approaches should be abandoned in favor of interdisciplinary and systemic approaches that best allow for understanding farmers' decisions in their specific context, and therefore provide a more solid basis for policy-making and interventions to promote safer pesticide use. The next three sections try to illustrate adopting such an approach through case studies from Iran, India and Colombia.

17.3 Pesticide Use and IPM in Iran

Chemical pesticide use has served as the dominant approach to pest control in Iran for over 60 years. In Iran, the estimated amount of total agrochemical pesticides used annually is 17–25 million liters. In addition, it is estimated that pests damage 42 % of agricultural products each year in Iran (Karamidehkordi and Hashemi 2010).

The use of pesticides is currently being seriously questioned as its negative impacts including pest outbreaks, pest resistance to pesticides, pesticide poisonings, and the threat to health and the environment have become evident in different parts of the country, particularly in provinces located on the southern coast of the Caspian Sea in northern Iran where about 60 % of the total pesticide consumption occurs (Heidari et al. 2007).

In general, the estimated amount of pesticides used each year in Iran is much more than is needed (Karamidehkordi and Hashemi 2010). The use of the insecticide diazinon on rice fields of Guilan Province, a Caspian Province, has been reported to be 5–10 times higher than the necessary amount (Allahyari et al. 2008). In addition, the frequency of overall pesticide applications in some fruits and vegetables may be as often as 6–12 times per season and almost 30 times per season in the Jirouft region (in the south-eastern part of the country) (Heidari et al. 2007).

According to Shahbazi et al. (2012), some outlawed organochlorine pesticides (OCP) (e.g., lindane and technical endosulfan) are still illegally used in rice, other cereals, and cotton cultivation (Norouzian 2000). Also, dicofol, a significant source of dichloro-diphenyl-trichloroethane (DDT), is still used in cotton cultivation and in forestry (Norouzian 2000). In a study conducted in 12 cities of Mazandaran Province, a Caspian province, 3.2 % of the authorized pesticides used were considered to be extremely dangerous, 11.8 % of these were classified as seriously poisonous, and 24.7 % were potentially dangerous (Yousefi 2008). In a more recent study aimed at surveying pesticides commonly used in Tehran and Isfahan, Dehghani et al. (2011) reported that 9.3 % of the pesticides used were highly hazardous and the remaining 58.5 and 32.2 % were moderately and less hazardous pesticides to human health, respectively.

Since 1994, the Iranian government has started a number of programs to reduce pesticide use; however, such initiatives failed to establish sustainable plant management systems at the farm level as most of them did not fully incorporate bottom-up participatory approaches (Heidari 2006).

According to Heidari (2006), in practice, no farms in Iran adopted the principles of IPM until 1999 when the Farmer Field School (FFS) approach was first introduced as part of a pistachio IPM project in Semnan Province which resulted in successfully empowering farmers to deal with many of their own problems, reducing production costs, and increasing income during two successive seasons. This project was conducted by the Iran National Plant Protection Research Institute in response to a request for help from the Semnan agricultural organization in controlling two surging pests on the main crops of Semnan Province, that is, psylla (*Agonoscena pistaciae*) on pistachio and melon fly (*Bactrocera cucurbitae*) on summer crops. The project successfully controlled the surging pest problems (Heidari 2006). Experiences with IPM/FFS projects in different parts of the country (Figs. 17.1 and 17.2) revealed that "IPM cannot be successful without active participation of the farmers" (Fathi et al. 2012; p. 20).

In general, even about a decade after the introduction of IPM/FFS in Iran (Table 17.1) by national and international institutions (Fig. 17.3)—FAO and the Global Environment Facility (small grants program)—IPM/FFS can still be

Fig. 17.1 Participants of FAO project on IPM/FFS for apple in Damavand County, Iran (photo Hossein Heidari)

Fig. 17.2 Participants of weekly meeting of UNDP GEF/SGP project on IPM/FFS for rice in Sooleh, Mazandaran Province, Iran (photo Hossein Heidari)

described as "a pilot project idea," although currently it is becoming a mainstream approach in Iran (Fathi et al. 2012).

17.3.1 Pesticide Use in Agriculture from the Iranian Farmers' Perspective

From a review of the relevant literature about Iranian farmers' perspective of pesticide use in agriculture, we can conclude that consideration of the Iranian farmers' perspective is very rare. In particular, almost all of those studies were conducted by researchers with a background in agricultural extension, without any contribution from relevant scientists with backgrounds in sociology, psychology, anthropology, and so on. In addition, there are currently extremely few, if any, studies that consider the farmers' perspective from an interdisciplinary point of view.

Table 17.1 Iran's national IPM/FFS program. (Source: Fathi et al. 2012)

Year	Number of FFS sites	Number of provinces	Number of crops
2004	5	2	4
2005	28	8	8
2006	91	15	10
2007	172	22	27
2008	252	29	37

Fig. 17.3 UNDP GEF/SGP project on training of rice IPM facilitators in Azbaran, Mazandaran Province, Iran (photo Hossein Heidari)

17.3.1.1 Farmers' Pesticide Use: Perceptions, Knowledge, Practices, Training Needs, and Health Effects

With regard to awareness, knowledge, and competence as important variables to adopt the safe use of pesticides and IPM technologies (Hashemi et al. 2012a, 2012b), most Iranian farmers lack basic knowledge of IPM, competence on pest management practices, and safe use of pesticides, according to studies conducted in different parts of the country. In a study conducted in Karaj in 2007, authors reported that most farmers lacked an acceptable knowledge of IPM (Hashemi et al. 2008) and most of them were not competent in basic pest management practices (Hashemi et al. 2009). In another study carried out in Zanjan Province in the northwest of Iran, Karamidehkordi and Hashemi (2010) reported that farmers had little awareness of non-chemical pest control methods (i.e., mechanical and biological techniques and natural enemies).

In a study conducted in Fars Province in southwest Iran in 2008, two distinct groups of farmers were revealed. One group of farmers clearly had a positive opinion about the efficacy of the current pesticide products (i.e., they felt that both current and older pesticides used are the same in relation to the level of active ingredients they have). On the other hand, the other group had a rather negative opinion of the efficacy of the current pesticide products (i.e., they felt that current pesticides are less effective than older pesticides they had used and that their efficacy decreases annually because they felt that companies deliberately dilute pesticide products to sell more pesticides) (Hashemi and Damalas 2011). As a result, one farmer from this group stated that "nowadays, current pesticides do not show adequate efficacy to control pests, and even if I wash my hands with pesticides, there will be no danger for my health" (Hashemi and Damalas 2011; p. 76).

Accordingly, many experts in Iran believe that the limited knowledge of Iranian farmers with regard as to how to use pesticides and how much pesticide to use is the main problem with pesticide use in Iran (Karamidehkordi and Hashemi 2010).

According to a study carried out in five provinces of Iran, 68 % of the farmers surveyed used no protection devices (e.g., coveralls, mask, gloves, etc). Further, 55 % of the farmers discarded the pesticide containers with no special care (Aghilinegad et al. 2008). In research which surveyed pesticide use among farmers in 2009, the authors reported that only 13 % of the farmers disposed of empty pesticide containers according to the pesticide label and also only 7 % of them were following the safety precautions on the label during pesticide use. In addition, about 60 % of the farmers stated that they were not using any special protective equipment when spraying pesticides and almost no one had received any special training in pesticide safety (Hashemi et al. 2012b). Results of similar studies conducted in other parts of the country confirm these findings (e.g., Ghasemi and Karami 2009; Karamidehkordi and Hashemi 2010; Shafiee et al. 2012).

In a study conducted to identify farmers' needs for pest management training, farmers showed different needs for future training on pest management because of their different levels of training already received and their different backgrounds. Farmers who had never attended a training workshop showed low levels of competence and consequently high levels of need for pest management practices training with regard to IPM principles. On the other hand, farmers who had participated in a workshop for pest management showed the highest level of competence for all three areas of pest management practices studied (i.e., pest identification, pesticide management, and IPM principles) (Hashemi et al. 2009).

According to a study conducted among vegetable growers by Shafiee et al. (2012), all of respondents reported health problems after routine pesticide use, including dizziness, cough, nausea, skin problems, poor vision, and stomach aches.

17.3.1.2 Pesticide Use and Risk Perceptions Among Farmers

Karamidehkordi and Hashemi (2010) report that 70 % of the farmers reported that pesticides have negative effects on human health. In addition, about 50 % of the respondents identified reported pesticide impacts on groundwater and non-pest insects. In another study, the majority of farmers reported that they consider current pesticides to be as harmful as older types of pesticides (60 %), whereas about 30 % of the farmers stated that they consider current pesticides to be harmless to human health compared with older types of pesticides (Hashemi and Damalas 2011). Pesticide use and farmers' risk perceptions of unsafe use of pesticides were explored in 2009 (Hashemi et al. 2012b). Three groups of farmers were revealed: the first group included 30.3 % of the farmers with the lowest perceived risk of unsafe use of pesticides; the second group, 63%, was the largest with an intermediate perceived risk of unsafe use of pesticides; and finally the last group, 16.7% of the farmers, perceived the highest degree of risk in the unsafe use of pesticides. In addition, this study found that there was not a simple and linear relationship between risk perceptions of unsafe use of pesticides and farmers' age, but farming experience and experience of pesticide-related adverse health effects in the past were the effective factors which lead to higher levels of perceived risk associated with the unsafe use of pesticides.

17.3.1.3 Safe Use of Pesticides: Determinants and Training Needs

Farmers' knowledge, attitudes, and practices of pest management were explored in a study conducted in four Iranian cities in Mazandaran Province (Arjmandi et al. 2012). Five categories of variables were considered as determinants of pesticide consumption: education, pesticide application technology, regulations, IPM implementation, and the price of pesticides.

Other research in Iran highlighted the role of cost of each pesticide product for farmers as the farmers' final criterion for the purchase and use of a specific product (Hashemi et al. 2012b). In addition, considering the fact that in Iran the price of the biological pesticides is much higher than that of the chemical pesticides, farmers normally do not tend to use these biological alternatives (Arjmandi et al. 2012). In Iran, pesticide subsidies were cut in 2009; therefore, this new situation will probably influence the behavior of farmers toward pesticide use (Hashemi et al. 2012b).

About 80% of Iranian farmers are not well-educated (either illiterate or undereducated) (Hashemi and Hedjazi 2011); some studies dealing with pesticide use among farmers revealed Iranian farmers' level of education as one of determinants of unsafe use of pesticides (e.g., Aghasi et al. 2010; Shafiee et al. 2012). In contrast, other studies have shown that there was no positive correlation between the farmers' level of formal education and their awareness of the side effects of the excessive use of chemical pesticides and farmers' personal safety in pesticide use (Karamidehkordi and Hashemi 2010; Arjmandi et al. 2012; Hashemi et al. 2012b).

Legislation and strong regulatory systems are necessary to ban or restrict use of dangerous chemicals and pesticides (Ecobichon 2001). The current regulations of the Iranian Plant Protection Organization go back to 1967 and do not cover components of environmental management of pesticide use in a comprehensive way. The regulations require revisions and amendments to include all environmental management of pesticide use (Arjmandi et al. 2012).

Hashemi et al. (2012a) focused on the three stages of pesticide handling (i.e., before, during, and after use) in pesticide safety training and compared the training needs of young farmers (up to 35 years old), middle-aged farmers (above 35 up to 50 years old), and old farmers (above 50 years old), according to a study conducted in 2009 (Hashemi et al. 2012a). The top training needs for the young farmers were mostly on measures or actions related to pesticide handling before use (i.e., "selecting appropriate pesticide products for a specific pest problem" and "defining the correct timing of application for a specific pest problem"). In contrast, the top training needs for middle-aged and old farmers were mostly on measures or actions related to pesticide handling during use (i.e., "providing first aid in case of sickness or poisoning by pesticides" and "discriminating degree of pesticide toxicity by the safety symbols").

17.3.1.4 Factors Affecting Farmers' Adoption of IPM

Veisi (2012) explored the determinants of farmers' adoption of IPM in the Iranian provinces of Mazandaran and Gilan, considering exogenous factors, farmer characteristics, farm characteristics, and the characteristics of innovations (IPM). The determinants with the highest effects on adoption behavior of IPM practices were "soil quality," "gender" (being male), and "level of knowledge." In Samiee et al. (2009), farmers' level of knowledge about IPM practices was found to be the most effective variable to explain the level of wheat growers' adoption of IPM practices.

17.4 Pesticide Problems and IPM in India

In India, insecticides are widely used in agriculture accounting for 64% of the total pesticide consumption (Peshin et al. 2009a). Insecticides are the main tool of pest management in cotton, vegetable crops, and rice (Peshin and Kalra 1998; Peshin et al. 2007, 2009b; Sharma 2011). Herbicides are commonly used in wheat and rice crops. The cotton crop accounted for about 50% of the total pesticide use before the introduction of transgenic cotton. Despite the implementation of many IPM programs in cotton, vegetable crops, and rice and widespread adoption of Bt cotton, pesticide use has increased from 37,959 tons in 2006–2007, to 55,540 tons (a.i.) in 2010–2011, corresponding to an increase of 46.31%. Prior to 2007–2008, pesticide use in Indian agriculture had decreased between 1990–1991 and 2006–2007 from 75,033 to 37,959 tons, a reduction of 49.41%. Pesticides continue to be the main plant protection tool in states like Punjab, Haryana, Andhra Pradesh, Maharashtra, Rajasthan, and Tamil Nadu, which consume 55% of the total pesticide use when taken together (Peshin et al. 2009a).

Pesticide-based pest management is a complex technology for farmers to efficiently adopt (Litsinger et al. 2009). It is a mix of software (consisting of a knowledge base) and hardware (consisting of inputs) technology. Hardware in terms of pesticides, and software in terms of selection of a right pesticide against a particular pest, right dosage, right dilution, and right time of application (Peshin et al. 2012). The hardware side of technology is dominant and is adopted faster than the software side (Roger 2003). The pesticide-based pest management requires higher levels of knowledge and greater skills on the part of farmers to select the right pesticide, pesticide dosage, and dilution (spray volume). Most pesticides are only toxic to specific pests, can be washed away by rain, can drift with wind, and are required to be placed on a specific part of the plant and must be diluted correctly (Nataatmadja et al. 1979; Litsinger et al. 2009).

17.4.1 Pesticide Use and Pest Problems in Punjab, India

The state of Punjab, comprising less than 1.5% India's land area, has been "the leader of the Green Revolution" in India. The rice yield increased from 1,035 kg/ha in 1960–1961 to 3,943 kg/ha in 2004–2005, and the wheat yield increased from 1,237 to 4,221 kg/ha during the same period (Anonymous 2006a). Punjab contributes 45% of the rice and 65% of the wheat to the production of these grains in India. In addition, the state is a major producer of milk, eggs, honey, fish, sugarcane, and cotton (PAU 1998). It has earned the name of "food basket of the country" and "granary of India." Punjab produces 2% rice, 3% wheat, and 2% of cotton of the world's production (Anonymous 2006b).

Pesticide use is also high (923 g/ha) (Agnihotri 2000). In cotton production, 2.580 kg of pesticide per hectare is applied to transgenic varieties and 6.440 kg/ha to non-Bt varieties (Peshin et al. 2007). In cotton, pest problems continued to increase inexorably resulting in reduced cotton productivity. Productivity initially increased from 269 kg/ha in 1960–1961 (pre–Green Revolution period) to 371 kg/ha in 1970–1971 (Green Revolution period) to as high as 502 kg/ha in 1994–1995 (post–Green Revolution period). The increased productivity was possible through the adoption of hybrid cultivars of cotton and increased fertilizer use and pesticides (insecticide) in the early years of their adoption. In the pre–Green Revolution era, the estimates of yield losses caused by pests in cotton were 18% (Pradhan 1964), and this figure jumped to over 50% in the post–Green Revolution era (Dhaliwal et al. 2004). This was due to: (i) the emergence and development of new pests such as spotted bollworm (*Earias vittella*), American bollworm (*Helicoverpa armigera*), and tobacco caterpillar (*Spodoptera litura*), (ii) the evolution of resistance in *Helicoverpa armigera* to insecticides, (iii) the resurgence of whitefly (*Bemisia tabaci*), and (iv) pest outbreaks of *H. armigera* in 1978, 1983, 1990, 1995, 1997, 2001, *B. tabaci* in 1995, and *S. litura* in 2003 (Dhawan et al. 2004). The farmers were caught on a "pesticide treadmill." The cost percentage of insecticide to total cost of cultivation increased from 2.1% in 1974–1975, 4.6% in 1979–1980, 11.9% in 1984–1985, 15.5% in 1989–1990, and then decreased to 13% in 1994–1995 (Dhaliwal and Arora 2001). In 1997–1998, productivity decreased to 220 kg/ha, and in 1998–1999 reached an all time low of 179 kg/ha. At the same time, the cost of insecticides as a percentage of the cost of cotton production increased to 21.21% in 1998–1999 (Sen and Bhatia 2004), reaching an all time high (50%) in the "pesticide hotspots" of Punjab (Bhathinda district) (Shetty 2004). The development of pest resistance to insecticides resulted in crop failures, with the cost of insecticides exceeding the other costs of production in 1998–1999.

The overuse of pesticides in Punjab has resulted in a change in the pest scenario, as up to 1970, the major pests of cotton were jassid (*Amrasca biguttula*) and pink bollworm (*Pectinophora gossypiella*). There were no pest outbreaks at that time. In 2001–2003, the major pests reported were jassid (*Amrasca biguttula*), whitefly (*Bemisia* tabaci), American bollworm (*Helicoverpa armigera*), and spotted bollworm (*Earias vitella*). Outbreak of American bollworm was reported in 1978, 1983, 1990, 1995, 1997, 1998, and 2001 (Dhawan et al. 2004).

17.4.2 Integrated Pest Management in Cotton

To overcome the negative effects of pesticide overuse in Indian agriculture, especially in the high productivity zone of the Northwest and the coastal regions covering 103 districts, numerous IPM programs were initiated, especially in rice and cotton, which accounted for 67% of total pesticide use prior to the introduction of Bt cotton. The Central Institute for Cotton Research (CICR), Nagpur, India, implemented an insecticide resistance management–based IPM (IRMIPM) program in 10 cotton-growing states (including Punjab) of India. The IRM approach is based on the premise that unless full-fledged efforts to understand all aspects of the resistance phenomenon are made, any attempt to implement IPM at field level would not bear results (Bambawale et al. 2004). The main focus of IRM program is on rationalizing insecticide use in cotton in the absence of availability of any effective bio-agents; this is presented within the full IPM context.

But the use of pesticides by farmers in cotton according to correct dosages, right timing, and application technology is not up to the accepted norms (farmers either apply an under-dosage or over-dosage) (Table 17.2). The farmers also did not apply the same dosage of a particular insecticide throughout the cropping season of cotton crop; they varied the dosage according to the crop stage and used a lower concentration for controlling young larvae of American bollworm (*H. armigera*) and increased the dosage for grown-up larvae. Under the Insecticide Resistance Management (IRM) program to prevent the build-up of resistance against insecticides, endosulfan was the recommended insecticide against jassid (*Amrasca bigutula*) but the farmers were reluctant to use it, as they felt intoxicated after its spray application (Peshin 2009). The Excel pesticide company was selling endosulfan as an IPM-compatible pesticide. The farmers were ahead of the scientists, because they had real-life experiences of the adverse effects with the use of endosulfan. In May 2011, the Supreme Court of India banned the production and sale of endolsulfan in the country. From their experiences with excessive use of insecticides in cotton, the farmers were knowledgeable about the resistance in insect pests. In local language (Punjabi) they termed it *Amli* (meaning pests having got inured to pesticides). The reasons cited by the farmers for the reduced pesticide use efficacy in cotton were development of resistance in insect pests (57%), excessive use of insecticide (36%), over/under dosage of insecticides (21%), tank mixing of different insecticides (13%), climate change (13%), spray equipment and spray technique (1%), and higher *H. armigera* infestation (3%) (Peshin et al. 2007).

Table 17.2 The adoption of correct and incorrect dosages of insecticides in cotton in Punjab. (Source: Peshin 2009)

Insecticide	IRMIPM villages (% farmers)	Non-IRMIPM villages (% farmers)
Alphamethrin 10EC		
i. Correct dosage (250 ml/ha)	29[a]	9[a]
ii. Higher dosage	81[a]	100[a]
N*	83	54
Cypermethrin 10EC		
i. Lower dosage	15[a]	0
ii. Correct dosage (500 ml/ha)	80[a]	100
iii. Higher dosage	9[a]	0
N	46	20
Cypermethrin 25EC		
i. Lower dosage	15	0
ii. Correct dosage (200 ml/ha)	8	0
iii. Higher dosage	77	100
N	13	12
Deltamethrin 2.8EC		
i. Correct dosage (400 ml/ha)	55	83
ii. Higher dosage	45	17
N	11	6
Fenvalerate 20EC		
i. Correct dosage (250 ml/ha)	10[a]	0
ii. Higher dosage	93[a]	100
N	41	14
β-cyfluthrin 0.25EC		
i. Correct dosage (500 ml/ha)	0	–
ii. Higher dosage	100	–
N	2	0
Lambda cyhalothrin 5EC[b]		
i. 1.200 ml/ha	100	100
N	3	1
Acephate 75SP		
i. Lower dosage	31[a]	52[a]
ii. Correct dosage (2 l/ha)	76[a]	57[a]
iii. Higher dosage	3[a]	5[a]
N	86	21
Chlorpyriphos 20EC		
i. Lower dosage	53[a]	54
ii. Correct dosage (5 l/ha)	51[a]	46
N	89	37
Dimethoate 30EC		
i. Correct dosage (625 ml/ha)	0	0
ii. Higher dosage	100	100
N	3	4
Ethion 50EC		
i. Lower dosage	40[a]	38[a]
ii. Correct dosage (2 l/ha)	67[a]	64[a]
iii. Higher dosage	5[a]	2[a]
N	92	55

Table 17.2 (continued)

Insecticide	IRMIPM villages (% farmers)	Non-IRMIPM villages (% farmers)
Monocrotophos 36SL[c]		
i. Lower dosage	22[a]	27
ii. Correct dosage (1.5 l/ha)	78[a]	46
iii. Higher dosage	11[a]	27
N	9	11
Profenophos 50EC		
i. Lower dosage	5	0
ii. Correct dosage (1.25 l/ha)	75	50[a]
iii. Higher dosage	20	67[a]
N	20	6
Quinalphos 25EC		
i. Lower dosage	16	0
ii. Correct dosage (2 l/ha)	75	100
iii. Higher dosage	9	0
N	32	5
Triazophos 40EC		
i. Lower dosage	40[a]	51[a]
ii. Correct dosage (1.5 l/ha)	64[a]	53[a]
iii. Higher dosage	18[a]	20[a]
N	121	59
Thiodicarb 75WP		
i. Correct dosage (625 ml/ha)	0	–
ii. Higher dosage	100	–
N	4	0
Endosulfan 35EC		
i. Lower dosage	35	20
ii. Correct dosage (2.5 l/ha)	58	60
iii. Higher dosage	7	20
N	57	15
Imidacloprid 17.8SL		
i. Correct dosage (100 ml/ha)	58[a]	63
ii. Higher dosage	50[a]	37
N	117	48
Acetamiprid 20SP		
i. Correct dosage (50 gm/ha)	11	7
ii. Higher dosage	89	93
N	46	15
Thiomethoxam 25WSC		
i. Lower dosage	4	0
ii. Correct dosage (100 gm/ha)	46	53
iii. Higher dosagne	50	47
N	24	19
Indoxacarb 15SC		
i. Lower dosage	4	6
ii. Correct dosage (500 ml/ha)	95	94
iii. Higher dosage	1	0
N	74	31

Table 17.2 (continued)

Insecticide	IRMIPM villages (% farmers)	Non-IRMIPM villages (% farmers)
Spinosad 48SC		
i. Lower dosage	8[a]	0
ii. Correct dosage (150 ml/ha)	33[a]	9
iii. Higher dosage	63[a]	91
N	52	22

– Decimals have been rounded up to nearest whole number
[a] Farmer applied different dosage of a particular insecticide for spraying on different occasions
[b] Not recommended by the Punjab Agricultural University
[c] Not recommended under IRM strategy
* N = The number of farmers out of a sample of 210 who have used a particular insecticide

17.5 Pesticide Use in the Colombian Andes[1]

17.5.1 Background and Research Problem

Human health and environmental effects of pesticide use are serious concerns among smallholder potato farmers in the Colombian Andes. Potato is one of the crops with the highest demand for fungicides and insecticides in Colombia (MADR 2006). The cultivation of potato is mainly concentrated in the Andean regions of Boyacá, Cundinamarca, and Nariño and is carried out by smallholders (MADR 2006). Smallholders in the region achieve an average yield of 14–15 t/ha, which has stayed constant in the last few decades (MADR 2006; Feola and Binder 2010c). Similar to many rural areas in the less developed countries, smallholders apply pesticides by means of a lever-operated knapsack sprayer and often wear inadequate personal protective equipment (PPE) (Cardenas et al. 2005; Ospina et al. 2008). Mostly carbamates (Carbofuran, Mancozeb, Methomyl), organophosphates (Metamidophos, Malathion), and pyrethroids (Cypermethrin) insecticides and fungicides are applied to the crop (details in Feola and Binder 2010c). In addition, smallholders in these regions were reported to overuse pesticides. Several studies showed that, as a consequence of such pesticide use practices, farmers in Boyacá and their environment are at risk because of exposure to pesticides (Leuenberger 2005; Cardenas et al. 2005; Ospina et al. 2008). Moreover, the negative economic consequences attracted the concern of governmental agencies; crop protection represents a significant share of the production costs for smallholders in this region (MADR 2001) and therefore more efficient pesticide use may not only reduce environmental and health risks, but also contribute to a more viable livelihood strategy.

Intervention programs in Boyacá often failed to achieve a durable and self-sustaining change from current pesticide use toward sustainable pesticide practices (e.g., Ospina et al. 2009). This is consistent with what has been observed in many

---
[1] An earlier and more extensive account of this research can be found in Feola and Binder 2010a, 2010b, 2010c, and Feola et al. 2012.

other contexts in poor countries (e.g., Orr 2003; Wyckhuys and O'Neill 2007)., and similar to those other contexts, this failure can be at least partially ascribed to the fact that policy-makers have only a limited understanding of how farmers conceptualize their SES and, consequently, of why farmers adopt certain pesticide use practices. Schöll and Binder (2009a, 2009b), for example, by using the structured mental model approach (Binder and Schoell 2010), showed that farmers and experts in Boyaca had divergent understandings of agricultural systems including the definition and importance of different capitals (i.e., human, physical, social, natural, and financial). Such a limited understanding does not translate into effective policies (Wyckhuys and O'Neill 2007). Therefore, sound knowledge was urgently needed to develop effective interventions for a transition toward a more sustainable pesticide use in Boyacá.

17.5.2 Goals

With reference to the study area of Vereda La Hoya in the region of Boyacá, this research aimed to: (i) uncover the behavioral dynamics underlying unsustainable pesticide use practices of smallholder potato farmers, and (ii) on this basis provide policy recommendations to foster a transition toward more sustainable pesticide use in this region.

17.5.3 Methods

The research was structured in three phases. Firstly, a theoretical framework was developed (see below) to allow for the understanding of farmers' behaviors as embedded in their specific SES (Feola and Binder 2010a). Secondly, data were collected through a survey ($N=210$) and statistical and econometric models of PPE and chemical pesticide use developed to identify influential factors and social dynamics (Feola and Binder 2010b, 2010c). Two practices were studied: PPE use and the chemical pesticide use. Finally, a dynamic behavioral model was developed and used to simulate alternative policies to achieve higher PPE use rates. This model was employed as a learning tool with local agriculture experts and policy-makers (Feola et al. 2012).

17.5.4 Theoretical Background

Most socio-psychological approaches to study farmers' behavior and decision-making fall short with respect to at least one of the following: (i) an explicit and well-motivated behavioral theory, (ii) an integrative approach, and (iii) understanding feedback processes and dynamics (Feola and Binder 2010a). The integrative agent-centered (IAC) framework (Feola and Binder 2010a), which was developed and ap-

17 From the Farmers' Perspective: Pesticide Use and Pest Control

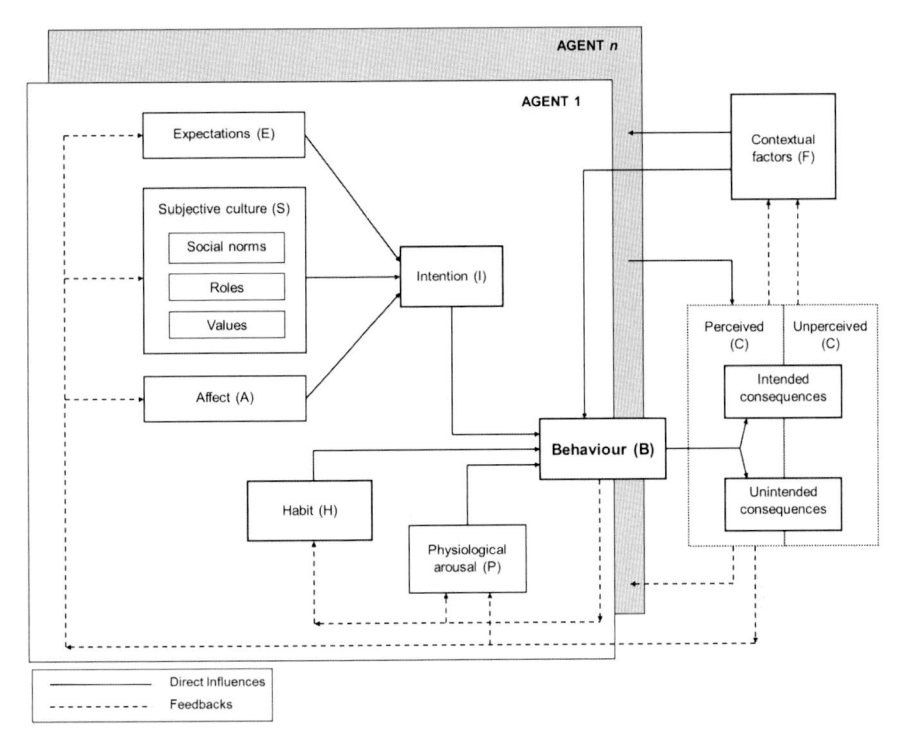

Fig. 17.4 The integrative agent centered (IAC) framework. The IAC framework provides a conceptual structure to understand social agents' behavior in their social–ecological systems by combining different behavioral drivers. It entails feedbacks and focuses on behavioral dynamics more than states and on the feedbacks among the determinants of a given behavior. (Source: Feola and Binder 2010a, with permission from Elsevier)

plied in this study, addresses simultaneously these three points and was developed to fill this gap. The IAC framework provides a conceptual structure to understand social agents' (i.e., farmers') behavior in their SES (i.e., agricultural systems).

The IAC framework is agent centered. It integrates and adapts Giddens' Structuration Theory (Giddens 1984) and Triandis' Theory of Interpersonal Behavior (Triandis 1980) to provide an understanding of farmers' behavior consistent with the perspective of agricultural systems as complex SES. It combines different behavioral drivers (i.e., rational expectations, subjective culture, affect, habit, and external factors) and, therefore, depicts a complex and potentially varied model of human behavior. It entails feedbacks, according to a circular, that is, systemic, conceptualization of human behavior. In addition, the IAC framework focuses on behavioral dynamics more than states and on the feedbacks among the determinants of a given behavior, and in particular between individual behavior and that of the system (Fig. 17.4).

In the framework (Fig. 17.1), an agent's (i.e., farmer) decision to enact a specific behavior (e.g., PPE use) is influenced by external and internal drivers. The former consists of contextual factors (i.e., facilitating conditions or barriers), whereas the latter includes habit (the frequency of past behavior), physiological arousal (the physiological state of the individual), and intention (Feola and Binder 2010a). The latter is determined by: (i) expectations (the beliefs about the outcomes, their probability, and their value), (ii) subjective culture (social norms, roles, and values), and (iii) affect (the feelings associated with the act). The behavior can have intended or unintended and perceived or unperceived consequences, which can feed back to the farmers. Only the perceived consequences, which are re-interpreted by the agent, feedback directly to farmers by influencing intention, affect, habit, and physiological arousal. The feedback processes can reinforce the current state or trigger change and can occur at different temporal levels (i.e., short- or long-term). Agents' interactions happen either directly or indirectly. The former depends on the agents' social network. The latter happens through the consequences of behavior, which can aggregate at the next highest hierarchical level, being perceived and reinterpreted by individual agents (Feola and Binder 2010a).

17.5.5 Results

With respect to the use of personal protective equipment (PPE), among the factors that influence this behavior, such as the cost of PPE and the ability to understand pesticide safety labels, there were two particularly important dynamics. Firstly, farmers tended to conform to the descriptive social norm, that is, the most common behavior observed in the peer group, thus reproducing the norm itself (reinforcing feedback; social level). Secondly, farmers tended to intermittently react to short-term pesticide-related adverse health effects by using more pieces of PPE more frequently, but disregarding PPE as the health effects loses relevance with time (balancing feedback; individual level). These behavioral dynamics were rendered, together with static factors, in the dynamic behavioral model that was used to simulate the effect of different combinations of policies on PPE use (Feola et al. 2012). The most effective simulated strategy was one that combined diversification of policies, long-term implementation, and intervention on structural aspects (i.e., descriptive social norm). Moreover, PPE use is influenced by the level of pesticide application (see below), farmers reacting to adverse health effects more frequently under more intense application levels.

Regarding the use of chemical pesticides, the results show that it is possible for smallholders in the region to achieve satisfactory productivity (average 13.6 t/ha) while applying insecticides and fungicides effectively, and consequently minimizing health and adverse environmental effects, and containing production costs (Feola and Binder 2010c). The analysis of the factors that influence farmers' pesticide use choices explains why the technical fix and approaches traditionally adopted by development agencies in the region might be bound to fail in Boyacá. These approaches focus on the short-term and assume the unsustainable practices are caused by a lack of knowledge. They do not address the specific social dynamics that induce ineffective pesticide use in the region, among which are conformity to social norms, market pressure for farmers to grow pest-vulnerable varieties, small parcels that hamper resource management, and the influence of pesticide producers and sellers on smallholders. Instead, the results suggest that a different approach is needed, in particular one that: (i) engages pesticide producers and sellers, (ii) facilitates new institutional settings such as farmer cooperatives, which support more efficient and less hazardous practices, and (iii) exploits social conformity in developing campaigns for sustainable practices (Feola and Binder 2010c).

17.6 Conclusions

We reviewed the potentials and limitations of different methodologies and approaches used in the literature to study pesticide use from the farmers' perspective. We contended that the reductionist paradigm's assumptions prevail in the current approaches and methodologies. This can result in creating an "unreal picture" of the farmers' perspective. In contrast with the narrow disciplinary approaches, we suggest adopting a more interdisciplinary approach with more potential to create a realistic and farmer-centered understanding of pesticide use.

Using three case studies from Iran, India, and Colombia, this approach was illustrated. In particular, drawing on studies currently available in the literature that look at pesticide use in agriculture from Iranian farmers' perspective, we found this area of scholarly research in nascent stages with a need for contributions from all relevant social scientists in an interdisciplinary and integrative way.

In addition, although there have been many efforts from both national and international supporters to encourage Iranian farmers to adopt safe use of pesticides and IPM practices, in practice many obstacles still prevent IPM from being a mainstream strategy for pest control in Iran. According to many studies currently available in the literature, Iranian farmers' attendance in educational courses on pesticide issues is highlighted as a critical need. Since such insights come from studies that are confined within narrow disciplinary boundaries, their recommendations may not be realistic enough when seen from a farmer's perspective. As such, other studies argue that Iranian farmers continue to use pesticides excessively and in an unsafe way even though they may be educated and aware of the hazardous effects of chemical pesticides. Their economic considerations and limited access to appropriate alternatives contribute crucially to choosing between pesticide products. Furthermore, farmers may not be interested in attending the classes provided by Iran's Ministry of Agriculture since they perceive that there is a wide gap between the "prescriptions" of the classes and the reality of their daily life. Even in cases where learning opportunities for farmers were provided in a more participatory and experiential way (FFS), some authors reported that Iranian farmers faced many obstacles such as lack of access to spraying tools and/or specific equipment needed to go through the pest management steps that they learned. The conclusion here is that there is a need to educate Iranian farmers about safe use of pesticides and other alternatives to pest control. We wish to suggest that this is not the only recommendation that needs to be made in every situation. This is consistent with results revealed in the case study conducted in the Colombian Andes which showed that more sustainable pesticide practices might result from diversified strategies.

The Indian case study showed that the farmers that have hands-on experience with pest management act rationally given their grasp of the relationship between cause and effect. Any IPM program and IPM technologies need to be modified by making farmers "partners" at the technology testing phase. Farmers' use of pesticides according to good agricultural practices is a complex technology. Researchers need to take into consideration farmers' perceptions about the technological attributes during the technology development process, rather than the technologists' predicting the adoptability in order to overcome innovation biases.

Finally, the Colombian case study illustrated how the IAC framework can be adopted to understand farmers' pesticide use practices, and thus help to define a policy agenda for triggering a transition toward a more sustainable pesticide use that goes beyond the search of "silver bullets" such as education. The IAC framework helps to understand the causes and meanings associated by farmers to selected pesticide use practices in the specific social and environmental context (i.e., social structures and the biophysical environment in SES) in which they take place, that is, the socially and environmentally adaptive value of those actions. It therefore also helps to overcome the rationality/irrationality discourse that often frames expert assessment of farmers' practices, such as not using PPE while applying chemical pesticides. It is on the basis of such a theory-based and integrative understanding that effective strategies and policies for a transition towards sustainable practices can be based.

Overall, this case study showed that while education and technological innovation are commonly claimed to be the way forward, more sustainable pesticide practices might result from different strategies. These include: (i) targeting the systemic processes which determine the actual social behavioral norms, (ii) diversification of measures to address different factors and processes co-influencing farmers, (iii) the involvement not only of farmers, but of other actors (e.g., pesticide producers) at the different levels of the agricultural system who influence farmers in symbolic and material ways, and (iv) strengthening institutional arrangements such as farmer cooperatives that scaffold best practices at the local level.

Acknowledgments The authors are grateful to Hossein Heidari, senior researcher at the Biological Control Research Department, Iranian Plant Pests and Diseases Research Institute, for providing us with some information needed for the case study "Pesticide use and IPM in Iran," funded by Tehran University Research Division. The research presented in the case study "Pesticide use in the Colombian Andes" was carried out within the project "Reducing human health and environmental risks from pesticide use: Integrating decision-making with spatially-explicit risk assessment models. The case of Vereda La Hoya, Colombia" led by Prof. C. R. Binder at the University of Zurich, and funded by the Swiss National Science Foundation.

References

Aghilinegad, M., Mohamadi, S. A., & Farshad, S. (2008). Impact of the pesticides consumption on agricultural health. *Research Journal of Shahid Beheshti University of Medical Science, 31*(4), 327–331 (in Farsi).

Aghasi, M., Hashim, Z., Moin, S., Omar, D., & Mehrabani, M. (2010). Socio-demographic characteristics and safety practices in pesticide applicators in Zangiabad area, Iran. *Australian Journal of Basic and Applied Sciences, 4*(11), 5689–5696.

Agnihotri, N. P. (2000). Pesticide consumption in India—An update. *Pesticide Research Journal, 12,* 150–155.

Allahyari, M. S., Chizari, M., & Homaee, M. (2008). Perceptions of Iranian agricultural extension professionals toward sustainable agriculture concepts. *Journal of Agriculture and Social Sciences, 4,* 101–106.

Anonymous. (2006a). *The Hindu: Survey of Indian Agriculture*. Chennai, India: National Press.

Anonymous. (2006b). *Agriculture at a Glance: 2005–2006*. Information Service, Punjab, India: Department of Agriculture.

Arjmandi, R., Heidari, A., Moharamnejad, N., Nouri, J., & Koushiar, G. (2012). Comprehensive survey of the present status of environmental management of pesticides consumption in rice paddies. *Journal of Pesticide Science, 37,* 69–75.

Atreya, K., Sitaula, B. K., & Bajracharya, R. M. (2012). Pesticide use in agriculture: The philosophy, complexities and opportunities. *Scientific Research and Essays, 7*(25), 2168–2173.

Bambawale, O. M., Patil, S. B., Sharma, O. P., & Tanwar, R. K. (2004). Cotton IPM at cross roads. In *Proceeding of International Symposium on Strategies for Sustainable Cotton Production: A Global Vision* (pp. 33–36). Dharwad, Karnataka, India: UAS.

Binder, C. R., & Schoell, R. (2010). Structured mental model approach for analyzing perception of risks to rural livelihood in developing countries. *Sustainability, 2,* 1–29.

Bruin, G., & Meerman, F. (2001). *New Ways of Developing Agricultural Technologies: The Zanzibar Experience with Participatory Integrated Pest Management*. Wageningen, Netherlands: Wageningen University.

Cardenas, O., Silva, E., Morales, L., & Ortiz, J. (2005). Estudio epidemiológico de exposición a plaguicidas organofosforados y carbamatos en siete departamentos colombianos, 1998–2001. *Biomédica, 25*(2), 170–180 (in Spanish).

Costanza, R., & Kubiszewski, I. (2012). The authorship structure of "ecosystem services" as a transdisciplinary field of scholarship. *Ecosystem Services, 1*(1), 16–25.

Damalas, C. A. (2009). Understanding benefits and risks of pesticide use. *Scientific Research Essays, 4,* 945–949.

Dehghani, R., Moosavi, S. G., Esalmi, H., Mohammadi, M., Jalali, Z., & Zamini, N. (2011). Surveying of pesticides commonly on the markets of Iran in 2009. *Journal of Environmental Protection, 2,* 1113–1117.

Dhaliwal, G. S., & Arora, R. (2001). *Integrated Pest Management: Concepts and Approaches*. Ludhiana, Punjab, India: Kalyani Publishers.

Dhaliwal, G. S., Arora, R., & Dhawan, A. K. (2004). Crop losses due to insect pests in Indian agriculture: An update. *Indian Journal of Ecology, 31*(1), 1–7.

Dhawan, A. K., & Peshin, R. (2009). Integrated pest management: Concept, opportunities and challenges. In P. Peshin & A. K. Dhawan (Eds.), *Integrated Pest Management: Innovation Development Process* (Vol. 1, pp. 51–82). Dordrecht, Netherlands: Springer.

Dhawan, A. K., Sohi, A. S., Sharma, M., Kumar, T., & Brar, K. S. (2004). *Insecticide Resistance Management on Cotton*. Ludhiana, Punjab, India: Punjab Agricultural University.

Ecobichon, D. J. (2001). Pesticide use in developing countries. *Toxicology, 160,* 27–33.

Evans, F. C. (1951). Symposium on viewpoints, problems, and methods of research in urban areas biology and urban areal research. *The Scientific Monthly, July,* 37–38.

Fathi, H., Heidari, H., Impiglia, A., & Fredrix, M. (2012). *History of IPM/FFS in Iran*. Tehran: IRIPP Publishers.

Feola, G., & Binder, C. R. (2010a). Towards an improved understanding of farmers' behaviour: The integrative agent-centred (IAC) framework. *Ecological Economics, 69*(12), 2323–2333.

Feola, G., & Binder, C. R. (2010b). Why don't pesticide applicators protect themselves? Exploring the use of personal protecting equipment among Colombian smallholders. *International Journal of Occupational and Environmental Health, 16*, 11–23.

Feola, G., & Binder, C. R. (2010c). Identifying and investigating pesticide application types to promote a more sustainable pesticide use. The case of smallholders in Boyacá, Colombia. *Crop Protection, 29*, 612–622.

Feola, G., Gallati, J. A., & Binder, C. R. (2012). Exploring behavioural change through an agent-oriented system dynamics model: The use of personal protective equipment among pesticide applicators in Colombia. *System Dynamics Review, 28*(1), 69–93.

Ghasemi, S., & Karami, E. (2009). Attitudes and behaviors about pesticides use among greenhouse workers in Fars Province. *Journal of Economics and Agricultural Development, 23*, 28–40 (in Farsi).

Giddens, A. (1984). *The Constitution of Society*. Cambridge, United Kingdom: Polity Press.

Hashemi, S. M., & Damalas, C. A. (2011). Farmers' perceptions of pesticide efficacy: Reflections on the importance of pest management practices adoption. *Journal of Sustainable Agriculture, 35*, 1–17.

Hashemi, S. M., & Hedjazi, Y. (2011). Factors affecting members' evaluation of agri-business ventures' effectiveness. *Evaluation and Program Planning, 34*(1), 51–59.

Hashemi, S. M., Mokhtarnia, M., Erbaugh, J. M., & Asadi, A. (2008). Potential of extension workshops to change farmers' knowledge and awareness of IPM. *Science of the Total Environment, 407*, 84–88.

Hashemi, S. M., Hosseini, S. M., & Damalas, C. A. (2009). Farmers' competence and training needs on pest management practices: Participation in extension workshops. *Crop Protection, 28*, 934–939.

Hashemi, S. M., Hosseini, S. M., & Hashemi, M. K. (2012a). Farmers' perceptions of safe use of pesticides: Determinants and training needs. *International Archives of Occupational and Environmental Health, 85*, 57–66.

Hashemi, S. M., Rostami, R., Hashemi, M. K., & Damalas, C. A. (2012b). Pesticide use and risk perceptions among farmers in southwest Iran. *Human and Ecological Risk Assessment, 18*(2), 456–470.

Heidari, H. (2006). *Farmers' Empowerment to Sustainable Agro-Ecosystem Management in Rice Fields Mazandarn Province, Iran*. Tehran: Barge Zeitoon Publishers

Heidari, H., Impiglia, A., Daraie, L., & Mirzaie, F. (2007). Farmer field schools deliver results in Iran (Integrated Pest Management). *Pesticides News, 76*, 8–10.

Jeffrey, P. (2003). Smoothing the waters. *Social Studies of Science, 33*(4), 539.

Jeyaratnam, J., & Chia, K. S. (1994). *Occupational Health in National Development*. Singapore: World Scientific Publishing Company Incorporated.

Kalaugher, E., Bornman, J. F., Clark, A., & Beukes, P. (2012). An integrated biophysical and socio-economic framework for analysis of climate change adaptation strategies: The case of a New Zealand dairy farming system. *Environmental Modelling & Software* (in Press).

Karamidehkordi, E., & Hashemi, A. (2010). Farmers' knowledge of IPM: A case study in the Zanjan Province in Iran. *Paper Presented in the Innovation and Sustainable Development in Agriculture and Food Symposium, ISDA 2010*. Montpellier, France. June 28–July 1.

Kesavachandran, C. N., Fareed, M., Pathak, M. K., Bihari, V., Mathur, N., & Srivastava, A. K. (2009). Adverse health effects of pesticides in agrarian populations of developing countries. *Reviews of Environmental Contamination and Toxicology, 200*, 33–52.

Leuenberger, M. (2005). *Environmental and health risk assessment of cultivation strategies in Tunja, Columbia*. Diploma thesis, Natural and Social Sciences Interface, ETH, Zurich, Switzerland

Litsinger, J. A., Libetario, E. M., & Canapi, B. L. (2009). Eliciting farmers knowledge, attitudes, and practices in the development of integrated pest management programs for rice in Asia. In R. Peshin & A. K. Dhawan (Eds.)., *Integrated Pest Management: Dissemination and Impact* (Vol. 2, pp. 119–274). Berlin: Springer.

Miller, T. R., Baird, T. D., Littlefield, C. M., Kofinas III, G., Chapin, F. S., & Redman, C. L. (2008). Epistemological pluralism: Reorganizing interdisciplinary research. *Ecology and Society, 13*(2), 46.

Ministerio de Agricultura y Desarrollo Rural (MADR). (2001). *Acuerdo de competitividad de papa en el Departamento de Nariño*, Ministerio de Agricultura y Desarrollo Rural, Bogotá, Colombia.

Ministerio de Agricultura y Desarrollo Rural (MADR). (2006). *La cadena de la papa en Colombia: Una Mirada global de suestructura y dinámica 1991–2005*. Ministerio de Agricultura y Desarrollo Rural, Bogotá, Colombia.

Nataatmadja, H., Tjitropranoto, P., Bernsten, R. H., Bagyo, A. S., & Hurun, A. M. (1979). Constraints to high yields in Subang, West Java, Indonesia. In *Farm Level Constraints to High Rice Yields in Asia* (pp. 97–121). Los Baños Laguna, Philippines: IRRI.

Norouzian, M. (2000). *List of Permissible Pesticides in Iran*. Tehran: IRIPP Publishers.

Oerke, E. C., & Dehne, H. W. (2004). Safeguarding production–losses in major crops and the role of crop protection. *Crop Protection, 23,* 275–285.

Orr, A. (2003). Integrated pest management for resource-poor African farmers: Is the emperor naked? *World Development, 31*(5), 831–845.

Ospina, J. M., Manrique, F. G., & Ariza, N. E. (2008). Salud, ambiente y trabajo en poblacionesvulnerables: los cultivadores de papa en el centro de Boyacá. *Revista de la Facultad Nacional de Salud Publica, 26*(2), 143–152 (in Spanish).

Ospina, J. M., Manrique-Abril, F. G., & Ariza, N. E. (2009). Intervención Educativasobre los Conocimientos y Practicas Referidas a los Riesgos Laborales en Cultivadores de Papa en Boyacá, Colombia. *Revista de Salud Pública, 11*(2), 182–190 (in Spanish).

PAU. (1998). *Punjab Agriculture 2020: Farmers and Farming in Punjab*. Ludhiana, Punjab, India: Punjab Agricultural University.

Peshin, R. (2009). *Evaluation of Insecticide Resistance Management Program: Theory and Practice*. New Delhi: Daya Publishing House.

Peshin, R., & Kalra, R. (1998). Integrated pest management at farmers' level. *Man and Development, 22,* 137–141.

Peshin, R., Dhawan, A. K., Vatta, K., & Singh, K. (2007). Attributes and socio-economic dynamics of adopting Bt-cotton. *Economic and Political Weekly, 42*(52), 73–80.

Peshin, R., Bandral, R. S., Zhang, W. J., Wilson, L., & Dhawan, A. K. (2009a). Integrated pest management: A global overview of history, programs and adoption. In R. Peshin & A. K. Dhawan (Eds.)., *Integrated Pest Management: Innovation-Development Process* (Vol. 1, pp. 1–50). Dordrecht, Netherlands: Springer.

Peshin, R., Dhawan, A. K., Kranthi, K. R., & Singh, K. (2009b). Evaluation of the benefits of an insecticide resistance management programme in Punjab in India. *International Journal of Pest Management, 55*(3), 207–220.

Peshin, R., Dhwan, A. K., Singh, K., & Sharma, R. (2012). Farmers' perceived constraints in the uptake of integrated pest management practices in cotton crop. *Indian Journal of Ecology, 39*(1), 123–130.

Pimentel, D. (2009). Pesticides and pest control. In R. Peshin & A. K. Dhawan (Eds.)., *Integrated Pest Management: Innovation-Development Process* (Vol. 1, pp. 83–88). Dordrecht, Netherlands: Springer.

Pradhan, S. (1964). Assessment of losses caused by insect pests of crops and estimation of insect population. In N. C. Pant, (Ed.)., *Entomology in India*, (pp. 17–58). Entomological Society of India, New Delhi.

Roger, E. M. (2003). *Diffusion of innovations* (5th ed.). New York: The Free Press.

Samiee, A., Rezvanfar, A., & Faham, E. (2009). Factors influencing the adoption of integrated pest management (IPM) by wheat growers in Varamin County, Iran. *African Journal of Agricultural Research, 4,* 491–497.

Schöll, R., & Binder, C. R. (2009a). System perspectives of experts and farmers regarding the role of livelihood assets in risk perception: Results from the structured mental model approach. *Risk Analysis, 29*(2), 205–222.

Schöll, R., & Binder, C. R. (2009b). Comparing system visions of farmers and experts. *Futures, 41*(9), 631–649.

Sen, A., & Bhatia, M. S. (2004). *State of the Indian Farmer: A Millennium Study-cost of Cultivation and Farm Income* (Vol. 14). New Delhi: Ministry of Agriculture, Government of India.

Shafiee, F., Rezvanfar, A., & Hashemi, F. (2012). Vegetable growers in southern Tehran, Iran: Pesticides types, poisoning symptoms, attitudes towards pesticide-specific issues and environmental safety. *African Journal of Agricultural Research, 7*(5), 790–796.

Sharma, R. (2011). *Evaluation of the integrated pest management farmer field school program in vegetable crops of Jammu region.* Ph.D dissertation, Sher-e-Kashmir University of Agricultural Sciences and Technology of Jammu, India.

Shetty, P. K. (2004). Social-ecological implications of pesticide use in India. *Economic and Political Weekly, 39,* 5261–5267.

Shahbazi, A., Bahramifar, N., & Smolders, E. (2012). Elevated concentrations of pesticides and PCBs in soils at the Southern Caspian Sea (Iran) are related to land use. *Soil and Sediment Contamination: An International Journal, 21*(2), 160–175.

Triandis, H. C. (1980). Values, attitudes and interpersonal behavior. In H. E. Howe Jr., & M. M. Page (Eds.)., *Nebraska Symposium on Motivation, 1979* (Vol. 27, pp. 195–259). Lincoln, Nebraska, USA: University of Nebraska Press.

van Huis, A. (2009). Challenges of integrated pest management in sub-Saharan Africa. In R. Peshin & A. K. Dhawan (Eds.)., *Integrated Pest Management: Dissemination and Impact* (Vol.2, pp. 395–417). Dordrecht: Springer.

Veisi, H. (2012). Exploring the determinants of adoption behaviour of clean technologies in agriculture: A case of integrated pest management. *Asian Journal of Technology Innovation, 20*(1), 67–82.

Wohl, R. (1955). Some observations on the social organization of interdisciplinary social science research. *Social Forces, 33,* 374–383.

Wyckhuys, K. A. G., & O'Neill, R. J. (2007). Local agro-ecological knowledge and its relationship to farmers' pest management decision making in rural Honduras. *Agriculture and Human Values, 24*(3), 307–321.

Yousefi, Z. (2008). A survey of pesticide toxicant utilization from Caspian Sea banks, Mazandaran Province, Northern Iran. *Environmental Justice, 1,* 101–106.

Chapter 18
Evaluation of Integrated Pest Management Interventions: Challenges and Alternatives

K. S. U. Jayaratne

Contents

- 18.1 Introduction ... 434
  - 18.1.1 Historical Background of IPM Interventions and Evaluation ... 435
  - 18.1.2 Current Context of IPM Evaluation ... 436
  - 18.1.3 Role of Evaluation in IPM Interventions ... 437
  - 18.1.4 Purpose of the Chapter ... 439
- 18.2 Issues of Evaluating IPM Interventions ... 439
  - 18.2.1 Lack of Theoretical Foundation for IPM Evaluation ... 440
  - 18.2.2 Use of Different Definitions for IPM ... 440
  - 18.2.3 Lack of Attention to Integrated Pest Management Evaluation ... 441
  - 18.2.4 Inadequate Allocation of Resources for Integrated Pest Management Program Evaluation ... 442
  - 18.2.5 Lack of Program Evaluation Knowledge ... 442
  - 18.2.6 Lack of Attention to Determining Evaluation Needs of Integrated Pest Management Stakeholders ... 443
  - 18.2.7 Lack of Utilizing Integrated Pest Management Evaluation Results ... 444
- 18.3 Challenges of Evaluating Integrated Pest Management Programs ... 444
  - 18.3.1 Planning Evaluation with Various Stakeholders ... 445
  - 18.3.2 Selection of Impact Indicators for Integrated Pest Management Evaluation ... 445
  - 18.3.3 Documentation of Long-term Impacts of Integrated Pest Management Interventions ... 446
  - 18.3.4 Managing to Conduct Integrated Pest Management Evaluations with Limited Resources ... 447
  - 18.3.5 Limited Knowledge Base for Evaluating Integrated Pest Management Interventions ... 447
  - 18.3.6 Empowering Integrated Pest Management Stakeholders to Evaluate their Programs ... 448
- 18.4 Alternatives for the Challenges and Issues of Integrated Pest Management Program Evaluations ... 448
  - 18.4.1 Stakeholder Identification and Coordination of Evaluation ... 448

K. S. U. Jayaratne (✉)
Department of Agricultural and Extension Education, North Carolina State University,
214 Ricks Hall, P.O. Box 7607, Raleigh, NC 27695, USA
e-mail: jay_jayaratne@ncsu.edu

18.4.2	Empowering Key Stakeholders and the Community Members to Conduct Evaluations	449
18.4.3	Identification of Evaluation Needs	450
18.4.4	Use of the Logic Model to Focus Evaluation	451
18.4.5	Use of an Evaluation Model to Conceptualize the Evaluation Holistically	453
18.4.6	Designs of Evaluation Studies Appropriate for Integrated Pest Management Program Assessments	462
18.4.7	Development of Evaluation Tools to Collect Data	463
18.4.8	Collecting, Analyzing, and Reporting Data	464
18.4.9	Utilization of Evaluation Results	464
18.5	Conclusion of Evaluating Integrated Pest Management Interventions	466
References		468

Abstract If properly conducted, evaluation can play a significant role in Integrated Pest Management (IPM) programming. Some issues and challenges associated with IPM program evaluation undermine the potential role evaluation can play in IPM programming. The major issues include lack of attention on the evaluation of IPM programs; inadequate resource allocation for evaluation; lack of evaluation knowledge among the stakeholders; lack of attention to identify the evaluation needs of the stakeholders; underutilization of evaluation results; lack of an IPM evaluation theory base; and the use of diverse definitions for IPM programs making it difficult to focus evaluations. The major challenges undermining the implementation of IPM evaluation are planning evaluations with various stakeholders of IPM; selection of impact indicators to reflect the broad benefits of IPM programs; documentation of long-term outcomes; conducting evaluations with limited resources; and empowering stakeholders to conduct evaluations. These issues and challenges are reviewed with the intention of finding alternatives for the development of IPM program evaluation as a useful tool. Stakeholder identification, incorporation of their evaluation needs, and empowering them to conduct evaluations; focusing evaluations by using logic models; application of an appropriate evaluation model, approach, and design to plan evaluations; and utilization of evaluation results for program improvement, accountability, marketing, advocacy, and policymaking are discussed as practical alternatives to improve the current situation of IPM evaluation.

Keywords IPM Program · Outcomes · Evaluation · Challenges · Alternatives

18.1 Introduction

Integrated pest management (IPM) as developed over the past four decades can be described as the practical application of all available pest control methods complimentary to each other for managing pests economically, environmentally, and socially in a desirable manner. Researchers and extension educators have made efforts to diffuse IPM technology as an effective strategy for controlling pests. Governments and international organizations such as the Food and Agriculture Organization of the United Nations (FAO) have taken initiatives to promote IPM as an effective strategy to control pests. With all of these efforts, IPM programming has become a valued strategy for combating pest problems worldwide. Over the last four decades, much of the attention has been paid to conducting IPM programs, evaluation was not considered an integral part of conducting IPM programs until very recently. Due attention and resources were not allocated for the evaluation of IPM programs. As a result, the IPM evaluation knowledge base is still in an early stage of development. Attention is needed to develop the theoretical foundation of IPM program evaluation. This chapter is intended to fill this knowledge gap by reviewing the major challenges and issues of evaluating IPM programs and formulating alternatives to improve the quality of IPM program evaluation.

18.1.1 Historical Background of IPM Interventions and Evaluation

After World War II, with the introduction of synthetic pesticides, chemical methods became the mainstream pest control method. Dichloro- diphenyl –trichloro-ethane, DDT, was widely used to control the malaria mosquito throughout the world. It took some time to realize the system-wide environmental pollution caused by DDT around the world. The Green Revolution pioneered by Norman Borlaug led to the breeding of high yielding crops resistant to insect pests and diseases. This plant breeding research opened a new approach to pest control; plant breeders used selected resistant lines and high yielding lines to build promising varieties resistant to pests. Plant breeding results of the Green Revolution rapidly reached many parts of the world during 1970s. Most of the farmland was used for these new varieties of crops. Farmers in America, Europe, and Asia were able to derive the benefits of the Green Revolution. Large tracts of farmland grew these new varieties as mono crops creating agro-ecological changes. Newly developed varieties performed well against pests for some time until new pests emerged threatening food production and farmers were compelled to use chemicals for controlling pests. Chemical companies promoted their products as effective means to control emerging new pests. Chemical control methods became the mainstream pest control method in many parts of the world causing environmental, economical, and social problems. Rachel Carson's book *Silent Spring*, published in 1962 initiated a public dialogue leading to the realization of the need to find alternative approaches to chemical pest control methods. This new movement encouraged scientists to develop IPM technology as an alternative approach to chemical pest control methods.

The concept 'integrated pest management' evolved over a period of time. This term and the acronym IPM did not appear in the literature until 1972 (Kogan 1998). IPM has become an important focus of research and extension efforts worldwide during the next 40 years (Ehler 2006). In the late 1970s, south Asian countries introduced the IPM approach to pest control in rice cultivation. Research and extension services worldwide promoted IPM. In the global context of the IPM movement, it is important to review the current status of IPM program evaluation to understand the assessment challenges and to find alternatives.

IPM program evaluation started in the United States (U.S.). The U.S. government Performance and Results Act of 1993 (GPRA 1993) enacted the requirement to report performance results of government funded programs. Most IPM programs were publicly funded and were required to report results under the GPRA. Researchers and extension educators started to evaluate IPM programs with the enactment of GPRA. Their major emphasis was to evaluate IPM program impacts and report in order to meet accountability requirements, but did not evaluate IPM programs holistically for program improvement. Many promoting IPM do not consider evaluation as an ongoing part of an IPM program implementation (Waibel 1999). Few studies have focused on the evaluation of the IPM implementation process (Wearing 1988; Whalon and Croft 1984). The evaluation of the IPM program process is important to identify implementation barriers and find alternatives.

18.1.2 Current Context of IPM Evaluation

Extension services around the world have promoted IPM programs since early 1980s. These IPM extension programs mainly focused on control methods such as biological control methods, the use of resistant varieties, agronomic practices, and mechanical methods, as alternatives to chemical control methods and emphasized the use of chemical control methods as the last resort for controlling pests. Due to this focus on reducing the use of chemicals for controlling pests, the effectiveness of IPM programs was assessed by estimating the cost savings on chemical pesticides used or the amount of pesticides applied. Traditionally, IPM evaluations focused mainly on a cost-benefit analysis. However, a cost-benefit analysis is inadequate to demonstrate the broad outcomes of IPM (Waibel et al. 1999). The focus for evaluating the effectiveness of IPM programs is still based on the reducing the amount of chemical pesticides applied (Bajwa and Kogan 2003; Horne and Page 2009; Olson et al. 2003). Understanding the reasons for the evaluation of IPM programs using this singular impact indicator is important for the development of effective evaluation approaches to document broad impacts of IPM programs.

The IPM program is a multi-faceted, systemic approach to managing pests economically, environmentally, and in a socially desirable manner. IPM as a systemic program means it approaches pest management as managing an ecosystem with minimal disturbance to maintain the natural balance of pests at economic threshold levels. IPM being multi-faceted implies it has economic, environmental, and social dimensions. Implementation of an IPM program encompasses many players at various levels. This multi-layer implementation includes farmers at the field level; extension educators and business representatives at the local level; extension specialists, extension administrators, researchers at the regional level; and policy makers at the state and national levels. Each of these players has a role to ensure the successful implementation of IPM programs. Even at the field level, farmers in a village are expected to follow IPM methods collectively in order to derive the best results. Because IPM is a system-wide pest management method and farms in a geographic area constitute the agricultural ecosystem, Kogan (1998) described three levels of implementing IPM by taking this agricultural system concept into account. The first level is at the crop field level targeting control strategies for a single species. The second level is at the crop community level targeting multiple pest interactions and the IPM control tactics. The third level is at regional agro-ecosystems level targeting multiple pests and their control agents with agro-ecosystems. Thus, the implementation of IPM programs involves various players at various levels and leads to the creation of economic, environmental, and social outcomes and impacts. This situation highlights the complexity of IPM program implementation and the need for reexamination of IPM program evaluations to understand assessment challenges and to find alternatives.

The potential beneficiaries of IPM programs include farmers, consumers, and many others in society. They derive diverse benefits from IPM programs which, unfortunately, are underreported (CIP 2010). Current evaluation approaches need to develop appropriate strategies for documenting the broad impacts of IPM programs. Until the full spectrum of IPM program benefits are assessed and documented, the general public and IPM's sources of public funding will not be fully aware of the real value of IPM programs. Public awareness is a prerequisite to gaining public as well as political support for IPM programs. Narrowly focused IPM evaluation on reduced levels of agrochemical usage undermines the reality of the broad benefits of IPM programs to society. Such results as currently reported may not address issues of value to the public. The evaluation of an IPM program should be capable of reflecting the reality of the economic, environmental, and social benefits of IPM programs to the society at large. How to conceptualize evaluations useful for program improvement and document broad outcomes of IPM programs is the focus of this chapter.

18.1.3 Role of Evaluation in IPM Interventions

The purpose of any program evaluation is to provide needed information for stakeholders (def. stakeholder—any individual who is interested in, benefits from, impacted by or involved in the IPM program) to make informed decisions about the program. The main role of evaluation is to contribute to IPM program improvement, not just to justify continued funding. The role of evaluation in analyzing the program context, resource allocation, the implementation process, and program outcomes is discussed in this section. Details of the application of these concepts in IPM evaluation will be discussed in Sect. 18.4.5. The results of these evaluations are useful for understanding the reality and effectiveness of IPM program planning, resource allocation, implementation, and outcomes.

The purpose of context evaluation is to help stakeholders understand the social, economic, and environmental factors justifying the IPM program in the given geographic area. Context evaluation finds answers to questions such as: What are the factors that led to the implementation of the particular IPM program in a given location? Was it due to a policy initiative at the national or state level? Was it due to research and extension programs in the area? Was it due to local needs and issues of pest management? Was it due to a combination of all or some of the above reasons? Finding answers to these questions is necessary to understand the reality that lead to the implementation of IPM programs in a given geographic location. Knowledge about the contextual situation is helpful in the planning and implementation of IPM programs tailored to the needs of a given situation. Some IPM programs focus on managing nonagricultural pests while most of the IPM programs are focused on managing agricultural pests. Some of the programs are top-down IPM program initiatives with policy directives while others are bottom-up, needs-driven initiatives. Some of the IPM programs are highly localized while others are regionally and/or nationally focused. Some IPM programs focus on a single crop while others have a multi-crop focus. Due to these complex foci of IPM programs, the context evaluation has a significant role to play in exploring the situation that lead to the creation of an IPM program in a given geographic location. Context evaluation is helpful for stakeholders to understand why the situation justified an IPM program in the given geographic location and align program objectives with the needs of the target audience.

The role of input or resource evaluation is to determine whether the best resource alternatives have been allocated for a given situation. The resources invested in IPM programs include financial resources, material resources, human resources, and time. Input evaluation is helpful for IPM stakeholders to determine whether the selected strategy is the best alternative in terms of using resources effectively for the given situation. The evaluation of the strategies used in the program is useful in maximizing the cost-effectiveness of IPM programming.

The process evaluation reveals what works and what does not work as planned. This information is helpful for finding alternatives and fixing problems. Process evaluation will prevent potential program failures. In this way, process evaluation can play a significant role in fine-tuning the program implementation process to achieve the desired cost effectiveness of IPM programming. The role of process evaluation of IPM programs is not yet up to the level it should be. Most of the focus of IPM program evaluation has been on outcome evaluation for the purpose of accountability.

The role of the outcome evaluation is to ascertain whether the IPM program is effective in terms of generating planned benefits to the target audience. Outcomes of IPM programs can be categorized as short-term outcomes, mid-term outcomes, and long-term outcomes. Long-term outcomes of IPM programs can be social, economic, and/or environmental benefits to the society. Some of the outcomes are tangibles and can be converted to monetary values easily. For example, prevented crop yield losses as a result of IPM can be estimated monetarily. However, some of the outcomes are indirect or intangible and difficult to convert into monetary values. For example, estimating the monetary value of improved biodiversity is challenging. Cost-benefit ratio is the most common outcome indicator used to evaluate outcomes of IPM programs. Reduced amount of chemicals used to control pests and the value of cost savings on pest control are other commonly used outcome

18 Evaluation of Integrated Pest Management Interventions

indicators. The impacts of IPM programs transcend the direct beneficiaries, such as farmers, and extend to others in society. Tracking these intangible impacts of IPM programs is critical to building public support for IPM programs. This can be achieved only by integrating evaluation concepts into IPM programming, allocating adequate resources to conduct systematic evaluations, and utilizing evaluation results effectively.

18.1.4 Purpose of the Chapter

Systematic evaluation provides answers to stakeholder questions enabling them to make informed decisions about their program. Some of the questions include:

- Why the IPM program is warranted for the given situation?
- What is the best alternative plan in terms of using resources effectively for the implementation of a warranted IPM program?
- Is the IPM program being implemented as planned? If it is not being implemented as planned, why not?
- What are the factors contributing to program improvement?
- What are the barriers hindering program implementation?
- Did the program generate planned outcomes and impacts? If so, what are those outcomes?
- How can the outcomes be attributed to the IPM program?
- How could the program be expanded?
- What can be done to sustain the IPM program as a solution to a given problem in a particular location?

Finding answers to these questions should be the focus of any systematic evaluation. The purpose of this chapter is to review IPM program evaluation issues and challenges and discuss alternatives for strengthening IPM evaluations.

18.2 Issues of Evaluating IPM Interventions

The review of the current situation of IPM program evaluations highlights the issues associated with IPM evaluations. The two major issues in evaluating IPM programs can be categorized as conceptual and practical. Lack of theoretical foundation for the evaluation of IPM programs and the use of diverse definitions for IPM are the major conceptual issues of evaluating IPM programs. Lack of attention to IPM evaluation; inadequate resources for evaluation; lack of evaluation capacity among IPM beneficiaries, extension educators, and researchers to conduct quality evaluations; lack of attention to determining evaluation needs of stakeholders; fragmented approach to evaluation; and lack of utilizing evaluation results can be considered as the major management issues of IPM program evaluation.

18.2.1 Lack of Theoretical Foundation for IPM Evaluation

Much of the IPM discussion is centered on program development and implementation. If there is any discussion about evaluation of IPM programs, it is primarily focused on evaluating the economic impacts of IPM programs which are reported as costs and benefits of IPM programs. Program costs and benefits are important parameters to measure the value of IPM programs but are not adequate to evaluate IPM programs holistically. Currently, IPM program evaluation theory is in an early stage of development and needs to begin with attention to the best approaches, models, and methods to conceptualize, plan, and implement IPM program evaluations and utilize evaluation results.

Contributions to the IPM evaluation theoretical foundation by emphasizing desirable approaches, models, and designs for the assessment of IPM programs are discussed in Sect. 18.4.

18.2.2 Use of Different Definitions for IPM

As Kogan (1998) described, the term Integrated Pest Management is often misinterpreted by professionals and laypeople without paying attention to what might be really meant by this term. IPM implementation is confused with the adoption of various definitions of IPM. IPM has many definitions in the literature (Ehler 2006). The Integrated Plant Protection Center (1996) at the Oregon University website lists 67 different definitions cited from the literature from 1959 to 1998. What IPM exactly means is still being debated (Sorby et al. 2003). This is because different groups define IPM differently depending on the context. The integrated use of pest control methods is first described as an alternative to chemical control methods by Stern et al. (1959). In the early days, IPM was described as the use of biological and other methods to reduce the dependency on chemical methods for pest control. With the dissemination of IPM as a pest management method in different parts of the world, different groups started to define it differently. These definitions have commonalities as well as some differences. Some of the definitions are narrowly focused while other definitions broadly encompass social, environmental, and economic aspects. Differences in IPM definitions may confuse IPM program planners as well as farmers and can make it difficult for someone to clearly understand the appropriate focus of IPM evaluations.

To address this issue, the Food and Agriculture Organization (FAO) of the United Nations appointed a panel of experts and adopted a broad definition in 1967. The FAO panel of experts used the term "integrated control" to describe IPM in 1967 and defined it as "a pest management system that, in the context of the associated environment and the population dynamics of the pest species, utilizes all suitable techniques and methods in as compatible a manner as possible, and maintains the pest populations at levels below those causing economic injury" (FAO 1967, p. 19). This definition has been widely cited in the literature (Kogan 1998). The current definition adopted by FAO describes IPM as "the careful consideration of all available pest control techniques and subsequent integration of appropriate measures that discourage the development of pest populations and keep pesticides and other interventions to levels that are economically justified and reduce or minimize risks to human health and the environment. IPM emphasizes the growth of a healthy crop with the least possible disruption to agro-ecosystems and encourages natural pest control mechanisms" (FAO 2012, p. 1). The current definition of FAO is broad and stresses the economic, social, and environmental aspects of pest control. An IPM program not based on a clear, broad definition may make the potential outcomes of the program difficult to conceptualize. The definition provides guidelines for program planners as well as evaluators to determine program boundaries and expectations. Therefore, it is important to follow a clear, broad definition for IPM for conceptualizing the program holistically and to plan evaluation approaches systematically.

18.2.3 Lack of Attention to Integrated Pest Management Evaluation

Until recently, IPM program planners mainly focused on program development and delivery and paid inadequate attention to program evaluation. Only limited evaluations have been conducted to determine the extent to which IPM programs have created sustainable results. Gathering evidence for the outcomes of IPM training workshops is not a regular practice (Charleston et al. 2011). Most of the IPM programs are funded through public funds. When public funds are shrinking, policy makers demand positive impacts of IPM programs to justify continuation and IPM program planners started to pay attention to program evaluation. The attention of IPM planners has mainly focused on assessing the outcomes of IPM programs to meet the expectations of policy makers such as reduced costs, increased profits, cost-benefit ratios, reduced levels of chemical pesticide use to showcase the value of IPM programs. As the FAO's IPM definition articulates, IPM is a systemic educational program and aims to impact the society, economy, and the environment. Therefore, it is important to determine whether the current impact indicators display the actual outcomes of IPM programming adequately. Beneficiaries of IPM programs are not limited to producers. Consumers are also an important beneficiary group of IPM programs. Current impact evaluations are mainly focused only on producers. The growing consumer demand for organically grown food is a result of food safety concerns over harmful residues of chemical pesticides. But little attention has been paid in evaluations to the benefits derived by the consumers from IPM programs. The National Roadmap for Integrated Pest Management in the U.S. stresses the need to evaluate national IPM programs by using broad performance measures including monetary, environmental, aesthetic, and health benefits (USDA 2004). However, review of the current evaluation practice of IPM programs reveals that insufficient attention has been paid to evaluating the IPM programming context and program implementation process. Without evaluating the implementation process and context of programming, outcome evaluation alone has little or no value for making programmatic decisions, especially program improvement decisions. This narrow focus on IPM outcome evaluation has contributed to overlooking the potential role that evaluation can play in program improvement and documentation of best practices. Therefore, IPM program evaluation has to be defined holistically using a systems approach at the IPM program planning stage to ensure the role that evaluation can and will play in educational programming. The roles evaluation can contribute to play are program improvement, accountability, marketing, decision making, and advocacy.

18.2.4 *Inadequate Allocation of Resources for Integrated Pest Management Program Evaluation*

Adequate resources are required to plan and implement systematic program evaluations. These resources include financial resources, human resources, material resources, and time. IPM program budgets are frequently prepared with little or no allocation for program evaluation because evaluation is not considered an important part of the program. Rather, program evaluation is an afterthought and conducted haphazardly with inadequate planning. IPM planners are not adequately aware of the role evaluation can play in program improvement and documentation of best implementation practices. However, policy makers' demands for IPM program results have prompted IPM program planners to begin to allocate some resources for program evaluation. Southern Regional Integrated Pest Management Competitive Grant Program in the U.S. allocated $ 1,580,000 for research and extension projects and $ 100,000 for IPM evaluation in 2012 (USDA 2012). These resources are mainly directed to evaluate the outcomes and impacts of IPM programs and address the 'so what?' question, namely what benefits do IPM programs provide to society? The use of evaluation as a tool for program improvement has not been considered when resources for IPM programs are initially allocated. When applying for funding, at least 10–15 % of the programming budget needs to be allocated to evaluate IPM programs systematically from the beginning of the program. Evaluation should be considered as an integral part of the IPM program and the required budget for evaluation should be built into the programming budget.

18.2.5 *Lack of Program Evaluation Knowledge*

Generally, farmers, extension educators, and researchers who plan and implement IPM programs do not have adequate knowledge and experience in program evaluation. Most of the time, IPM programs are delivered and no one has the knowledge to evaluate those programs (CIP 2010). Without adequate knowledge about IPM program evaluation, planning and conducting an evaluation systematically is difficult; this situation has become a major hindrance to IPM program evaluation.

IPM program planners do not have adequate training opportunities to develop their evaluation knowledge and skills due to a lack of resources available for evaluation training. When the extension educators who implement IPM programs do not have necessary evaluation knowledge and skills, their ability to conduct quality evaluations is limited. Therefore, building evaluation capacity among the farmers, extension educators, and researchers who conduct IPM programs is a necessary step to improve the quality of IPM evaluations. Understanding the need for evaluation training and providing learning opportunities as a part of IPM programming can address this issue if resources for this training are available.

18.2.6 Lack of Attention to Determining Evaluation Needs of Integrated Pest Management Stakeholders

The key stakeholder groups of IPM programs include policy makers, administrators, researchers, extension educators, agricultural industry workers, farmers, and consumers. These diverse groups will likely have different evaluation needs and expectations. Incorporation of the diverse needs of key stakeholder groups is important to ensure the usefulness of evaluations for these groups to make informed decisions about IPM programs. Incorporation of stakeholder needs into IPM evaluation will contribute to their support for evaluation and the effective use of evaluation results. Currently, no systematic approach exists to incorporate the evaluation needs of key stakeholders of IPM programs when planning program assessments. The IPM program evaluations are driven by the needs of policy makers and funding agencies most of the time. The evaluation needs of these two groups are mostly limited to program outcomes and impacts. These outcome evaluations are generally focused on the determination of the benefits derived by farmers from IPM programs but farmers are only one group of IPM program beneficiaries.

Consumers and others in the society are passive but important beneficiaries of IPM programs. Even though consumers are important beneficiaries, their needs are not incorporated into IPM programming or evaluation. Generally, IPM programs have overlooked the need for educating consumers about IPM programs and their impacts on consumers. As a result, consumers are not aware of the value of IPM programs and are unable to make informed decisions at the market place when buying agricultural commodities produced with IPM practices since IPM labeling on agricultural commodities is rare. Consumer resistance to pesticides and pesticide residues has become a growing issue due to increased health concerns (Hollingsworth and Coli 2004; Misra et al. 1991; Task Force 2003; Wearing 1988). If consumer needs are incorporated into IPM programming and evaluation, a consumer demand for agricultural commodities produced with IPM practices will develop, assuming that commodities have an IPM label. As a result, farmers will be able to ask a premium price for their agricultural commodities produced with IPM practices. If this situation can be created, it will be the best incentive for farmers to adopt and continue their IPM practices. Educating farmers as well as consumers is essential for promoting and sustaining IPM as an economically, socially, and environmentally desirable pest management strategy.

Identification and incorporation of the evaluation needs of various stakeholder groups of IPM programs is essential in order to plan an evaluation useful for these groups to make informed decisions relating to IPM programming. For example, policy makers may want to know what policies are working to promote IPM; administrators may want to know whether IPM programs are generating intended results for the funds spent; extension educators may want to know the best strategies to disseminate IPM programs; farmers may want to know whether IPM programs have increased their income and profit; consumers may want to know whether they receive any health benefits; and the community may want to know whether their environmental conditions have improved. This discussion highlights the need for planning an evaluation with the inputs from various stakeholders to ensure that their evaluation needs are met by a IPM program assessment.

18.2.7 Lack of Utilizing Integrated Pest Management Evaluation Results

Program evaluation is worthwhile and meaningful only if key stakeholders plan to utilize evaluation results for making programmatic decisions. Currently, evaluation results are used for the justification of resources spent in IPM programming. The use of evaluation results for improving programs is minimal. IPM programs can be improved by considering context, resources, and process evaluation results. Evaluation results can be used for IPM program improvement if the evaluation needs of the key stakeholders of IPM programs are incorporated into the evaluation plans; that adequate resources to conduct evaluations systematically are available; and evaluation results are shared with key stakeholders in a timely manner. These stakeholder concerns include but are not limited to program improvement, accountability, marketing, advocacy, and policy making. Utilization of evaluation results is helpful for preventing program failures and contributing to the maximization of cost effectiveness.

18.3 Challenges of Evaluating Integrated Pest Management Programs

IPM programs are complex in terms of the type of stakeholder groups involved, the variety of applications, and the diversity of outcomes and impacts. Evaluating diverse outcomes in various IPM programs to meet the specific information needs of diverse stakeholders is a challenging task. These challenges include planning evaluations with various stakeholder groups to meet their information needs; selection of outcome indicators to reflect the reality of the benefits of various IPM programs;

documentation of the long-term outcomes of IPM programs; meeting the evaluation needs of various stakeholder groups with limited resources available for IPM evaluations; planning and implementation of evaluation with a limited knowledge base available for IPM program assessment; and empowering grassroots level stakeholders to conduct evaluations. This section will review each of these challenges and discuss strategies to manage them.

18.3.1 Planning Evaluation with Various Stakeholders

There are multiple stakeholder groups in IPM programming. These stakeholders include but are not limited to farmers, consumers, extension educators, industry people, researchers, administrators, and policy makers. These diverse stakeholders have different responsibilities, expectations, and evaluation needs related to IPM programming. For example, farmers are responsible for implementing IPM practices at the field level and certainly want to know that IPM is saving them money and increasing farm profits. Extension educators are responsible for planning and implementing IPM programs at the community level and expect to determine the overall outcomes and ways to improve educational programming. Stakeholders' evaluation needs and expectations can vary with the context of the IPM program. For example, the context of the use of IPM as a strategy to overcome pesticide resistance in a rice farming area is different from using IPM as a solution to improve current environmental quality of a farming area where the pest resistance to chemical pesticides is not yet a problem. The context of the first scenario requires alternative pest control methods to combat the pest issue. The context of the second scenario does not require alternative pest control methods. These two scenarios represent two different contextual situations.

Due to the diversity of programming contexts, the challenge is to identify the evaluation needs and expectations of various stakeholder groups. However, understanding and incorporation of evaluation needs and expectations of the stakeholders of IPM programs are essential to ensure the usefulness of the evaluation for them. Stakeholder expectations and needs must be incorporated as a practical strategy to ensure stakeholders' active participation in evaluation of IPM programs and utilization of evaluation results.

18.3.2 Selection of Impact Indicators for Integrated Pest Management Evaluation

Impact indicators are reasonable and practical measurements used to gauge the value of program results. Selection of impact indicators to reflect the true nature of the results and benefits of IPM programs is challenging due to variety of outcomes and different evaluation expectations of stakeholders. IPM programs encompass economic, environmental, and social benefits. Some of the economic benefits include reduced cost of production; increased income and profits, and cost-benefit ratios. Some of the environmental benefits are reduced levels of environmental pollution due to reduced amount of chemical pesticides used; reduced incidence of pest outbreaks; improved quality of the environment such as water and air quality; increased bio-diversity; and sustainability of agro-eco systems. Some of the social benefits include farmers' increased knowledge about managing bio-diversity in agro-ecosystems; increased public interest to support IPM; establishment of favorable policies to manage agro-eco systems; reduced incidences of health hazards created by chemical pesticides, and increased awareness of consumers about the advantages of IPM. Documentation of all of these outcomes is a challenging task. Selection of a few of the highest priority and practical impact indicators to evaluate economic, environmental, and social outcomes is the best strategy to demonstrate the public value of IPM programs.

18.3.3 Documentation of Long-term Impacts of Integrated Pest Management Interventions

Depending on the time taken to manifest program results, IPM program outcomes can be categorized into three groups: short-term, medium-term, and long-term outcomes. Short-term outcomes of IPM programs can be observed generally in less than 3 months and often demonstrate learning achieved. Since IPM is a knowledge intensive pest management approach (Chung et al. 1999; SP-IPM 2008), the success of diffusing IPM technology depends on the effectiveness of the educational strategies used. If the educational strategies are effective, IPM program participants will gain new knowledge, change attitudes, acquire new skills, and aspire to adopt IPM as a practical approach to pest management. All of these are short-term outcomes. The IPM target audiences need time to understand IPM concepts, change their minds and attitudes about pest management, develop skills, and mentally assess pros and cons of IPM and decide to use IPM technology. That is why short-term outcomes will take about 3 months to manifest. These short-term outcomes are not tangible, but can be measured easily by using educational program evaluation methods to document changes in participants' knowledge, attitudes, skills, and aspirations related to IPM.

If the IPM educational efforts are effective, the target audience will apply what they have learned to manage pests in their fields. Some of these methods include the use of pest resistant varieties, field sanitation, crop rotations, the use of mechanical methods, crop scouting, and judicious application of selective chemical pesticides. If the IPM program is successful, application of these practices by IPM program participants can be observed as the medium-term outcomes of IPM programs. These outcomes take 3–12 months to manifest. Medium-term outcomes are observable and demonstrate practice changes implemented. For example, a number of farmers performing crop scouting can be observed and recorded.

If the IPM program participants are practicing integrated control methods to manage pests, then the long-term outcomes or program impacts will begin to manifest. Generally, the long-term outcomes will take more than 6–12 months to manifest. Some of these long-term outcomes include reduced cost of production, increased income, increased profits, reduced dependency on chemical pesticides, increased water quality, increased biodiversity in agricultural fields, fewer incidences of pest outbreaks, fewer incidences of increasing pest resistance to chemical pesticides, reduced incidence of health hazards caused by pesticides, increased community support for IPM programs, and increased consumer demand for the commodities produced with IPM practices. Some of these long-term outcomes are relatively easy to measure while others are more difficult to measure. For example, recording reduced cost of production is relatively easy compared to recording the increased quality of the environment. The real challenge with documentation of long-term outcomes is the separation of the contributions made by IPM programs from the contributions possibly attributable to other things happening in the community. For example, if the water quality is improving, it can be due to IPM program results as well as many other things happening in the community including environmental changes as well. When impact indicators are selected for long-term outcomes, it is important to consider the practicality and the relative strength of the indicator to use it for documenting impacts.

18.3.4 Managing to Conduct Integrated Pest Management Evaluations with Limited Resources

Resources available for IPM evaluations are very limited which makes conducting evaluations challenging. Expecting more resources for IPM program evaluation with contracting economies around the world is unrealistic. IPM evaluators should be prepared to face this reality with practical alternatives. Some of this work can be completed with community volunteers. Using volunteers and community resources can be considered as practical options to augment the resources available for evaluation.

18.3.5 Limited Knowledge Base for Evaluating Integrated Pest Management Interventions

Available literature for planning and conducting IPM program evaluations is limited. Building evaluation capacity among the IPM planners is a challenge when the available knowledge base for IPM evaluation is limited and has resulted in a lack of attention paid to IPM evaluation in the past. This chapter will contribute to filling the information gap of IPM evaluation.

18.3.6 Empowering Integrated Pest Management Stakeholders to Evaluate their Programs

IPM program planners, extension educators, and farmers often lack adequate knowledge to conduct quality evaluations. This lack of knowledge blocks the full potential of evaluation to be a practical tool for building effective IPM programs. For example, the use of process evaluation as a strategy to improve program implementation is often lacking since stakeholders lack knowledge about program evaluation. Empowering stakeholders to conduct evaluations includes increasing their knowledge and capacity to identify evaluation needs, organizing community resources to plan and conduct evaluations, and utilize evaluation results. Empowering stakeholders is difficult due to the limited knowledge base available for IPM evaluation and the lack of evaluation knowledge and capacity among the stakeholders, especially among the farmers, extension educators, and researchers. Building evaluation capacity among the farmers and extension educators is essential to improve the quality and frequency of evaluation because of their involvement in planning, conducting, and utilizing evaluations. Building evaluation capacity of farmers is very challenging due to their busy schedules. This situation indicates the need for conceptualizing practical educational approaches to empower grassroots level stakeholders of IPM programs.

Section 18.4 describes alternative strategies for the issues and challenges discussed in Sect. 18.2 and 18.3 of this chapter.

18.4 Alternatives for the Challenges and Issues of Integrated Pest Management Program Evaluations

Finding practical solutions as alternatives to challenges and issues of IPM evaluations is necessary to tap the full potential of evaluation as a useful tool for program improvement and accountability. This section discusses evaluation concepts with the intention of formulating alternatives to address the challenges and issues of IPM program evaluations. First, stakeholders of IPM programs and their evaluation information needs will be discussed. Second, application of effective approaches, models, and designs for evaluation of IPM programs will be discussed as alternatives. Finally, data collection, analysis, and utilization will be discussed.

18.4.1 Stakeholder Identification and Coordination of Evaluation

Understanding who wants what information, why they need that information, and what decisions will they be able to make with that information is the basic premise for planning a useful evaluation. "The utilization-focused evaluator keeps these questions front and center throughout the design process" (Patton 1997, pp. 189–190). That is why stakeholder identification is considered the most important step for planning a useful evaluation. The most common stakeholders of IPM programs include farmers, consumers, communities, extension educators, researchers, extension and research administrators, agriculture industry personnel, and policy makers. The type of stakeholder groups involved in IPM programs may vary with the location and the focus of the program so that not all the aforementioned stakeholder groups will be involved in every IPM program. However, most of the IPM programs will have farmers or a target audience and extension educators.

Evaluation should be planned as a part of the IPM program during its development stage (CIP 2010) in order to make use of the evaluation to guide the programming process. For this purpose, at the inception of an IPM program, evaluation needs should be identified through stakeholder input as the IPM program develops in the given geographic location. Key stakeholders are the potential users of evaluation results for making programmatic decisions. This includes extension educators and community leaders in a given geographic location. Generally, community leaders are progressive, successful, and innovative farmers respected by the community. Identification of key stakeholders in the given geographic location is the most important step in planning IPM evaluation.

Identification of key stakeholders and the development of partnerships with them are the most important steps for planning, implementation, and evaluation of IPM programs. Once the key stakeholders are identified, a working relationship with them should be developed so their evaluation needs can be identified. Patton (1982) highlights the importance of collaborating with stakeholders to form an evaluation team for tailoring the evaluation planning process to find answers to their evaluation questions. Coordination of evaluation planning activities with the key stakeholders will ensure their support and that they buy into the IPM program planning and evaluation process.

18.4.2 Empowering Key Stakeholders and the Community Members to Conduct Evaluations

Stakeholder empowerment to conduct IPM evaluation means enabling them to plan and conduct meaningful evaluations for their own benefits. Enabling stakeholders to conduct meaningful evaluations includes organizing the group for the evaluation task, building their evaluation knowledge and capacity, and guiding them to plan and implement an evaluation that will provide information for them to make programmatic decisions. IPM is a community education program and its success depends on the extent to which the stakeholders are educated about the application of integrated pest management methods. Similarly, if the stakeholders are educated to be knowledgeable about evaluation, they will be able to engage in evaluating their IPM program effectively. They will be able to plan, conduct, and utilize the evaluation.

There are different approaches to evaluation. Empowerment evaluation creates an environment conducive for program improvement and self-determination of program participants. Empowerment evaluation is more democratic and invites community members to actively participate in the evaluation process. Community members jointly identify needs and set evaluation priorities (Fetterman 2001). Due to these qualities, the empowerment evaluation approach is suitable for mobilizing key stakeholders and community members to conduct IPM evaluations.

According to Fetterman (2001) there are three steps in empowerment evaluation. The first step is defining what the stakeholders are expected to accomplish with the IPM program. By defining stakeholder expectations for the program, they will be able to map out what activities and program processes should be in place. The second step is identifying and prioritizing the most important program activities for accomplishing the expectations of the IPM program. Extension educators and program participants (stakeholders) will review how well each of the program activities are progressing and record the progress on a grading scale. One may use 1(low) to 9 (high) scale with 5 being moderate for recording participants' observations about program activities. This rating will help stakeholders to understand the current situation of the program implementation process. The third step is drawing their action plans to overcome weaknesses and achieve expected outcomes of IPM programs. Extension educators and IPM program participants will jointly work to evaluate the evidence as to whether the program is moving in the desired direction and achieving anticipated results. The focus of empowerment evaluation is on "program development, improvement, and lifelong learning" (Fetterman 2001, p. 6). An evaluator's role in empowerment evaluation is mainly teaching and facilitation. An evaluator is expected to educate IPM program stakeholders and facilitate the evaluation process with them. The evaluator functions as a coach with the first time IPM program stakeholders conducting evaluations. When the stakeholders are educated and actively engaged in evaluation for the first time, they will be empowered with necessary knowledge, skills, and capacity to conduct program evaluations on their own in the future. Empowerment evaluation contributes to building and sustaining the evaluation capacity within the community.

18.4.3 *Identification of Evaluation Needs*

The identification of stakeholder needs is important to ensure the usability of evaluation results for making programmatic decisions. The identification of evaluation needs can be accomplished by having a focus group interview, a qualitative research technique used to discern common views of a group, with key stakeholders. The moderator facilitates the discussion using a pre-planned set of questions to ascertain what the group wants to accomplish. The IPM stakeholders are guided to determine what information they need and why they need that information. Someone other than a stakeholder should record the interview. The group should be small enough for everyone in the group to express their needs and views about the point that is being discussed, but large enough to relate diverse needs and view points of the target community (Krueger 1988). That is why the ideal size of a focus group is considered to be 10–12 people. If the stakeholder group is very diverse, two or more focus group interviews may be required to ensure that all views and needs of the target group have been gathered for planning evaluation. The focus group selection should include all stakeholder groups of the target community. The questions for the focus group interview should be developed in advance and should be effective enough to gather all the necessary information for planning evaluation. Focus group interview questions should be open ended questions. Questions leading to 'yes' and 'no' answers should be avoided. The following list of questions provides an example for conducting a focus group interview with a group of IPM farmers to identify their evaluation needs.

- Please tell us why do you want to start this IPM program in your community?
- What would you like to know about the resources allocated for this program?
- Why do you need that information about the resources allocated?
- What would you like to know about planning this IPM program?
- Why do you need that information about program planning?
- What would you like to know about the IPM program activities implemented?
- Why do you need that information about program activities implemented and scheduled?
- What are the results/benefits that you expect from this IPM program?
- Tell us what do you expect to do with the information about the results of IPM program?
- What are the other comments, concerns, and needs you have for the evaluation of this IPM program?

Responses to focus group interview questions should be analyzed and summarized to determine the needs of stakeholders for planning an IPM evaluation. The evaluation information needs of stakeholder groups must be identified so that the information can be used to tailor an evaluation useful for the target audience.

18.4.4 Use of the Logic Model to Focus Evaluation

Focusing an evaluation means determining what information there is to collect from whom at what stage of the program for meeting the evaluation needs of stakeholders. IPM programs are highly dynamic and this dynamic nature should be incorporated into evaluation plans for measuring program success (Waibel 1999). The logic model is the best tool for achieving this planning task. The logic model proposes that if an antecedent takes place then the consequence of the antecedent will result. It depicts the logical linkages of program resources or inputs with program outputs or program activities and program results or outcomes. Before starting to develop the logic model for any IPM program, it is necessary to understand the program logic or the theory behind the program to make rational linkages between antecedent and its consequences. "The legitimacy of evaluative statements depends on explicitness with regard to assumptions about causality" (Jiggins 1999, p. 25). Program theory tells the cause and effect relationship of the program. For example, the use of varieties resistant to common pests in a given location will contribute to overcome the dependency on chemical pesticides for pest control. The logical statement is that if resistant varieties are used, then there will be fewer incidents of pest problems and chemical pesticides will not be necessary. There is research evidence to support this causal relationship. Program logic should be based on research evidence to ensure its utility. The following steps must be addressed before drawing the logic model for focusing evaluation:

- Determine the major issue or context leading to the adoption of an IPM program in the given geographic location.
- Identify the program objectives relating to the issue/context of the program.
- Determine the resources required for the implementation of the program.
- Understand the activities planned to accomplish the program objectives with the assigned resources.
- Clearly identify the target audience/beneficiaries of the IPM program.
- Determine the potential outcomes of the planned activities. This should be done based on available program theories.
- Categorize the rational sequence of potential outcomes into short-term, medium-term, and long-term outcomes.
- Determine other external factors which may impact the program. Some of those external factors have a positive impact while others may have a negative impact on the program.

After gathering this information one will be able to draw the logic model for the IPM program.

The following case study is used as an example to demonstrate how to use this information for the development the logic model.

Case Study: Rice farmers in a village were attempting to control insect pests using a wide spectrum of pesticides for the last few years. They noticed that previously effective insecticides were no longer effective with the target insects and they looked for other more effective pesticides. In some years, they have had to control pest outbreaks with other more effective pesticides. This continuous cycle of applying a broad spectrum of insecticides to control insects in their rice fields significantly increased their costs of production and reduced their profits. Farmers became frustrated with this continuous battle with pests and discussed this situation with the extension educator in the area. The extension educator explained the IPM concept to the farmers. The farmers agreed to follow IPM as an alternative to their current practice. The extension educator worked with farmers to develop program objectives and pool community resources for planning and implementing the IPM program. Their main objectives were to reduce the cost of production; increase profits; reduce the dependency on chemical pesticides for controlling pests; and prevent pest outbreaks. They pooled $ 2,000 for implementing the program. The extension educator planned to use the farmer field school approach as the extension strategy to educate all 100 rice farmers in the village in IPM concepts. The extension educator and the farmers agreed that if the farmers participated in the educational program, the farmers would be knowledgeable about IPM concepts and skillful in applying those concepts. If they apply IPM practices in their fields across their community, they would be able to derive the intended outcomes. This information is used as an example to develop the logic model (Fig. 18.1).

The logic model provides a visual map of what will happen from beginning to the end of the programming process within a given context. The model links outcomes with outputs and inputs. This visual map is useful for planning the evaluation. The logic model shows the data collecting points for various indicators and is not only useful in focusing evaluation but also in planning the IPM program. The model describes what resources are needed to accomplish the planned objectives and what activities are needed to achieve the planned outcomes.

18.4.5 Use of an Evaluation Model to Conceptualize the Evaluation Holistically

More documentation of the diverse outcomes of IPM programs are needed (Thrupp 1999). Therefore, it is important to select an appropriate evaluation model to conceptualize the evaluation plan holistically. IPM programs are systemic community education programs. The systemic nature of IPM programs highlights the need for a holistic evaluation model to conceptualize the evaluation of IPM programs. There are different evaluation models available in the literature for conceptualizing program evaluations. These evaluation models include (but are not limited to) Stufflebeam's (1983) CIPP Evaluation model, Bennett's (1975) Hierarchy Model, Rockwell and Bennett's (1994) Targeting Outcomes Model, Kirkpatrick's (1995) Training Evaluation Model, and Jacobs's (1988) Five-tier Model. Of these models, Stufflebeam's (1983) CIPP evaluation model provides a comprehensive framework for planning IPM program evaluations. The term CIPP is an acronym for Context, Inputs, Process, and Product or outcomes of a program. Generally, context, inputs, process, and product evaluation "respectively ask, What needs to be done? How should it be done? Is it being done? Did it succeed" (Stufflebeam 2007, p. 1)? Later the CIPP model was further expanded into an evaluation checklist of 10 components by Stufflebeam (2007). These 10 components include contractual agreement with stakeholders, context evaluation, input evaluation, process evaluation, impact evaluation, effectiveness evaluation, sustainability evaluation, transportability evaluation, meta-evaluation, and the final evaluation report.

According to Stufflebeam (2007), the contractual agreement means the evaluator is expected to enter into an agreement for the evaluation work to be done with the key stakeholders. By signing the evaluation agreement, both sides will be fully aware of what to expect from the evaluation. This is the major benefit of signing an agreement. Context evaluation refers to evaluation of the background environ-

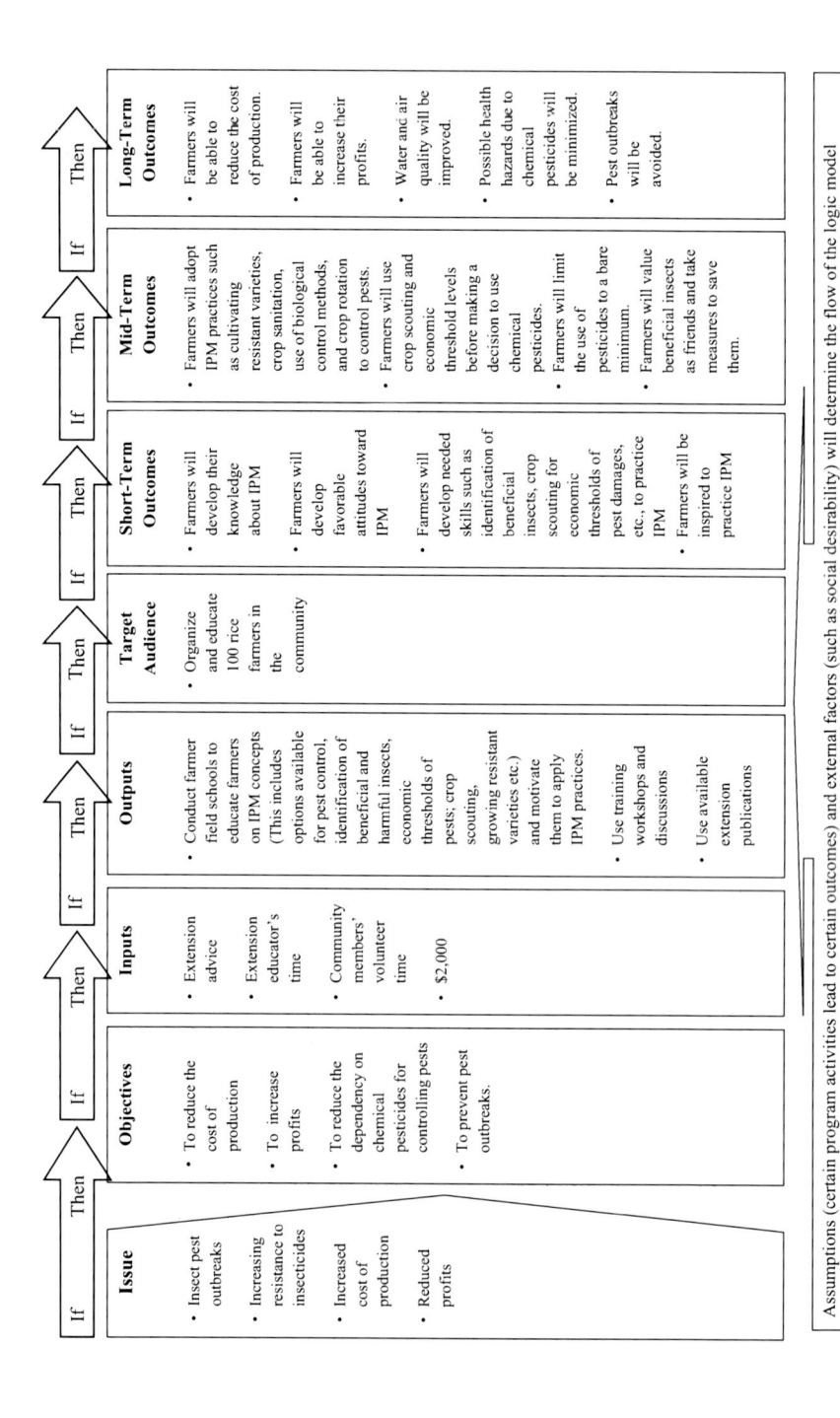

Issue	Objectives	Inputs	Outputs	Target Audience	Short-Term Outcomes	Mid-Term Outcomes	Long-Term Outcomes
• Insect pest outbreaks • Increasing resistance to insecticides • Increased cost of production • Reduced profits	• To reduce the cost of production • To increase profits • To reduce the dependency on chemical pesticides for controlling pests • To prevent pest outbreaks.	• Extension advice • Extension educator's time • Community members' volunteer time • $2,000	• Conduct farmer field schools to educate farmers on IPM concepts (This includes options available for pest control, identification of beneficial and harmful insects, economic thresholds of pests; crop scouting, growing resistant varieties etc.) and motivate them to apply IPM practices. • Use training workshops and discussions • Use available extension publications	• Organize and educate 100 rice farmers in the community	• Farmers will develop their knowledge about IPM • Farmers will develop favorable attitudes toward IPM • Farmers will develop needed skills such as identification of beneficial insects, crop scouting for economic thresholds of pest damages, etc., to practice IPM • Farmers will be inspired to practice IPM	• Farmers will adopt IPM practices such as cultivating resistant varieties, crop sanitation, use of biological control methods, and crop rotation to control pests. • Farmers will use crop scouting and economic threshold levels before making a decision to use chemical pesticides. • Farmers will limit the use of pesticides to a bare minimum. • Farmers will value beneficial insects as friends and take measures to save them.	• Farmers will be able to reduce the cost of production. • Farmers will be able to increase their profits. • Water and air quality will be improved. • Possible health hazards due to chemical pesticides will be minimized. • Pest outbreaks will be avoided.

Assumptions (certain program activities lead to certain outcomes) and external factors (such as social desirability) will determine the flow of the logic model

Fig. 18.1 An example of the logic model for IPM programming

ment the program encounters at the given location. Input evaluation assesses the alternative options to determine the best option for achieving desired results. Process evaluation means evaluation of the program implementation process against planned options to determine needed improvements. Impact evaluation includes the evaluation of the benefits of the program to ascertain whether the benefits justify the resources invested in the program. Effectiveness evaluation determines the significance of outcomes in terms of meeting the needs of target beneficiaries. Sustainability evaluation assesses the extent to which program results are integrated into the community and continue in the future. Transportability evaluation determines to what extent this program is capable of being replicated in other locations. Meta-evaluation means reflective assessment of the overall evaluation process to understand the extent to which it follows the evaluation standards. The final evaluation report includes writing and communicating evaluation findings with key stakeholders (Stufflebeam 2007). This model is a comprehensive framework because it focuses on program context, inputs, process, impacts, effectiveness, sustainability, and transportability assessments with the emphasis on program improvement, sustainability, and replication. The CIPP model checklist aims to achieve long-term sustainable development of programs. Therefore, the CIPP model checklist is appropriate for conceptualizing the IPM program evaluation. It provides a framework to evaluate the IPM program from beginning to the end. The next sections briefly discuss how to apply Stufflebeam's CIPP evaluation checklist for conceptualizing the IPM program evaluation.

18.4.5.1 Signing the Contractual Agreement

If the IPM evaluation is performed by an outside evaluator, it may be appropriate to sign an agreement with the key stakeholders in the IPM program to clearly spell out what to expect from the evaluation. However, most IPM programs are community education programs carried out by extension educators or researchers and evaluated jointly by farmers, extension educators, and researchers. In this situation, it is not necessary to sign an agreement. Instead, it is appropriate to discuss, clarify, and determine evaluation expectations of the key stakeholders so that evaluation can be tailored to meet their informational needs.

18.4.5.2 Context Evaluation

Programming context entails the locale in which the program is taking place. Careful analysis of the context in which the program operates is an important step for planning any evaluation (Wholey et al. 2004). Understanding the context of the IPM program is necessary to determine whether the program is effective or not (Waibel 1999). The IPM programming context is mainly determined by the socioeconomic environment and the agro-ecological environment of the program. The program context changes over the course of program implementation. The context evaluation will find answers to the question, is this IPM program the best alternative for the given situation? If the programming context changes then it is necessary to evaluate the suitability of the program in the new context to make necessary adjustments to the IPM program. Context evaluation assesses the situation that warranted the IPM program in the given geographic location, community assets available for the program, and the factors contributing to the program. The context evaluation is helpful in determining whether the program goals and objectives are aligned with the needs of the target audience of the IPM program and achievable within the available resources. If there is any discrepancy between program goals and the needs of the target beneficiaries of the IPM program, it is important to revise program goals to ensure the usefulness of the program to stakeholders. For example, if the IPM program goal is to reduce the cost of production by reducing the use of unnecessary chemical pesticides among the rice farmers in a rural village, then at the end of the program it is necessary to evaluate the context to ascertain the situational changes. If most of the farmers have reduced the use of chemical pesticides, then the continued application of the IPM program in that farming community requires realignment of goals with the changing needs of the target audience. The changing needs of the farmers may be maintaining pest populations at economic threshold levels, maintaining sustainability or enhancing biodiversity. Parallel to these changing needs, it is important to realign the goals and objectives of the program to ensure relevance of the IPM program to the target audience. Alignment of the program goals and objectives with the needs of the target audience is essential to retain their full participation in the program. Context evaluation is useful for accomplishing this task.

18.4.5.3 Input Evaluation

Input evaluation assesses available program plans with the resources assigned for the program to determine the best plan within the resources assigned. IPM is a community education program and there are many ways to plan educational strategies to teach IPM concepts to the target audience within the allocated resources. The challenge of the program planner is to select the best effective educational strategy under the given socio-cultural conditions and resource limits. Resources include money, staff and volunteer time, materials, equipment, facilities, etc. Input evaluation in IPM programs should not be limited to economic terms but to use socio-cultural values of the target audience when assessing the appropriateness of program plans and strategies. For example, teaching women farmers by the local male extension educator may not be culturally appropriate. Bringing an outside female extension educator may be more expensive but might be more culturally appropriate. This highlights the need for assessing program plans and educational strategies to determine whether those are the best alternatives for the socio-cultural conditions of the target audience and the resource limits of the IPM program. Plans and strategies should be assessed for their technical appropriateness and political viability to make sure those are the best alternatives for the given situation. For example, if the target audience is illiterate, written extension materials are not useful. Pictures, demonstrations and hands-on learning activities may be effective alternative strategies for educating them although these educational aids also help educate literate audiences. Input evaluation is helpful to determine the best program strategies to educate the target audience on IPM concepts within the given socio-cultural conditions and resource limits. If the input evaluation is conducted by the extension educator and the key leaders of the farming community, they need to discuss and compare the planned strategy with all possible extension strategies to verify the superiority of the planned option within their resource limits. If the input evaluation reveals that there are better options than the planned strategies, then IPM program stakeholders will be able to use the input evaluation information to revise the original plan for achieving desired results. This way, input evaluation will prevent the use of inappropriate strategies and contributes to maximizing the cost effectiveness of IPM programming.

18.4.5.4 Process Evaluation

The success or failure of many IPM programs is mainly determined by the program implementation process. "However, there is little evidence that IPM (as originally envisioned) has been implemented to any significant extent in American Agriculture" (Ehler 2006, p. 2). This statement highlights the need for assessing the IPM programming process. Process evaluation assesses the course of program implementation to ascertain whether the program is being implemented as envisioned at the planning stage. Process evaluation will find answers to questions such as:

- Did the program receive planned resources?
- Did the program get adequate staff and volunteers to implement it?
- Did the program staff and volunteers use allocated funds for the planned work?
- Is the program reaching the target audience and serving their needs?
- Are educational methods effective?
- What is the level of learner satisfaction?
- What are the problems and issues in implementing the program?
- What are the strengths and weaknesses of the program implementation?
- What are the constraints and challenges of program implementation?
- What are the best alternatives for the challenges within the available resource limits?
- How should the implementation process be modified to achieve the desired results?

Process evaluation results are helpful in fine-tuning the program implementation process by eliminating weaknesses and overcoming barriers.

Process evaluation should be built into the program planning so that, at the end of each program activity, that activity will be evaluated to learn from success and failures. Process evaluation is meaningful only if process evaluation results are utilized to fix program weaknesses. This is possible if the plan has incorporated the

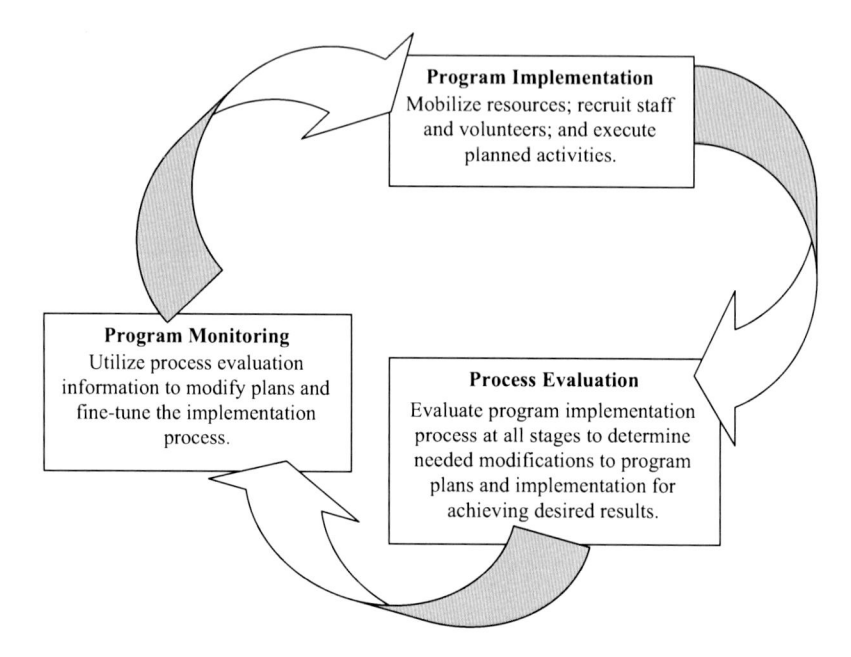

Fig. 18.2 Incorporation of process evaluation into program monitoring

monitoring process into program implementation. At the monitoring step, process evaluation data and information will be reviewed by the key stakeholders and necessary adjustments will be made based on the process evaluation results as shown in Fig. 18.2. Program monitoring followed by process evaluation will contribute to maximizing the cost effectiveness of IPM programs by preventing possible program failures.

18.4.5.5 Impact Evaluation

Impact evaluation assesses the extent to which the target audience is impacted by the program as planned. The first step is to determine whether the program reached the target audience as planned; the major target audiences of most IPM programs are farmers. Consumers are also important beneficiaries of IPM programs. The major target audience of IPM programs varies with the scope of the program. For example, it may be an IPM program focused on rice farmers in a district or vegetable farmers in a village. Geographic scope and the type of crop targeted are the major criteria that determine the target audience of any IPM program. Determining the extent to which the program reached the target audience is the first step in impact evaluation. The reasons for any discrepancy between the planned target and the actual attainment in reaching the target audience should be explored to improve the program. For example, if the program achievement is below expectation, finding reasons for the gap will be helpful for program staff to fix the problem. If the achievement is exceeding the target, finding the reasons contributing to success will be useful to further expand the program.

The second step of impact evaluation is to determine the extent to which the target audience benefited from the program. The logic model can be used to map the type of impacts each of these beneficiary groups will derive from IPM programs.

The following questions will be useful for focusing the impact evaluation:

- What are the geographic boundaries of this IPM program?
- Who are the major beneficiaries of this IPM program?
- What are the planned outcomes of the IPM program?
- To what extent did the program materialize planned outcomes for the target audience?
- If there is any discrepancy between the planned and realized impacts, what are the reasons for discrepancy?

The third question highlights the need for determining impact indicators for planned outcomes. The impact indicator is a reasonable and useful measure of intended outcomes. When impact indicators are determined, it is important to work with the key stakeholders to identify which indicators are more meaningful and practical. Table 18.1 summarizes the potential outcomes and useful impact indicators for IPM program outcomes.

It is easier to document short-term outcomes compared to mid-term outcomes and long-term outcomes. Higher level impact evaluations demand more time and resources than those of lower level impact evaluations. Mid-term and long-term outcomes provide strong evidence for the success of IPM programs. Practicality and the type of evidence needed will determine what level of impact evaluation is appropriate for a given situation. Available resources and time will determine what level of impact evaluation is practical.

The short-term outcome evaluation is usually practical. The short-term outcome indicators are appropriate for assessing the immediate outcomes of IPM educational programs. A change in attitudes is an important prerequisite for changing the mindset of farmers to embrace IPM as a viable alternative for chemical pest control. Until their mindset is changed, it is difficult to achieve desirable results by conducting IPM programs. If program participants developed positive attitudes toward IPM, they will actively participate in IPM educational activities, gain knowledge, develop skills, and aspire to apply IPM practices as a practical means to control pests. Changes in participants' attitudes, knowledge, skills, and aspirations are useful impact indicators for measuring short-term outcomes of IPM programs. Participants' levels of aspiration reflect their readiness to apply what they learned and become convinced about and that they will be taking charge of as a result of the program. Level of aspiration is a useful impact indicator to measure the success of educational activities (Jayaratne 2010) presented under the IPM program.

Mid-term outcome indicators are useful to determine the extent to which the program participants have adopted IPM practices for controlling pests. Most of the mid-term impact indicators are recording actual practice changes by program

Table 18.1 Impact indicators of integrated pest management programs

Type of Outcomes	Impact Indicators
Short-term Outcomes (Changes in learning)	
1. Changes in participants' IPM knowledge	Number of participants who improved their IPM knowledge
2. Changes in participants' IPM skills	Number of participants who improved their IPM skills such as identification of beneficial insects, estimation of economic threshold levels, etc.
3. Changes in participants attitudes toward IPM	Number of participants who developed favorable attitudes toward IPM
4. Changes in levels of aspiration	Number of participants intending to practice IPM
Mid-term Outcomes (Changes in Practices)	
1. Adoption of IPM practices	Number of farmers adopted IPM practices such as cultivating resistant varieties, crop sanitation, mechanical pest control methods, crop rotations, biological methods, etc.
2. Adoption of crop scouting and economic threshold levels	Number of farmers practicing crop scouting and economic threshold levels before making decisions to apply pesticides
3. Reduced level of pesticide usage	Number of farmers who stopped or reduced use of chemicals to control pests Reduced amount of pesticides used
4. Increased attention to managing the ecosystem	Number of farmers identifying beneficial insects and taking measures to save them
Long-term Economic Outcomes	
1. Reduced cost of production	Amount of money saved on pest control
2. Increased income	Amount of income increased by practicing IPM
3. Increased profit	Amount of profit increased (benefit/cost ratio)
Long-term Social Outcomes	
1. Reduced health hazards	Reduced number of health hazards caused by pesticides
2. Increased public awareness of IPM	Increased number of consumers demand for foods produced under IPM practices
3. Public support for IPM	Number of pro-IPM legislations/policies adopted
Long-term Environmental Outcomes	
1. Increased water quality	Reduced amount of pesticide residues present in waterways
2. Increased biodiversity	Reduced incidences of pest outbreaks as a result of natural balance Reduced incidences of building resistance to pesticides

participants (Table 18.1). Some of the mid-term indicators record the immediate results of desired practice changes by program participants. For example, reducing the amount of pesticide used is the immediate result of practicing IPM. If there is a discrepancy between the targets and actual adoption of IPM practices, it is important to understand the reasons for the discrepancy in order to find alternatives.

The long-term outcome indicators determine the extent to which IPM programs contribute to improve the economic, social, and environmental conditions of the target population. If properly coordinated with the program participants, long-term economic outcome data can be collected at a reasonable cost. When collecting and interpreting long-term outcome data, the evaluator must determine whether the long-term outcome is a result of the IPM program or is due to other events taking place in the area. For example, if the reduced level of pesticide residues in water ways is observed, it is necessary to ascertain the observed condition is not because of other events such as reduced levels of farming activities in the area and that it is primarily due to the IPM program implemented in the area. Long-term outcome data provides convincing evidences about the value of IPM programs.

Short-term and mid-term outcome data can be collected at a reasonable cost by planning and coordinating the data collection with program participants and staff. Collecting long-term outcome data is costly and needs a lot of planning and coordination with participants and program staff. The use of multiple indicators for measuring important outcomes will provide broader evidence by compensating the weaknesses of one indicator with the strength of another indicator (Rossi et al. 2004).

18.4.5.6 Effectiveness Evaluation

According to Stufflebeam (2007, p. 8) effectiveness evaluation determines the "quality and significance" of outcomes. Effectiveness evaluation finds answers to the question of whether the IPM program is effective in terms of meeting the economic, social, and environmental expectations of the target audience. Intended and unintended outcomes of IPM programs should be assessed with the stakeholders to ascertain whether the positive outcomes exceed negative outcomes and meet their expectations. Meeting stakeholder expectations of the program is significant to qualify IPM as an effective strategy for controlling pests.

18.4.5.7 Sustainability Evaluation

Sustainability evaluation determines the extent to which the IPM program has been integrated into the community for continuation (Stufflebeam 2007). The continued application of IPM is decided upon by the participants based on the program's ability to meet their expectations which can only be determined by the IPM educators having a dialogue with the program participants. This dialogue is necessary to understand participants' judgments about the value of the program. It is necessary to assess the extent to which the program participants and key stakeholders are in favor of continuing the program to determine the sustainability of the program. The greater the IPM program outcomes meet the participants' expectations and the participants come to value the program, the more likely they will continue the IPM program. Sustainability evaluation will determine the extent to which program participants have embraced the IPM as a useful strategy for managing pests. Outcome data and information must be made available to program participants to enable them to make informed judgments about the value of continuing their IPM program.

18.4.5.8 Transportability Evaluation

The transportability evaluation assesses the extent to which the program could be replicated in similar locations (Stufflebeam 2007). If an IPM program is conducted with rice farmers in a village and the rice farmers in a neighboring village are getting interested in the results of that IPM program, there is a great chance to expand the IPM program into the neighboring village particularly if the neighboring village has comparable conditions. The transportability evaluation determines to what extent program replication is possible and provides useful information and data for extension educators to understand the necessary conditions for replicating the IPM program in similar locations.

18.4.6 Designs of Evaluation Studies Appropriate for Integrated Pest Management Program Assessments

The appropriate design for a given evaluation is determined by the nature of the evaluation, available resources, and the level of rigor needed for the evaluation. The nature of the evaluation may be exploratory evaluations, outcome evaluation, or assessing causal relationships. Qualitative methods such as focus group interviews and case studies are exploratory by nature and appropriate for exploratory evaluations. Focus group interviews are conducted with a selected group of 10–12 people to understand the situation from their perspectives and experiences. The interview is recorded and analyzed for content so a summary can be prepared. The case study technique, an exploratory evaluation, uses observations, interviews and available records to gather needed information to assess the situation. The case study technique is a time-consuming, qualitative research technique.

Quasi-experimental studies are practical and appropriate for outcome evaluations. Randomized experimental designs are appropriate for assessing causal relationships. Randomized designs provide more rigorous evidence than that of any other design but demand more time and resources. Sometimes randomized designs are impractical due to social and ethical limitations such as assigning individual farmers randomly for a control (no IPM program) or a treatment (of IPM programming). The design of evaluation studies needs to take into account the two factors—rigor and the practicality of the design. Evaluation design should be rigorous for relying on evaluation results. Rigorous evaluations demand more resources. Practical considerations of coordinating the evaluation task with the target audience, available resources, and human subject protection regulations limit the available design options (Rossi and Freeman 1993). When these two factors are taken into account, quasi-experimental designs are considered more practical than other designs for conducting IPM evaluations.

There are many quasi-experimental designs. Of them, 'before and after' design and the 'nonequivalent group design', are considered the two most practical designs for conducting IPM evaluations. 'Before and after' design is the most commonly used design in extension evaluation. When using the 'before and after' design, evaluation variables are measured before and after the IPM intervention. Before and after measurements are compared for changes. The changes in measuring variables are considered as the outcomes of the IPM program. However, results of this method can be biased. Sources of potential biases are called threats to internal validity (Campbell and Stanley 1966). The most common threats to internal validity of using 'before and after' design for evaluating IPM programs are history, seasonality, attrition, and statistical regression. History means anything other than the IPM program that can alter the measuring variables of IPM program. For example, if the advertising budgets of pesticide companies were reduced at the time of IPM program and it contributed to reduced levels of pesticide usage of farmers, then the reduced levels of pesticides could not be attributed to the IPM program completely. Seasonality refers to the outcome variation caused by seasonal variation. For example, if the weather is unfavorable for pest infestation then the measured changes after the program could not be attributed to the IPM program completely. Attrition is the systematic dropout of IPM participants during the program. For examples, if the large scale farmers dropped out of the program, the results will be biased toward the small scale farmers. Statistical regression means the tendency of gravitating data toward the mean.

Nonequivalent group design uses two comparable groups of participants. One will receive the IPM program and the other group will be used as the control site. Participants at the control site will not receive IPM programming. Measuring variables will be recorded before and after implementing the program at the IPM site and the control group site. This method is useful to control the effects of history and the seasonality. However, it is not possible to assure the comparability of two different groups. This is the major weakness of nonequivalent group design.

18.4.7 Development of Evaluation Tools to Collect Data

Evaluation data are collected from primary sources and secondary sources. Primary sources of data are collected from participants, consumers, and extension educators. Valid and reliable survey tools are needed for this purpose, one source for developing such survey tools is the article, "Evaluation research: Methodologies for evaluation of IPM programs" (Peshin et al. 2009). If an already developed survey instrument or a newly designed instrument is used, first determine the purpose of the evaluation and what information needs to be collected (Basarab Sr. and Root 1992). Defining the purpose and the focus of the evaluation is the most important step in determining what questions to include in the survey (Colton and Covert 2007). The following guidelines are helpful tips in designing a useful evaluation tool:

- Determine the evaluation needs of the IPM stakeholders.
- Determine the type of data needed before formulating the survey questions.
- Determine the levels of education of the target audience so that the survey questions are understood.

- Draft questions to collect needed data.
- Keep the number of questions to a minimum.
- Develop necessary scales to record the situation of measuring variables.
- Provide clear instructions on how to complete the survey.
- Check with a group of IPM program educators to confirm the content validity of the survey.
- Pilot test the survey with a comparable group of 10–15 people to ensure the reliability of the instrument.
- Use the pilot test results to make necessary changes to the survey.
- Finalize the survey and use it for data collection.
- If the target audience is illiterate, then someone will have to interview participants and complete the survey.

The secondary sources of data are collected from existing reports such as agency reports. For example, collecting sales data for pesticides from the agrochemical store in the area or collecting water quality data from the water quality monitoring agency to assess the changing situation.

18.4.8 Collecting, Analyzing, and Reporting Data

Collecting accurate data from IPM participants is a challenging task. Generally, participants are reluctant to provide evaluation data such as yield and income. Collecting accurate data requires that the evaluator first build trust with the program participants. One way to build trust is to explain the reasons for collecting evaluation data. If the participants are educated and empowered to actively engage in the evaluation, they could become partners in the evaluation and provide accurate information and data. Valid and reliable survey tools are needed to collect data from participants.

When composing the evaluation report, the key to effective communication to the stakeholders is the use of the simplest statistics for data analysis and the appropriate level of language. For example, if the report is targeting the general public, it is appropriate to use simple statistics such as percentages and avoid technical terms to help them understand the report. It should be concise to help them understand it easily.

18.4.9 Utilization of Evaluation Results

An evaluation is meaningful and worthwhile only if IPM program stakeholders utilize evaluation results for making informed decisions about necessary changes to the program. Evaluation results can be utilized for program improvement, accountability, marketing, advocacy, and policy development. Utilization of evaluation results can be achieved mainly by working with stakeholders to plan useful evaluations (Rossi et al. 2004).

18.4.9.1 Program Improvement

The information in the evaluation gathered from context, input, process, and results can be used to improve the IPM program.

Context evaluation information can determine the extent to which IPM program objectives are aligned with the needs of the target audience. Any discrepancy between the needs of the target audience and the program objectives can be used to align program objectives with the actual needs of the target audience. Context evaluation can be used to determine the extent to which the evaluated program is based on the broad notion of IPM and whether the IPM concept has been misused. If the program is deviating from the quality standards of the broad definition of IPM, then it can be discussed with the stakeholders and fixed at the outset. Additionally, context evaluation reveals what assets are available in the community to build the program and what problems could prevent the success of the program.

Input evaluation determines the extent to which the program plan is using the best cost-effective strategies and ensures the program is using the best cost-effective strategies.

Process evaluation determines to what extent actual implementation is progressing as planned. Process evaluation information reveals any discrepancy between the plan and implementation of program activities and the reasons for any discrepancy. This information can be used to fix problems and fine-tune the implementation process as discussed in Sect. 18.4.5.4 and displayed in Fig. 18.2.

Results evaluation determines the extent to which the program realized outcomes as planned and the reasons for any discrepancy between the planned expectations and the realized outcomes. By knowing the reasons for discrepancy, program staff and key stakeholders will be able to find alternatives for fine-tuning the program.

18.4.9.2 Accountability

Historically, accountability is the driving force for evaluation of IPM programs. Funding agencies of IPM programs demand accountability. A variety of evaluation data can be used for this purpose. These data include number of educational activities and lesson plans developed and presented (output data), number of target audiences reached, and program outcomes. It is important to use this accountability information for the justification of the resources invested in the program. Long-term economic outcomes are more appealing evidence for accountability than any other evaluation data.

18.4.9.3 Marketing

Outcome evaluation data can be used to market IPM programs to potential audiences. Long-term economic outcome data and long-term environmental outcome data provide convincing evidence for someone to seriously consider applying IPM as a practical strategy to control pests. When extension educators are planning to diffuse IPM technology, outcome evaluation data should be used to educate audiences about the benefits of practicing IPM. Marketing IPM programs with real facts enable extension educators to gain and retain the trust and support of audiences for implementing IPM programs in new locations.

18.4.9.4 Advocacy and Policy Development

IPM programs are mostly supported by public funds. Establishment of favorable policies is a prerequisite to ensure adequate funding for IPM research and extension in a country or region. Public support for IPM is an important determinant for the establishment of favorable policies. Impact evaluation data could be used to educate the public and advocate IPM to policymakers to gain their support for the establishment of favorable policies for the expansion and sustenance of IPM programs.

18.4.9.5 Meta-Evaluation for Continuous Improvement

The term meta-evaluation refers to the critical assessment of an evaluation itself to ascertain the extent to which it abides by the standards of sound evaluation. The Joint Committee on Standards for Educational Evaluation (2011) published The *Program Evaluation Standards* (3rd ed.) as guidelines for conducting sound evaluations. Meta-evaluation is important to identify shortfalls and find ways to improve the evaluation practice of anyone engaging in IPM program evaluations.

18.5 Conclusion of Evaluating Integrated Pest Management Interventions

IPM technology is an important strategy for managing pests. Much time and resources have been invested in research to develop IPM technology. Extension educators have spent their time and resources to develop and deliver educational programs for disseminating IPM technology. With these research and extension efforts, still the adoption of IPM technology by the stakeholders has not reached its full potential. Problems associated with the IPM technology transfer process are the major reason for low IPM program adoption levels (Wearing 1988). This situation highlights the need for the evaluation of the IPM programming process for finding alternatives to disseminate IPM technology effectively.

The review of IPM programming reveals that many IPM programs were implemented with little or no attention to evaluate programs systematically. As a result, most IPM program outcomes were not documented or publicized which kept policy makers, farmers, consumers, and the general public unaware of the real value of

IPM programs. If the farmers are not aware of the full benefits of IPM, they will not adopt IPM as a viable technology. If the consumers are unaware of the benefits of IPM, they will not be able to make an informed decision at the market place to demand foods produced under IPM practices. If the general public and the policymakers are unaware of the benefits of IPM, they will not support funding the promotion of IPM. This situation signifies the need for documenting the broad outcomes of IPM programs and using that information to educate policy makers, farmers, consumers, and the general public. Educating these key stakeholder groups about the benefits of IPM is necessary for the diffusion and sustenance of IPM as a viable strategy for managing pests.

Little or no attention has been paid to context evaluation of IPM programs. The context evaluation of IPM programs is helpful to align objectives of an IPM program with the contextual factors that call for IPM technology in a given geographic location. To the extent IPM program objectives are aligned with the pest control needs of a geographic location, the target audience of the program will be able to relate the rationale of adopting IPM technology for their situation. This will contribute to the dissemination of IPM technology. Special attention should be paid to context evaluation for aligning program objectives with the situation.

The review of IPM programming further reveals that the program process evaluation better known as the formative evaluation is lacking. Formative evaluation will be helpful to detect problems, weaknesses, and strengths of the program implementation process. Lacking process evaluation, IPM programs are being conducted without having systematic implementation reportage in place to detect and fix program implementation shortfalls. This is a major drawback for achieving the cost effectiveness of IPM programming. Due attention should be paid to improve the program process evaluation and utilize evaluation data to improve IPM programs.

Currently, some issues and challenges associated with IPM programs limit the potential role of evaluation. These issues and challenges should be properly addressed for tapping the full potential of evaluation as a useful tool for making informed decisions in program improvement, accountability, marketing, advocacy, and policy development. Alternative strategies discussed in Sect. 18.4 can be used to address these issues and challenges for improving the quality of IPM evaluations. Due attention and adequate allocation of resources for evaluation; integration of evaluations as an integral part of the programming process; and building evaluation capacity among the extension educators, researchers, and farmers are the major steps needed to improve the quality of IPM program evaluation.

Currently, IPM evaluation results are utilized mainly for accountability to policy makers. IPM evaluation results must become a tool for making programmatic decisions. Incorporation of stakeholders' evaluation needs is a practical step to change this situation. If the evaluation is providing needed information for stakeholders to make programmatic decisions, then they will use that information. Properly planned, sound evaluations of IPM programs generate information useful for improving programs, accountability, marketing, advocacy, and policy making.

Meta-evaluation is an important step to review the evaluation practice of IPM programs for the improvement of evaluation. Individuals and groups engaged in the evaluation of IPM programs should critically assess their own practices by reviewing the standards of evaluation to ensure the soundness of their evaluations. Evaluation standards provide guidelines for conducting a sound evaluation. Those who engage in IPM evaluation should pay due attention to meta-evaluation for improving the quality of IPM evaluation practice.

References

Bajwa, W. I., & Kogan, M. (2003). Integrated pest management adoption by global community. In K. M. Maredia, D. Dakouo, & D. Mota-Sanchez (Eds.)., *Integrated Pest Management in the Global Arena* (pp. 97–107). Cambridge, United Kingdom: CABI Publishing.

Basarab, Sr., D. J., & Root, D. K. (1992). *The Training Evaluation Process.* Boston: Kluwer Academic Publishers.

Bennett, C. (1975). Up the hierarchy. *Journal of Extension, 75*(2). http://www.joe.org/joe/1975march/1975-2-a1.pdf. Accessed 12 July 2012.

Campbell, D., & Stanley, J. C. (1966). *Experimental and Quasi-experimental Designs for Research.* Chicago: Rand McNally.

Charleston, K., Miles, M., & Brier, H. (2011). IPM workshops for growers and consultants—Lessons for R, D, and E. *Extension Farming Systems Journal, 7*(2), 86–91. http://www.csu.edu.au/__data/assets/pdf_file/0008/199151/EFS_Journal_vol7_n02_17.pdf. Accessed 28 Dec 2012.

Chung, D. K., Pincus, J., Rola, A., & Widawsky, D. (1999). Assessment of household and village level impacts of IPM. In H. Waibel, G. Fleischer, P. E. Kenmore, & G. E. Feder (Eds.)., *Evaluation of IPM Programs: Concepts and Methodologies* (pp. 9–12). Hannover, Germany: Institute for Economics in Horticulture. http://www.ifgb.uni-hannover.de/fileadmin/EUE_files/PPP_Publicat/Series/PPP08.pdf. Accessed 15 July 2012.

(CIP) International Potato Center. (2010). *Introductory Guide for Impact Evaluation in Integrated Pest Management (IPM) Programs.* http://cipotato.org/publications/pdf/005514.pdf. Accessed 25 Aug 2012.

Colton, D., & Covert, R. W. (2007). *Designing and Constructing Instruments for Social Research and Evaluation.* San Francisco: Jossey-Bass.

Ehler, L. E. (2006). Integrated pest management (IPM): Definition, historical development and implementation, and the other IPM. *Pest Management Science, 62*(9), 787–789. doi:10.1002/ps.1247.

FAO. (1967). *Report of the First Session of the FAO Panel of Experts on Integrated Pest Control* (Rome, 18–22 Sept 1967), Rome: Food and Agriculture Organization of the United Nations.

FAO. (2012). *Integrated Pest Management.* http://www.fao.org/agriculture/crops/core-themes/theme/pests/ipm/en/. Accessed 20 June 2012.

Fetterman, D. M. (2001). *Foundations of Empowerment Evaluation.* Thousand Oaks, California, USA: Sage Publications, Inc.

Government Performance and Results Act of 1993 (GPRA). (1993). http://www.whitehouse.gov/omb/mgmt-gpra/index-gpra. Accessed 20 July 2012.

Hollingsworth, C. S., & Coli, W. M. (2004). Consumer response to IPM: Potential and challenges. In O. Koul, G. S. Dhaliwal, & G. W. Cuperus (Eds.)., *Integrated Pest Management: Potential, Constraints, and Challenges* (pp. 255–264). Cambridge, United Kingdom: CABI Publishing.

Horne, P., & Page, J. (2009). Integrated pest management dealing with potato tuber and all other pests in Australian potato crops. In J. Kroschel & L. A. Lacey (Eds.)., *Integrated pest management for the Potato Tuber Moth, Phthorimaea Operculella (Zeller)—A Potato Pest of Global Importance* (pp. 111–117). Tropical Agriculture 20, Advances in Crop Research 10. Weikersheim: Margraf Publishers. http://www.google.com/url?sa=t&rct=j&q=&esrc=s&source=web

&cd=2&ved=0CD4QFjAB&url=http%3A%2F%2Fwww.vgavic.org.au%2Fpdf%2Fr%26d_IPM_potato_review.pdf&ei=wNbULDJBYaQ9QSNhYCQDA&usg=AFQjCNFEXLtEmtFzZldQR_S1oPFPLfVGoA&bvm=bv.41248874,d.eWU&cad=rja. Accessed 20 Dec 2012.

Integrated Plant Protection Center. (1996). *Compendium of IPM Definitions*. Oregon State University, Corvallis, Oregon, USA. http://www.ipmnet.org/ipmdefinitions/defineIII.html#90%27s. Accessed 20 June 2012.

Jacobs, F. (1988). The five-tiered approach to evaluation: Context and implementation. In H. Weiss, & F. Jacobs (Eds.)., *Evaluating Family Programs* (pp. 37–68). Hawthorne, New York, USA: Aldine de Gruyter.

Jayaratne, K. S. U. (2010). Practical application of aspiration as an outcome indicator in extension evaluation. *Journal of Extension, 48*(2). http://www.joe.org/joe/2010april/tt1.php. Accessed 28 Dec 2012.

Jiggins, J. (1999). Evaluation of IPM extension: Institutional aspects. In H. Waibel, G. Fleischer, P. E. Kenmore, & G. E. Feder (Eds.)., *Evaluation of IPM Programs: Concepts and Methodologies* (pp. 25–29). Hannover: Institute for Economics in Horticulture. http://www.ifgb.uni-hannover.de/fileadmin/EUE_files/PPP_Publicat/Series/PPP08.pdf. Accessed 15 July 2012.

Kirkpatrick, D. L. (1995). Evaluating training programs: The four levels. *Human Resource Development Quarterly, 6*(3), 317–320.

Kogan, M. (1998). Integrated pest management: Historical perspectives and contemporary developments. *Annual Review of Entomology, 43*, 243–270. doi:0066-4170/98/0101-0243.

Krueger, R. A. (1988). *Focus Groups: A Practical Guide for Applied Research*. Newbury Park, California, USA: Sage Publications, Inc.

Misra, S. K., Huang, C. L., & Ott, S. L. (1991). Consumer willingness to pay for pesticide-free fresh produce. *Western Journal of Agricultural Economics, 16*(2), 218–277.

Olson, L., Zalom, F., & Adkisson, P. (2003). Integrated pest management in the USA. In K. M. Maredia, D. Dakouo, & D. Mota-Sanchez (Eds.)., *Integrated Pest Management in the Global Arena* (pp. 249–271). Cambridge, United Kingdom: CABI Publishing.

Patton, M. Q. (1982). *Practical evaluation*. Newbury Park, California, USA: Sage Publications, Inc.

Patton, M. Q. (1997). *Utilization Focused Evaluation* (3rd ed.). Thousand Oaks, California, USA: Sage Publications, Inc.

Peshin, R., Jayaratne, K. S. U., & Singh, G. (2009). Evaluation research: Methodologies for evaluation of IPM programs. In R. Peshin, & A. K. Dhawan (Eds.)., *Integrated Pest Management: Dissemination and Impact* (Vol.2, pp. 31–78). Netherlands: Springer. doi:10.1007/978-1-4020-8990-9-2.

Rockwell, K., & Bennett, C. (1994). *A Hierarchy for Targeting Outcomes and Evaluating Their Achievement*. http://citnews.unl.edu/TOP/english/. Accessed 12 July 2012.

Rossi, P. H., & Freeman, H. E. (1993). *Evaluation a systematic approach* (5th ed.). Newbury Park, California: Sage Publications, Inc.

Rossi, P. H., Lipsey, M. W., & Freeman, H. E. (2004). *Evaluation a Systematic Approach* (7th ed.). Thousand Oaks, California, USA: Sage Publications, Inc.

Sorby, K., Fleischer, G., & Pehu, E. (2003). Integrated pest management in development: Review of trends and implementation strategies. *Agriculture & Rural Development Working Paper 5*, World Bank, Washington D.C. http://130.203.133.150/viewdoc/summary?doi=10.1.1.201.6369. Accessed 20 Feb 2013.

SP-IPM. (2008). Incorporating Integrated Pest Management into national policies. *IPM Research Brief No. 6*. SP-IPM Secretariat, International Institute of Tropical Agriculture (IITA), Ibadan, Nigeria. http://www.spipm.cgiar.org/c/document_library/get_file?p_l_id=17830&folderId=18484&name=DLFE-87.pdf. Accessed 20 Feb 2013.

Stern, V. M., Smith, R. F., van den Bosch, R., & Hagen, K. S. (1959). The integrated control concept. *Hilgardia, 29*(2), 81–102.

Stufflebeam, D. L. (1983). The CIPP model for program evaluation. In G. F. Madaus, M. Scriven, & D. L. Stufflebeam (Eds.)., *Evaluation Models: Viewpoints on Educational and Human Services Evaluation* (pp. 117–141). Boston: Kluwer Nijhof Publishing.

Stufflebeam, D. (2007). *CIPP Evaluation Model Checklist* (2nd ed.). http://www.wmich.edu/evalctr/archive_checklists/cippchecklist_mar07.pdf. Accessed 20 June 2012.

Task Force. (2003). *Integrated Pest Management: Current and Future Strategies*. Ames, Iowa, USA: Council for Agricultural Science and Technology.

The Joint Committee on Standards for Educational Evaluation. (2011). *The Program Evaluation Standards: A Guide for Evaluators and Evaluation Users* (3rd ed.). Thousand Oaks, California, USA: Sage.

Thrupp, L. A. (1999). People, power, and partnerships to sustain IPM impacts: Assessing advances in the adoption of agro-ecological alternatives. In H. Waibel, G. Fleischer, P. E. Kenmore, & G. E. Feder (Eds.)., *Evaluation of IPM Programs: Concepts and Methodologies* (pp. 30–35). Hannover, Germany: Institute for Economics in Horticulture. http://www.ifgb.uni-hannover.de/fileadmin/EUE_files/PPP_Publicat/Series/PPP08.pdf. Accessed 15 July 2012.

USDA. (2004). *National Roadmap for Integrated Pest Management*. Washington, D.C.: United States Department of Agriculture. http://www.csrees.usda.gov/nea/pest/pdfs/nat_ipm_roadmap.pdf. Accessed 25 Jan 2013.

USDA. (2012). *Regional Integrated Pest Management Competitive Grant Program, Southern Region: FY 2012* request for applications. United States Department of Agriculture, Washington, D.C. http://www.google.com/url?sa=t&rct=j&q=&esrc=s&source=web&cd=5&ved=0CEoQFjAE&url=http%3A%2F%2Fwww.csrees.usda.gov%2Ffunding%2Frfas%2Fpdfs%2F12_ripm_south.pdf&ei=umUJUZTQCIec9gSu9YHYCQ&usg=AFQjCNE0r1-gF0_jrJLzgALntLDkAIyx7Q&bvm=bv.41642243,d.eWU&cad=rja. Accessed 25 Jan 2013.

Waibel, H. (1999). Policy perspective of IPM evaluation. In H. Waibel, G. Fleischer, P. E. Kenmore, & G. E. Feder (Eds.)., *Evaluation of IPM Programs: Concepts and Methodologies*. Hannover: Institute for Economics in Horticulture. http://www.ifgb.uni-hannover.de/fileadmin/EUE_files/PPP_Publicat/Series/PPP08.pdf. Accessed 15 July 2012.

Waibel, H., Fleischer, G., Kenmore, P. E., & Feder, G. E. (1999). *Evaluation of IPM Programs: Concepts and Methodologies*. Institute for Economics in Horticulture, Hannover, Germany. http://www.ifgb.uni-hannover.de/fileadmin/EUE_files/PPP_Publicat/Series/PPP08.pdf. Accessed 15 July 2012.

Wearing, C. H. (1988). Evaluating the IPM implementation process. *Annual Review of Entomology, 33,* 17–38. http://www.annualreviews.org/doi/pdf/10.1146/annurev.en.33.010188.000313. Accessed 17 July 2012.

Whalon, M. E., & Croft, B. A. (1984). Apple IPM implementation in North America. *Annual Review of Entomology, 29,* 435–470. http://www.annualreviews.org/doi/pdf/10.1146/annurev.en.29.010184.002251. Accessed 17 July 2012.

Wholey, J. S., Hatry, H. P., & Newcomer, K. E. (2004). *Handbook of Practical Program Evaluation* (2nd ed.). San Francisco: Jossey-Bass.

Index

A

ACCase inhibitors, 283, 288, 289, 292, 294, 296
Agriculture, 7–14, 17–24, 27–31, 35, 37, 48, 56, 65, 75–80, 90, 93, 95, 100–105, 109, 112–118, 128, 131–134, 142, 143, 158–162, 169–176, 178, 184–188, 190, 203–208, 214, 215, 230, 237, 239, 255, 259, 260, 274, 276, 306, 308, 312, 324, 332–334, 357, 410–414, 418, 420, 424, 427, 449
Agroecosystems, 54, 206–208, 270–272, 277
ALS inhibitors, 283, 285, 286, 288, 296, 348
  in wheat, 294
Alternative pest control, 131, 445
Alternatives, 15, 76, 78, 105, 131–133, 135, 369, 395, 417, 427, 435–439, 447, 456, 457, 460, 465, 466
  for the challenges and issues of integrated, 448
  pest management program evaluations, 448
Andean potato weevil (APW), 250
  management, plastic barriers for, 259

B

Biological control, 13, 27, 31, 89, 90, 104, 122, 134, 162, 179, 186, 215, 255, 258, 270, 298, 299, 315, 316, 357, 358, 365, 375–378, 428, 436
  and unintended consequences of, 276
  food web and ecosystem perspective, 271
  identifying risks, 276
  in Grasslands and Rangelands, 274
  key, strategy, 274
  non-target effects, 276
Biological weed control, 160, 314
  advances in, 315

C

Carbon dioxide emission, 121, 247
  in pesticide application, 120
Challenges, 160, 168, 203, 228, 282, 314, 403, 412, 436, 467
  of evaluating integrated pest management, 444
  programs, 104, 105, 205, 403, 433, 435, 437, 439, 444, 445, 448, 457
China, 3, 7, 30, 32, 35, 61, 102, 104, 187, 246, 296, 360, 406
  development of IPM in, 31, 34
  pesticide consumption and environment, 33
  impact in, 33
Corn, 11, 12, 48, 55, 58, 64, 128, 129, 133–135, 142–152, 155–161, 173, 179, 181, 286–290, 294–297, 305–309, 315, 318–320, 333–339, 342–349
Costs, 50, 53–59, 63–65, 77, 83, 93, 112, 128, 131–135, 142, 151, 152, 160, 173, 180, 181, 188, 190, 203, 215, 247, 250–252, 259, 273, 275, 298, 305, 324, 333, 346–348, 358, 361, 362, 365, 367, 371, 375, 379, 413, 419, 423, 426, 440, 441, 452
Cotton, 7, 9, 11, 12, 23, 24, 27–32, 35, 55–58, 75, 128, 129, 134, 173, 179, 181, 186, 187, 206, 228–230, 233, 234, 237, 239, 290, 297, 298, 308, 333–340, 342–347, 394, 407, 413, 418–420
Cover crops, 116, 132, 145, 158, 160, 282, 299, 305, 313, 314
Crop loss assessment, 203
  elements of, 204
Crop losses, 48, 53, 55, 57, 58, 105, 129–132, 176, 180–183, 189, 190, 203, 205–208, 214–217, 229, 233, 236, 240, 249, 250, 262, 270, 276, 305, 403

D. Pimentel, R. Peshin (eds.), *Integrated Pest Management*,
DOI 10.1007/978-94-007-7796-5,
© Springer Science+Business Media Dordrecht 2014

Crop protection, 75, 78–80, 82, 85, 92–94, 157, 169, 173, 175, 188, 203, 204, 216, 229, 270, 323, 395, 397, 399, 405, 406, 412, 423

Crops, 11, 14, 29, 32, 51–59, 61, 76–85, 88, 91, 94, 103–106, 110, 115, 128–135, 143, 145, 155, 160, 161, 169, 173, 176–181, 186–190, 204, 206, 208, 215, 216, 228–230, 237, 240, 256–259, 270, 275, 276, 297, 305–310, 313–318, 332–335, 423, 435

Cultural weed control, 130

D

Denmark, 7, 17, 19, 21, 105, 128, 155, 157

E

Economic benefits, 10, 13, 64, 215, 445
   from reduced pesticide use, 131

Economic damages, 203

Energy inputs, 116, 117, 146, 150, 151, 161, 162
   in pesticide application in New Zealand, 117–121
   in spraying, 112

Environment, 7–10, 31, 63–65, 76, 79, 81, 85, 95, 131, 142, 143, 161, 169, 173–175, 180, 184, 187–190, 204, 215, 247, 249, 262, 271, 272, 282, 304, 305, 337, 349, 357, 358, 368, 383, 394–396, 410, 413, 423, 440, 441, 446, 447, 450

Environmental problems, 55, 64, 188, 333

Europe, 13, 14, 17, 19, 49, 77, 78, 159, 178, 182, 187, 205, 255, 282, 286, 287, 294, 399, 435

European Union(EU), 14, 15, 23, 24, 75, 80, 173
   pesticide reduction strategy, 76

Evaluation, 9–12, 151, 203, 204, 254, 260, 435–459, 461–466

F

Farmer capacity building, 260
   for arthropod management, 261
   for pathogen management, 245, 259, 261

Floral biodiversity, 305, 320

Food Quality Protection Act (FQPA), 357

Foods, 51, 62, 90, 130, 162, 172, 177, 207, 324, 467

Food safety, 94, 168, 172, 176, 182
   case studies, 189
   current trends in the use of agrochemicals and, 174, 175
   integrated pest management and, 170, 189
   nutrition and food quality, 171
   through eco-friendly pesticides, 182

Food sovereignty
   pesticides, 176

Framework Directive, 81
   European Union, 14

Fruit crops, 359, 361, 365, 374

G

Genetically modified (GM), 175, 332, 333, 349

Glyphosate, 12, 13, 61, 108, 175, 282, 283, 289, 294–299, 306–309, 333, 337, 339–348, 350
   in mid-west USA, 290
   in Roundup Ready crops, 290, 292

H

Herbicide, 10–15, 21–23, 29, 57, 58, 78, 90, 105, 108, 109, 112, 118, 134, 135, 146, 155, 173, 175, 270, 282–299, 306–309, 316, 317, 321, 332–337, 340–350
   application timings, 310
   use pattern in United States, 306–308
   use reduction in agronomic crops, 318
   use reduction in corn, 319

Herbicide resistance, 283, 296

Herbicide resistant crops, 297–299, 307, 308

Herbicide resistant weeds, 282–284, 291–299, 333, 335

Herbivory, 206

I

India, 23, 24, 27–30, 55, 61, 170, 175, 183, 186–189, 205, 208, 228–239, 402, 411, 412, 418–420

IPM programs since 1993
   operational research project, 24
   pesticide use in Indian agriculture, 28, 418

Insecticide use, 11–15, 24, 27, 29, 53, 86, 108, 129, 133, 134, 179, 215, 259, 360, 365–369, 373, 420
   patterns in blueberries, 370
   patterns in peaches, 362

Insect pests, 30, 54, 78, 90, 129, 130, 161, 169–175, 184, 185, 188, 206, 228–237, 249, 250, 261, 263, 306, 308, 323, 357, 368–372, 376, 382, 420, 435, 452

Integrated pesticide management, 2, 37

Integrated pest management (IPM), 116, 178, 308, 434
   in cotton, 7
   European pesticide policy and, 75

Index

food safety and, 172
impacts, 249, 394, 412
pesticides and, 2, 7, 8, 75, 81, 394
program, 7, 8, 81, 105, 208, 357
Integrated weed management (IWM), 298, 299, 315
Integrative approach, 424

L

Late blight, 247–250, 253–256, 260, 262
  risk assessment for insects, 261
Leaf miner fly (LMF), 250, 263
  control, 257, 258
Livestock, 94, 101, 106, 110, 145, 146, 161, 162, 177, 183, 189, 270, 274, 275, 298, 345

M

Maize, 77, 82, 101, 173, 175, 183, 184, 187, 189, 206, 215, 229–233, 237, 282, 346
Mechanical weed control, 19, 104, 152, 160, 317
Monetary losses, 231, 237

N

Netherlands, 8, 17, 160, 247
New Zealand, 102, 103, 106–109, 111, 114, 118, 159, 316
  agriculture in, 100, 101
  energy consumption in pest control, 112
  energy inputs in pesticides application, 117
  pesticide consumption in, 108, 122
Non-chemical weed control, 321

O

On-farm management
  cultural management practices, 256
  fungicides, use of, 254
Outcomes, 13, 24, 75, 78, 345, 401, 426, 436–447, 450–455, 459, 461, 463, 465–467
Outreach, 380, 381

P

Pest control, 54, 64, 65, 75, 78, 95, 106, 119, 128, 131, 132, 135, 160, 170, 180, 183–190, 208, 214, 215, 249, 357, 365, 371, 394, 396, 403, 410, 427, 435, 438, 441, 452, 459, 467
  chemical methods, 102, 105, 435, 440
  in Iran, 412, 427
  methods, 7, 85, 89, 102–105, 111, 112, 169, 172, 205, 216, 412, 434, 435, 440, 445

non-chemical methods, 89, 102
total energy outputs in, 119, 120
Pesticide resistance, 55, 92, 105, 169, 181, 207, 445
  in pests, 54
Pesticides, 6–20, 23, 28–29, 33, 34, 37, 48–63, 75, 76, 81, 82, 90, 92, 107–110, 119, 120, 129–135, 143, 157, 158, 169–180, 185–187, 191, 206, 208, 215, 216, 262, 323, 347, 366, 375–377, 382, 394, 397, 403, 407, 410, 441, 447, 464
  externalities of, 170, 171, 177
  and farmers, 256, 417
  health effects of, 10, 49, 50, 105, 169, 182, 252, 415, 417, 426, 445, 452, 453
Pesticide use, 4, 13–17, 23, 28, 29–34, 49, 50, 53–55, 63–65, 75, 77–79, 90–93, 106, 109, 118, 120, 129–132, 135, 142, 161, 169, 180–183, 190, 250–252, 262, 270, 335, 365, 366, 374, 375, 395, 402, 413, 419, 420, 441, 460
  in agriculture, 37, 56, 76, 80, 105, 128, 133, 176, 178, 208, 308, 411–418
  in Colombian Andes, 423–427
  and IPM in Iran, 412
Pest management, 75, 78, 79, 81, 85, 104, 105, 116, 170–179, 182–191, 205, 208, 216, 230, 239, 249, 256, 257, 261, 264, 270, 271, 274, 308, 314, 357, 359, 363, 364, 366–370, 373, 374, 383, 395, 396, 405–407, 412, 415–418, 420, 428, 434–438, 440–449, 462
  costs, 64, 65, 173, 181, 203, 259, 273, 358, 361, 362, 365, 367, 371, 375
  in Florida, 374, 375, 377, 380–382
Potato tuber moths (PTM), 250, 264
  complex, 258
Public health, 63–65, 95, 128, 135, 142, 158, 175, 190
  effects, 49–52, 76, 177, 188

R

Reduced pesticide use, 19, 21, 105, 128, 131, 132, 161, 190, 323, 374, 395, 403, 420
Reduced-risk practices, 367, 369
Residues, 33, 51, 52, 58–60, 79, 93, 94, 105–107, 130, 145, 171–180, 183, 186, 187, 252, 272, 314, 320, 382, 403, 406, 441, 443
Resistance management, 284, 334, 342–345, 347, 362, 372, 404, 420
  strategies, 283, 299, 366, 380, 405
Resistant weeds, 308, 348

AC case inhibitor (A/1), 283, 284, 288–291, 294
ALS inhibitor (B/2), 285–291
cases, 284, 285–291, 294
glyphosate (G/9), 289, 290
triazine (C1/5), 286–290
Responsible use, 405
of chemical pesticides, 395, 408
of crop protection products, 399
of pesticides, 394, 395, 401, 406, 407

S

Safe use of pesticides, 79, 169, 411, 416, 417
and IPM practices, 427
and IPM technologies, 415, 428
determinants and training needs, 417, 418
Socio-economic impact
of biotic constraints in potato production, 247–253
Soil organic matter, 152
and biodiversity, 157
Soybeans, 53, 58, 64, 128, 133–135, 143–146, 155, 159–161, 180, 290, 297, 307, 317, 333–346
Stewardship, 397
Survey, 33, 59, 77, 85, 90–92, 106, 189, 228, 233, 250, 259, 260, 284, 338–347, 365, 376, 377, 380, 406, 424, 463, 464

Sustainable agriculture, 14, 64, 157, 191, 215, 270, 314, 323, 397
Sustainable use, 14, 15, 76, 81
Sustainable Use Directive (SUD), 76
Sweden, 8, 17, 21, 105, 128, 142, 161

T

Thematic Strategy on the Sustainable Use of Pesticides (TSSP), 75
Training, 79, 80, 90, 94, 256, 257, 264, 395, 396, 404, 406, 407, 415–417
in IPM principles, 403
materials, 260, 364, 405, 407
programs, 21, 27, 105, 183, 260, 359, 377, 399, 401–405, 418, 443
re-enforcement of, 406, 407

U

United States of America (USA), 8

W

Water Framework Directive (WFD), 80
Weed control, 135, 142, 146, 160, 215, 273, 274, 283, 292, 294, 297–299, 306, 310, 317–322, 332–334, 340, 343, 349
acceptable levels of, 317
broad spectrum of, 308, 313, 335
methods, 82, 105, 112, 282, 308, 315, 339, 346